Birkhäuser

Moon Duchin • Olivia Walch
Editors

Political Geometry

Rethinking Redistricting in the US with Math, Law, and Everything In Between

Birkhäuser

Editors
Moon Duchin
Department of Mathematics
Tufts University
Medford, MA, USA

Olivia Walch
University of Michigan–Ann Arbor
Falls Church, VA, USA

ISBN 978-3-319-69160-2 ISBN 978-3-319-69161-9 (eBook)
https://doi.org/10.1007/978-3-319-69161-9

This book is published under the imprint Birkhäuser, www.birkhauser-science.com by the registered company Springer Nature Switzerland AG
The registered company address is: Gewerbestrasse 11, 6330 Cham, Switzerland

Contents

Preface

This unusual book was originally conceived as the proceedings of an unusual conference. In 2016-2017, a kitchen-table collective of mathematicians formed the Metric Geometry and Gerrymandering Group, and in July 2017 we planned a Geometry of Redistricting Workshop at Tufts University. Speakers came from many fields and vocations—scholars, technologists, organizers, and litigators—and the conference drew so much interest that we had to hold it at a theater off campus. More than four years later, the collective is now a Lab, and the Lab is engaged in the decennial redistricting in some way in over a dozen states. Meanwhile, the conference proceedings evolved into the ambitious boundary-flouting volume you see here.

What is "political geometry"? For starters, it's a riff on "political geography," an established academic discipline that looks at the spatial dimensions of elections and governance. Where there's space, there's shape, and mathematicians might have something useful to say. And it's not just math, geography, and political science: there's software, graph algorithms, policy, civil rights, history, political philosophy, and of course law in the mix. This book was designed to serve up a multi-disciplinary buffet, with both traditional fare and fresh fusions.

WHAT'S IN THIS BOOK

We've divided this book into five broad parts, coarsely chunked by domain but with lots of overlaps and cross-talk: Political Thought, Law, Geography, Math and Computer Science, and "On the Ground." Each chapter is written by a different author, or set of authors, drawing from a huge variety of backgrounds and perspectives. We've added smaller "Explainers" in a few key places that treat important topics in a stand-alone fashion. (They are marked with colored strips on the page corners so they're easy to flip to.) We've also sprinkled interviews with a range of practitioners and theorists throughout.

OK, BUT REALLY, WHAT'S IN THIS BOOK

We start the book off with an introduction and overview from *Moon Duchin* (Chapter 0). She'll cover some of the basics: Can you use shape to define a gerrymander? What about judging from results themselves, where representation is out of whack with the vote balance? If not either of those, what can you do? This chapter surveys the lay of the land in data-driven redistricting.

Chapter 1, "Explainer: Compactness by the numbers," builds on the Introduction, defining scores commonly used to judge the shape of a plan and identifying some of their basic shortcomings.

Then we're off to the Parts:

POLITICAL THOUGHT

Traditionally, the identification of gerrymandering has been the province of political science.

In Chapter 2, "Measuring partisan fairness," *Mira Bernstein* and *Olivia Walch* provide a mathematical view on some metrics of fairness from the political science literature.

But what do we even mean by fairness? In Chapter 3, "Concepts of Representation," we interview four political thinkers—philosophers Elizabeth Anderson, Ryan Muldoon, and Brian Kogelmann and political scientist Claudine Gay—on what fairness in representation looks like.

In Chapter 4, "Redistricting: Math, systems, or people?", *Keith Gåddie* gives a high-level take on "the redistricting problem" from his perspective as a political scientist and veteran redistricting expert.

Chapter 5, "Political geography and representation," goes in-depth on a particular conundrum raised throughout this Part: We know that geography matters in elections, but how? Here, *Jonathan Rodden,* a political scientist who specializes in political economics and geography, teams up with data scientist *Thomas Weighill* to take a look at how political geography is reflected in districting outcomes, particularly addressing questions of size and scale.

LAW

Then we turn to the law.

In Chapter 6, "Explainer: A brief introduction to the VRA," *Arusha Gordon* and *Doug Spencer* provide the reader with key background on the Voting Rights Act and its relevance to redistricting.

Leading voting rights scholar *Ellen Katz* jumps off from there in Chapter 7, "Race and Redistricting," where she surveys the law of racial gerrymandering with a detailed look at the intertwining roles of Congress and the Courts.

Chapter 8, "Law, computing and redistricting in the 1960s," brings a historian's perspective to the book. In it, *Alma Steingart* looks back to the moment that the U.S. Supreme Court introduced its One Person, One Vote standard, which made computing a permanent part of the redistricting scene.

Finally, in Chapter 9, "The law of gerrymandering" *Guy-Uriel Charles*, an expert in race and constitutional law, and *Doug Spencer*, who teaches both law and policy, examine the parallels between racial and partisan gerrymandering law.

GEOGRAPHY

As a discipline, geography spans from philosophy of place to technologies of space.

Chris Fowler is a geographer who studies cities, planning, and neighborhood change. In Chapter 10, "Race, space, and the geography of representation," he reminds us not to take demographic distributions for granted, but to put them in historical and social context.

In Chapter 11, "The elusive geographies of communities," *Garrett Dash Nelson*, a historical geographer and the curator of maps at the Boston Public Library, takes a close look at community and regionalization.

Chapter 12, "Explainer: Communities of interest," *Heather Rosenfeld* teams up with Moon for a practical primer on what constitutes a "COI" where redistricting is concerned.

In Chapter 13, "Geography as data," geographer and data scientist *Ruth Buck* comes together with *Lee Hachadoorian*, whose work uses geospatial technology in urban and demographic analysis, to give us a close look at geo-electoral data and the software that wrangles it.

MATH AND COMPUTER SCIENCE

To a mathematician, the redistricting problem can feel like a playground where any idea from math might find fruitful application.

In Chapter 14, "Three applications of entropy," *Larry Guth*, *Ari Nieh*, and *Thomas Weighill* test this out by taking the math/physics idea of entropy and seeing where it fits. They end up describing three use cases: 1) how different are two plans? 2) how much does a map split counties? 3) how segregated is a city?

Of course, entropy isn't the only hammer in the toolkit. In Chapter 15, "Explainer: Measuring clustering and segregation," *Moon Duchin* and *James Murphy* tackle metrics of spatial patterning in a completely different way, examining a construct that geographers call "Moran's I."

In Chapter 16, "Redistricting algorithms," computer scientists *Amariah Becker* and *Justin Solomon* give a big, big picture overview of how computing can bear on the redistricting story. It's notoriously hard to compare algorithmic strategies against each other because implementation can be very finicky, and the goals of different researchers don't line up perfectly in the first place. But they go for it anyway, and it makes for some very illuminating comparisons.

Next comes Chapter 17, "Random walks," by mathematicians *Daryl DeFord* and *Moon Duchin*. This chapter looks at one of those algorithmic strategies, Markov chain sampling, in closer detail.

ON THE GROUND

We close with the voices of practitioners.

Megan Gall, Karin Mac Donald, and *Fred McBride* all have political science training but now work hands-on with maps and data in the field. In Chapter 18, "Making maps," they talk about their experiences drawing in the real world.

Nate Persily is a law professor who has frequently been appointed by courts to draw the lines when the primary parties can't agree. In Chapter 19, we interview him about his experiences.

The longer we spend in the redistricting world, the more our focus moves from flagging the rulebreakers to upgrading the rules. Chapter 20, "Explainer: Ranked choice voting" acts as a primer on one policy change that may be able to do just that.

Next is a redistricting story very close to MGGG's home in Boston—in Chapter 21, *Iván Espinoza-Madrigal* and *Oren Sellstrom* tell us about the voting rights suit that they filed on behalf of plaintiffs in Lowell, Massachusetts—the first to base a voting rights claim on a coalition between Asian and Latino voters. As we write, the lines are being drawn (by Nate Persily!) for a brand new city council structure.

After that is Chapter 22, "Explainer: Race vs. Party," a brief look at how race and party preference intertwine in voting patterns, and how this has played out in several recent cases.

We close with Chapter 23, "The state of play in voting rights," from *Kristen Clarke,* who has held many civil rights law positions across government and nonprofit organizations, and *Arusha Gordon,* an attorney who is her former colleague at the Lawyers' Committee for Civil Rights Under Law (LCCR). They leave us with a wide-angle view of where voting rights stand today, and point to possible locations of battlefields to come.

Who's the audience for this book? Great question.

We have tried to frame the book to be simultaneously engaging to community organizers, math-curious high school students, philosophers, programmers, and election lawyers. We hope it's suitable to assign in a political science class and to ground a data science curriculum. Not every chapter is written at the same level, and they're certainly not written in the same voice. We think this keeps it true to its interdisciplinary conference roots: a symposium of sorts, bringing a lot of different people together to share tools and ideas. There are illustrations (many by Olivia) throughout the book, which provides one kind of throughline. There are also sidebars, often written in the voice of the editors, that can be skipped without cost to the exposition but should provide more depth or color in strategic places.

For readers, we hope there is something for every taste. You will get the most out of this buffet by giving all the dishes a try. The authors have made a real effort to make their flavors accessible but complex.

Code-heavy chapters have a corresponding GitHub repo (github.com/political-geometry). We hope that both the code and the book overall can be public resources in the long and wild mapping wars that are already underway.

EDITORS' ACKNOWLEDGMENTS

MD: I wandered into my current obsession with redistricting through teaching a class on abstract voting theory (Mathematics of Social Choice), primed by two happy coincidences. One, I had attended a session of the American Political Science Association a few years earlier and I chanced on a talk that introduced me to the idea of "compactness." Two, one of my closest and oldest friends, Kristen Clarke, is a prominent civil rights attorney, who was just telling me that voting rights litigation was in need of a new generation of experts. So I had election systems, district shape, and civil rights on my mind. I teamed up with two multi-talented, multi-curious mathematician friends, Ari Nieh and Mira Bernstein, and the *Metric Geometry and Gerrymandering Group* was born. We recruited the brilliant and demonically productive computer scientist Justin Solomon (who'd read about us on Reddit!) not long after. I thank Kristen, Mira, Ari, and Justin first for the life-changing inspiration and collaboration.

Five years later, this is my main research focus, and I run a Lab that is deeply engaged in public mapping, plan evaluation, and scholarship in data science for civil rights. And while this book was in press, Kristen became the brand new Assistant Attorney General of the United States—the top civil rights attorney in the land!

Kristen and Moon in 2017 at the first MGGG workshop

MD thanks: Daryl DeFord, Jordan Ellenberg, Aidan Kestigian, Thomas Weighill, Sam Gutekunst, John Powers, Sarah Cannon, Max Hully, Ruth Buck, Parker Rule, JN

Matthews, Gabe Schoenbach, the rest of the truly fantastic MGGG crew, Joe Auner, Alan Solomont, Alex Swift, Charles Stewart, Steve Ansolabehere, Rebecca Green, Joey Fishkin, Sarah Pinto, Amahl Bishara, Eitan Hersh, Gary King, Mike Hopkins, David Fisher, Bernard Fraga, Malia Jones, Bernie Grofman, Jonathan Cervas, Anita Earls, Paul Smith, Bill Cooper, Blake Esselstyn, Ben Williams, Sam Wang, Hannah Wheelen, Andrew Blumberg, Dick Engstrom, Karen Saxe, danah boyd, Xiao-Li Meng, Max Palmer, Jowei Chen, Wendy Cho, Yan Liu, David Hoffman, Leah Aden, Nina Perales, Yurij Rudensky, Michael Li, Kathay Feng, Dan Vicuña, Rick Pildes, Larry Tribe, Andrew Chin, Peter Aronow, Ellen Veomett, Nick Stephanopoulos, Ruth Greenwood, George Cheung, Heather Villanueva, Colin Cole, Howie Fain, Pedro Hernandez, Kenny Easwaran, Laurie Paul, Nate Persily, Zach Schutzman, Assaf Bar-Natan, Nestor Guillen, Tyler Jarvis, Ben Fifield, Harrison Bray, Guy Charles, Doug Spencer, Megan Gall, Karin Mac Donald, Fred McBride, Larry Guth, all the other book authors, James Whitehorne, Simson Garfinkel, Cynthia Dwork, Aloni Cohen, Peter Wayner, Tomiko Brown-Nagin and the Radcliffe Institute, Sarah Reckhow, Hakeem Angulu, Maira Khan, Tyler Piazza, Oliver York, 85 VRDI participants, Danny Goroff, Henry Cohn and Microsoft Research, Justin Levitt, Sam Hirsch, Jessie Amundsen, Charlie Parekh, Wendy Underhill, Jamie Chesser, Anna Dorman, Jackson Skeen, Michael Sarahan, Seth Rothschild, Youlian Simidjiyski, Dustin Mixon, Diane Souvaine, Hugo Akitaya, Phil Klein, Matt Scheck, Nathan Foster, Richard Barnes, Bridget Tenner, Wes Pegden, Maria Chikina, Jonathan Mattingly, Greg Herschlag, Christy Graves, Jim Greiner, Eric Maskin, Greg Warrington, Robert Vargas, Dave Wasserman, John Urschel, James Murphy, Ari Stern, Justin Moore, Kristine Jones, Beth Malmskog, Jeanne Clelland, Diana Davis, David Savitt, the staff of the Tisch College of Civic Life, Lucy Millman, Heather Rosenfeld, Heather Kapplow, Floor van de Velde, and especially Olivia Walch and Amanda Gunn.

Alfonso Gracia-Saz is a rock-star educator, crack square-dancer, and hilarious and loving human who helped develop the Educator materials for the original MGGG workshop cycle. He tragically died in the COVID epidemic earlier this year. I love and miss you, Alfonso.

OW: My happy coincidence is that I only met Moon at MGGG's first conference in 2017, when I went to draw a comic about it, and not a few years earlier, when we were both at the University of Michigan and independently going to occasional poker games at Dick Canary's house. If I had met her then, I would definitely have lost a lot of money to her.

OW thanks the authors, Mira Bernstein, Thomas Weighill, Heather Rosenfeld, and (of course) Moon Duchin. She also thanks David Renardy and Matt Jacobs for their love and support.

We thank The Norman B. Leventhal Map & Education Center for their kind permission to use the historical maps included in this volume, as well as Mapbox for permission to use screenshots from the Districtr tool.

The original version of this book was revised: "The original HTML version of the book was inadvertently published with the incorrect author names attributed to the chapters. This has now been amended in each chapter". The correction to this chapter is available at https://doi.org/10.1007/978-3-319-69161-9_25.

Chapter 0

Introduction

MOON DUCHIN

You've probably been hearing about gerrymandering lately. As I write in 2021, the U.S. Supreme Court has heard cases from Wisconsin, Maryland, Texas, Virginia, and North Carolina in its last two terms, if only to give back equivocal answers. Meanwhile, in Pennsylvania, plaintiffs pressed a partisan gerrymandering suit and prevailed in state court in 2018, followed by a frenzied few weeks with new proposed maps flying around, and finally a brand-new set of congressional districts, with sweeping effects for the Congressional delegation.[1] In North Carolina, multiple state-level cases fared better than the federal one, and the state put new congressional and legislative districts in place by the 2020 election, one tick before they would have had to be redrawn anyway in the new Census cycle.

What was at stake in these cases? Apparently quite a few seats, for one thing. Pennsylvania's new map coincided with a major shift in its congressional delegation, from 13–5 Republican control to a 9–9 split. Was that responsive to shifts in the vote, or a mere function of the carefully re-drawn lines?

Gerrymandering, or agenda-driven line-drawing, is a practice (and an anxiety) as old as the Republic. In a country that vests power in elected representatives, there will always be skirmishes and scrapping for control of the process, and in a system like our House of Representatives where winner-takes-all within a geographical district, the delineation itself is a natural battleground.

[1] Full disclosure: I got a front-seat view of Pennsylvania's districting reboot as a consulting expert for Governor Tom Wolf.

0.1 U.S. ELECTORAL DISTRICTS 101

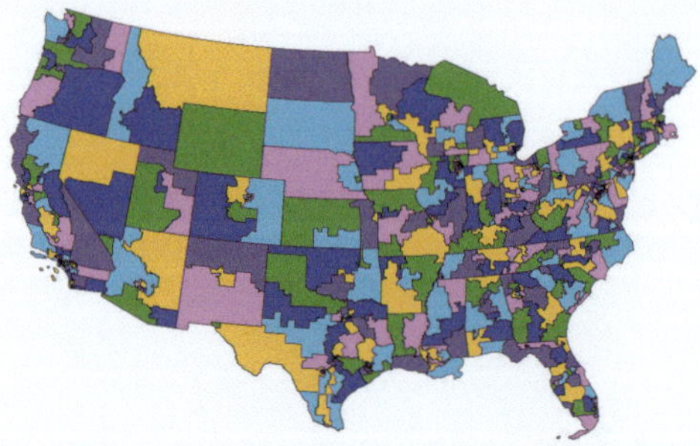

- The U.S. House of Representatives has 435 members, ever since 1911. They must be elected one per district by common practice that was made official in law in the 1960s.[a] The picture shows the 432 in the continental U.S. circa 2019.
- State legislatures currently have 7308 members, all elected from districts. (Over 1000 of them are elected from multi-member districts.)
- Many thousands more elected representatives sit on city councils that are elected from districts—New York and Chicago have the two largest city councils, with 51 and 50 districts respectively.
- And then there are school districts, county commissions, ambulance districts, water boards, executive commissions, and more. The U.S. Census Bureau conducted a count of local governments in 2017, enumerating around 90,000 across the country.[b]
- "One person, one vote" jurisprudence from the Supreme Court (from the 1960s onward) tells us that electoral districts should be population-balanced within their jurisdiction—so zones to elect a school board must have nearly the same population, even if zones defining school attendance need not.
- We usually use *plurality* or "first past the post" voting in districts—i.e., the single candidate with the most votes wins. There are exceptions, like the multi-member legislative districts mentioned above, and the many local elections that use "at-large" schemes to fill several seats at once.
- We have two major parties, but the parties have shifted significantly over American history and certainly might continue to do so. For elections where candidates run with a party ID, there is often a primary several months in advance to pick the nominee in each party before the inter-party competition in the general election.
- Incumbency advantage is enormous. U.S. House races happen every two years, and for instance in 2016 only five incumbents ran for re-election but lost in the primary; another eight lost in the general election; and 380 were re-elected, for an overall success rate in the neighborhood of 97%.[c]

[a] Apportionment Act of 1911, Pub.L. 62–5; 2 U.S.C. §2c "no district to elect more than one Representative"

[b] https://www.census.gov/programs-surveys/cog.html

[c] Brookings Institute, *Vital Statistics on Congress.*

I am a mathematician with a background in geometry (shape) and dynamics (systems in motion). I have a long-standing investment in civil rights work and social movements. I'm also invested in social studies of science, and I like to think about how scientific argument circulates in politics, policy, and law—how technical expertise acts in the social sphere. This comes with a healthy skepticism of scores and metrics that promise to take a *complicated* thing and make it *simple*. So gerrymandering is an irresistible problem for me. It's all about peeling back layers of intuition about shapes, numbers, and power.

Today the primary image of gerrymandering centers on party politics, but the long history of manipulative redistricting has been driven by many other agendas, like back-room deals to make safe seats for incumbents or to dice up a district to stick it to a hated rival. And it's impossible to understand the current context or the bulk of the jurisprudence without contending with the history of schemes to suppress the political power of racial minorities, especially Black and Latino voters—not a practice of the past, but one that's even arguably on the rise in places where new demographic formations are visible. All of these flavors of gerrymandering have in common their basic structure: draw the lines to arrange pluralities for one set of voters and dilute the influence of the other voters.

This is stubbornly difficult to identify. People think they know gerrymandering by two hallmarks: bizarre shapes and disproportionate outcomes. But neither one is reliable.

1 HOW (NOT) TO SPOT A GERRYMANDER

1.1 BIZARRE SHAPES

We think crazy shapes tip us off to moustache-twirling gerrymanderers for a few reasons. The simplest is that we can easily imagine that the district line had to veer around wildly to include *this* pocket of people, but not *that* one. This seems especially likely if a district has been made to narrowly favor one party's voters in election after election. Another reason—if we expect that different kinds of people with shared community interests tend to clump together—is that jagged lines may indicate that an unspoken agenda has dominated over the contours of neighborhoods and communities. Finally and possibly most persuasively, we may worry that those who draw the lines just have too much detailed control over outcomes. Wildly winding boundaries flaunt the power of the pen.

The 1812 episode that gave us the word "gerrymander" sprang from this same pile of intuitions. The name is derived from Elbridge Gerry, governor of Massachusetts at the time. Gerry has quite a Founding Father pedigree—member of Congress, James Madison's vice president, a major player at the U.S. Constitutional Convention—so it's remarkable that he's mainly remembered in connection with nefarious redistricting. The "Gerry-mander," or Gerry's salamander, was the curvy state Senate district in Boston's North Shore that was allegedly drawn to favor one party, Gerry's Democratic-Republicans, over the rival Federalists (see Figure 1). A woodcut political cartoon ran in the Boston Gazette in 1812 with wings and claws and fangs suggestively added to the contours of the district to heighten its appearance of reptilian contortions—Figure 2 shows a Salem Gazette adaptation the next year.

So the idea that eccentric shapes are red flags for wrongdoing is old. And just as old is the idea that close-knit districts promote democratic ideals. Even before the notorious Gerry-mander, James Madison had written in the Federalist Papers (1787) that "the natural limit of a democracy is that distance from the central point which will just permit the most remote citizens to assemble as often as their public functions demand"—in other words, districts should be *transitable* to promote the possibility of deliberation. The new federalist model would knit these together: the United States was to be a republic built from these districts, serving as its constituent democracies. Forming districts of manageable size would ensure that the representatives have knowledge of "peculiar local interests" to be conveyed to the legislature (Fed. 14, 37, 56). So here, shape is in the mix but only as a correlate of function. In 1901, a federal apportionment act marked the first appearance in U.S. law of the vague desideratum that districts should be composed of "compact territory." That word *compact* then proliferated throughout the legal landscape of redistricting as a districting criterion, but almost always without a definition.[2]

Figure 1: Democratic-Republican Thomas Jefferson (left) and Federalist Alexander Hamilton (right) disagreeing over the compactness of this district. (Reenactment.)

Going back to the 1810s, the language from the Original Gerrymander broadsides is instructive. In the Salem Gazette, the *democratic* sins of the district are that it "cut up and severed this Commonwealth" much like "the arbitrary deeds of Napoleon when he partitioned the territories of innocent nations to suit his sovereign will." The *geographic* sins are those of its "peculiarities" of shape: three towns too far north, a town from a separate county "tacked on," and so on.

[2]Apportionment Act of 1901, 31 Statute 733. For a precise definition, see Chapter 1.

SALEM GAZETTE.

VOLUME XXVII.] SALEM, FRIDAY, APRIL 2, 1813. [No. 27.

THE GERRY-MANDER,
OR
ESSEX SOUTH DISTRICT FORMED INTO A MONSTER!

FEDERALISTS! FOLLOWERS OF WASHINGTON!

AGAIN behold and shudder at the exhibition of this terrific Dragon, brought forth to swallow and devour your Liberties and *equal* Rights. Unholy party spirit and inordinate love of power gave it birth;—your patriotism and hatred of tyranny must by one vigorous struggle strangle it in its infancy. The iniquitous Law, which cut up and severed this Commonwealth into Districts, is kindred to the arbitrary deeds of NAPOLEON when he partitioned the territories of innocent nations to suit his sovereign will. This Law inflicted a grievous wound on the Constitution,---it in fact subverts and *changes our Form of Government*, which ceases to be *Republican* as long as an *Aristocratic* HOUSE OF LORDS under the form of a Senate tyrannizes over the People, and silences and stifles the voice of the *Majority*.

When Tyranny and arbitrary Power thus make inroads upon the Rights of the People, what becomes the duty of the citizen? Shall he submit quietly and ignominiously to the decrees of Usurpers? Are the citizens of this Republic less jealous of their rights than their ancestors? Will you, then, permit a Party to disfranchise the People,---to convert the Senate Chamber into a Fortress in which ambitious office-seekers may entrench themselves and set at defiance the frowns of the People? No,---this usurping Faction must be dislodged from its strong-hold.

Arise, then, *injured Citizens!* Turn out! turn out! let Monday Next be the day of your Emancipation---by one manful Struggle reclaim your usurped Rights---and frown into obscurity those audacious men who unblushingly boasted---"*We have secured the Senate for ten years, and should have been Fools if we had not done it.*" Prove on election day that the Folly of their men is equal to their want of honesty and contempt of the People. Elect patriots who will be loyal to the Constitution, and faithful to the interests of the State,---men who will harmonize with our illustrious STRONG,---and who will not sell our Drafted Soldiery, like Hessians, to any Foreign Power.

requiſition their utmoſt cunning to deviſe means expreſsly to accommodate their own views and to promote their own intereſt, ſhould ſtill be unwilling to adhere to the policy preſcribed by themſelves.

Such a man would have to learn, that violence and ſubordination muſt be employed to accompliſh what diſhoneſty and trick had failed to ſecure. It is in fact the ſettled maxim of ſome, who, if the teſtimony of their own profeſſions is to be credited, have much love for the people and a ſacred regard for their rights, to contemn all legal proviſions, which impede their ambitious deſigns, however juſt and neceſſary to ſecure the peace and order of ſociety; to execute by force whatever they are prohibited to do by law, and to aggrandize themſelves without reſpect to juſtice or decency. Upon no other principle can we account for the conduct and activity of thoſe, who, at a late election in this town, ſcrupled not to bring to the poll, not only alien reſidents, who were known never to have been naturalized, not only a certain clergyman (the learned Comentator of Politics for this vicinity) whom they had, in common with the reſt of his brethren, diſqualified to vote in the choice of town officers, and thus to involve him in the turpitude of their tranſgreſſions, and to multiply the cauſes of repentance, (unleſs he ſo far believes, "that the end would ſanctify the means," that there would be *non cauſa penitentiæ*); but even to ſeduce citizens from *other* towns, to inveſt the minority with a power, which they would be ſure to abuſe. He, who promiſes freedom to the captive, when he increaſes the number and weight of his chains, acts as conſiſtently, as he does, who declaims much and loud about the liberties of the people and the rights of his country, when all his actions evince, that he regards neither the one nor the other, any farther, than he diſcovers a direct advancement of his own intereſt. It is greatly to be deſired, that *honeſt* men would ſometimes ſuffer themſelves to be influenced as well by objects of *viſion*, as *hearing*, and compare *actions* with *profeſſions*; and then let judgment decide upon the worth of character, and the ſafety of repoſing confidence. CIVIS.

MR. PRINTER,—I am a native of *Great-Britain*, and ſome years ago was naturalized at *Gottenburg*; but for ſeveral months I have been engaged in trade in this State. I have ſeen very preſſing notices of the Marſhal for alien enemies to report themſelves forthwith While I remain here I wiſh to comply with the laws and ordinances of the country, and feel no enmity towards the government or people. A peace or change of times would deſtroy the profits of my preſent ſpeculations, and in that caſe I ſhould probably ſoon depart. Moreover, I do not find that the U. S. of America are at war with my adopted country. Now I will thank any of your correſpondents to inform we what is proper or expected from one in my circumſtances; as there are many in the country, and more expected, ſimilarly ſituated. A. B.

"UNDER A SIGNATURE."

Is the correct, and was once the universally received phrase. But a spirit of innovation and hyper-criticism directed by gross misconception has engendered and brought into too general use the awkward and absurd term, "*over a signature.*"—To show how mischievous tolerated abuses are, we would refer to the following from the Boston Gazette. "*The Road to Peace.*

"Messrs. Russell & Cutler—I have read a very sensible, but small tract [compose your muscles, gentle reader] *over the above title.*" &c.

Now, the tract, "*over*" this title is worthy of all praise for its correctness and candor, and may be read *without offence* and certainly with advantage, *by all parties.* It is peculiarly calculated for *distribution*—it *repels no reader*, whatever be his politics or character.

Given *over* my hand and seal, March, 1813.

THE GOOD OLD WAY.

DARKNESS VISIBLE.

Mons. GALLATINI has directed the Lights near the Chesapeake and Nantucket to be extinguished, so that our Enemies may be left in the dark; and of course it will be perfectly easy to place Torpedoes under the Fleet and blow it out of our waters.—How perfectly in character is it for our present rulers to *extinguish* instead of illumining, and to scatter *darkness* where their predecessors diffused *light.* It is the sport of these Proclamation-men, like Mad Tom,

"To play at bowls with sun and moon,
"And shade them in eclipses."

It is expected that every man who does not wiſh to be drafted as a conſcript and ſent out of the State to fight the battles of the Southern Nabobs—every Farmer who wiſhes to prevent his farm and ſtock from being ſeized and ſold to pay the enormous taxes which are ſhortly to be laid on them, will turn out on the firſt Monday of April next, and ſupport the Peace Ticket.—New Hampſhire has ſet a bright example; let us follow it. You all know you were proſperous and happy under Waſhington; you are now wretched and miſerable under Madiſon.—You cannot, you will not give your ſuffrages for men who are in favor of the preſent war, unleſs you are unjuſt to yourſelves, and unfaithful to your country.—*Auguſta Herald.*

Salem,
FRIDAY, APRIL 2, 1813.

SOUTH with a Vengeance!

The Senatorial Diſtrict made and called "ESSEX SOUTH *Diſtrict*," by a Gerry-mander Legiſlature, has the following *peculiarities*:

1. It contains three Towns (Saliſbury, Ameſbury and Haverhill) more *northerly* than any other town in the County.

2. An Eaſt and Weſt line drawn through the centre of the County leaves Saliſbury, Ameſbury, Haverhill, Methuen, a large part of Andover and a ſmall part of Middleton [ſix] on the *Northern* ſide.

3. The only towns wholly in the *ſouthern* ſection of the county that the diſtrict contains, are Danvers, Lynnfield, Salem, Marblehead and Lynn [five.]

4. The *northernmoſt* town of *another county* is tacked on it.

☞ There neighbor, take a piece of chalk and ſee if it does not come out ſo.

Meaſures for Defence have been adopted in all the ports threatened by the Britiſh Squadron. At Wilmington a Committee of Safety has been appointed. The Honorable Meſſrs. BAYARD and HORSEY, (U. S. Senators from Delaware) and other zealous Federaliſts conſtitute this Committee. Theſe are the Patriots on whom the country muſt rely when the hour of extreme peril approaches. Our democratic rulers have ſufficient talents to plunge the nation in miſery and calamity, but none but Federaliſts can reſcue the nation and ſave its "drowning Honor." The deceaſed Heroes WASHINGTON, LINCOLN, KNOX, WAYNE, MORGAN and HAMILTON,—the Soldiers of Liberty who atchieved our Independence,—were zealous and decided *Federaliſts* to the laſt hour of their lives;—and even WASHINGTON was calumniated and traduced as a Britiſh Tory by the Democratic party,—which party he ſolemnly declared in a letter to a friend was "A CURSE TO THIS COUNTRY." This ſame party is now ſcourging the country with War, and the People *feel* the truth of the words of WASHINGTON.

The Right of Suffrage.

ALL men will agree in the importance of having the right of ſuffrage properly defined by law; and moſt men, when uninfluenced by paſſion, will acknowledge the neceſſity of enforcing the law with the utmoſt ſtrictneſs, and can correctly calculate the immenſe benefits which would reſult from ſuch a courſe.—Theſe are the natural dictates of the underſtanding and of conſcience. But man in *theory* and in *action* is not the ſame being. In the diſpaſſionate moments of ſpeculation, he liſtens to the ſuggeſtions of his reaſon, and approves of the admonitions of his underſtanding; but in practice he acknowledges no law but convenience, and feels no ſatisfaction for his exertions except in triumph and victory.—To a perſon, whoſe aſtoniſhment had not been ſo often excited, till he had become incapable of new ſurpriſe, by the moral phenomena which are continually occurring in the political world in theſe days of ſtrange deluſion and of diſaſters; it would appear incredible, that a party of men, after they had had the opportunity of putting in

Figure 2: The O.G. (Original Gerrymander)

There's no question that the outline of a crooked reptile is doing a lot of work on your intuition. If this feels like an "eyeball test," that's exactly what it is, and it's a major driver in redistricting to this day. Thirty-seven states have some sort of shape regulation on the books, and in almost every case (at least until the map goes to court!) the eyeball test is king.

But the problem is that the eyeballed outline of a district tells you a very partial, and often very misleading, story. Take Alabama's 1st district (Figure 3), bounded to the south by the jagged Gulf coast and to the north by a pair of rivers.

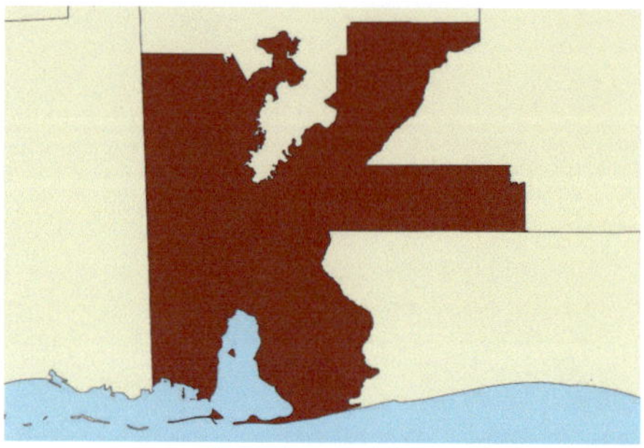

Figure 3: Alabama's 1st district.

The parts of its boundary that are not dictated by physical geography tend to follow county lines fairly faithfully. And county lines may be tortuous themselves, but you wouldn't want to punish a district for following them! (In fact, many states have rules telling you that district lines should follow county lines "to the extent practicable.") This spotlights a balancing act that is both real and often scapegoated: districters quite often will claim that other districting rules and principles forced horrible shapes on them. The plausibility of this claim.... varies, to say the least.

For instance, Figure 4 shows a pair of notorious districts, one from North Carolina and one from Maryland. Party politicians on both sides of the aisle claimed (and may have believed!) that the shape of NC-12 was forced on them by the Voting Rights Act. And at least one leading political figure asserted (but surely did not believe) that MD-3 had to look this tortured to hit a precise population number.[3]

Certainly there can be benign reasons for ugly shapes. Even more importantly, districts that are plump and squat and symmetrical to the eye offer no real seal of quality. For example, in the 2018 redistricting scrum in Pennsylvania, the state Supreme Court invalidated the 2011 Congressional plan and asked for a new one. Interestingly, the court order named a number of metrics that had to be calculated for any new plan, including five somewhat different scores of shape to be reported for each district, without specifying what role all those numbers would have in their decision. First crack at a new plan goes to the legislature, which had the opportunity to commission a new plan and to pass it as a bill. They didn't end up voting on it, but the Republican leaders Mike Turzai (House) and Joe Scarnati

[3]John T. Willis, the Democratic party stalwart who chaired the redistricting subcommittee, said: "It's a very complex situation, and population is the No. 1 driving characteristic. ...All of our congressional districts don't deviate by more than one person." See E. Batista, "Politics Makes Strange Bedfellows, Even Stranger Congressional Boundaries," perma.cc/P6Z4-S2NL.

Figure 4: Examples of scapegoating: North Carolina's 12th district and Maryland's 3rd, circa 2013.

(Senate) floated an alternative plan on Twitter before filing it with the court—so I'll call it "the Twitter plan." The Twitter plan achieved glowing compactness scores, under all five formulas specified by the court, relative to the much-mocked 2011 enacted plan that it was aimed at replacing. But the court found that despite its more pleasing forms, the plan locked in the same extreme partisan skew as its predecessor. So in Pennsylvania, you can get extreme performance with innocent shapes (see Figure 5).

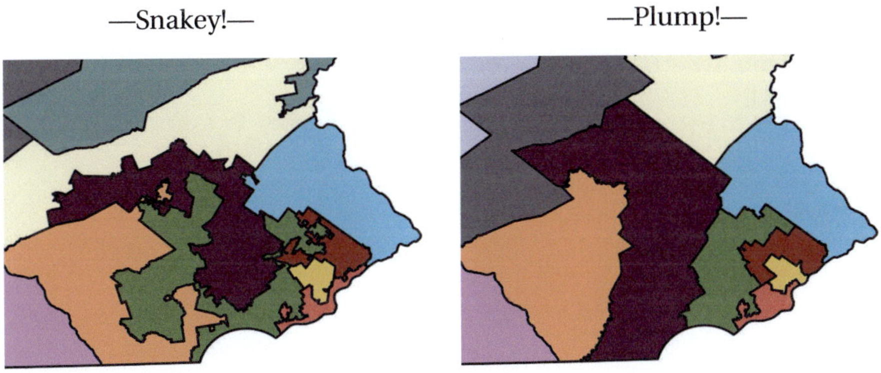

Figure 5: Philadelphia-area inset of the 2011 enacted Congressional plan (left) and the replacement map proposed in 2018 by Republican legislative leaders (the "Twitter plan," right). The replacement *looks* great, but both plans only have 4 out of 18 Democratic-majority districts when laid over the 2016 Senate vote pattern, which was nearly equal between the two major parties. (Compare Figure 12.)

In fact, some of the reason why shapes were often so flagrantly ugly in the past is not that horrible contours were strictly needed for more extreme partisan performance, but that the right kinds of pressure were not yet in force to rein them in.[4] What's more, the Twitter plan is not the exception—even strong shape imperatives may fail to constrain. Under scrutiny, line-drawers can often lock in all the advantage afforded by an ugly plan while keeping the shapes nice and plump.

[4]The era of shape-based legal invalidation really began in the 1990s with the so-called Shaw Line of Supreme Court cases (see Chapter 7), when the court grumbled about—but still began to engage in—"endless beauty contests" about district appearance.

0.2 THE RULES OF REDISTRICTING

Redistricting is made extremely complicated by a patchwork of rules that are typically unranked and often fuzzy to boot. Here's a quick primer on the "traditional redistricting principles" across the country, plus some that are less traditional but still make appearances.

Equal population—*Districts within a polity should all have very close to the same population.* The standard way to count is to use the Decennial Census numbers, which is one of the reasons why the Census is so important. This rule applies to the whole nation, and these days any two Congressional districts within a state will most often have a zero- or one-person difference in their Census count![a]

Contiguous—*Each district should be a single connected component.* You may be surprised to hear that only around 30 states require this property by law. This rule is mostly straightforward except when you're building from units that are themselves disconnected, or where there are water crossings to consider.

Compact—*The districts should be reasonably shaped.* ...Whatever that means! Language varies on this one, but for the most part it's a matter of the eyeball test. At least 37 states reference this principle.

Voting Rights Act—*The districts must not undercut the opportunity for minority communities to elect candidates of choice.* This has been a federal law on the books since 1965 and has a formidable (and formidably complicated) legal history and practice.

Communities of interest—*Groups with significant shared interests should be strategically placed in order to boost their voice in government.* While it's one of the most conceptually important, this principle is especially open-ended. Shared interests could be about industry, environment, or culture, and groups are sometimes better served by being kept together and sometimes by forming a significant part of multiple districts. More states will take concrete steps toward COI consideration in the 2021 redistricting cycle than in any previous cycle.

Political boundaries—*Counties, cities, and other relevant jurisdictions should not be split among multiple districts when there is a way to keep them whole.* In some states, this is phrased as a preference that district lines should follow political boundary lines.

Units—Some states prescribe which building-block pieces plans should be assembled from. For instance, Louisiana and New Mexico mandate whole precincts in legislative plans, and Iowa requires that counties be kept whole in congressional districts.

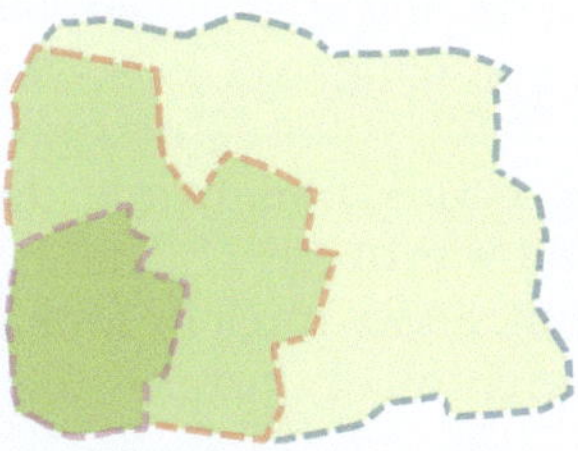

Nesting—Eight states currently require the state House districts to nest inside the state Senate districts two-to-one, and two additional states require three-to-one nesting.

Incumbency—In some states, there is a rule on the books that implies that *new maps should avoid pairing incumbents to run against each other.* (Pairing incumbents also goes by the colorful name "double-bunking"!) In other states, the rules forbid having the redistricters consider incumbency at all.

Partisan properties—A handful of states have rules indicating that there is a priority on the creation of *competitive* districts or districts that react *responsively* to changes in voter opinion, and numerous states have considered adopting language of that kind. Several other states forbid considering partisan data in the redistricting process.

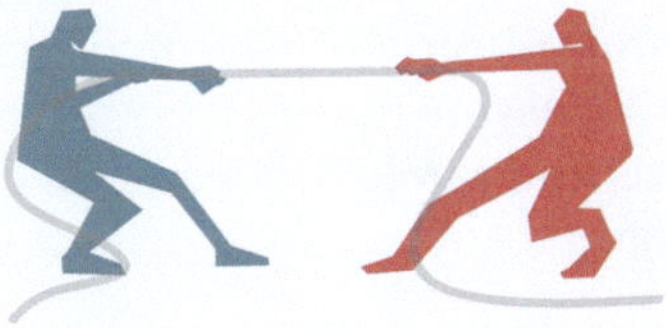

[a]ncsl.org/research/redistricting/2010-ncsl-redistricting-deviation-table.aspx

1.2 DISPROPORTIONATE OUTCOMES

So district shape will not do the trick on its own. How about if we cut out the middleman and get right to the bottom line, studying the extent to which the representatives match the electorate. Many people hold the strong intuition that disproportions give prima facie evidence of abuse. That is, a group with 30% of the votes *would have gotten* 30% of the seats, if the lines had not been rigged.

But not so fast. Let's zoom in on a particular case to understand some of the root causes of disproportionate outcomes. We'll look at a subgroup that reliably has over 1/3 of the votes but is locked out of even 1/9 of the Congressional representation: Republicans in my home state of Massachusetts.[5] This is a situation where even

[5]Let me flag at the outset that it's hard to directly measure people's party preferences for Congress

if your heart expects or desires a proportional outcome, the structure gods are cruel—it can't be done.

If you consider the elections for President and U.S. Senate held in MA since the year 2000, the Republican share of the statewide vote is most often between 30 and 40%, averaging over 36%. Since that's well over a third of the vote and we have nine seats to fill, you might expect a fair map to send three Republicans to the House in each cycle; meanwhile, the last time a Republican won *any* MA Congressional district was in 1994.[6] That is thirteen straight election cycles of total Republican lockout. So we must be looking at a vicious gerrymander that denies Republicans their rightful opportunity districts, right?

Except the mathematics here exonerates the Bay State line-drawers. The Bush-Gore election in 2000 is a great example. There is literally no way to put together a subset of the state's 351 towns making up enough population for a district—no matter now disconnected and scattered—that preferred Bush. That sounds like a paradox, but it's easily explained. Though Bush won 35.2% of the statewide vote, only 32 towns preferred Bush outright, making up under 3% of the state population. Preferences were very flat around the average, and there just aren't enough Bush-majority towns to anchor a district, no matter how cleverly you group them.[7]

The state started reporting more granular precinct-level results just after that, giving us an opportunity to see that the pattern held up in Massachusetts all through that Census cycle. Kenneth Chase, the Republican challenger to Ted Kennedy in 2006, cracked 30% of the statewide vote. But once again the districting numbers don't shake out for Chase voters. It is mathematically impossible to create a single district-sized grouping of *precincts* that preferred Chase; this is a realistic redistricting setting because precincts are typically preserved whole in Massachusetts legislative plans and rarely broken up in congressional plans. Chase voters simply were not clustered enough for a district to give them access to representation.

The problem is that even though Republican voters are nearly a third of the state, they are also about a third of every town and a third of every precinct—and a third of every household, as far as I know!—so no combination of units can combine to form a Republican majority, even if you throw niceties like compactness and contiguity to the winds. And this phenomenon carries over to any group in the numerical minority. You need a certain level of nonuniformity in the distribution for districting to offer even a theoretical opportunity to elect. The takeaway is that districts are ineffective if a minority is dispersed.

---

from Massachusetts, because the races are so often uncontested, as five out of nine seats were in 2016. Also, like many states, Massachusetts votes Democratic for national office but loves its Republican governors. But really I just want to make a point about the consequences of certain distributions of *votes*, so we can look at statewide elections for federal office—Senate and President—to understand that.

[6]If there are nine districts, each has about 11% of the state population—since you just need a plurality to prevail, you should be able to control a district with just 6% of the statewide vote. (The apportionment for Massachusetts dropped from 10 to 9 during this timeframe, but I will stick with 9 districts to simplify the discussion.) So Republicans routinely get more than *six times* the vote support needed to control a district.

[7]Exercise for the enterprising reader: collecting units in order of Republican vote share is a less effective greedy strategy than going by Republican margin per capita. You can find an appropriate sorting lemma in Duchin et al. [1].

So as a system of representation, districting doesn't start out looking like it will provide strong guarantees for minority groups (which includes Republicans in Massachusetts). And that's only looking at the population shares in the units and not at their spatial arrangement, which often compounds the difficulty. Even the mildest constraints on shape, like requiring that each district be one connected piece, make it harder to convert scattered votes for a minority-preferred candidate into representation. In a fairly ironic turn, this means that minority groups with the most strongly segregated geographic patterns—like racial groups historically targeted by discriminatory housing policy—may be in the *best* position to leverage the system of districts to secure representation. Spatially dispersed groups have no hope. Suddenly it looks unreasonable to expect that representative democracy can make good use of winner-take-all districts.

We have identified a problem with the system: districts beget disproportion. Let's look to mathematics to measure the extent of this problem, and to try to understand some of the mechanisms that cause it.

2 THE UNIVERSE OF POSSIBILITIES

We want to understand how districts might be able to provide some minority representation within a majoritarian paradigm. More broadly, we just want to understand what they can and can't do. It would be enormously useful to be able to survey all of the possible districting plans that satisfy some basic constraints, and then reason from there.

Mathematicians like to ease into a hard problem by first abstracting to a "toy problem"—an ultra-simplified model that helps illuminate structural issues. So instead of directly tackling the question at hand (what are all the ways to divide the geographical units in a state into nearly equipopulous districts?) we'll start much simpler: redistricting a small square grid with homogeneous units. This is already hard, as it turns out.

2.1 NUMBER

Maybe I've made some progress in convincing you that neither weird shape nor glaring disproportion gives a sure stand-alone indicator of gerrymandering. If you want to evaluate whether an election result should make you distrust the districts, you should really be comparing the plan to *other possible ways of districting the same jurisdiction.* The catch is that studying the universe of possible plans becomes an intractable problem.

When you think about "big data," you probably think of space exploration and medical imaging. It may come as a surprise that the humble math problem of how many ways to cut up a square pie belongs in the same conversation.

Think of a simple 4×4 grid, and suppose you want to divide it into four "districts" of equal size, 4 units each. The only requirement is that the districts should be contiguous. If we imagine the regions on a chessboard and we interpret contiguity to mean that a rook (traveling vertically and horizontally) should be able to visit the whole district, then there are exactly 117 ways to do it, summarized in Figure 6.[8] I'll denote rook-contiguous partitions of the $n \times n$ grid into k equal districts as the $n \times n \to k$ problem, for short. A cleverly programmed laptop can generate the $4 \times 4 \to 4$ solutions instantly.

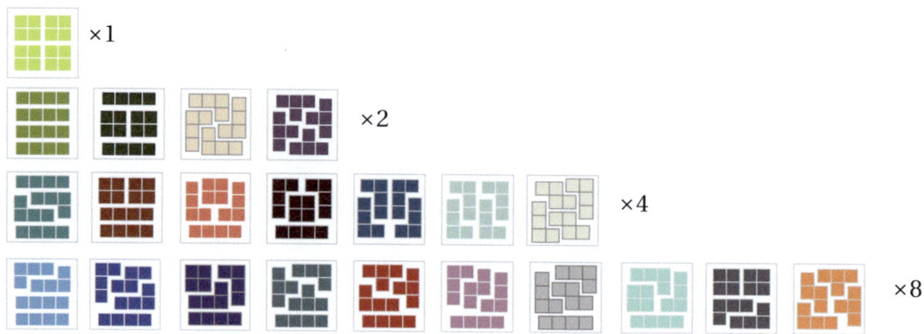

Figure 6: The 4×4 grid has 117 four-district plans—start with the 22 types shown here and apply rotations and reflections to get the full list. Try for yourself! No matter how you rotate or reflect the first plan, it looks the same (so it only contributes $\times 1$ to the ultimate list), but each plan on the next row is one of a pair of variants (so they contribute $\times 2$).

But to my surprise—forgive me, I'm trained as a theorist—I've learned that it's not obvious how to get even a high-performance machine running the best known algorithms to count all the possible configurations in a reasonable amount of time. At the time of writing, our best methods can handle $7 \times 7 \to 7$ in seconds and $8 \times 8 \to 8$ in minutes, but the 9×9 is a much more formidable computing task and the 10×10 is out of reach. Now try 18 districts built from Pennsylvania precincts!—it's not only a far bigger problem (9059 units) but has a more complicated connection topology of the units, with no symmetry to exploit. Forget about getting an answer during a 10-year census cycle; this complete enumeration calculation almost certainly can't be done before the heat death of the universe.

One reason for that is that the contiguity and balance constraints are stubbornly nonlocal, meaning that if you just look in one small neighborhood you can't be sure that a district is globally connected or that it's the right size. And these requirements have a lot of bite: unconstrained, there are roughly $4^{16}/4! = 179$ million ways to label 16 grid squares as belonging to district 1, 2, 3, or 4. Balance (insisting that each district is of equal size) cuts it down to $\binom{16}{4}\binom{12}{4}\binom{8}{4}/4! = 2.6$ million. Contiguity without balance cuts it down to 62,741. And both together leave you with just 117. So "brute force" algorithms that have to check all possible labelings just don't scale. This seems to call for a clever idea and not just the determination to search exhaustively.

Unfortunately, the problem doesn't reduce in a nice way: knowing the full answer for smaller grids gets us nowhere at all with the $n \times n$. (In math-speak, the problem lacks *recursive* structure.) So to find the very large number of valid partitions, you're searching blindly in an exorbitantly larger ambient space.

[8]If corner adjacency is permitted—so-called "queen contiguity"—the number jumps to 2620.

2.2 CLUSTERING

Since we can't simply build out all the plans, we will need to start understanding what *features* of the problem have important consequences for the measurements we care about. If we are trying to divide a population of two types into districts, it really matters how that population is laid out over the area we are dividing. Let's call this the *political geography*. We've already seen that political geography doomed the hapless Republican voters of Massachusetts—they were too uniformly distributed across the units (towns or precincts) to secure representation. They were not clustered enough.

On the other hand, conventional wisdom in redistricting carries the strong view that 21st century Democrats are disadvantaged by *excessive* clustering. "Democrats pack themselves!" as the slogan goes—because the votes are densely arranged in dense cities, even party-blind redistricting tends to create wastefully high Democratic percentages in urban districts, causing inefficient *packing* (shares far higher than needed) in parts of the map and *cracking* (shares just below the winning threshold) in others. But the math is actually subtle.

If too little clustering is bad and too much clustering is bad, is there is a sweet spot? Let's explore in a toy grid. Below I've represented four configurations in a 6×6 grid, each with one-third of units marked with a club suit (Figure 7).

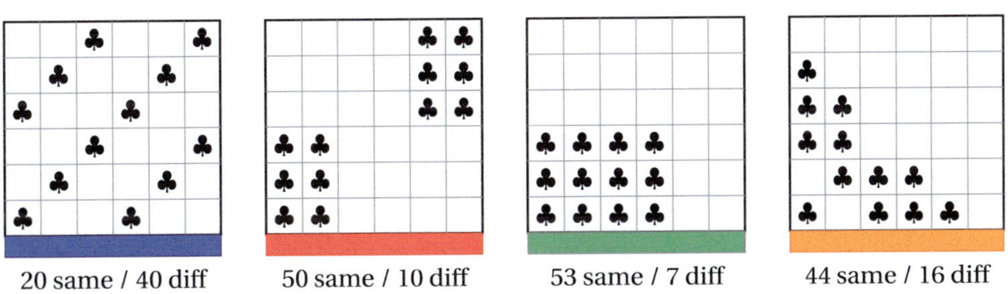

Figure 7: Spatiality matters! For each of these different ways of arranging 12 clubs voters, I've used a "same/different" count to measure clustering. by noting how many pairs of neighbors have matching or different symbols; for instance, 44 pairs of neighbors in the orange grid have the same marking (club-club or blank-blank) while 16 neighbor pairs are different (club-blank). In network science, this kind of same/different statistic is called *assortativity*. This captures something (but not everything) about the geometry of the configuration.

I can try partitioning these into six same-sized "districts" and see how much the layout matters, even while the vote share stays constant. Try it for yourself—some of these symbol layouts give you greater control of the outcome than others. Some spatial arrangements make it possible to lock out the clubs voters from representation entirely; in other arrangements, it's possible to overshoot proportional representation. For instance, I can shut out the blue grid's clubs voters by drawing vertical-stripes districts. The best I can do if I'm trying to *maximize* their representation, on the other hand, is to draw a plan that gives them two districts out of six, and that's not so easy to find. This world of possibility is almost disjoint from the one afforded by the political geography in the orange grid!

How the distribution of clubs votes relates to district outcomes is surprisingly subtle. But in this small example, it's the *most* clustered arrangement (green) that is in line with proportional representation (1/3 of the votes tending to earn 1/3 of the seats), and this is way better than the outcomes I should expect of a typical layout. We can see histograms summarizing *all* the possible ways of districting these grids in Figure 8. The very best layout possible for 12 clubs voters is the one shown in orange—the expectation is actually slightly super-proportional!—but others with a similar clustering score are not as advantageous. The spatial effects are stubbornly multidimensional; political geography is not captured in a single clustering score.[9]

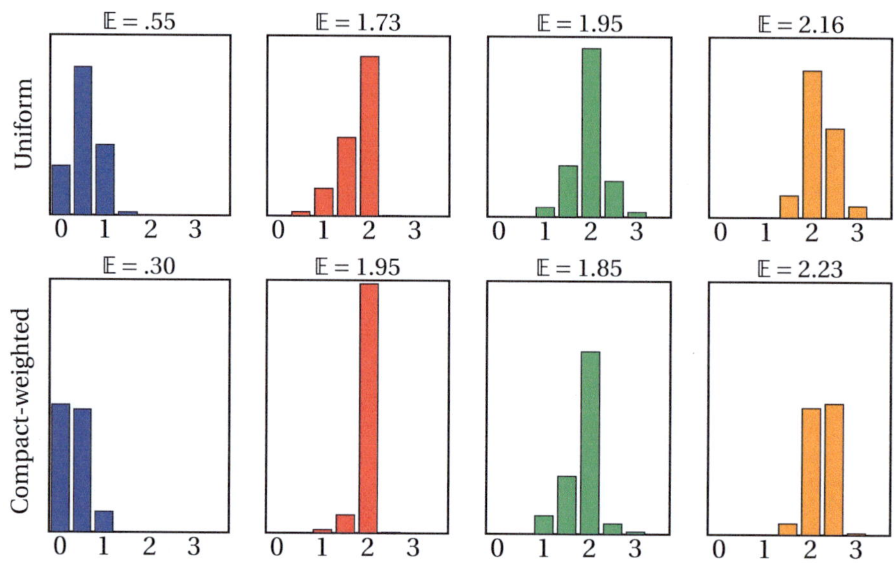

Figure 8: How much can I gerrymander? This plot shows how many seats would be won by the clubs party for every single way of districting the grid—there are 451,206 contiguous plans in all. (I gave clubs credit for .5 of a district if it got three out of six votes.) The top row is uniform: all plans are weighted equally, so for instance the most common outcome on the red grid is that two out of six seats are held by the clubs party. The bottom row shows the exact same set of possibilities, but where plans are weighted according to compactness—plump plans get heavier weight and snakey plans contribute more lightly to their histogram bars. (To be precise, this uses *spanning tree weighting*, which will be explained a bit further below.) So if there's a preference for choosing compact plans, the two-seats outcome becomes overwhelmingly likely on the red grid.

We were able to unearth considerable complexity in the problem by completely enumerating the plans for the $6 \times 6 \to 6$ districting problem. Now consider that I'm not able to construct all the plans even for a 9×9 grid, and I can't even *count* all the possibilities for a 10×10 grid or reasonably *estimate* the possibilities for Pennsylvania's precincts. How can I assess the consequences of the "political geography" to disentangle gerrymandering from the neutral consequences of districting?

[9]Or at least not *this* clustering score (meaning assortativity, or the "same/diff" count shown above), or any that I have seen or tried—see Chapter 15 for more discussion of spatial statistics. It would take a lot of space to provide enough examples to make this point fully, but you can play with spatial effects yourself at mggg.org/metagraph.

2.3 SAMPLING

This sounds like a hopeless state of affairs. We're trying to evaluate one way of cutting up a state, but without any measure of the size of the universe of alternatives, let alone a catalog of its properties to compare against. This sounds like groping around in a dark, infinite wilderness.

The good news is that even universes that can't be definitively mapped can often be effectively explored with random sampling. You don't need to talk to every American to conduct a good poll; you can use statistics from a representative sample to understand the wider universe. To do this well, you'll need to think about weighting and sample size. (We'll return to this below.)

There is a bevy of sampling techniques you might use for redistricting (Chapter 16, Chapter 17). Instead of profiling those, let's stay broad. What it is to be a representative sample in any context?

Building a Sample

Step 1. Come up with relevant categories or types;

Step 2. Construct a raw sample that encounters all relevant types;

Step 3. Re-weight the raw sample to reflect the population you want to represent.

Let's stick with polling to illustrate some of the issues in play. There are a lot of ways to fail as a pollster! Suppose my ultimate goal is to get a sample of intended voters that is representative of the electorate. If my whole poll is conducted by cell phone calling, then I will entirely miss some kinds of people—those who don't have cell phones, or those who don't pick up from an unknown number. If a lot of people hang up on me when they hear my first question, I'll have too few responses from a certain type of voter. In order to counteract the over-representation and under-representation in my raw sample (relative to the electorate), I need to do work to come up with relevant categories, such as "Angry White Guy Who Thinks Coronavirus Is A Hoax" (AWGWTCIAH). I will then need a sense of how much of the electorate is made up of AWGWTCIAH so that I can counteract the skew in my sample relative to the universe I want to represent. That lets me re-weight my raw sample so that AWGWTCIAH voting preferences are in balance.

If you are thinking "Well, I don't know any AWGWTCIAH!"—yes, that is kind of my point. A uniform distribution on your friends and family, or even a uniform distribution on the voting-age public, is not going to give you a sample that represents the electorate. It's easy to miss that a lot of fundamental conceptual work happens in Step 1 and Step 3. It's also easy to forget that if Step 2 fails, so that you *never* encounter certain types, it can't be corrected by re-weighting.

These reweighting ideas are crucially important in redistricting, because there is a type of silly and unreasonable districting plan that wants to dominate your sample if you let it! Namely, there is an over-supply of plans that are so wild and snakey and flagrantly noncompact that they look like fractals and put the original gerrymander to shame (Figure 9).

0.3 BUILDING RANDOM SAMPLES OF PLANS

Markov chain Monte Carlo, or MCMC, is an industry standard across scientific domains for impossible search tasks such as ours. It's a tool capable of decoding ciphertext, probing the properties and phase transitions of liquids, finding provably accurate fast approximations for hard computational problems, and more. (Persi Diaconis's classic 2009 survey [3] estimates that 10–15% of statistical work in science and engineering is driven by MCMC, and the number has probably only gone up since then.)

Essentially, the strategy of MCMC for sampling a collection of objects is just to take a *random walk* in the universe of objects and see what you see. In our case, you can start at any districting plan and make a random transformation to obtain another, then iterate as many times as you like. Then you can compare a proposed plan to the *ensemble* that you encountered on your random walk. It turns out that for many problems where solutions are hard to construct exhaustively, you can still sample quite efficiently if you have a well-designed engine for making those iterative alterations. You're building out what you need from a starting point, using chains of elementary moves.

So you take a million, a billion, or a trillion steps and look at the aggregate statistics. There's mathematics in the background (ergodic theory, to be precise) guaranteeing that if you random-walk in a suitable space for long enough, you'll hit a probabilistic steady-state. This allows you to collect a sample whose properties are representative of the overall universe, typically far before you've encountered every alternative.

I've been involved in developing a family of samplers called "ReCom" (or recombination) that are powered by large moves in the space of plans, and for which we have a good approximate description of how their ensembles ultimately distribute. Heads up: recombination-style samplers do not weight all districting plans equally, *and that's a good thing!* Weighting all plans equally would tend to put far more weight on noncompact districting plans than on plausibly compact ones, just because of how many more ways there are to be snakey than plump.

ReCom works by fusing two whole districts at every step, choosing a district skeleton called a *spanning tree*, then finding a place to snip the tree that leaves two population-balanced pieces behind. Once they converge, ReCom samplers draw from (approximately) the distribution that weights plans according to the number of spanning trees of their districts. (A different elementary move would target a different distribution; you can think of this as the *distributional design* of the random walk.) This distribution is precisely the "compact-weighted" one that produced the club-suit statistics in the bottom row of Figure 8.[a] The spanning tree distribution nicely blends visual compactness, in that it favors fat over spindly districts, and functional compactness, because it favors plans that have more connections *within* relative to connections *between* the districts.

Today the cutting-edge scientific questions concern better theoretical guarantees and convergence diagnostics, as well as efficient implementations. And together with all this, we must keep building persuasive ways of talking about it!

[a] This distribution gives the most compact partitions of the grid over 11 million times the weight of the least compact ones. Chapter 17 and particularly Sidebar 17.8 provide the ingredients to compute this for yourself!

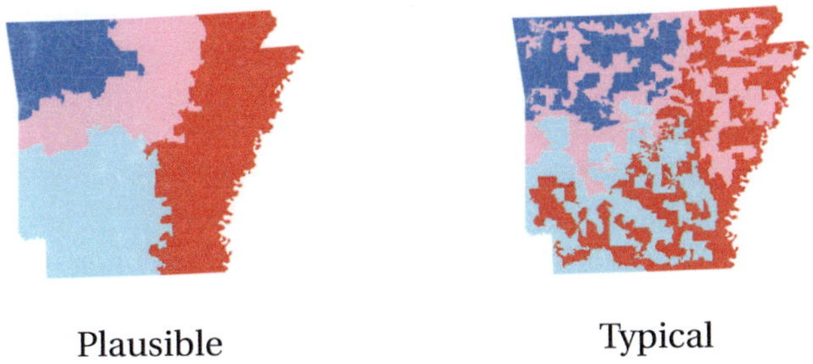

Figure 9: If you don't re-weight, your plans will look crazy! In a uniform sample where all contiguous and population-balanced plans are counted equally, over 99% of plans will look wildly noncompact. If you try to fix this by just setting a worst allowable compactness score, then 99% of plans will be at the worst level you allowed and you will essentially never see compact ones.

So finding good ways to sample while maintaining reasonable compactness is the name of the game—we need to over-sample or up-weight the compact plans in order to collect an ensemble that is plausible for the redistricting application, and we would like to do so by an elegant mechanism that leaves us able to explain why and by how much we weight some more than others. In my lab we've developed a method that weights plans in direct proportion to the compactness of their districts, and only that—no hidden factors. (See Chapter 17.) Then other plan criteria can be layered in. A uniform sample of all plans is both intractable and not that valuable; a *representative* sample of plausible, valid plans is the goal.

3 THEORY MEETS PRACTICE

What happens when we scale up from a small grid to, say, Pennsylvania? That is, how would plausible redistricting plans tend to look if drawn without a partisan agenda? Let's be careful: that's a really different question from locating partisan fairness, let alone living up to our highest ideals of representative democracy. And it allows us to stay in *descriptive* mode (what do plans made with no election data look like in partisan terms?) rather than making *normative* declarations (this is how party spoils "should" be distributed).

3.1 THE DATA

First, real-world modeling demands the collection and preparation of data. And, as always, there's a whole story to be told about that. Throughout this book and all of the debates about how best to study gerrymandering, there's a fundamental need for a certain kind of data that is very hard to get.

Americans vote in *precincts*, which are geographical units for election administration that usually have a population of a few hundred to a few thousand. (Typically, each precinct has a single polling place, but it's sometimes several-to-one in either direction.)

For instance, the math department at Tufts is in a small building that has two entrances, each in different precincts. The main entrance is in one precinct (population 3902 in the 2010 Census), and the back entrance is in another (population 3567). Election results are typically reported in cast vote totals per precinct. So the MA Secretary of State website tells us that the front-door precinct went 1270–120 for Clinton over Trump in 2016, while the back-door precinct went 1013–167. What's more, those precincts are in different cities and different Congressional districts! The front door is represented by Katherine Clark and the back door is represented by Ayanna Pressley.

0.4 THE PRECINCT PROBLEM

At the Voting Rights Data Institute in 2018, a team led by Ruth Buck, Katie Jolly, and Katya Kelly put dozens of students to work on figuring out the precinct boundaries from the 2016 elections in Ohio. We called all 88 counties to ask the simple question: can you send us a map of your precincts? 46 counties sent shapefiles, which are the industry standard digital format for spatial data. 27 counties had PDF maps. 8 counties sent paper maps (including highway maps with marker and tape!). 7 had nothing. We spent hundreds of person-hours digitizing and georeferencing the maps to build a statewide shapefile.[a] And that was for one year's elections!

So whenever you see highly granular color-coded maps of election results, think twice about the accuracy of the dataset. (I'm looking at you, New York Times![b])

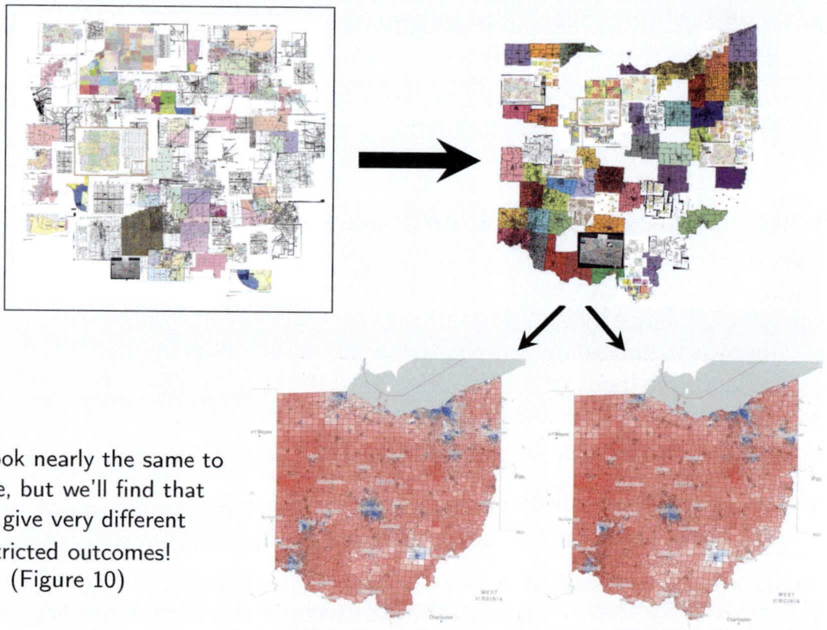

These look nearly the same to the eye, but we'll find that they give very different districted outcomes! (Figure 10)

President 2016 vote
45.74% D / 54.26% R

Senate 2016 vote
39.04% D / 60.96% R

[a]The artistic rendering here is by Emilia Alvarez. See github.com/mggg-states for details.
[b]"An Extremely Detailed Map of the 2016 Election," 25 July, 2018.

Since the precincts are where the votes are reported, you really need to know where they are located to analyze the impact of districting lines. But believe it or not, in most states in the country, *nobody knows where all the precinct boundaries are at any given moment.* That's because local election officials—usually county officials, but in some states like MA and WI it's actually town officials—have the authority to administer elections and to change the precincts, and in many states they have no reporting requirements, so even the secretary of state is not kept abreast of changes.[10] For the rest of the chapter, we'll restrict ourselves to states where it is reliably reported, or where someone has painstakingly assembled it.

3.2 POLITICAL GEOGRAPHY

Data in hand, we'll churn out a party-blind sample of plans and see how they cut up the votes into representation. This gives us a capsule summary of the consequences of political geography. For the following three sets of three states, we've built 50,000 random plans that are compact, contiguous, and population-balanced for each state. Each of these states had a U.S. Senate election in either 2016 or 2012; I have laid the random plans over the voting pattern for the Senate election and for the Presidential election the same year. For each plan in the ensemble, I then report the same summary statistic: how many districts have more Democratic than Republican votes? The outputs show us the "mere" mathematical consequences of single-member districts interacting with recent American political geography.

What can we see at a glance? I see two forces here, sometimes aligned and sometimes in opposition.

- A tendency for districted systems to underrepresent minorities, in this case minority party preferences;
- A Republican advantage.

In the states that lean Republican overall (Figure 10), these point in the same direction and combine to create a world of districting plans that favors Republicans by several seats, so that a proportional outcome (marked by the red line) is rarely observed under blind districting. This is most dramatic in Ohio's Senate pattern, where even a 25% (4 out of 16) showing for Democrats is rare, despite their 39% vote share. It is fundamentally important to note that this does not mean that one couldn't draw a plan with six Dem seats—on the contrary, it is sometimes quite easy for a person to draw by hand a plan with properties never observed in an ensemble. But in the world neutrally constructed by the rules, if Ohioans cast votes in the pattern that they did for Rob Portman against Ted Strickland, but it would probably take more than the lifetime of the candidates for Team Strickland to draw a map from a neutral process that would give him a proportional showing.

[10] I'm really not exaggerating. See for instance this court filing by the Pennsylvania legislative leaders (perma.cc/ECU9-PATG), or this statement by the Virginia Division of Legislative Services (perma.cc/KE85-3A95).

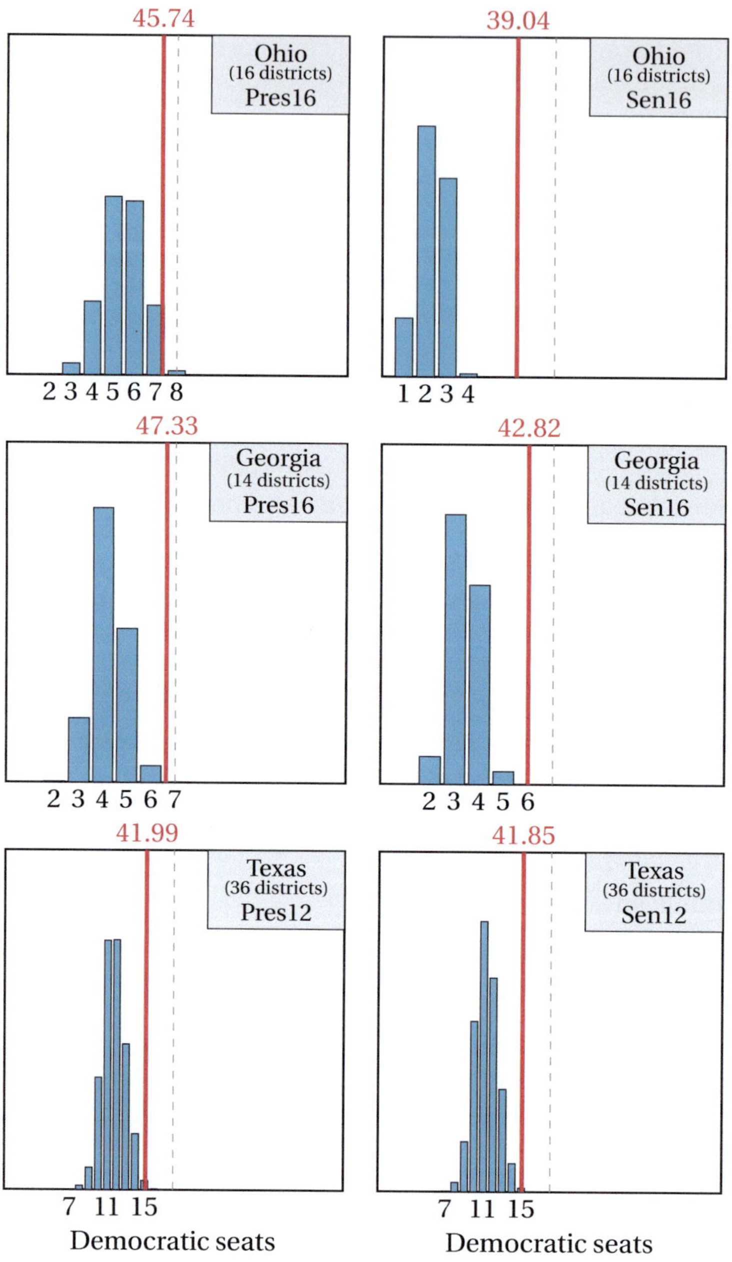

Figure 10: **Consequences of political geography**: how many Democratic-majority districts would result from "blind" redistricting (using no partisan data) in Republican-leaning states? The box at top right tells you what state, how many districts, and which statewide vote pattern is used in the background. The dotted line is placed at 50–50, while the red line marks the statewide Democratic vote share in that election. The histogram shows a neutral ensemble of 50,000 compact, contiguous, population-balanced plans made without partisan data. For example, we observe that a proportional outcome for Democrats in Texas Congress would be 15 seats for either 2016 vote pattern (see red line), but random plans (in blue) very rarely achieve that.

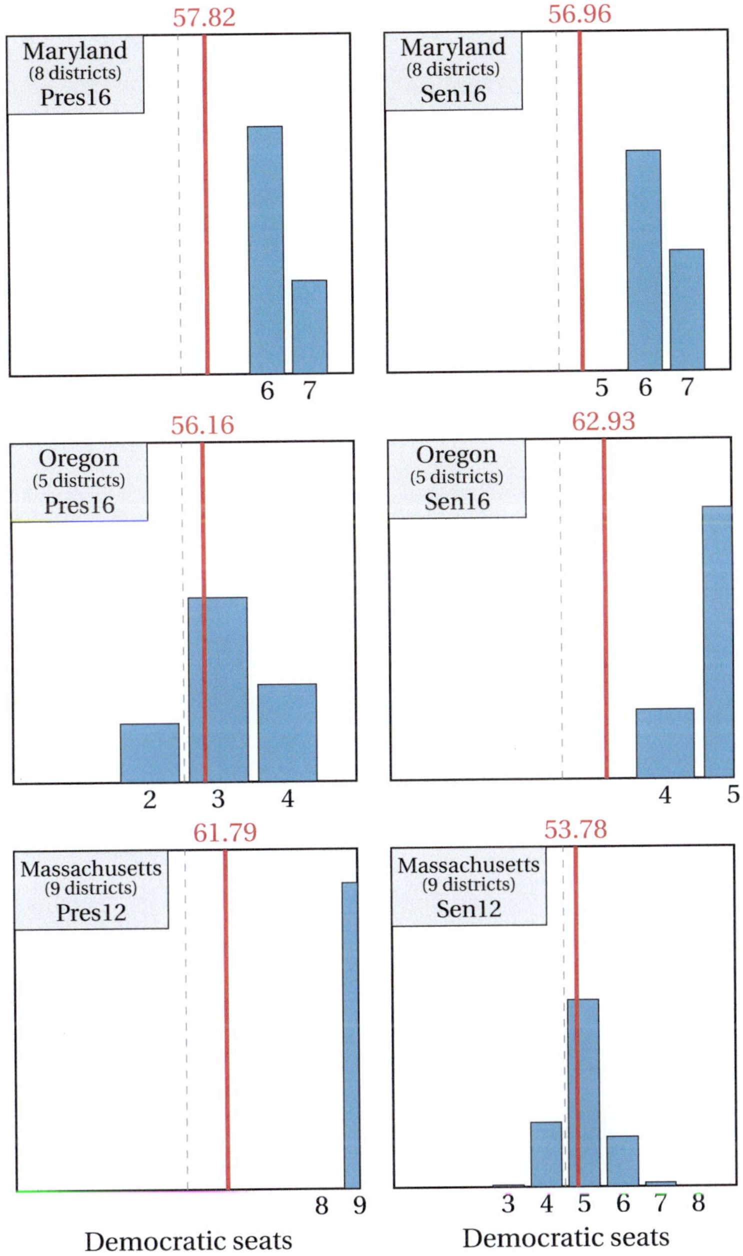

Figure 11: **Consequences of political geography**: how many Democratic-majority districts would result from "blind" redistricting (using no partisan data) in Democratic-leaning states? The box at top left tells you what state, how many districts, and which statewide vote pattern is used in the background. The dotted line is placed at 50–50, while the red line marks the statewide Democratic vote share in that election. The histogram shows a neutral ensemble of 50,000 compact, contiguous, population-balanced plans made without partisan data. For example, we observe that more than half the plans in the ensemble give a proportional outcome in Massachusetts with respect to Senate voting, but no plans get close to proportionality for the Presidential vote pattern. This squares with the earlier observation that elections with two-thirds/one-third vote preferences in Massachusetts produce a district lockout for the minority party.

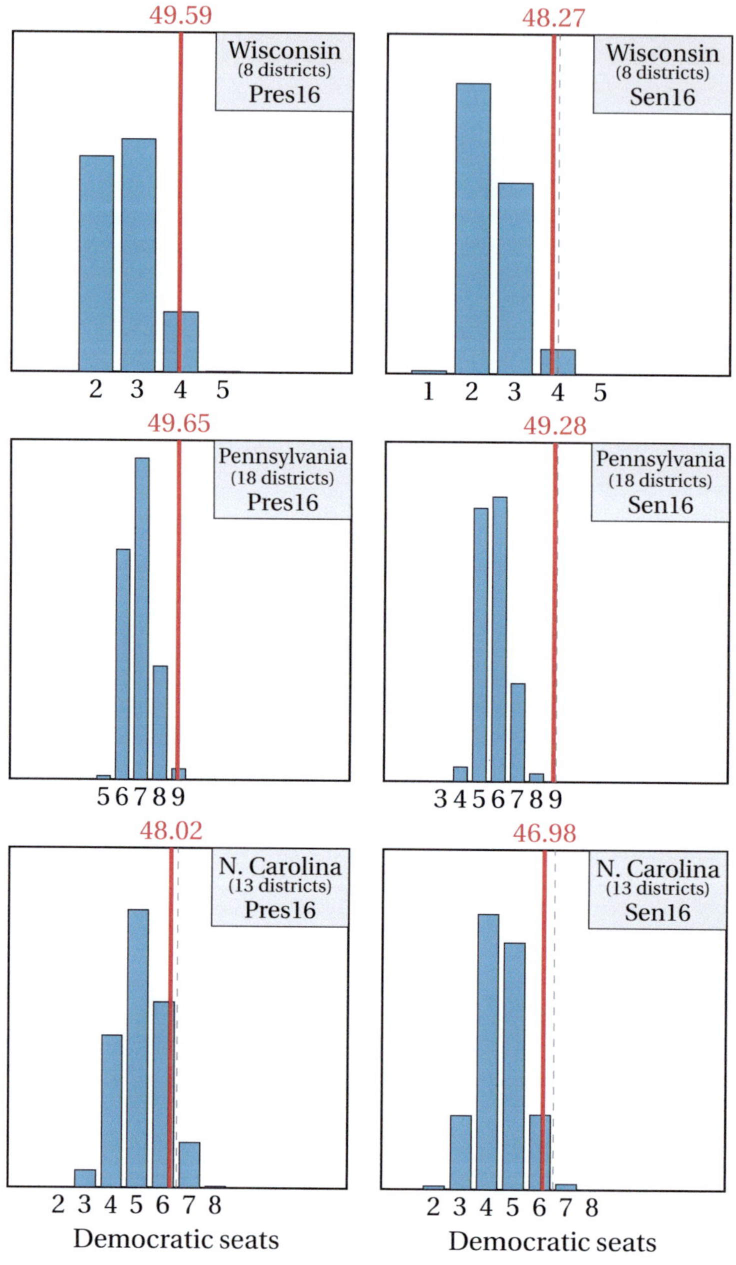

Figure 12: **Consequences of political geography**: how many Democratic-majority districts would result from "blind" redistricting (using no partisan data) in very close states that were litigated as Republican gerrymanders? The box at top right tells you what state, how many districts, and which statewide vote pattern is used in the background. The dotted line is placed at 50–50, while the red line marks the statewide Democratic vote share in that election. The histogram shows a neutral ensemble of 50,000 compact, contiguous, population-balanced plans made without partisan data. All of these examples show a tilt by several seats in the Republican direction, with respect to proportionality.

The outcomes in Democratic-leaning states (Figure 11) are consistent with a tension between these forces; look at Oregon/Pres16 and Massachusetts/Sen16, for instance, where a small but substantial lead for the Democratic candidate has not reaped the same benefits for Democrats as it did for Republicans in the earlier set of examples.

In the last set (Figure 12), we see that razor-thin statewide margins do not induce histograms centered around a 50-50 delegation. Since the races are so close to even, the explanation must be spatial, having to do with where the votes fall and not just with their overall balance. This, then, is the political geography in action.

Why the anti-minoritarian outcomes? Once this is observed, it's not that hard to explain the math: having 1/3 of a population, say, does not translate to having greater than half of a random sample 1/3 of the time. This produces consistent structural disadvantage for minority groups in districted systems. Smaller groups can only be rescued by spatial concentration, which lets a global minority appear as a local majority. These effects were previewed with the club-suit voters of Section 2 above—only the most clustered configurations gave them ready access to good representational outcomes.

The precise causes of the partisan lean are harder to diagnose. For one thing, our discussion of the benefits of clustering is directly at odds with the conventional wisdom about Democratic disadvantage in contemporary American districting, where many authors have blamed *overconcentration* for Democratic shortfalls. (Namely, through the heavy Democratic lean of urban areas.) Personally, I have not yet heard a specific mechanism proposed for city effects that is amenable to modeling. So I think this remains one of the tantalizing open questions in the field: how does the human and political geography of many U.S. states in the 21st century interact with the mathematics of districts? And how do the features combine to produce such a gravitationally tilted playing field for the two major parties?

3.3 REASONING FROM ENSEMBLES

As promised, we've built a fairly powerful descriptive account of redistricting. Using hard-won datasets and carefully designed sampling techniques, we've set ourselves up to understand the neutral tendencies when districts divide up territory, letting the chips fall where they may against American voting patterns.

From here, there are several normative moves we can make—that is, we have some choices about what we'll regard as fair. One simple idea is to flag outliers as *unfair*. For instance, the wild-looking Pennsylvania plan from 2011 gives four Democratic seats when you lay it over the Senate 2016 voting pattern, putting it in a tail of the curve, and indeed in the tail that's most favorable to the party in charge. And the plumper and lovelier Twitter map? Also four.

It should not be surprising that all of three of the purple states from Figure 12 saw legal challenges on the basis of partisan imbalance (though Wisconsin's focused on legislative rather than congressional districting); the world is tilted in a Republican direction, and Republican legislatures still chose from the "far side of the bell curve" in each case.[11] They leaned in to their structural advantage.

But let's remember that taking up a negative norm (like "don't be an outlier") does not commit you to a positive norm (like "you must look perfectly typical"). The fact that 7 seats and 6 seats are at the peak of the bell curves for those election patterns in Pennsylvania can perhaps help us pull apart the effects of careful partisan design from the consequences of the system itself. As such, it might help us flag a gerrymander.[12] But are those neutral plans our best choices for a healthy democracy?[13]

4 ADDING THINGS UP

Proportionality along party lines is a ready benchmark, but it's not the only one that we might select. There are many alternatives, such as wanting every sizeable bloc and region to be reflected in the legislature; ensuring rotation of control by engineering for competition and responsiveness; promoting continuity in representation when it has proven effective; avoiding gridlock by buffering a small vote advantage to a more secure governing advantage; excluding violent extremists; and on and on. We can certainly achieve some suite of good-government goals in a districted system, especially when all the extrinsic factors are favorably aligned—but our chances of that are better when the lines are drawn by those who are well informed about the alternatives, and whose incentives align with legitimate societal goals.

We should never confuse neutrality with fairness, in any context, especially when the process itself is up for grabs. That is, we have to keep thinking and debating about our representational ideals and requirements—and seeing how best to draw districts that measure up. At some point that might even be too big of a strain, making alternative voting systems more attractive. We never signed up for blind districting, and even districting itself wasn't graven on tablets as a fundamental form of American government. As the properties of districting become clearer, we should keep asking whether and how we can get good outcomes, whatever we take good outcomes to be.

The sum total of some years of obsessive study of this stuff has left me with some big-picture views, many of which will be explored (or challenged!) by the authors in this volume.

[11] Relative to (Pres16,Sen16) voting, the legislatures' enacted plans give these Democratic party outcomes: Wisconsin (2,2), Pennsylvania (6,4), and North Carolina (3,3). The state courts ultimately issued replacement plans: Pennsylvania (8,5) and North Carolina (5,5). The Wisconsin case was federal and hit a dead end.

[12] As my friend and collaborator Jordan Ellenberg memorably put it not so long ago, "The opposite of gerrymandering isn't proportional representation; the opposite of gerrymandering is not gerrymandering." *The Supreme Court's Math Problem*, March 29, 2019. perma.cc/R6VJ-CDPM

[13] This question takes on a whole different fairness dimension applied to racial and ethnic minorities instead of mutable party preferences; see [2] for an extended discussion of the merits of race-blind redistricting.

4.1 THINGS I DON'T BELIEVE IN

1. **Presenting any single number as a metric of fairness.** Given all of the complexity of balancing multiple objectives, you might say that the premise that fairness could be captured in a single statistic is implausible on its face. After years of studying the various attempts to score fairness, I can tell you that it's unsupportable on closer review as well. This does not doom our ability to handle it well. We have a system for deciding when there's been a murder, and we're comfortable with the fact that this requires complex, interdisciplinary evidence that could come from chemistry, ballistics, psychology, etc. We don't expect it all to be captured in a "murder score."

2. **Especially any single number with a prescribed ideal.** People often feel that a non-gerrymandered map would have such-and-such a property in a state of nature. For instance, we've spilt a lot of ink on (a) the nearly universal intuition that neutral districting attains rough proportionality, and (b) the finding that it does not. It might be just as tempting to believe, for instance, that a non-gerrymandered map would have a roughly equal number of "wasted votes" for each party.[14] Likewise, if a certain election had a statewide average of 60% for one party, we might expect that a truly neutral outcome would have the same number of districts over 60% as under 60%.[15] But districted plurality outcomes are a highly geometrized and nonlinear affair, and there's no reason at all for these ideals to obtain in the absence of gerrymandering. Instead, if you want to know if a score is good, there's just no reasonable way around comparing to the others that are actually achievable in your particular setting.

3. **Redistricting as optimization.** When many technical people take up the redistricting problem, they import a paradigm that I think is unhelpful, which is to look for an *objective function* (some sort of grand unified score of goodness) and seek a plan that *optimizes* it. But redistricting never has been—and shouldn't be—a literal matter of finding the best map, even if you *did* have a reliably informative score in hand. The whole project of drawing good territorial districts is about capturing community while working with an eye to representational balance, and this means it's properly a human and holistic affair. We wouldn't and shouldn't be satisfied that one plan is chosen over another qualitatively very different one because its score is better in the fifth decimal place. Computational redistricting can help people to understand tradeoffs and generate novel alternatives, but it works best as assistance rather than outsourcing.

4. **Redistricting as a game.** While computer scientists reach for optimization, economists often reach for game theory as a paradigm to make sense of fairness. Here, the state might be regarded as a resource that should be divided "fairly" among game-playing agents.[16] But this typically sets up redistricting as an adversarial game between the political parties, and all the other qualities that make a plan matter fall away to ancillary status.

---

[14] This is the efficiency gap test.

[15] This is roughly what is called the mean-median test.

[16] There is really some beautiful math in this direction. See Pegden–Procaccia–Yu, Landau, etc. [4, 5]

5. **Confusing rules of thumb for laws of nature.** When things are hard to model, it's totally reasonable, even essential, to use rules of thumb. But often a shortcut gets elevated to a method and the simplifications that made it useful turn into conceptual baggage. From "uniform partisan swing" to "the seats-votes curve" to a "vote index" that captures the sum total of a state's likely voting behavior, this field is full of those. Interdisciplinarity can sometimes provide needed conceptual independence from standard constructs, making space for novel ideas.

6. **Gerrymandering tests only a computer can pass.** (Or even worse, gerrymandering tests that only your own algorithm can pass!) This is a clear risk when computational indicators are elevated as badges of fairness. For instance, imagine that we are presented with a human-generated string of 0s and 1s and we're trying to judge if they were generated without a bias towards one kind of digit. That is really different from asking if they are statistically indistinguishable from binary seqeunces made from random number generators. It's well known that people are bad at imitating coinflips: a human trying to be fair is much more likely to write 01001010111011001010 than 01111100011100100001 even though they are equally likely for a perfect uniform coinflip process. Anomaly detection is important, but it's a whole lot less useful if it devolves into mere *human detection*. To avoid this, it's essential that gerrymandering tests be "ground-truthed" on plans that were made without gerrymandering intent.

7. **"Democrats pack themselves."** As I've argued above, the conventional wisdom around urban disadvantage is underdeveloped, and identifying mechanisms for this should be a big research question in the coming years. Meanwhile, the good news is that we can model the effect, and identify its magnitude.

8. **The tyranny of the median.** So you've got an ensemble of alternatives! What do you do with it? Demanding that a plan should fall right in the middle of the distribution to be deemed fair is a little bit like demanding that a coin should have exactly 50 heads in 100 flips. If you get 54 heads, we shouldn't conclude a slight pro-heads bias; that is well in the normal range.

4.2 THINGS I DO BELIEVE IN

1. **Arguing from alternatives.**
2. **Ranges, not ideals.**
3. **Using quantitative information to tell a qualitative story.**
4. **Mathematical modeling of rules and their consequences.**
5. **Community.**
6. **Humans.**

5 CONCLUSION: WHAT'S NEXT?

In working on redistricting, I have tried to look for the places where mathematical thinking and modeling can make an intervention that helps people to understand and clarify their representational goals. This has also led me to identify some contexts in which districts are just not the right tool for the job, and to study other systems of election (particularly ranked choice voting) that may better promote shared goals and ideals of fairness.

In the meantime we can work to build better districts. Since the last big decennial redistricting, quite a few states have shifted the way they draw the lines: voters in Colorado, Michigan, Missouri, Ohio, Utah, and Virginia all approved redistricting reform at the ballot box. As we've watched new independent or bipartisan commissions get spun up all over the country, they differ in the kinds of help they are seeking. Some of them are sticking with old-school consultants, or even dueling consultants (Democratic and Republican). Others are calling in the kinds of analysis you've seen introduced here, hoping to see where their plans fall in the world of possibility. At the same time, Congress is debating voting bills that contain elements of redistricting reform. Then there's the looming inevitability of litigation. In all of these settings, fancy algorithms and shiny metrics will be leveraged for litmus tests and beauty contests just as much as for measured analysis.

The spirit of this chapter, and I hope of this book as well, is to use the best available tools and perspectives from many fields to help us understand and improve this enigmatic, deeply American, and now widely mistrusted electoral device: plurality districts.

How should we govern ourselves? Who should represent us? How should we elect? To answer these questions requires reflection, not (just) calculation. With law and computing and geography and political thought colliding productively, and with reform energy all over the country, the 2021 redistricting cycle stands to be an innovative and exciting one.

ACKNOWLEDGMENTS

Warm thanks to Parker Rule, Olivia Walch, Zach Schutzmann, Emilia Alvarez, Talia Blum, Emmett Witkovsky-Eldred, Zach Wallace-Wright, Ari Nieh, and Dylan Thurston for contributing supporting material and/or helpful comments to this introduction. Thanks also to Scientific American; the starting point for this chapter appeared as *Geometry v. Gerrymandering* in the November 2018 issue.

REFERENCES

[1] Moon Duchin, Taissa Gladkova, Eugene Henninger-Voss, Ben Klingensmith, Heather Newman, and Hannah Wheelen, *Locating the Representational Baseline: Republicans in Massachusetts.* Election Law Journal 18 (2019), 388–401

[2] Moon Duchin and Doug Spencer, *Models, Race, and the Law.* Yale Law Journal Forum **130** (2021), 744–797.

[3] Persi Diaconis, *The Markov chain Monte Carlo revolution.* Bulletin of the American Mathematical Society 46 (2009), 179–205.

[4] Wesley Pegden, Ariel D. Procaccia, and Dingli Yu. "A partisan districting protocol with provably nonpartisan outcomes." arXiv preprint arXiv:1710.08781 (2017).

[5] Zeph Landau, Oneil Reid, and Ilona Yershov. "A fair division solution to the problem of redistricting." Social Choice and Welfare 32.3 (2009): 479-492.

1 *Explainer: Compactness, by the numbers*

MOON DUCHIN

A few years of hands-on work analyzing redistricting has left me convinced that "compactness" is over-emphasized as a cure for gerrymandering. But on the other hand, there is always room to augment the list of metrics for measuring it. The tools of 20th century geometry (let alone 21st!) have been slow to enter the conversation. Let's review the scope of common compactness scores and introduce a new one.[1]

SCORING SHAPES

The history of shape metrics in redistricting is itself long and winding. As we'll see, there are dozens of possible scores that have been proposed in the academic and legal literature, and most of them leverage (literally) ancient mathematics. Still, the most-used scores bear the names of their 20th century popularizers: Reock, writing in 1961 [10]; Schwartzberg, writing in 1966 [11]; and Polsby–Popper, writing in 1991 [8].

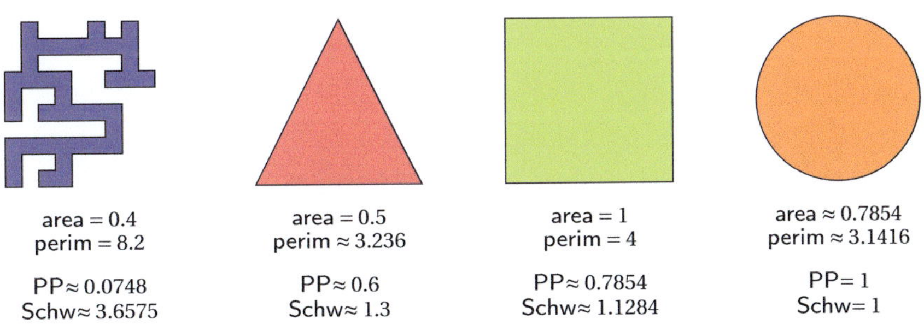

Figure 1: Shapes. If you're simply trying to measure the regularity or area-efficiency of a shape, a standard measure since antiquity has been to compare area to perimeter. The circle is the unique shape that maximizes enclosed area for a fixed perimeter.

The **Polsby–Popper score** of a region is *the ratio of the region's area to the area of a circle with the same perimeter*; bigger is better and 1 is ideal. The **Schwartzberg score** is *the ratio of the region's perimeter to the perimeter of a circle with the same area*. This time smaller is better and 1 is ideal. It's clear from the description that both of them report an ideal score when the region is a circle; it's a classical fact that scores like this *only* report ideal scores for the circle (see Figure 1). Brushing off your high school geometry and writing some formulas, these scores read as follows.

[1] This short treatment draws on ideas from Duchin and Tenner [4], which is written with an audience of political geographers in mind.

M. Duchin, O. Walch (eds.), *Political Geometry*, https://doi.org/10.1007/978-3-319-69161-9_1

$$\mathsf{PP}(\Omega) := \frac{4\pi \cdot \mathsf{area}(\Omega)}{\mathsf{perim}(\Omega)^2}; \qquad \mathsf{Schw}(\Omega) := \frac{\mathsf{perim}(\Omega)}{\sqrt{4\pi \cdot \mathsf{area}(\Omega)}}.$$

But wait—this format makes it clear that $\mathsf{Schw}(\Omega) = \mathsf{PP}(\Omega)^{-1/2}$. Since one score is simply the other score raised to a power, it is immediate that, although specific numerical values will differ, Schwartzberg and Polsby–Popper assessments must rank districts from best to worst in precisely the same way.[2]

But here is an interesting historical quirk. Because Joseph Schwartzberg worried that there was no way (with 1966 technology) to accurately measure perimeters of districts, he also proposed a notion of *gross perimeter*, which comes from a partial discretization, placing points along the boundary and approximating the perimeter by the successive distances between those points [11]. As a result of engineers taking this suggestion literally, software like Maptitude for Redistricting uses this alternative perimeter in the computation of a Schwartzberg score but not in the computation of a Polsby–Popper score, which of course can break the scores' monotonic relationship.[3]

Another collection of scores is based on comparing the district to some idealized relative (Figure 2). For instance, if $\overline{\Omega}$ is the circumcircle of Ω—or a bounding box, or the convex hull, or some other comparison figure—then we can define a score by the proportion of the area filled in by the shape; that is, we compute $\mathsf{area}(\Omega)/\mathsf{area}(\overline{\Omega})$. Of course, redistricting is about people, not acres and trees, so we might prefer to count population rather than land area.

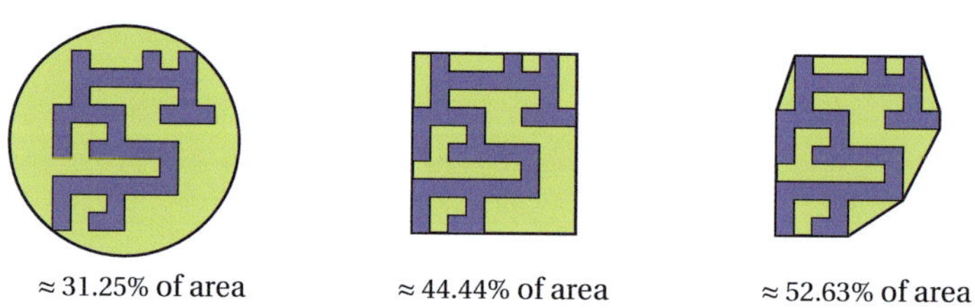

Figure 2: Comparing a shape to its circumcircle (left), bounding box (middle), or convex hull (right—intuitively called the "rubber band enclosure"). For each of these relative area scores, the ideal would be to fill in 100% of the comparison figure.

The **Reock** score, **convex hull** score (AKA "minimum convex polygon"), and **population polygon** score are, respectively,

[2]This is because for positive values of x and y, we have $x > y \iff x^{-1/2} < y^{-1/2}$. Therefore, a higher (and thus better) PP score corresponds to a lower (and thus better) Schw score.

[3]Maptitude is so dominant in the industry that all but one brief about a proposed remedial plan in the Pennsylvania case simply included a printout of the Maptitude report to describe the compactness of the plan (as well as for other metrics like county splitting). The only exception was written by me! [5] This is a good reminder of how thoroughly software mediates our interaction with the quantifiable side of redistricting.

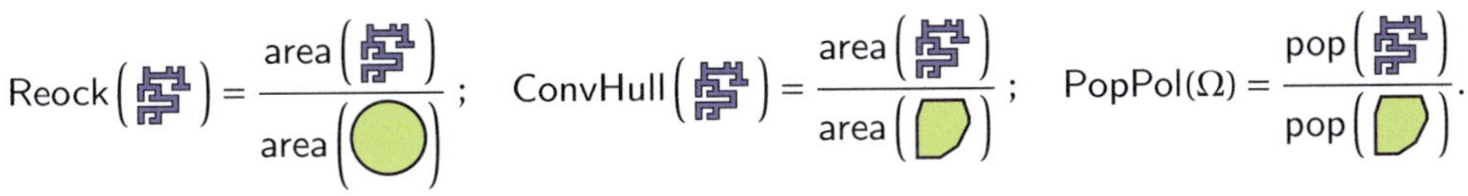

And now we have met the five scores that were cited by the Pennsylvania Supreme Court in the 2017–2018 redistricting challenge: Polsby–Popper, Reock, Schwartzberg, convex hull, and population polygon. Every party that submitted a remedial map was required to report these five scores for all 18 Congressional districts in the proposed map.

CONTEXT AND AGGREGATION

Armed with scores, we are ready to go! Here's a district: is it good or bad?

Well, it's got one very straight edge, which might be good. But it's pretty elongated rather than plump, so that might be bad. It's got some pretty "thin necks," which seems bad. And it's pretty windy and erratically formed, which is probably bad.

Our mystery district turns out to be Maryland's 6th (Figure 3), which was successfully challenged in district court as a pro-Democratic gerrymander in the case that eventually became Benisek v. Lamone. But its sins, such as they are, are not primarily geometric. MD-6 looks much more benign overlaid on a precinct map of the state.[4]

One more major problem demands attention: most of the zoo of compactness scores consists of individual district metrics. Even if it's clear how to compare two shapes head-to-head, how do you compare a set of 8 or 18 scores? For instance, we heard above that the Pennsylvania Supreme Court required petitioners to report 5 scores for each of the 18 districts in their plans. That means that each plan is assessed by 90 numbers, making it totally-not-obvious how to compare one plan to the next! Most petitioners reported the *average* of the 18 individual scores in each

[4] Caveat: the district is not actually made of whole precincts, but its precinct-level approximation, made using geospatial approximation based on population [6], is depicted here.

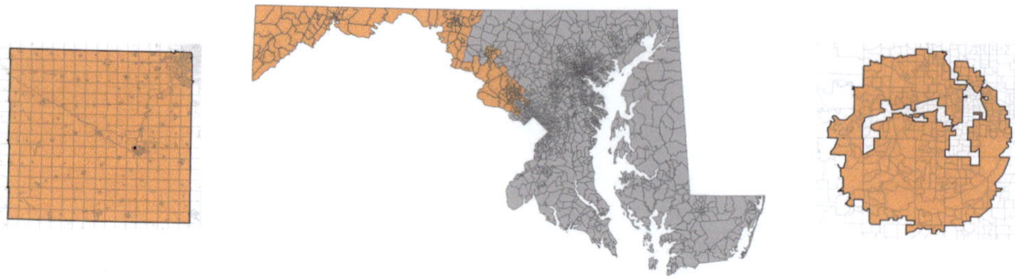

Figure 3: Left: You can build a square district when the units cooperate! But imagine trying to do that from the Maryland precinct units shown in the middle. Right: unreasonable districts can have good scores—in this case, a stellar Reock score. (Image credit: Amy Becker.)

metric, but it's rather unclear that that's reasonable. If you have a really terrible district, is it actually balanced out by a nice plump district somewhere else in the state?[5] The Pennsylvania remedial plan reported an average Polsby–Popper score of 0.33 across its 18 districts. Coincidentally, that's the same average Polsby–Popper score of the enacted Congressional plan in Minnesota across its 8 Congressional districts.[6] Given all the differences in the natural landscape, number of districts, shapes of the units, and all the rest, this does not feel like a particularly meaningful comparison.

BUT WHAT DO THE NUMBERS TELL US? AND WHAT ELSE CAN YOU DO?

Four out of the five scores mentioned above are pure shape scores, with no reference at all to the units, the population, or the particularities of the state and its districting setup—they just use area, perimeter, and some kind of old-school (we're talking *millennia*-old) plane geometry. You could compute all those scores for a Rorschach blot or a coffee stain just as easily as for a voting district. Even the population polygon score, which does make reference to the people and where they live, is still *contour-based*, in the language of Duchin and Tenner [4], which means that it is determined by the outline of the district on a flat plane. All contour-based scores will be sensitive—and sometimes highly sensitive—to things that don't matter for district quality, like the choice of projection from the Earth to a plane (see Chapter 13 and Bar-Natan et al. [1]) and the measurement precision of the winding boundaries [2]. Most will be majorly impacted by other electoral irrelevancies, like the assignment of unpopulated areas to one district or the next. On the other hand, these contour-based scores are insensitive to the physical geography (mountains, rivers, and other features of the natural and built environment) and to the units that were actually available to the districter as building blocks. It is not a great state of affairs when your metrics are heavily impacted by irrelevant factors but

[5]To drive this point home, consider that most of the scores are valued between 0 and 1. So if you raise them to any $\alpha > 0$, you get a new score between 0 and 1, but where averaging behaves differently!

[6]See `https://www.gis.leg.mn/redist2010/Congressional/C2012/reports/compactness.pdf` for the Minnesota report and `https://www.pubintlaw.org/wp-content/uploads/2017/06/attachment-1.zip` for the Pennsylvania remedial plan files and report.

not impacted at all by important features of the problem you are studying.

If we want to modernize the geometry of district shape, we should strive to build approaches that are (a) keyed to the geographical units, and (b) designed for ease of comparison to relevant alternatives. At the same time, any reasonable score should (c) comport with the all-important "eyeball test," both because of public optics and because the case law tells us that this matters [7]. Bonus points if the score is (d) attached to a mathematical formulation with good theory and good algorithms behind it. And finally a quantitative approach will always succeed better if it is (e) simply motivated and easily described.

There will not be a perfect compactness score for the 21st century, but there will be new ideas.[7] Throughout this book you will hear about the *cut edges* score of a districting plan. First, choose a redistricting setting (like Maryland Congressional districts or South Carolina House districts) and fix the units of the problem (like precincts or census blocks that you'll be using to build districts). Then you simply score a plan by the number of pairs of units that were adjacent in the state but were severed from each other by the division into districts (see Figure 4). This is directly based on the geographic units and doesn't care at all about how jagged the units themselves are. It doesn't need any averaging or summing to give you a whole-plan score, so it's set up very well for within-state comparisons. It does a good job of tracking with visual district appearance, as you can see in Figure 4. And it is extremely natural from a mathematical point of view: in combinatorics terms, it is the size of the cut-set for the graph partition. A host of theorems and algorithms exist that reference and leverage this notion.[8] Finally, I think it does pretty well on the simplicity scale: It measures the number of neighbors that are separated when cutting out the plan with scissors!

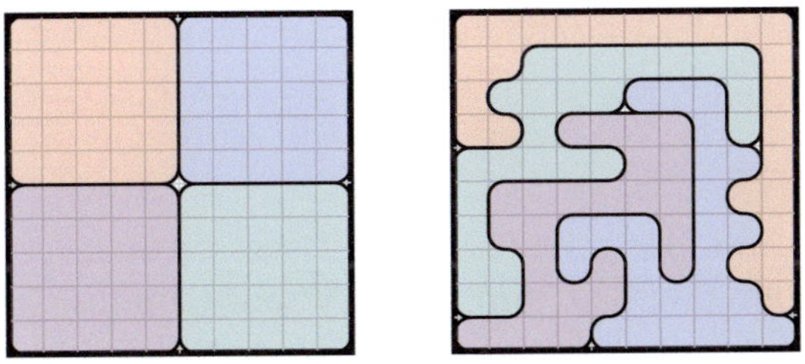

Figure 4: Cut edges as a measure of compactness. The plan on the left has just 20 cut edges, whereas the plan on the right cuts far more (73 out of the 180 edges in the grid). (Reproduced from DeFord et al. [3].)

[7] Having been asked about this point in a recent deposition, I'd like to be very clear that I'm proposing discrete geometry to *complement* traditional contour-based geometry, giving another vantage point on the efficiency or complexity of district shapes. Also, as you'll learn in this book, discrete geometry undergirds the powerful algorithms used to explore the universe of districting plans.

[8] To name just a few settings for cut-sets: the *max flow–min cut* problem is one of the foremost mathematical models of the 20th century, and important algorithms like Karger's algorithm were built to find minimum cuts. The *Cheeger constant* in geometry relates the sizes of cut-sets to the sizes of the separate pieces of the partition; there are theorems connecting the values of this constant h to various notions of curvature, expansion, and to spectral properties. Multiple authors, including Clelland et al., Procaccia–Tucker-Foltz, and Tapp, show evidence that the cut edge count correlates closely with the spanning tree count discussed in Chapter 17.

No score is going to do the hard, human, deliberative work of finding fairness in representative democracy. But there are still best practices for designing metrics, and they call for thinking about *the work that you want the score to do for you.* Your score should track the distinctions that you are most interested in flagging, and not superfluous ones. Being attentive to sensitivity, robustness, and the incentives created by a score—or the flip side of incentives, namely gameability—will help you to troubleshoot and improve it. And you won't get anywhere without trying your new metric out on real data. These are sound guidelines for *critical modeling* that will be valuable throughout the study of redistricting, and beyond.

WHERE TO READ MORE

In addition to the references cited in this chapter, those interested in learning more should check out the following resources:

- Stephen Ansolabehere and Maxwell Palmer. *A Two Hundred-Year Statistical History of the Gerrymander.* Ohio State Law Journal, vol. 77, 2016: 1–23.
- Daryl DeFord, Hugo Lavenant, Zachary Schutzman, and Justin Solomon. *Total Variation Isoperimetric Profiles.* SIAM Journal on Applied Algebra and Geometry 3.4, 2019.
- Aaron Kaufman, Gary King, and Mayya Komisarchik. *How to Measure Legislative District Compactness If You Only Know It When You See It.* American Journal of Political Science 3 (2017): 55-58.
- Richard G. Niemi, Bernard Grofman, Carl Carlucci, and Thomas Hofeller. *Measuring Compactness and the Role of a Compactness Standard in a Test for Partisan and Racial Gerrymandering.* The Journal of Politics 52 (4) 1990: 1155–81.

REFERENCES

[1] Assaf Bar-Natan, Elle Najt, and Zachary Schutzmann, *The Gerrymandering Jumble: Map Projections Permute Districts' Compactness Scores.* Cartography and Geographic Information Science, Volume 47, Issue 4, 2020, 321–335.

[2] Richard Barnes and Justin Solomon, *Gerrymandering and Compactness: Implementation Flexibility and Abuse.* Political Analysis, Volume 29, Issue 4, October 2021, 448–466.

[3] Daryl DeFord and Moon Duchin. *Redistricting Reform in Virginia: Districting Criteria in Context.* Virginia Policy Review, Vol. 12 No. 2 (2019), 120–146.

[4] Moon Duchin and Bridget Tenner. *Discrete Geometry for Electoral Geography,* preprint 2018. `https://arxiv.org/abs/1808.05860`

[5] Moon Duchin. *Outlier Analysis for Pennsylvania Congressional Redistricting.* Technical report, 2018. Expert report submitted in *LWV v. Pennsylvania.*

[6] MGGG Redistricting Lab, *MAUP: A geospatial toolkit for redistricting data.* GitHub repository. `https://github.com/mggg/maup`

[7] Richard H. Pildes and Richard G. Niemi. *Expressive Harms, "Bizarre Districts" and Voting Rights: Evaluating Election-District Appearances after Shaw v. Reno.* Michigan Law Review, Vol. 92, No. 3 (Dec., 1993), 483–587.

[8] D.D. Polsby and R.D. Popper, *The Third Criterion: Compactness as a Procedural Safeguard Against Partisan Gerrymandering.* Yale Law & Policy Review Vol. 9 (1991), 301–353.

[9] Ariel D. Procaccia and Jamie Tucker-Foltz. *Compact Redistricting Plans Have Many Spanning Trees,* Preprint. 2021. `https://arxiv.org/abs/2109.13394`

[10] Ernest C. Reock. *Measuring Compactness as a Requirement of Legislative Apportionment.* Midwest Journal of Political Science, Vol. 5, No. 1 (Feb., 1961), 70–74.

[11] Joseph E. Schwartzberg. *Reapportionment, Gerrymanders, and the Notion of "Compactness."* Minnesota Law Review, Vol. 50 (1966), 443–452.

I

Political thought

Chapter 2

Measuring partisan fairness

MIRA BERNSTEIN AND OLIVIA WALCH

CHAPTER SUMMARY

What does fairness in the context of redistricting look like? Can you identify a gerrymander based on election results alone? In this chapter, two mathematicians examine a variety of metrics that have been proposed in the courts as tools to detect partisan gerrymandering and to quantify its effects. The takeaway is that most of these metrics can lead to counterintuitive results. They are also unstable: slightly different conditions can yield markedly different outcomes. Fundamentally, these metrics share a common problem: you cannot interpret them without the context of what is "normal" for a particular state, based on the geographic distribution of its voters.

INTRODUCTION: WHAT IS A FAIR MAP?

In this chapter, we focus on **partisan gerrymandering**, in which boundaries of electoral districts are manipulated to give advantage to a political party. Given a **districting plan**, or electoral map, we'd like to be able to tell whether it has been gerrymandered and to quantify the advantage gained by the benefiting party. This issue arises frequently in gerrymandering lawsuits, and scholarship on this topic has been greatly influenced by decisions of the U.S. Supreme Court. While our focus will be on quantitative methods, we will include pointers to the necessary legal context along the way.

To begin with, any discussion of advantage requires a baseline: advantage relative to what? What would a districting plan that does *not* give any extra advantage to either side look like?

Finding the baseline turns out to be an extremely complex and controversial question. There are two fundamentally different ways one can define fairness in redis-

M. Duchin, O. Walch (eds.), *Political Geometry*, https://doi.org/10.1007/978-3-319-69161-9_2

tricting (as in many other contexts). One possibility is to focus on the map-making *process*: we could declare a map fair if it was drawn based on nonpartisan considerations, without the intent to hurt or benefit any party.[1] The other approach is to define fairness based on *results*: we could say that a districting plan is fair if it leads to fair electoral outcomes, i.e., outcomes that are consistent with some abstract notion of justice or equality.

In *Davis v. Bandemer*, the Supreme Court ruled that a districting plan is unconstitutional only if it violates both definitions of fairness: it must be motivated by "discriminatory intent" *and* have a "discriminatory effect" on voters of one party [1]. Over the years, the Court has considered several possible standards for assessing whether a plan's effect is discriminatory: **proportionality** (*Davis v. Bandemer*, 1986), **partisan symmetry** (*LULAC v. Perry*, 2006), and **low efficiency gap** (*Gill v. Whitford*, 2018). As we shall see, the Court had good reason to be skeptical of all three proposals. When elections are based on geographically defined districts, the outcomes depend on voter geography, and any standard that does not take this into account is bound to run into trouble.

It is only in the last few years that mathematical and technological advances have enabled researchers to evaluate the effects of a districting plan in a way that takes the geographic distribution of voters into account. This has led to the development of a new test of partisan gerrymandering, the **extreme outlier standard**, which takes on the task of disentangling the effects of gerrymandering from the effects of geographical districts. Unfortunately, even though this kind of work was in evidence in the last partisan gerrymandering case to come before the Supreme Court (*Rucho v. Common Cause*, 2019), the Court split along usual partisan lines to rule that partisan gerrymandering should not be adjudicated by the federal courts at all.

But of course, the fight against partisan gerrymandering is not over: it continues in state courts, state legislatures, and grass-roots efforts around the country. As we strive for fair maps, through both legal and political means, it is important to understand just how complex and contradictory the ideal of fairness can be. Quantitative methods can help us to contend with some of this complexity.

Ultimately, the question "what is a fair map?" is a philosophical one: it requires a normative choice (what *should* be done), and such choices cannot be dictated by any mathematical analysis. But mathematics can guide our decision-making by helping us to understand the implications of the choices we make; and, once we have chosen, it can help us to design more effective ways to implement our choices.

1 PROPORTIONALITY

When we give talks on gerrymandering to nonspecialists, we often start by asking the listeners a simple question:

[1] In this chapter, by "fairness" we mean exclusively *partisan* fairness. A map drawn without a partisan agenda could still be unfair in other ways—for instance, as a racial gerrymander.

> "Suppose Party *A* got 55% of the votes in your state. The state legislature has 100 seats. Ideally, how many of these seats should go to Party *A*?"

Invariably and unanimously, the answer is 55%. People are fine with small deviations like 52% or 58%, but by the time you get above 63% or so, pretty much everyone agrees that this is not ideal. In other words, most people's intuitive concept of fairness is **proportionality**: they want the number of seats that each party gets to be proportional to the number of votes it receives.

Because most people think of fairness in terms of proportionality, discussions of gerrymandering in the popular press are often framed in these terms as well. Here is a typical example from *The Washington Post*:

> "The 2012 election results give some sense of the extent of [Pennsylvania's] gerrymander. That year, Democratic candidates for the state's 18 U.S. House seats won 51 percent of their state's popular House vote. But that translated to just 5 out of 18, or a little more than one-quarter, of the state's House seats."[2]

There is, in fact, plenty of evidence that Pennsylvania's 2011 districting plan was a Republican gerrymander. Yet the disproportional results cited in *The Washington Post* cannot be used to prove this or to measure the advantage this gerrymander gave to Republicans. Counterintuitively, our system for electing representatives — dividing a state into districts, then choosing one representative per district — is not set up for proportionality. It is a system in which elections that are "fair" in the sense of process (no gerrymandering or partisan intervention of any kind) are highly *un*likely to produce proportional results.

To better understand what's going on here, we'll first introduce some formalism for discussing elections mathematically and then look at two real elections as case studies.

1.1 VISUALIZING VOTES AND SEATS

A districting plan (together with a procedure for choosing a representative in each district) forms an **electoral system**: a method for converting voter preferences into a choice of representatives. The results of every partisan election for a representative body (Congress, Parliament, state assembly, etc.) are often summarized in two sets of numbers: the fraction of **votes** that each party gets and the fraction of **seats** that each party wins. The fraction of votes can be calculated by simply aggregating the votes for each party across the state, but you might also want to calculate the average vote share across districts. These two values coincide when turnout is equal across all districts. (See Sidebar 2.1 for further discussion.)

For the remainder of this chapter, we will assume that every election involves only two parties. We arbitrarily choose one of the parties (in our real-world examples, it will always be Republicans) and look at the election from their point of view:

[2] *Ingraham, Christopher,* "How Pennsylvania Republicans pulled off their aggressive gerrymander", *The Washington Post,* 6 February, 2018

- We denote the Republican **vote share**—the fraction of the two-party vote that the Republican party received statewide in the election—by V.
- The Republican vote share in each of a state's N districts is represented by the vector $(v_1, v_2, \ldots, v_N)$, and the **average district Republican vote share** is denoted by $\overline{V}$.
- We denote Republican **seat share**—the fraction of available districts in which Republicans got more votes than Democrats—by S.

We can now visualize the results of a single election as a point in the **seats–votes plane**.[3] For instance, in Figure 1 we plot the results of the 2016 Congressional election in four states: Minnesota, Maryland, Ohio, and Michigan. It is natural to put vote share (V or $\overline{V}$) on the x-axis and S on the y-axis, because we usually think of vote share as an input into the electoral system that we are examining, while seat share is its output. Our goal is to understand how our system "converts" votes to seats.[4]

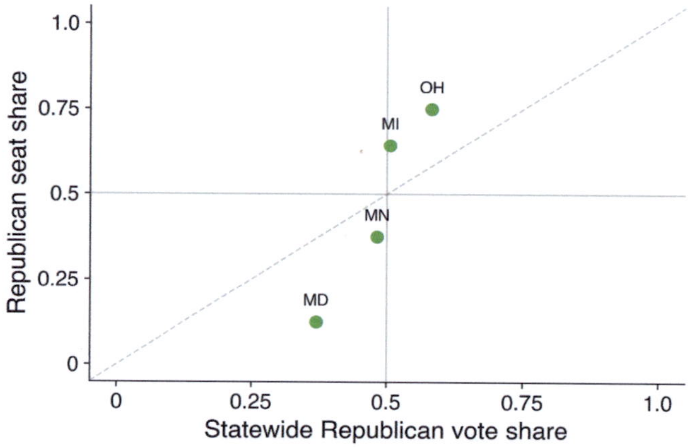

Figure 1: The 2016 Congressional election in four states, plotted on the seats–votes plane. A point in the upper-right quadrant ($V > \frac{1}{2}$, $S > \frac{1}{2}$) is an election where Republicans won a majority of the statewide vote and a majority of the seats (as in Ohio and Michigan). A point in the lower-left quadrant ($V < \frac{1}{2}$, $S < \frac{1}{2}$) is an election where Republicans won a minority of the statewide vote and a minority of the seats (as in Maryland and Minnesota).

If our electoral system promoted proportionality, we would expect most elections to cluster near the **line of proportionality**, defined by $S = V$. But this is not what happens! Figure 2 shows the fraction of Congressional districts in the 2012 and 2016 elections where Republicans won the majority of the Presidential vote share, for all states with six or more districts. The points are indeed clustered in a linear pattern, but not around the line of proportionality: the line of best fit has slope 2.6.

---

[3]All data, unless otherwise specified, taken from the MIT Election Science + Data Lab [2, 3].

[4]In reality, both V and $\overline{V}$ are affected by the districting plan, since districting lines can influence voter choices in many subtle ways (e.g., suppressing turnout in very safe districts, changing the availability of incumbents, etc). But as a first approximation, we can think of each party's vote share (V or $\overline{V}$) as an expression of the true underlying preferences of the voters.

The fact that the slope is greater than 1 is often termed a **winner's bonus**. For instance, in this case, you might say that there is an extra 2.6% of seat share for each additional percent in the winner's vote share. But reporting only the slope hides the fact that 11 out of the 52 elections plotted here buck the pattern completely by awarding more seats to the losing party. Such elections fall in the upper left and lower right quadrants.

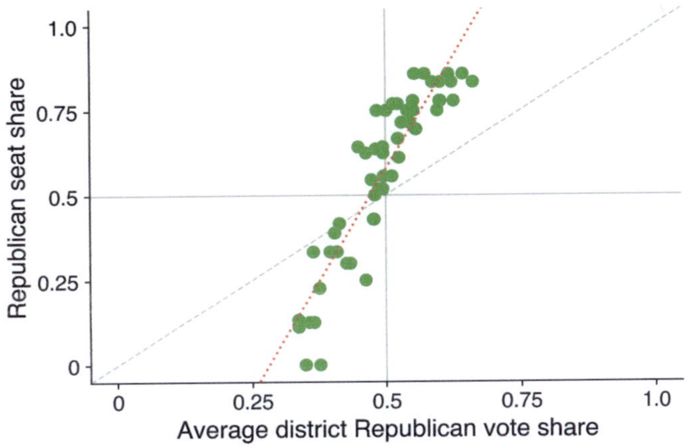

Figure 2: Data from the 2012 and 2016 elections for all 26 states with at least six Congressional districts. (Taken from The Daily Kos [4].) The x-axis shows average district Republican vote share ($\overline{V}$) in the Presidential race, while the y-axis shows the fraction of Congressional districts in which the Presidential election results had more Republican than Democratic votes. The red line is the line of best fit, slope 2.6. The reason for using Presidential election results is explained in Sidebar 2.2.

2.1 V VS. $\overline{V}$

You might wonder why we make the distinction between V, the statewide Republican vote share, and $\overline{V}$, the average district Republican vote share. The simple reason is that some authors use V, while others use $\overline{V}$. While V and $\overline{V}$ are typically close, they can be different in important ways. For instance, in Michigan's 2016 Congressional elections, V was very slightly greater than 0.5, while $\overline{V}$ was slightly less than 0.5. This changes the quadrant where Michigan appears on the seats–votes plane.

What is the conceptual difference between the two (beyond simply the way they are calculated)? Here's one way to think about it: V reflects the number of votes cast in the election, full stop. $\overline{V}$ is a version of V that's normalized by district turnout, making it so that every district is weighted the same, regardless of how many people turned out on election day.

In this chapter we use V for the majority of our figures, since we're often talking about "fairness" in an abstract sense, and V feels like the simpler and more elegant choice in this context. Others have argued for using $\overline{V}$ in all calculations of partisan metrics, to control for the effects of differential turnout across districts [5]. Here, we use $\overline{V}$ when the available data only contain district vote shares and not raw votes, and in each of the figures we've tried to be clear about which one we're using (or said "vote share" when the two quantities are the same).

2.2 MODELING UNCONTESTED RACES

Figure 2 uses the votes cast in each district in the 2012 and 2016 *Presidential* elections to derive the average district vote share and seat share. Why didn't we use the actual results from the 2012 and 2016 Congressional races?

The problem is that almost half of the 26 states shown in Figure 2 (12 states in 2012 and 10 in 2016) had at least one **uncontested district**: a district in which one of the two parties did not field a candidate. In these districts, the Republican vote share in the Congressional election would have been either 0% or 100%. Clearly this number does not reflect the true partisan preferences of the district's voters, so we don't want to use it in computing V or $\overline{V}$ for the state. Instead, we use the Republican vote share in the Presidential election from the same year, which tells us how many voters in the district preferred Republicans to Democrats in a different context. Of course, there are many reasons why the Presidential vote share may differ from the Congressional vote share, so this is only a very rough estimate—but it's certainly better than 0% or 100%.

If we want a more precise estimate, we can look at recent election outcomes for all the districts in the state and use these data to construct a statistical model for how Republican vote share tends to vary from election to election. Ideally, the model would include all the major factors that might affect a district's Republican vote share, such as: the year of the election, which office it is for, whether one of the candidates is an incumbent, etc. For example, a simple model might indicate that, in our state, incumbents tend to get a ~5% bonus in Congressional elections; and that once you account for this bonus, the Republican vote share in each district tends to be ~2% higher for Congressional elections than for Presidential ones. Then if an uncontested district with a Democratic incumbent had a 30% Republican vote share in the 2016 Presidential election, we would predict that its Republican vote share in the 2016 Congressional election would have been about $30\% + 2\% - 5\% = 27\%$.

To check the quality of our model's predictions, we can perform a standard modeling test by splitting the data from contested districts (for which we know the true Republican vote share) into *training data* and a small amount of set-aside *test data*. We base our model only on the training data, then see how well it predicts the Republican vote share in the test data. If the model does reasonably well, then we can feel justified using it to impute what the Republican vote share would have been in uncontested districts as well.

Political scientists use these kinds of models all the time, both to estimate vote share in uncontested districts and to address more general questions like "Has incumbency advantage in the US increased or decreased over time?" See Gelman and King [6] for a broad overview of imputation strategies and Mayer [7] for a concrete example: an expert witness report in *Whitford v. Gill* with a detailed description of its imputation model.

1.2 WHY DISTRICTS DON'T PRODUCE PROPORTIONALITY (USUALLY)

The following two scenarios illustrate why we should not expect elections under our electoral system to produce proportional outcomes.

Example 1: Competitive districts. Competitive districts, where the two parties have approximately equal support, are generally considered good for democracy: since neither party can count on an easy win, both candidates have to work hard to gain their constituents' vote. However, competitive districts can be terrible for proportionality. In theory, if all the districts in a state are closely contested, then a small swing in overall preferences can drive big deviations from proportionality.

Consider Minnesota, where Republicans in 2016 won 38% (3 out of 8) of the Congressional seats with 48% of the overall statewide vote (V). Three of Minnesota's eight Congressional districts (#1, #2, and #8) were extremely competitive, won by a margin of less than 2% (Table 2.1). With so many competitive districts, Republicans in Minnesota could easily have gotten anywhere from 25% to 62.5% of the Congressional seats (2–5 districts) with essentially the same statewide vote share.

District	Total votes	% Republican	Winner	Margin of victory... in %	in votes
1	335,595	49.6%	D	0.8%	2,547
2	341,285	51.0%	R	1.9%	6,655
3	392,313	56.9%	R	13.7%	53,837
4	324,332	37.3%	D	25.4%	82,266
5	330,617	24.4%	D	51.2%	169,297
6	358,395	65.7%	R	31.4%	112,375
7	330,516	47.5%	D	5.0%	16,628
8	356,185	49.7%	D	0.6%	2,009
Statewide	2,769,238	48.2%	3R/5D		

Table 2.1: Results of the 2016 Congressional election in Minnesota, with the margin of victory in percentage points and votes.

Example 2: A dispersed minority. You may have already read about Massachusetts (Chapter 0), where Republicans consistently get about 35% of the vote in Presidential elections. If Congressional voting followed a similar pattern, then a proportional outcome would give the state 2–4 Republican representatives (out of 9 or 10). Yet Massachusetts has not sent a Republican to Congress in over 20 years. Why?

The reason why Massachusetts does not have any majority-Republican districts is not gerrymandering, but the geographic distribution of its voters. Republican voters in Massachusetts are a significant minority almost everywhere, but a majority almost nowhere. For instance, in 2012, Mitt Romney (R) outperformed Barack Obama (D) in only about 15% of the state's precincts, and even in these "red precincts," his average vote share was only 54.5%.[5] Moreover, the red precincts did not form

[5]Code at: `https://github.com/political-geometry/partisan-fairness`

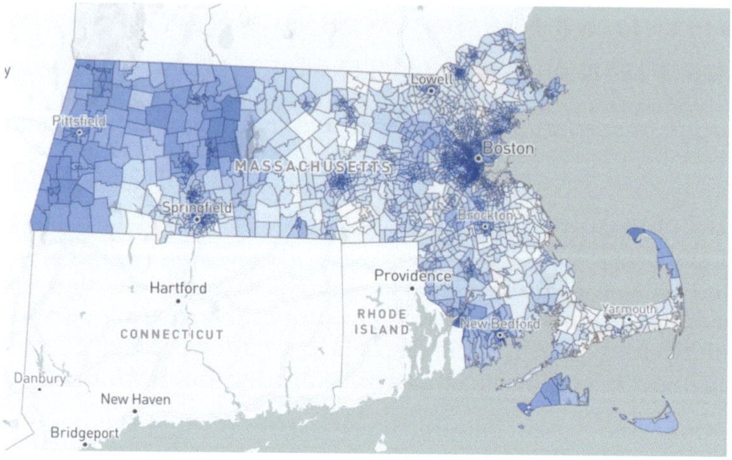

Figure 3: Voting patterns in Massachusetts precincts in the 2008 Senate election—one of the races shown in Duchin et al. [8] to produce no feasible Republican districts. Darker blues indicate a higher Democratic percentage of the two-way vote. Majority Republican precincts, shown in shades of red, are almost entirely absent from the plot, even though the two-way Republican vote share in the state as a whole was nearly 1/3 (31.9%). Image taken from the MGGG Districtr tool (`https://districtr.org/`).

a contiguous region: any Congressional district that included them would also have had to include some "blue precincts", many of which Obama won by a huge margin (getting 70% of the vote on average). It would have been challenging to create a majority-Republican district in Massachusetts in 2012, even if you were trying to gerrymander *for* the Republicans.[6]

Compare this to New York, where the two-party statewide Republican vote share tends to be only slightly higher than in Massachusetts (38.2% vs 35.3% in the 2016 Presidential election). Unlike Massachusetts, New York does have some majority-Republican clusters, and this has a noticeable effect on election results. From 2012 to 2020, Republicans consistently won about one-third of the state's Congressional seats (6 to 9 out of 27, depending on the year). New York's Congressional districts were drawn by the judiciary [9, 10], so are unlikely to have been gerrymandered for either party. Even if we suppose that the magistrate who worked on the maps was trying to "gerrymander" for proportionality, the fact that she was able to do so is in itself significant. The difference between New York and Massachusetts serves as a striking illustration of the extent to which a party's possible seat share in our electoral system is contingent on the geographic distribution of its voters.

If we want election outcomes to be proportional in a districted system, we generally would need to engage in what might be called "benign gerrymandering." In particular, to be confident in a proportional outcome, we would need to make many of our districts safe for one party or the other. We might allow ourselves a few competitive districts, but we definitely can't afford too many, since, in com-

[6]With some elections, it is *impossible* to make a majority-Republican district in Massachusetts out of small building blocks like towns or precincts, even if you abandon the requirement for contiguity. For a map corresponding to an election where a Republican district in MA would have been impossible to construct, see Figure 3. For a more detailed analysis of the obstacles to constructing a majority-Republican district in Massachusetts, see Duchin et al. [8].

petitive districts, small swings in voter preference can lead to large swings in the outcome. Thus, we arrive at the same conclusion that we have already hinted at in our Minnesota example: in the U.S. electoral system, competitiveness and (assured) proportionality are fundamentally incompatible.

It does not have to be this way. Almost half of the world's democracies elect their representative bodies using some method of **proportional representation**—that is, an electoral system that is designed to ensure proportionality with respect to political parties. (See Sidebar 2.3 and Chapter 20.) In such systems, competitiveness and proportionality do not undermine one another. Proportional representation does have some drawbacks (see, for example, King and Browning [11]), but one of its chief advantages is that it accords with most people's—including most Americans'—normative standard of fairness. If our goal is proportional outcomes, then we need to change not just the way we draw our districts, but our electoral system as a whole. In this chapter, however, our goal is to detect and quantify partisan gerrymandering within the context of our current system. And for these purposes, as we have just demonstrated, proportionality is simply the wrong metric.

THE SUPREME COURT WEIGHS IN

When partisan gerrymandering came to the Supreme Court in *Davis v. Bandemer*, the plaintiffs argued that intentional deviation from proportionality in redistricting violated the Equal Protection Clause. But the Court ruled decisively that "the mere lack of proportional representation [is not] sufficient to prove unconstitutional discrimination," citing essentially the reasons that we have outlined above:

> "If all or most of the districts are competitive...even a narrow statewide preference for either party would produce an overwhelming majority for the winning party... This consequence, however, is inherent in winner-take-all, district-based elections" [1].

Crucially, the Court in *Bandemer* did not say that proportional representation was *un*fair. In fact, in an earlier case (*Gaffney v. Cummings*, 1973), the Court had explicitly recognized proportionality as a legitimate goal that a state might pursue in designing its districting plan, even if this required making the district sizes slightly imbalanced. The *Bandemer* decision explains this apparent inconsistency as follows:

> "To draw district lines to maximize the representation of each major party[7] would require creating as many safe seats for each party as the demographic and predicted political characteristics of the State would permit... We upheld this "political fairness" approach in *Gaffney v. Cummings*, despite its tendency to deny safe district minorities any realistic chance to elect their own representatives. But *Gaffney* in no way suggested that the Constitution requires the approach...adopted in that case" [1].

[7] That is, to ensure proportionality by party (our footnote).

2.3 PROPORTIONALITY BY DESIGN

The Framers of the Constitution frequently used the phrase "proportional representation" in their deliberations. To them, it denoted the principle that "equal numbers of people ought to have an equal number of representatives" [12]. Writing during the American Revolution, John Adams put it slightly differently in *Thoughts on Government*: "equal interests among the people should have equal interests in [a representative assembly]" [13].

Nearly a century later, John Stuart Mill, in *Considerations on Representative Government*, pointed out what seems obvious in retrospect: equal representation by geographic region does not ensure equal representation by any other trait [14]. Mill called for true "proportional representation of all minorities," writing, "I cannot see... why people who have other feelings and interests, which they value more than they do their geographical ones, should be restricted to these as the sole principle of their political classification."

But how does one achieve equal representation of all interests? Must every group be represented in Congress according to its proportion in the population? If 24% of Americans are Catholic, must 24% of Congressional representatives be Catholic? And if 8% of Americans are Catholics who also believe in UFOs,[a] must 8% of representatives hold the same combination of beliefs?

This is where political parties come in. In principle, parties can form around any group of people with a common agenda who want their views represented in government. By choosing a party, a voter can explicitly designate which of her "feelings and interests" should form the primary basis of her "political classification." Instead of proportionality by geography, the Mill view then supports proportionality by party. Today, many people use the phrase **proportional representation (PR)** to denote an electoral system in which each party's seat share is structurally guaranteed to be (roughly) equal to its vote share.

There are many different PR systems in use around the world. The simplest one is **party-list PR**: each person votes for a party, and each party is allocated a number of seats proportional to the number of votes it receives. Party-list PR is the most common system across nations, but there are also more complex systems, which combine both proportional and local representation. For example, Germany uses a system called **mixed-member proportional representation (MMP)**, in which every person votes both for a local district representative and a (possibly different) political party. The district representatives account for about half the seats in the Bundestag, and the remaining seats are allocated by party in such a way as to make the overall results proportional. The system of electing one representative per district is used almost exclusively in Great Britain and its former colonies.

In 2017, and again in 2019, Representative Don Beyer (D-VA) introduced a House resolution called the Fair Representation Act,[b] which would require Congressional elections to be conducted using a ranked choice voting system that promotes proportionality. Congressional representatives under the Fair Representation Act would be elected locally, but districts would be larger than they are now: each district would elect 3–5 representatives. A similar system for national legislative elections has been used in Ireland since the 19th century.

Ranked choice voting (including the multi-winner version that promotes proportionality) is introduced in Chapter 20. For more on the mechanics, advantages, and disadvantages of different electoral systems, a good place to start is the ACE Electoral Knowledge Network (`aceproject.org`).

[a] Gallup poll, August 2019 (https://news.gallup.com/poll/266441/americans-skeptical-ufos-say-government-knows.aspx)

[b] Don Beyer, *Let's change how we elect the House of Representatives*, The Washington Post, 27 June, 2017.

In other words, the Court's message was: if you think that fairness hinges on proportionality, that's your business — go ahead and gerrymander for proportionality. You can even pass a state or Federal law requiring it. However, proportionality by party cannot be a constitutional standard of electoral fairness, because our district-based electoral system, which was in use at the time of the Constitutional Convention, fails this standard by its very nature.

On the question of what does constitute a "discriminatory effect", the Court offered only some very general guidelines:

> "Unconstitutional discrimination occurs only when the electoral system is arranged in a manner that will consistently degrade a voter's or a group of voters' influence on the political process as a whole" [1].

The justices knew that this was too vague to be actionable, but the majority wanted to leave the door open for a more precise "arithmetic" standard that might be found in the future:

> "We are not persuaded that there are no judicially discernible and manageable standards by which political gerrymander cases are to be decided" [1].

Thus, the *Bandemer* decision, while inconclusive, was an invitation to keep looking.

2 PARTISAN SYMMETRY

As it happens, just around the time of *Gaffney v. Cummings* in the 1970s, political scientists had begun to develop a new set of statistical tools for analyzing elections in nonproportional systems [15]. By the 1990s, a cadre of top political scientists and statisticians— most notably, Andrew Gelman, Bernard Grofman, and Gary King—had built support for the idea that any reasonable definition of fairness in redistricting (or, more generally, in electoral systems that were not designed for proportionality) should be based on **partisan symmetry** [16].

The basic idea of partisan symmetry is that, in a fair voting system, if one were to swap the parties' vote shares, their seat shares should also swap. Of course, when we say "swap vote shares", we are not imagining that individual Republican voters would turn into Democrats overnight and vice versa. So to make this idea usable, we will need to approximate more realistic swings in overall voter preferences.

For instance, suppose that, in the first election held under a given districting plan, Democrats get 52% of the votes and 65% of the seats. In the next election, Republicans get 51% of the votes and 67% of the seats. These results are not proportional: in each case, the majority party secured a seat share well above its vote share. But the results are roughly *symmetric*: the size of the "winner's bonus" was approximately the same in both cases, so the districting plan does not appear to give either party an inherent advantage.

This kind of analysis appears to require comparing the results of more than one election. And this is a problem: we don't want to wait through several election cycles (and hope for just the right shift in the voters' preferences) before we can judge whether a given districting plan is sufficiently symmetric. So we need a modeling assumption that allows us to predict, from the results of a single election, what would happen if such a shift occurred.

2.1 UNIFORM PARTISAN SWING AND SEATS–VOTES CURVES

Uniform partisan swing (UPS) is a model for how voters' partisan preferences change over time: its core assumption is that the change is closely linked across different parts of the state. For example, if we know that Democrats are becoming more popular in one region, then, no matter what's driving this trend, we assume that it affects the rest of the state in the same way. Of course, we don't expect Republican strongholds to suddenly switch sides, but we do expect them to become a little *less* Republican.

To make this precise, let us formulate a linear UPS model.[8] Suppose we observe an election in a state with N districts, with statewide Republican vote share V and district Republican vote shares $v_1, \ldots, v_N$. In a second election, suppose the observed vote share has changed from V to $V' = V + \delta$. (For instance, $\delta = 0.05$ if the overall Republican share has gone up from 0.47 to 0.52.) Under the UPS model, we assume that all the individual district vote shares have also changed by the same amount:

$$v_i' = v_i + \delta \text{ for } i = 1, \ldots, N.^9$$

This allows us to predict the new Republican seat share S': it is just the fraction of the v_i' that are greater than 0.5.

Thus, under the UPS assumption, we can use the results of a *single* election under a districting plan D to predict the number of seats that Republicans would win for *any* Republican vote share V'. In other words, UPS allows us to think of a districting plan D, along with a single election outcome, as specifying a *function* $V \to S(V)$ for converting vote share to seat share.[10] It is important to remember that this is just a prediction based on an assumption of how voter preferences change. In most cases, however, the assumption turns out to be fairly accurate. (For a detailed look at how well the UPS model holds up compared to real data, see Katz et al. [5].)

The graph of the function $S(V)$ is called the **seats–votes curve.**[11] Figure 4 shows

[8]Why specify linear here? Because there is another way of formulating UPS that swings the odds, rather than the votes themselves. Swinging the votes, as in linear UPS, can yield unpleasant edge conditions, like having to add 2% Democratic vote share to a district that is 99% Democratic (which you would resolve by setting the vote share to 100%). Swinging the odds neatly avoids headaches like that. Unfortunately, nobody really uses it. We've included code for it in the GitHub repository that goes along with this chapter.

[9]If v_i' ends up outside the interval $[0,1]$, we round it to 0 or 1.

[10]Variations on this construction of the function $S(\overline{V})$, including a stochastic version of the UPS model, can be found in Katz et al. and King [5, 17].

[11]Confusingly, the term "seats–votes curve" has been used over the years to refer to many different relationships between votes and seats. In early works, it often denotes the best-fit curve for electoral data from multiple elections [11, 15]. These days, however, it almost always refers to the curve derived from a single election based on the UPS model, as described above. See Katz et al. [5] for more on the formulation we discuss.

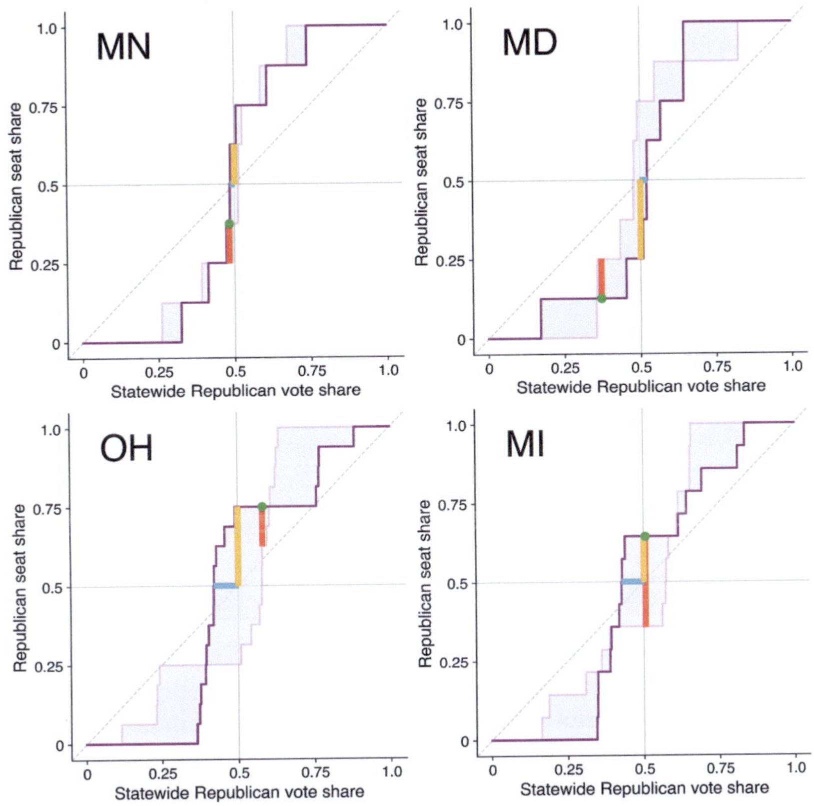

Figure 4: Seats–votes curves and symmetry scores for four 2016 Congressional elections. The seats–votes curve is dark magenta, and its 180° rotation is light magenta. The area of the gray region between the two curves is $\int_0^1 |S(V) - (1 - S(1 - V))|\, dV = 2\int_0^1 |\beta(V)|\, dV$. The actual election is marked with a green dot, and the distance between the two curves at that point, $2\beta(V^\circ)$, is shown in red. $\beta(0.5)$ is shown in yellow and the approximate mean–median score in blue.

the seats–votes curves for the four elections in Figure 1, with the observed election outcome marked by a green dot. The distinctive stair-step shape of the curves is a consequence of the fact that seats are whole numbers, so seat share changes abruptly as rising δ pushes each $v_i + \delta$ past the 0.5 threshold.

We can now give a more precise definition for the concept of partisan symmetry. For every possible value of V, we want $S(V)$—the modeled Republican seat share—to equal the Democratic seat share when *their* vote share is V. Since Democrats get vote share V when the Republican vote share is $1 - V$, we get the following.

Definition 1. *A districting plan satisfies the* **partisan symmetry standard** *if* $S(V) = 1 - S(1 - V)$ *at every* V.

This definition has a nice geometric interpretation. Switching the role of Democrats and Republicans corresponds to reversing the directions of both the V-axis and the S-axis in the seats–votes plane. This double reflection is equivalent to a 180° rotation around the point $(0.5, 0.5)$. So the partisan symmetry standard is upheld in the political science sense if and only if the seats–votes curve is symmetric about $(0.5, 0.5)$ in the familiar, geometric sense. Note that every symmetric curve must pass through the point $(0.5, 0.5)$, which suggests this point as a good reference for assessing the symmetry of a districting plan.

2.2 MEASURING (A)SYMMETRY

Of course, we don't expect any districting plan to be exactly symmetric, so if we want to use partisan symmetry as a standard of fairness, we need to specify how much *asymmetry*, or bias, we are willing to tolerate. In other words, we need some way of quantifying a plan's deviation from the ideal of symmetry.

Following Katz et al. [5], we define the **partisan bias** of a districting plan at each vote share V by the formula

$$\beta(V) = \frac{S(V) - (1 - S(1 - V))}{2}.$$

If we regard the average of the curve and its $180°$ rotation as a symmetrization, then $\beta(V)$ measures the distance from the curve to its symmetrization at each value of V.

But how do we reduce the bias of the curve as a whole to a single number? Several different metrics, or **symmetry scores**, have been proposed in the political science literature:

- The seats–votes curve is derived from an actual election with Republican vote share $V°$. If we want to summarize a districting plan's partisan bias, one possibility is simply to use $\beta(V°)$.
- We can compute the average value of $|\beta(V)|$ over the entire interval $[0,1]$: as an integral, this is

  $$\int_0^1 |\beta(V)|\, dV = \frac{1}{2}\int_0^1 |S(V) - (1 - S(1 - V))|\, dV,$$

  i.e., half the area between the curve and its rotated copy.[12]

  Since the extremes of the seats–votes curve usually correspond to unrealistic scenarios, we may choose to restrict the integral to a smaller interval around 0.5, such as $[0.4, 0.6]$. We could also remove the absolute value and compute $\int \beta(V) dV$, if we want the sign of the integral to reflect which party gains the advantage (and are OK with the possibility that positively and negatively counted areas might cancel). We refer to this summary score, and all its variants, as the **β-average** (signed or unsigned).
- Recall that any perfectly symmetric curve must go through the point $(0.5, 0.5)$. So one way to summarize a plan's deviation from symmetry is to measure

[12]The idea of using area to measure deviation from an ideal might remind you of Gini coefficients in economics, and some authors have called this score the "partisan Gini" [18, 19].

State	$\int_{0.4}^{0.6} \beta(V)dV$	$\int_0^1 \lvert\beta(V)\rvert dV$	$\beta(V^\circ)$	$\beta(0.5)$	Mean–median score
MN	0.005	0.017	0.062	0.125	0.019
MD	−0.022	0.05	−0.062	−0.25	−0.018
OH	0.036	0.078	0.094	0.25	0.087
MI	0.021	0.051	0.143	0.143	0.082

Table 2.2: Symmetry scores for the 2016 Congressional election in four states. From left to right: β-average on the interval $[0.4, 0.6]$, with area favoring Democrats counted as negative; unsigned β-average on the interval $[0, 1]$, with all area counted as positive; $\beta(V^\circ)$, the bias at the actual vote share; $\beta(0.5)$, the bias at $V = 0.5$; and the mean–median score, the approximate horizontal distance from the curve to $(0.5, 0.5)$. For all scores except the unsigned β-average, positive values indicate a Republican advantage while negative values correspond to a Democratic advantage.

how far its seats–votes curve lies from this central point, either vertically or horizontally. The vertical deviation is just $\beta(0.5)$. This corresponds to the counterfactual that asks what the seat outcome would have been if the vote had been exactly evenly split.

- The horizontal distance between the seats–votes curve and the point $(0.5, 0.5)$ estimates how much Republicans could fall short of half of the vote while still winning at least half of the state's seats. Note that half of the seats is particularly important if you are analyzing a state legislature, where controlling the majority is significant.[13] This measure of partisan asymmetry is approximately equal to the difference between the median and the mean of the district vote shares, or the **mean–median score**. (See Sidebar 2.4.)

Symmetry scores for each of the 2016 Congressional elections in Figure 1 are shown in Table 2.2.

INTERPRETATIONS AND LIMITATIONS

Partisan symmetry scores take the seats–votes curve and reduce it to a smaller, ideally more illustrative set of numbers. But reducing the information in this way can have some undesirable consequences.

- **Inconsistency:** Having so many different ways of summarizing the partisan symmetry of a districting plan means that they can produce contradictory results. In particular, most of these are signed measures, where in our convention a positive result indicates a bias toward Republicans. For geometric reasons, $\beta(0.5)$ and the mean–median score always have the same sign, but they can have very different magnitudes.[14] And the three seat-based measures of symmetry—$\beta(V^\circ)$, $\beta(0.5)$, and β-average—can theoretically have any combination of signs. This can make it difficult to find a consistent interpretation of these scores.

[13] For states with an even number of districts, the seats–votes curve intersects the line $S = 0.5$ not at a single point, but in a line segment. In this case, we measure the distance from (0.5,0.5) to the midpoint of the line segment.

[14] Some have observed that under certain circumstances this sign can tell a counterintuitive story; see DeFord et al. [19].

- **Instability:** Two elections with similar results can lead to very different partisan symmetry scores for the same districting plan. For instance, Figure 5 shows the seats–votes curves for Congressional elections in Minnesota from 2012 to 2018. If the UPS assumption held perfectly, the curves would look exactly the same: only the green dot corresponding to the actual results would change position along the curve. In reality, the curves do look broadly similar, so the UPS assumption seems not too far off. Yet even slight shifts in the curve can result in very different summary scores: for instance, in Minnesota in 2012, all the metrics suggest a highly symmetric plan. In 2014, however, $\beta(V^\circ)$ jumps from zero to -0.06 (suggesting a pro-Democratic bias), and then up to 0.06 in 2016 (suggesting a pro-Republican bias). So while the seats–votes curve as a whole is not sensitive to small deviations from UPS, the summary scores can change drastically, leading to qualitatively different conclusions.
- **Plateaus and "firewalls":** The scores we discussed are ill-equipped to capture the long flat plateaus that can arise in seats–votes curves. For example, look at Ohio and Maryland, where the actual election lies in the middle of a particularly long, flat part of the curve. In Maryland, as long as the Republican vote share stays between roughly 25% and 45%, the results will be the same — one seat for Republicans, seven for Democrats. This is certainly suggestive of gerrymandering, yet it is not captured by Maryland's small mean–median score and relatively small β-average on the interval $[0.4, 0.6]$.
- **Unrealistic counterfactuals:** It seems strange to declare a districting plan "fair" on the basis of a hypothetical situation that is extremely unlikely to occur. Again, we can think of Maryland. Why should it matter what the seats–votes curve looks like near $V = 0.5$ or $V = 0.6$, if the actual Republican vote share in Maryland hasn't gone above 0.4 in over a decade?

In fact, unrealistic counterfactuals pose a problem not just for partisan symmetry summary scores but for the symmetry standard as a whole.

In any situation where one party has a much higher statewide vote share than the other, a partisan mapmaker could start with the most ruthless districting plan they can think of, then make cosmetic adjustments to some of their own party's safe districts to game the scores, e.g., pulling $\beta(0.5)$ and mean–median to zero. It would cost them nothing, since all the adjustments would be in the unrealistic region of the curve. For this reason, advocates of the symmetry standard generally advise limiting it to states where both parties have a realistic chance of winning a majority [20].

Partisan symmetry scores reduce the information contained in the seats–votes curve, but the seats–votes curve itself is already a very reduced representation of the state: it depends only on the vector of votes in each district and contains no information about the state's geography and demography. Yet (as you'll hear again and again in this book), spatiality matters. We'll come back to this later in the chapter.

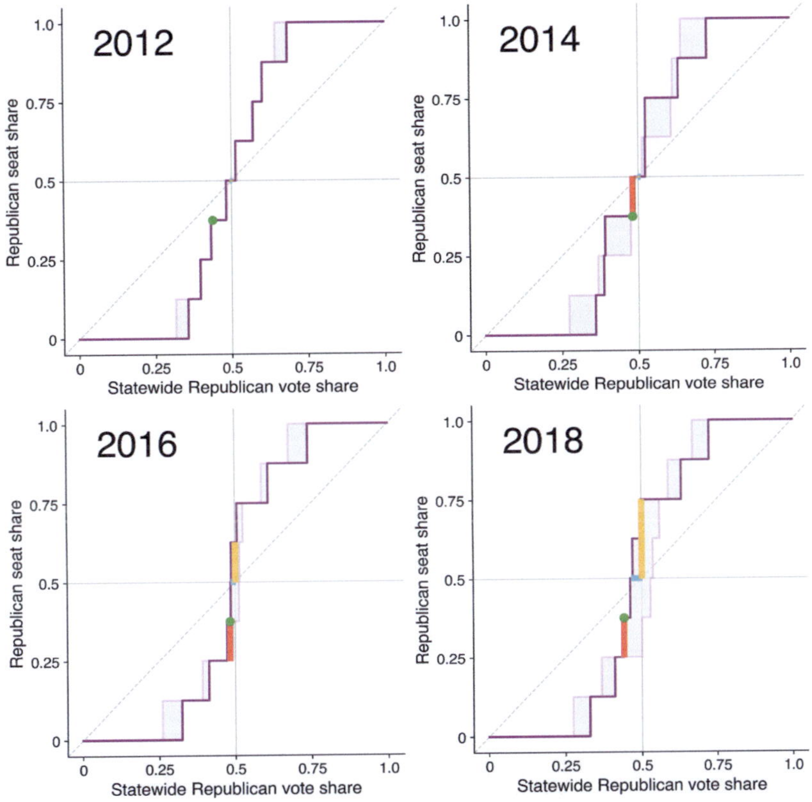

Figure 5: Seats–votes curves for Minnesota Congressional elections from 2012 to 2018, with summary scores as in Figure 4. As before, the magenta line is the seats–votes curve and the pale magenta line is the 180° rotation of that curve. The mean–median score is approximated by the blue line, $\beta(0.5)$ is the yellow line, and $2\beta(V^\circ)$ is the red line. There is no red line in 2012, because $\beta(V^\circ)$ is zero, and there are no yellow lines in 2012 and 2014 because the seats–votes curve in those years goes through the point (0.5, 0.5). While the curves themselves are broadly similar, some of the scores change significantly from year to year: for instance, $\beta(V^\circ)$ jumps from 0 in 2012 to -0.06 in 2014 to +0.06 in 2016.

THE SUPREME COURT WEIGHS IN AGAIN

After *Bandemer*, the next two partisan gerrymandering cases to reach the Supreme Court were *Vieth v. Jubelirer* (2004) and *LULAC v. Perry* (2006). In *LULAC*, several prominent political scientists submitted an *amicus* brief, proposing partisan symmetry as a new standard of fairness in redistricting [16].

The four liberal justices on the Court seemed content with the symmetry standard. The four conservative justices were not going to be convinced by any standard: they believed the Court should have nothing to do with partisan gerrymandering at all. So the decision was up to Anthony Kennedy, the swing justice. Kennedy agreed with the liberal justices that the Court should continue to hear partisan gerrymandering cases. In his *Vieth* opinion, he had expressed the hope that "new technologies may produce new methods of analysis that make more evident the precise nature of the burdens gerrymanders impose on the representational rights of voters and parties" [21]. But in *LULAC*, he rejected all the standards of fairness under consideration by the Court, including partisan symmetry.

Kennedy's objection to the symmetry standard was not one outlined in the previous section. He had a more fundamental concern:

> "The existence or degree of asymmetry may in large part depend on conjecture about where possible vote-switchers will reside" [22].

In other words, what bothered Kennedy was that there was no way to measure partisan bias without relying on a statistical model (such as uniform partisan swing). It's not just that he doubted the validity of the UPS assumption; he was wary of using any model at all:

> "Even assuming a court could choose reliably among different models of shifting voter preferences, we are wary of adopting a constitutional standard that invalidates a map based on unfair results that would occur in a hypothetical state of affairs" [22].

So, after all was said and done, the Court ended up back where it started: without a Constitutional standard of fairness in redistricting, but still holding out hope that such a standard might be found.

2.4 THE MEAN–MEDIAN SCORE

Suppose we observe an election where the district Republican vote shares are $v_1, \ldots, v_N$. We define the **mean–median score** of the districting plan under which the election was held to be

$$M = \text{median}\{v_i\} - \text{mean}\{v_i\}.$$

The mean–median score has been proposed as a measure of partisan asymmetry [18, 23, 24, 25] because, under the equal turnout assumption ($V = \overline{V}$), it coincides with the horizontal distance between the point (0.5,0.5) and the seats–votes curve derived from the election. (See Sidebar 2.5 for a discussion of the equal turnout assumption.)

We show that the two measures are equal in the case when N is odd; the even case is proved similarly. Assuming equal turnout, each point on the seats–votes curve corresponds to an election with statewide Republican vote share $V+\delta = \overline{V}+\delta$ and district vote shares $v_1+\delta, \ldots, v_N+\delta$. Since N is odd, there is some district m for which $v_m = \text{median}\{v_i\}$. When $v_m+\delta < 0.5$, Republicans will lose district m and at least half of the remaining districts; thus, $S(V+\delta) < 0.5$. Similarly, when $v_m+\delta > 0.5$, Republicans will win district m and at least half of the remaining districts; thus, $S(V+\delta) > 0.5$. We conclude that the seats–votes curve intersects the line $S = 0.5$ precisely when $v_m+\delta = 0.5$. The statewide Republican vote share at this point is

$$V+\delta = \overline{V}+0.5-v_m = 0.5-M.$$

Thus, M is the horizontal distance from the intersection point to (0.5, 0.5), as required.

3 THE EFFICIENCY GAP

The *LULAC* case made one thing clear: if you wanted to fight gerrymandering in the Supreme Court, you had to devise a measure of fairness that would appeal to Justice Kennedy. "Symmetry" might be a good conceptual hook, since Kennedy didn't shut the door on it completely.[15] But whatever measure you came up with could not involve any counterfactuals or hypotheticals.

In 2014, law professor Nicholas Stephanopoulos and political scientist Eric McGhee came out with an influential article about a new metric, called the **efficiency gap (EG)**, which they believed would do the trick [26]. The efficiency gap proposes to measure unfairness by comparing the number of votes "wasted" by each party in an election. A fair map is defined to be one in which EG = 0 because the parties waste the same number of votes—a kind of symmetry.[16] Once you have the full vote data, computing EG for any given election is very simple, and at first glance no hypotheticals are required.

In 2015, a team of lawyers including Stephanopoulos used the new EG measure to challenge the districting plan in Wisconsin. The case, *Whitford v. Gill*, made national headlines: for the first time since *Bandemer*, a federal court sided with the plaintiffs and declared a districting plan to be an unconstitutional partisan gerrymander. The Wisconsin plan's high efficiency gap was not the only basis for the decision, but it was a significant part of the plaintiffs' argument.

As *Whitford v. Gill* headed to the Supreme Court (becoming *Gill v. Whitford* in the process), there was a lot of excitement about the efficiency gap, both in the legal community and in the popular press. But many political scientists were skeptical. Indeed, as we shall see in this section, when you translate the requirement EG = 0 into the language of votes and seats, you get some very uncomfortable results.

3.1 DEFINITION AND EXAMPLES

For the purposes of defining EG, a vote is considered to be "wasted" if it does not contribute to the election of a candidate. This includes all votes cast for a losing candidate, as well as votes for a winning candidate in excess of the 50% required to win.

To see how this works in practice, we return to our running example of Minnesota's Congressional race in 2016 (Table 2.3).

Let's go through the calculation of wasted votes in District 1:

- First, we compute how many votes were required for a party to win in District 1. The total number of voters in the district (Republicans and Democrats) was 335,595. So to win, a party had to get just over half of those votes: 167,798.

[15]What Kennedy actually said about partisan symmetry was that he did not "altogether discount... its utility in redistricting planning and litigation" [22]. The four liberal justices were much more explicit in their support, but they were not the ones that needed to be convinced.

[16]Of course, this is a different use of the word "symmetry" than the technical sense described in Section 2 [5]. But it seems plausible that when Justice Kennedy wrote that "symmetry" may be useful in redistricting litigation, he was not wedded to the technical definition either.

District	Votes for R	Votes for D	Total votes	Needed to win	Wasted by R	Wasted by D
1	166,524	169,071	335,595	167,798	166,524	1,273
2	173,970	167,315	341,285	170,643	3,327	167,315
3	223,075	169,238	392,313	196,157	26,918	169,238
4	121,033	203,299	324,332	162,166	121,033	41,133
5	80,660	249,957	330,617	165,309	80,660	84,648
6	235,385	123,010	358,395	179,198	56,187	123,010
7	156,944	173,572	330,516	165,258	156,944	8,314
8	177,088	179,097	356,185	178,093	177,088	1,004
Total	1,334,679	1,434,559	2,769,238		788,681	595,935

Table 2.3: Vote data from 2016 Congressional election in Minnesota, with the number of votes wasted by each party. Importantly, with only small changes to the votes in Districts 1, 2, and 8, these numbers can change dramatically.

- The Democrats in District 1 got 169,071 votes and won the election. But as we just saw, only 167,798 of these votes were necessary to win. We say that the remaining 1,273 Democratic votes were wasted, since they did not contribute to the election of a Democrat.
- The Republicans lost District 1, so none of their votes contributed to electing a Republican. We therefore say that all 166,524 Republican votes in District 1 were wasted.

In this particular case, the losers happened to waste more votes than the winners, but it can also go the other way. In District 5, the Democrats wasted more votes than the Republicans, even though they won the election: they had such a large margin of victory that their “extra” votes outnumbered *all* the votes cast by Republicans.

Note that the language of wasted votes provides a useful way of quantifying the intuition that gerrymandering can be accomplished by “packing” and/or “cracking” voters of the opposing party. When many voters of Party A are packed into a few districts, Party A will win those districts by huge margins, causing it to waste more votes than Party B (as in Minnesota’s District 5). On the other hand, when a block of Party A voters is cracked, this usually involves creating several districts with safe-but-low margins of victory for Party B. Now the winners waste fewer votes than the losers, once again giving Party B an advantage in terms of wasted votes.

After calculating the wasted votes in each district individually, we add up all the votes wasted by Republicans and Democrats in the entire state and call these quantities W_R and W_D respectively. We can see that, by this definition, many more Republican votes were wasted in Minnesota in 2016 (788,681 to 595,935). Another way of putting it is that Republican voters were unable to use their votes as *efficiently* as Democrats. This is where the name “efficiency gap” comes from.

Definition 2. *Suppose we have two parties, A and B. The efficiency gap favoring Party A, for a given pattern of votes, is defined as the difference in wasted votes, divided by the total number of votes:*

$$\mathrm{EG} = \frac{W_B - W_A}{T}.$$

Dividing by T means that we are measuring the difference in wasted votes relative to the total number of votes cast. This seems reasonable: for instance, a difference of 100 wasted votes would be quite significant in an election with only 1,000 voters, but barely noticeable in an election with millions of voters, like Minnesota's.

The sign of EG indicates the direction of the advantage. If $\text{EG} > 0$, then Party B wasted more votes than Party A, so we conclude that the districting plan was tilted in favor of A. If $\text{EG} < 0$, we conclude that the plan was tilted in favor of B. Of course, in reality, you never get EG exactly equal to zero. Based on their analysis of previous elections, Stephanopoulos and McGhee suggest 0.08 as a reasonable margin of error [26]. In other words, any plan with $|\text{EG}| < 0.08$ should be considered fair enough, whereas a plan with $|\text{EG}| > 0.08$, while not necessarily a gerrymander, should be seen as a cause for concern.

Working from Table 2.3, we find that the efficiency gap for the 2016 Congressional election in Minnesota is

$$\text{EG} = \frac{W_D - W_R}{T} = \frac{595,935 - 788,681}{2,769,238} = -0.0696.$$[17]

Since this is below the threshold of 0.08 proposed by Stephanopoulos and McGhee, the EG standard would lead us to conclude that Minnesota's districting plan is probably *not* a partisan gerrymander.

3.2 SOME ISSUES WITH THE EFFICIENCY GAP

Let's do a little sensitivity analysis. In Table 2.3, notice that the biggest difference in wasted votes between the winner and the loser occurs in the most competitive districts (#1, #2, and #8). What would have happened if a small number of voters in these districts had changed sides?

- If just 2400 Democrats in Districts #1 and #8 had switched their votes, it would have been enough to give the Republicans narrow victories in both districts. Most of the wasted votes in those districts would then belong to Democrats. We would get $\text{EG} = 0.18$, signaling a huge Republican gerrymander.
- If 3400 Republicans in District #2 had switched their votes, the district would have gone to the Democrats. Most of the wasted votes in that district would then belong to Republicans, and we would get $\text{EG} = -0.19$, now signaling an even more egregious Democratic gerrymander.

In other words, while the actual vote results do not flag Minnesota's districting plan as problematic, a shift of a few thousand votes (out of nearly 3 million) could make it look like *either* an egregious Republican gerrymander *or* an egregious Democratic gerrymander.

This example shows that, in the presence of competitive districts, EG is extremely volatile and thus arguably useless as a measure of unfairness. Stephanopoulos

[17] We identify Party A with Republicans and Party B with Democrats, so that, as usual, a positive score corresponds to a Republican advantage and a negative score to a Democratic advantage.

and McGhee are aware of this problem: they recommend not using high EG as an indicator of gerrymandering if small changes in competitive districts could make it fall below the 0.08 threshold. Still, it should give us pause that such a straightforward and reasonable-sounding metric turns out to behave so unreasonably in this case.

What about Massachusetts? For reasons that will become clear soon, we will use data from 2000 rather than 2016. Right away, we run into a problem: five of the state's ten Congressional districts were not even contested by Republicans in 2000. As we saw in Sidebar 2.2, we cannot reasonably include the votes from those districts in our calculation of EG, yet we can't just ignore these districts either.

The usual approach to uncontested districts is to use a statistical model to estimate what the election results in those districts *would have been* if both parties had fielded a candidate. This is the approach that Stephanopoulos and McGhee recommend as well. But notice that, if we take this route, we are basing our EG calculations on a *counterfactual*—exactly what Justice Kennedy protested against and what EG was explicitly designed to avoid. Thus, the claim that EG does not rely on statistical modeling turns out to be inaccurate for elections with uncontested districts (which are extremely common).[18] The truth is that almost all analyses of voting data have some statistical assumptions underlying them. You can't avoid "hypotheticals"; you can only de-emphasize them and hide them in the background.

Getting back to Massachusetts: to avoid dealing with uncontested districts, let's use data from the 2000 Presidential election instead of the Congressional one.[19] In 2000, a total of 1,616,487 people in Massachusetts voted for the Democrat (Gore) and 878,502 voted for the Republican (Bush), for $T = 1,616,487 + 878,502 = 2,494,989$ two-party votes in all.[20] Republicans were a minority in every district, so we know that *all* their votes were wasted: $W_R = 878,502$. Democrats needed just over half the votes in each district to win that district, Therefore, up to rounding, the total number of votes required to win all the districts was $T/2$. The rest of the Democratic votes were wasted: $W_D \approx 1,616,487 - \frac{2{,}494{,}989}{2} \approx 368,992$. Thus, the efficiency gap for Massachusetts under these assumptions is $\mathrm{EG} = \frac{W_D - W_R}{T} = \frac{368{,}992 - 878{,}502}{2{,}494{,}989} = -0.20$, indicating a massive tilt in favor of Democrats. This accords with our intuition that Republicans in Massachusetts were very inefficient at turning votes into seats: their 35% vote share got them no seats at all.

But if there is any unfairness to Republicans here, it has nothing to do with the districting plan. As we touched on in Section 1, there are some elections that could not yield a Republican district in MA no matter how the districts were drawn. It turns out that the 2000 Presidential election was one of them [8]. Thus, our calculation shows that, for this election, *every* possible districting plan in Massachusetts would have resulted in $\mathrm{EG} = -0.20$. The efficiency gap fails for the same reason

[18]For instance, the Wisconsin districting plan that was challenged in *Whitford v. Gill* (the court case in which EG was first introduced) had 99 districts, of which as many as 49 were uncontested in some of the elections under scrutiny.

[19]Note that this does not get around the issue of hypotheticals. We are just using an extremely rudimentary statistical model: "In a district where both parties field a Congressional candidate, the Republican vote share is roughly the same as it is in the concurrent Presidential election." This model would be too crude to use in a serious analysis, but it is good enough for our illustrative purposes here.

[20]We continue to ignore votes for other candidates, even though in the 2000 Presidential election, the third-party vote share in Massachusetts was unusually high — about 6.4% of the total.

proportionality does: it detects unfairness that is arguably real, but has nothing to do with gerrymandering.

Here is a more surprising result: what if Republicans in Massachusetts were not a 35% minority, but a 20% minority, still dispersed across the state? The Democrats would once again win all the seats, so for any districting plan, W_R would be 20% of T (since all Republican votes would still be wasted). W_D would be $80\% - 50\% = 30\%$ of T (since all Democratic votes above the 50% needed to win in each district would be wasted). Thus, the efficiency gap would be $\text{EG} = \frac{W_D - W_R}{T} = 0.3 - 0.2 = 0.1$. Paradoxically, we would conclude that the districting plan is unfair to Democrats, even though they won all the seats in the state. In other words, the EG standard calls for a 35% minority to get more than 0 seats, but for a 20% minority to get *fewer* than 0 seats!

To be fair to Stephanopoulos and McGhee: they are aware of all these issues with EG, and they propose a number of safeguards, i.e., contexts in which a court should not interpret a high EG score as a sign of gerrymandering.[21] But when a method that seemed so simple turns out to have so many exceptions and counterintuitive results, you can't help but wonder: what is really going on here? Clearly we need a better understanding of what the efficiency gap is actually measuring.

SEATS AND VOTES AGAIN

Recall that we were able to calculate the efficiency gap for Massachusetts (and the hypothetical "20% Republican" version of Massachusetts) knowing almost nothing about the distribution of voters in districts. For the 20% version, all we used were the Republicans' (fictional) statewide vote share ($V = 0.2$) and seat share ($S = 0$). It's not hard to check that we could have used the same approach to calculate EG for the actual 2000 election, with $V = 0.35$ and $S = 0$. Of course, Massachusetts is a particularly straightforward case, but it turns out that, with some elementary algebra, we can approximate EG for *any* election with a simple expression in terms of V and S. We just need to make one simplifying assumption, **equal turnout**: that is, we assume that there are T/N voters in each of the N districts. (See Sidebar 2.5 for what happens when this assumption does not hold.)

To compute EG, we need to know W_R and W_D. To find W_R, let's first figure out how many votes the Republicans did *not* waste. By the equal turnout assumption, it takes $\frac{T}{2N}$ votes to win each district. Republicans won SN districts, which required a total of $SN \cdot \frac{T}{2N} = \frac{ST}{2}$ votes. The remainder of the VT Republican votes were wasted. This gives $W_R = VT - \frac{ST}{2} = \frac{2V-S}{2} \cdot T$. An analogous calculation shows that $W_D = \frac{1-2V+S}{2} \cdot T$. Thus,

$$\text{EG} = \frac{W_D - W_R}{T} = \frac{1-2V+S}{2} - \frac{2V-S}{2} = S - 2V + \frac{1}{2}.$$

We can rewrite this as $EG = \left(S - \frac{1}{2}\right) - 2\left(V - \frac{1}{2}\right)$. In other words, the efficiency gap

[21] Unfortunately, these caveats are routinely ignored by people who cite EG in their writing. Journalists, in particular, often report EG scores uncritically, lending the appearance of scientific precision to many spurious claims about gerrymandering.

standard ($EG = 0$) locates fairness not on the diagonal $S = V$ but on the line with slope 2 passing through $(0.5, 0.5)$. The permissible zone where $|\mathrm{EG}| < 0.08$ is simply a narrow band around this line.

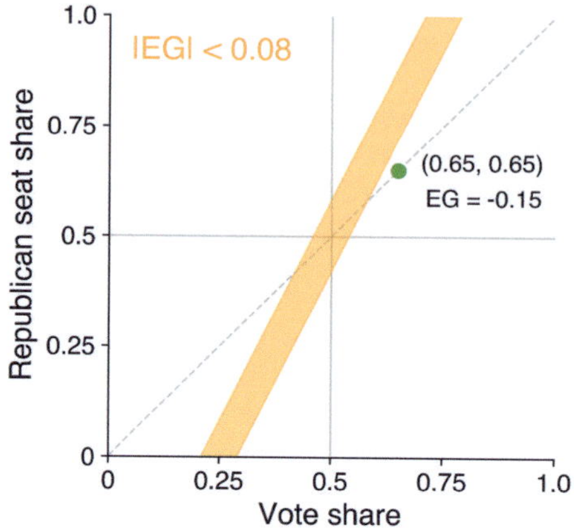

Figure 6: The yellow band marks the region in the seats–votes plane where $|\mathrm{EG}| < 0.08$ under the equal turnout assumption. Note that the point $(0.65, 0.65)$ lies outside this region. Although such an election would accord with most people's intuitive definition of fairness (proportionality), its efficiency gap of -0.15 would indicate a Democratic gerrymander. (Figure adapted from Duchin [27].)

Notice that the permissible zone in Figure 6 leaves out most elections with proportional outcomes. For example, a districting plan under which Republicans got 65% of both votes and seats would be suspected of being a Democratic gerrymander. When we said in Section 1 that our electoral system tends not to produce proportional outcomes, that was a factual, empirical statement — but the efficiency gap standard has turned it into a normative judgment. Instead of saying that proportionality is not required for fairness, it actually implies that proportionality is unfair! [22]

THE SUPREME COURT DOES NOT WEIGH IN

We have discussed the efficiency gap at great length because, for a few years after its publication in 2015, it was all the rage in anti-gerrymandering circles. Due to its central role in the early stages of *Gill v. Whitford*, the first major partisan gerrymandering case since *LULAC*, it got a lot of attention in the popular press, and numerous scholarly articles were published analyzing, critiquing, and tweaking it [28, 29, 30, 31, 32].

[22] In practice, as the proponents of EG point out, a plan that produced proportional results would be safe in court no matter what its efficiency gap was. It takes both "discriminatory effect" and "discriminatory intent" for a redistricting plan to be declared unconstitutional. Since the Supreme Court has already declared proportionality a legitimate goal in *Gaffney v. Cummings*, no one could accuse the designers of a proportional plan of discriminatory intent.

But this does not address the deeper issue. The purpose of a quantitative standard of fairness is not to replace our intuitive notion, but to formulate it more precisely. Given how strongly our intuition associates fairness with proportionality, a measure that labels proportionality *unfair* can hardly be said to be a measure of fairness at all.

In the end, the Court did not issue a formal opinion on using EG as a measure of fairness. Instead of being decided on the merits, *Gill v. Whitford* was sent back to the lower courts to deal with issues of legal standing.

2.5 THE EQUAL TURNOUT ASSUMPTION

In any real election, turnout across districts will certainly not be exactly equal. The districts across a state have very nearly equal total population at the beginning of the Census cycle, but the number of people who actually turn out to vote is bound to vary. For instance, as you can see in Table 2.3, district turnout in Minnesota in 2016 ranged from a minimum of 324,332 (District 7) to a maximum of 392,313 (District 3).

How well does the simplified EG expression $S - 2V + \frac{1}{2}$ approximate the original EG formulation in real elections? In theory, it could be very far off, but in practice, the two values are usually quite close. The table here shows the comparison for the 2016 Congressional elections, for all states with 8 or more Congressional districts in which every district was contested by both parties.

State	EG	$S-2V+\frac{1}{2}$
MD	−0.11	−0.12
MI	0.15	0.13
MN	−0.07	−0.09
MO	0.06	0.04
NC	0.19	0.2
NJ	0.06	0.0
OH	0.11	0.09
TN	−0.01	−0.02

The largest discrepancy, 0.06, occurs in New Jersey, where the variation in district turnout in 2016 was extraordinarily high: minimum 167,070, maximum 334,038.

In general, turnout tends to be lower in Democratic areas than in Republican areas [33]. Correspondingly, in recent elections, most states had lower average turnout in Democrat-won districts than in Republican-won districts, sometimes by a significant ratio [30]. This observed tendency causes the equal turnout assumption to overestimate W_R, underestimate W_D, and therefore (usually) underestimate EG, as can be seen in the table above.

While the simplified $\mathrm{EG} = S - 2V + \frac{1}{2} = 0$ formula calls for half the votes to secure half the seats, this is not true of $\mathrm{EG} = \frac{W_D - W_R}{T} = 0$. The implications can be rather counterintuitive. Ellen Veomett in [30] derives a more sophisticated formula for EG in terms of S, V, and the turnout ratio, and then applies this formula to a hypothetical 50-50 election with the same turnout ratio as the actual 2016 Congressional election in Texas. It turns out that, to receive a score of $\mathrm{EG} = 0$, such an election would have to award Democrats 60% of the seats!

4 ENSEMBLES AND OUTLIERS

So far, we have considered three standards of fairness in redistricting: proportionality, partisan symmetry, and equality of wasted votes (EG). Each of these standards has its advantages and disadvantages, but they all share the same fundamental flaw: they attempt to set an *absolute* baseline of fairness, without taking the specifics of a state's distribution of voters into account.

Proportionality and EG both require a particular relationship between seat share and vote share ($S = V$ and $S = 2V - \frac{1}{2}$ respectively). But as we have seen again and again, in our district-based electoral system, the same vote share can legitimately lead to many different seat shares. The more sophisticated partisan symmetry standard avoids this pitfall, but it too fails the test of geography: as we will demonstrate later in this section, certain kinds of asymmetry in the geographic distribution of voters will naturally lead to plans with asymmetric seats–votes curves.

Given the myriad potential arrangements of voters in a state, it is hard to imagine any absolute standard, no matter how sophisticated, that would be satisfied by all neutrally drawn districting plans under *any* population distribution. And as long as a measure of fairness is unable to distinguish between the effects of gerrymandering and the effects of geography, mapmakers will use this as an excuse, rendering the standard toothless.

THE SUPREME COURT WEIGHS IN (FOR THE LAST TIME?)

How do we deal with this seemingly intractable problem? One possibility is to give up and declare the problem unsolvable. It is impossible to decide whether a districting plan is fair unless we first agree on a definition of fairness. And since there does not appear to exist an absolute, all-purpose standard of fairness, we are stuck. This is the view of the Supreme Court in its most recent (and perhaps final) partisan gerrymandering decision, *Rucho v. Common Cause* (2019):

> "Federal courts are neither equipped nor authorized to apportion political power as a matter of fairness. It is not even clear what fairness looks like in this context. Deciding among... different visions of fairness poses basic questions that are political, not legal. There are no legal standards discernible in the Constitution for making such judgments. And it is only after determining how to define fairness that one can even begin to answer the determinative question: 'How much is too much?' [34]"

Based on this reasoning, the Court in *Rucho* declared partisan gerrymandering to be **nonjusticiable**: "outside the courts' competence and therefore beyond the courts' jurisdiction" [34].

The *Rucho* decision emphasizes that the Court "does not condone excessive partisan gerrymandering" and acknowledges that "[e]xcessive partisanship in districting leads to results that reasonably seem unjust" [34]. It urges the states and

Congress to address this problem via the political process, the first step of which is to decide what kind of fairness they want.

4.1 INTRODUCING ENSEMBLES

Everything we have seen in this chapter so far seems to confirm the Court's opinion in *Rucho*: our electoral system does not lend itself to a one-size-fits-all standard of fairness. But recent advances in statistics and computer science have given us a new way of tackling the challenges of geography: **ensemble-based analysis**. Here, the baseline of fairness is defined empirically, by creating a collection (or **ensemble**) of districting plans for the state and looking at its properties in the aggregate.

The first step is to have a computer randomly generate a large number of "eligible" nonpartisan districting plans: plans that satisfy all state and Federal laws (population equality, contiguity, compactness, the Voting Rights Act, etc.) and use no partisan data in their construction. For each eligible plan, we then use precinct-level results from a recent election to figure out which of its hypothetical districts Republicans would have won. Thus, we can report what the Republican seat share S would have been under each of our eligible plans.

This is where the state's voter geography enters the picture. For instance, in a Massachusetts election where it is provably impossible to create a Republican district [8], all the randomly generated plans will have $S = 0$. But in New York, which has a similar Republican vote share but has some more concentrated Republican areas, the number of Republican seats under each plan will depend on how the plan happens to chop up those areas. In general, our ensemble of randomly generated plans will produce a range of values for S, reflecting the possibilities and the probabilities for this particular state, with this particular geography.

Finally, we compare the plans in our ensemble to the enacted or proposed plan. Was the seat share within the reasonable range for S across the ensemble of alternatives or is it a statistical outlier? If it is outside the reasonable range, does it deviate in the direction of proportionality, or does it exacerbate the controlling party's advantage? This leads to the **extreme outlier standard**: unless the mapmakers can provide some other explanation for their plan's deviation from the norm, we can reasonably conclude that the plan was the result of partisan gerrymandering. We can then quantify its effect on the disadvantaged party by comparing it to the nonpartisan plans in our ensemble.[23]

As an example of ensemble analysis, consider North Carolina's 2016 Congressional districting plan (the subject of *Rucho v. Common Cause*). The official set of criteria used in the plan's construction included the usual desiderata of population

[23]What if the mapmakers do offer an alternative explanation? Suppose they claim that their plan is different from all the plans in our ensemble because they were trying to achieve some additional innocuous goal X. Then we can simply restrict our ensemble to only those plans that also achieve goal X (and create more such plans if necessary). If the districting plan under examination is still an outlier, then we know it's not because of goal X. Once all other possible explanations are gone, the only one that remains is partisan gerrymandering: we know the intent and we can quantify the effect. The court can decide how much is too much, but at least we have identified precisely the quantity that we were looking for.

equality, contiguity, and compactness, minimizing the number of counties split between districts, and adhering to the Voting Rights Act. But the list of criteria also included "partisan advantage": the mapmakers were instructed to try to "maintain the current partisan make-up of North Carolina's congressional delegation" [35].

To analyze how much advantage Republicans actually got from the resulting plan, researchers at Duke University randomly generated 66,544 nonpartisan plans that performed at least as well as the enacted plan on the criteria of population equality, compactness, and minimizing split counties. To comply with the Voting Rights Act, they ensured that each plan in their ensemble had at least two districts that were majority African-American.[24] The histogram in Figure 7 shows the number of seats that Republicans would have won in the 2016 Congressional election under each of these 66,544 random plans [36].

The statewide Republican vote share in this election was 53%. Most of the randomly generated plans give Republicans eight seats, but even if Republicans had won nine seats, they could still reasonably claim that their plan did not excessively disadvantage Democrats: after all, a random nonpartisan plan would have a 30% chance of producing the same result.[25]

But under the North Carolina districting plan, Republicans in 2016 won 10 of the state's 13 seats. Among the computer-generated nonpartisan plans, only 1% produced such an extreme result.

This has two implications. First, it gives strong evidence that North Carolina's districting plan was, in fact, an intentional partisan gerrymander: the odds of obtaining such a plan using only nonpartisan considerations are tiny. In the specific case of North Carolina, this evidence is unnecessary, since we already knew that "partisan advantage" was one of the official criteria for creating the plan. However, in situations where the gerrymanderers are better at hiding their tracks, the fact that ensemble analysis can produce evidence of *intent* to gerrymander could prove very useful.

Second, we can now quantify the *effect* of the gerrymander on the outcome of the 2016 election. Nine Republican seats would have been a high but still plausible outcome under a nonpartisan plan. Since the Republicans actually got ten seats, we can say with high confidence that they gained at least one seat, or 8% in seat share, as a direct consequence of the gerrymander.

The beauty of the extreme outlier standard is that it does not require us to agree on a normative definition of fairness. Its baseline of fairness is just the absence of clear partisan gerrymandering, i.e., the requirement that a plan's partisan metrics should not look radically different from an ensemble of neutral, nonpartisan plans.

[24] The actual requirements of the Voting Rights Act are more complex, but since North Carolina has had two majority-minority districts for years, it is reasonable to assume that a plan with two such districts would be in compliance. For more on the Voting Rights Act in the context of redistricting, see Chapter 6 and Chapter 7.

[25] Note that a plan with 9 Republican seats would have an efficiency gap of

$$S - 2V + 0.5 = 0.69 - 2(0.53) + 0.5 = 0.13,$$

if we used the equal turnout assumption formula. So the EG standard would end up flagging 30% of randomly generated plans as Republican gerrymanders!

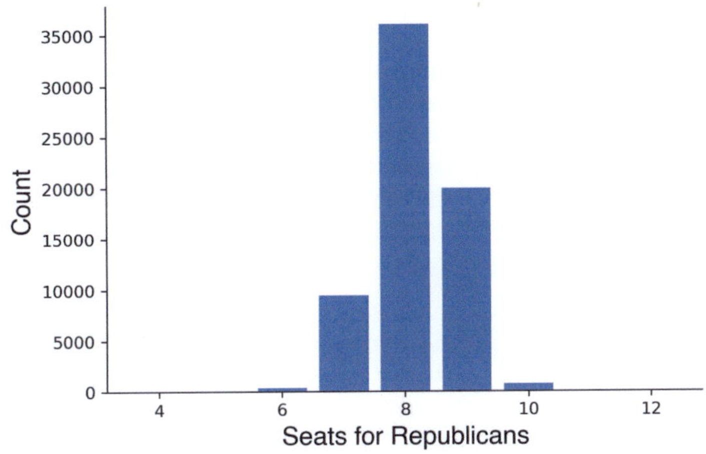

Figure 7: Number of Congressional seats (out of 13) that would have been won by Republicans in North Carolina's 2016 Congressional election under 66,544 randomly generated neutral plans. The actual number of seats won was 10. Figure based on data from Herschlag et al. [36].

This sounds simple, but it relies on our ability to produce a representative sample of eligible plans satisfying all the legal requirements. Until researchers had developed the statistical and computational tools to attempt this, we had no way of knowing what "absence of gerrymandering" looks like with a particular state's geography.

Ensemble analysis is an active area of current research. There are many important technical details to work out about how the set of eligible maps is generated; different research groups use somewhat different approaches. You can find more information on the technical aspects of ensemble analysis in Chapter 16 and Chapter 17.

4.2 APPLICATIONS OF ENSEMBLES

We can use ensemble analysis to verify empirically some of the theoretical results mentioned earlier in this chapter.

First, consider proportionality (or lack thereof). In Figure 7, most of the plans in the ensemble give Republicans 62% of the seat share (8 out of 13 districts) even though their vote share was only 53%. The proportional outcome would have been 7 districts ($S = 0.53$), but this happens in less than 14% of the plans in the ensemble.

The introduction to this book, as well as several other chapters, feature similar histograms for many other elections.[26] In most of them, proportional outcomes are unlikely. The expected winner's bonus differs from state to state and from election to election: sometimes the most likely seat share is extremely far from proportionality, sometimes quite close. This bears out our claim that there is no one "correct" seat share for any given vote share.

[26]Note that the histograms in Chapter 0 show seat share and vote share for Democrats rather than Republicans.

In all the histograms in Chapter 0, the party that got the majority of the statewide vote also gets a majority of the seats under most plans in the ensemble. But in general, even this basic result is not guaranteed! Figure 8(A) reproduces the histogram for Pennsylvania's 2016 Presidential election, where the statewide Republican vote share was 50.4%. Figure 8(B) shows what would happen if we shifted the vote by 0.5% toward Democrats, using uniform partisan swing. In this hypothetical election (which is extremely close to the real one), Republicans would have gotten less than half the vote, yet all 50,000 plans in the ensemble still lead to $S \geq 0.5$ (9+ seats out of 18), and 96% of the plans lead to $S > 0.5$.[27]

The 2016 election in Pennsylvania also poses a problem for the partisan symmetry standard. Figure 8(C) shows the seats–votes curves for all the plans in the ensemble superimposed on each other. The mean value of S for each $\overline{V}$ is shown in black. This "average seats–votes curve" for the ensemble is highly asymmetric by any measure of symmetry, as are almost all the individual seats–votes curves. Something about the geographic distribution of voters in Pennsylvania gives the Republicans a huge advantage, making it very unlikely that a random neutral districting plan will pass the symmetry standard.

What is going on here? One possible explanation is the frequently cited demographic trend of "Democrats packing themselves." Figure 8(D) illustrates this phenomenon at the level of precincts, the smallest geographic units for which election data are available. (Intuitively, you can think of each precinct as a neighborhood.) The histogram shows many Pennsylvania voters living in extremely Democratic precincts (Republican vote share $< 10\%$), but almost none living in extremely Republican precincts (Republican vote share $> 90\%$). Similar trends have been observed in other elections across the country, even in states that lean strongly Republican overall [37].

The same asymmetry that we see at the extremes of the histogram also exists in the state as a whole. In 2016, in Pennsylvania's majority-Republican precincts, the overall margin of victory for Republicans was 16%. In majority-Democratic precincts, the overall margin of victory for Democrats was 21%. Let's call this difference in the margins of victory **differential packing**, and denote it by d_{precinct}. For Pennsylvania in 2016, $d_{\text{precinct}} = 0.21 - 0.16 = 0.05$, indicating that, at the precinct level, Democrats were somewhat more packed than Republicans.

Does differential packing in precincts extend to districts? Not necessarily: if the precincts were distributed completely at random across the state, all such local differences would get diluted. But intuitively, nearby precincts tend to be similar to each other, so we would expect the trend to persist at larger spatial scales as well. This might plausibly lead to a tendency for Democratic districts to be more packed,

[27] This is a good opportunity to revisit the quote from *The Washington Post* in Section 1, regarding Pennsylvania's 2012 Congressional election. That year, Republicans won 13 out of 18 seats, with a statewide vote share of just over 49%. These results were taken by the author of the article as proof of "an aggressive gerrymander". Now compare this outcome to the hypothetical election in Figure 8(B), with a similar statewide Republican vote share. A plan that gave Republicans 13 seats would indeed be flagged as a gerrymander, but a plan with just one fewer Republican seat would not be. Intuitively, a seat share of 67% (= 12/18) for a minority party would probably still look suspicious to the author of the article and to most readers. But ensemble analysis shows that, given Pennsylvania's voter geography, it would not be an indication of gerrymandering.

on average, than Republican ones.

Ensemble analysis confirms this empirically. For a given districting plan, we can define d_{district} to be the difference between the average margin of victory in Democratic districts and the average margin of victory in Republican districts. Across all the districting plans in our Pennsylvania ensemble, the mean of d_{district} is 0.044. Differential packing at the district level in Pennsylvania is, on average, less extreme than at the precinct level, but not by much.

In a state with $\overline{V} = 0.5$, a positive value of d_{district} automatically implies $S > 0.5$: if Democratic districts are won by larger margins, then there must necessarily be fewer of them. In both Figure 8(A) and (B), the overall Republican vote share is extremely close to 0.5 and d_{district} is significantly greater than 0 for most plans in the ensemble. From these two facts alone, we can conclude that most of the plans will give the majority of Pennsylvania's seats to Republicans, though to what extent, we can't say. (For a much deeper dive into the geography and vote spatiality of Pennsylvania, see Chapter 5.)

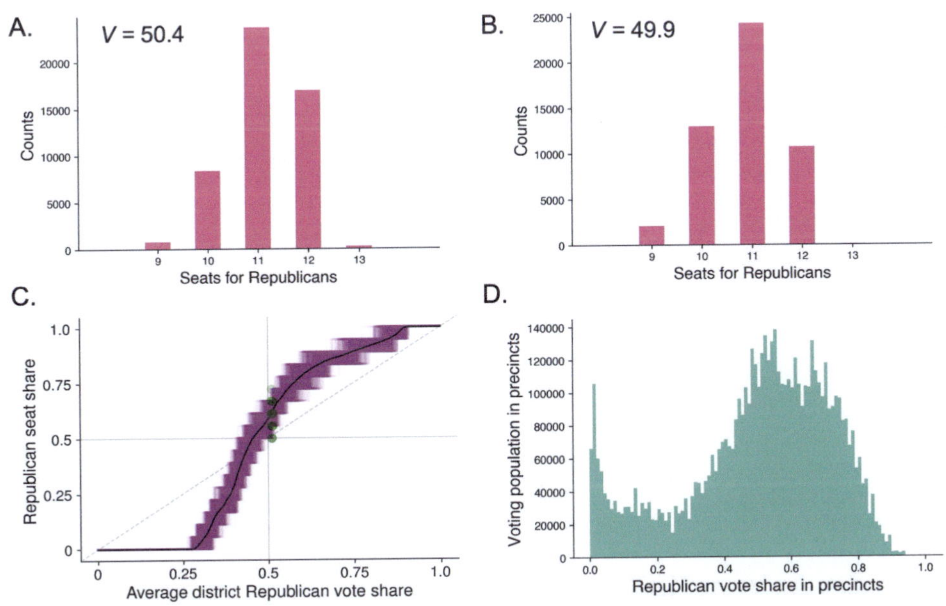

Figure 8: Analysis of the 2016 Presidential election in Pennsylvania. (A) Number of Congressional seats (out of 18) that would have been won by Republicans under 50,000 randomly generated plans. (B) Number of seats that would have been won by Republicans if the statewide Republican vote share decreased by 0.5% (assuming uniform partisan swing). (C) Seats–votes curves for 50,000 randomly generated plans. The black curve shows the mean value of S for each vote share. (D) Total number of voters in precincts at each Republican vote share.

Finally, let us revisit the efficiency gap through the lens of ensembles. For their report on redistricting criteria in Virginia [38], researchers at MGGG generated two ensembles of neutral plans: one with 11 seats (for Virginia's Congressional delegation) and one with 40 seats (for the state Senate). Figure 9 shows the number of seats won by Democrats and the distribution of EG for each plan of ensemble.[28]

We see immediately that the EG distributions look like noisier versions of the seat distributions. This is unsurprising, given the formula $EG \approx S - 2V + \frac{1}{2}$ that we derived in Section 3. Here V is fixed: it is the actual Republican vote share in the 2017 election for Attorney General. So under the equal turnout assumption, each plan's EG is just a function of its seat share S. The noise in the EG histograms reflects slight variations in district turnout across the plans in each ensemble, but it is negligible.

In general, we expect a plan that is an outlier in seats to be an outlier in EG and vice versa. Thus, in the context of ensemble analysis, the efficiency gap offers very little new information.

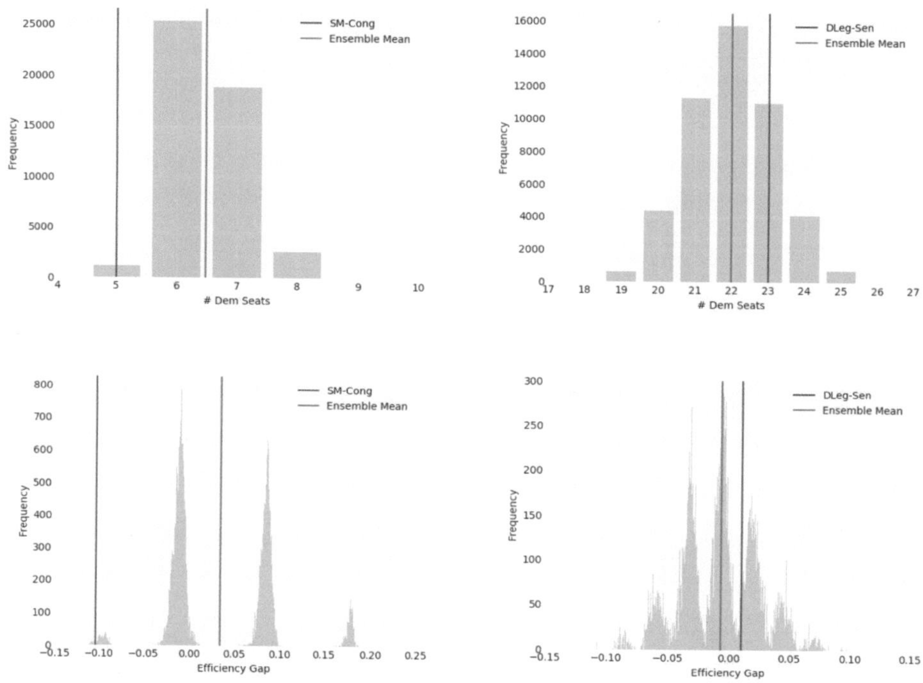

Figure 9: Histograms of seats and efficiency gaps from ensembles in Virginia, using data from the 2017 election for Attorney General. The plots on the left correspond to ensembles with 11 seats, and the plots on the right correspond to ensembles with 40 seats. The red lines show results from Virginia's actual districting plans: the special master's Congressional map (SM-Cong, left) and the legislature's state Senate map (DLeg-Sen, right). The green lines show the ensemble means. Taken from Figure 5 of DeFord and Duchin [38].

THE STATE SUPREME COURTS WEIGH IN

So far, the extreme outlier standard has been validated by the Supreme Courts of Pennsylvania and North Carolina, in *League of Women Voters v. Commonwealth of*

[28]Since all analyses in DeFord and Duchin [38] are in terms of Democratic rather than Republican vote share, $EG > 0$ here corresponds to an advantage for Democrats.

Pennsylvania (2018) and *Common Cause v. Lewis* (2019) respectively. In each case, the Court heard from several expert witnesses who had used different methods to generate their ensembles, but arrived at the same conclusion: the plan in question was an extreme outlier, constructed with clear partisan intent and conferring a significant, quantifiable advantage on the mapmakers' party. Both Courts concluded that the plan under consideration caused sufficient damage to the disadvantaged party to violate the Free Elections Clause of their respective state Constitutions.

Neither Court chose to set any specific numeric threshold for how extreme an outlier a plan has to be in order to be deemed unconstitutional. Perhaps this was because, in both cases, the Republican seat share of the enacted plans was so far out of the range of the nonpartisan ensembles that it left no room for doubt. In the next redistricting cycle, there are likely to be many more cases brought in different states, and some of them might be less clear-cut. We may even see proponents of different versions of the extreme outlier standard arrive at different conclusions because of their differences in methodology. A lot of legal and mathematical details remain to be worked out. Nonetheless, based on the experience of the last few years (and the resounding endorsement of U.S. Supreme Court Justice Elena Kagan, albeit writing for the dissent [39]), it seems clear that the extreme outlier standard is well on the way to becoming the principled, widely accepted quantitative standard of fairness in redistricting that the Supreme Court in *Rucho* has assured us cannot possibly exist.

5 CONCLUSION: DEBATING FAIRNESS

We began this chapter by asking a seemingly simple question: "What is a fair map?" We discovered that this is not a simple question at all: the idea of fairness is extremely complex and multifaceted, and our voting system isn't set up to accommodate any absolute definition of it. This is what makes it so hard to detect gerrymandering and to measure its effects. Did an election have a disproportionate result because of gerrymandering... *or* because the districts were highly competitive, as in Minnesota? Is the seats–votes curve asymmetric because of gerrymandering... *or* because of differential packing, or both?

Ensemble analysis circumvents this problem by establishing a baseline grounded in the actual geography of a state. Instead of measuring deviation from an abstract ideal, it allows us to measure deviation from neutrality. As a new and powerful way to measure the actual effects of gerrymandering, it should prove invaluable in redistricting reform around the country.

But as the Supreme Court reminds us in *Rucho*, "deciding among... different visions of fairness poses basic questions that are political, not legal" [34]. While absolute standards of fairness are not good for deciding whether gerrymandering has actually occurred, they are still very much relevant at the political level. The people of a state (or of the United States) can decide on any definition of fairness that reflects their values (such as proportionality or partisan symmetry) and pass a law accordingly. Of course, voter geography might make it difficult for mapmakers to comply with this law. But drawing maps that strive to comply with such laws is something that the Supreme Court has previously declared constitutional, and it is up to the people of each state to decide whether careful partisan tuning is in accordance with their values.

A more serious problem is that many of the properties that voters might want from a districting plan are incompatible with one another. Foremost among these, as we have seen, are competitiveness and proportionality. Arizona's State Constitution requires mapmakers to strive for competitive districts [40]; Ohio's Constitution requires mapmakers to strive for proportionality by party [41]. Most likely, Arizona voters would not declare themselves to be against proportionality, nor Ohio voters against competitive districts. Yet as we saw in Section 1, under our current electoral system, competitiveness and (assured) proportionality are fundamentally incompatible.

This might be an argument for change that goes beyond the way we draw our district lines. If proportionality—or some other ideal that is elusive to districters—is truly what we want, maybe it's the voting system that should change, and not the standard we use to judge it.

ACKNOWLEDGMENTS

The authors thank Doug Spencer for his help on the law content of the chapter.

REFERENCES

[1] Davis v. Bandemer. 478 U.S. 109 (1986).

[2] MIT Election Data and Science Lab. U.S. House 1976–2018, 2017.

[3] MIT Election Data and Science Lab. U.S. President 1976–2016, 2017.

[4] Daily Kos Elections' presidential results by congressional district for 2016, 2012, and 2008. Retrieved from https://www.dailykos.com/stories/2012/11/19/1163009/-Daily-Kos-Elections-presidential-results-by-congressional-district-for-the-2012-2008-elections. Data also linked at https://tinyurl.com/dailykos2018.

[5] Jonathan N. Katz, Gary King, and Elizabeth Rosenblatt. Theoretical foundations and empirical evaluations of partisan fairness in district-based democracies. *American Political Science Review*, 114(1):164–178, 2020.

[6] Andrew Gelman and Gary King. A unified method of evaluating electoral systems and redistricting plans. *American Journal of Political Science*, pages 514–554, 1994.

[7] Analysis of the Efficiency Gaps of Wisconsin's Current Legislative District Plan and Plaintiffs' Demonstration Plan. Kenneth R. Mayer, at 46 tbl.10, Whitford v. Gill, 218 F. Supp. 3d 837 (W.D. Wis. 2016) (No. 3:15-cv-00421-bbc), 2015 WL 1009101.

[8] Moon Duchin, Taissa Gladkova, Eugene Henninger-Voss, Ben Klingensmith, Heather Newman, and Hannah Wheelen. Locating the representational baseline: Republicans in Massachusetts. *Election Law Journal: Rules, Politics, and Policy*, 18(4):388–401, 2019.

[9] U.S. Judge is named to help guide redistricting. *The New York Times*, Feb. 22, 2012. Retrieved from https://www.nytimes.com/2012/02/22/nyregion/federal-judge-is-named-to-help-oversee-redistricting-in-new-york.html.

[10] New Congressional Lines Imposed by Federal Court. *The New York Times*, Mar. 12, 2012. Retrieved from https://www.nytimes.com/2012/03/20/nyregion/judges-impose-new-congressional-map-for-new-york.html.

[11] Gary King and Robert X Browning. Democratic representation and partisan bias in congressional elections. *American Political Science Review*, 81(4):1251–1273, 1987.

[12] James Madison. Notes of Debates in the Federal Convention of 1787, 9 June, 1787.

[13] John Adams. Thoughts on Government, Applicable to the Present State of the American Colonies. In a Letter from a Gentleman to His Friend. Philadelphia: John Dunlap, 1776. Retrieved from https://www.masshist.org/publications/adams-papers/index.php/view/PJA04dg2.

[14] John S. Mill. Considerations on Representative Government. London: Parker, Son, Bourn. 1861. Retrieved from https://www.gutenberg.org/files/5669/5669-h/5669-h.htm.

[15] Edward R. Tufte. The relationship between seats and votes in two-party systems. *American Political Science Review*, 67(2):540–554, 1973.

[16] Brief for Professors Gary King, Bernard Grofman, Andrew Gelman, and Jonathan N. Katz as Amici Curiae in Support of Neither Party. League of United Latin American Citizens v. Perry, 548 U.S. 399 (2006) (no. 05-204). Retrieved from https://gking.harvard.edu/publications/brief-amici-curiae-professors-gary-king-bernard-grofman-andrew-gelman-and.

[17] Gary King. Representation through legislative redistricting: A stochastic model. *American Journal of Political Science*, pages 787–824, 1989.

[18] Bernard Grofman. Measures of bias and proportionality in seats-votes relationships. *Political Methodology*, 9(3):295–327, 1983.

[19] Daryl DeFord, Natasha Dhamankar, Moon Duchin, Varun Gupta, Mackenzie McPike, Gabe Schoenbach, and Ki Wan Sim. Implementing partisan symmetry: Problems and paradoxes. *Political Analysis*, to appear, 2021.

[20] Bernard Grofman and Gary King. The future of partisan symmetry as a judicial test for partisan gerrymandering after *LULAC v. Perry*. *Election Law Journal*, 6(1):2–35, 2007.

[21] Vieth v. Jubelirer. 541 U.S. 267 (2004) (Kennedy A. concurring opinion).

[22] League of United Latin American Citizens v. Perry. 548 U.S. 399 (2006).

[23] John F. Nagle. Measures of partisan bias for legislating fair elections. *Election Law Journal*, 14(4):346–360, 2015.

[24] Michael D. McDonald and Robin E. Best. Unfair partisan gerrymanders in politics and law: A diagnostic applied to six cases. *Election Law Journal*, 14(4):312–330, 2015.

[25] Brief for Heather K. Gerken, Jonathan N. Katz, Gary King, Larry J. Sabato, and Samuel S.-H. Wang as Amici Curiae in Support of Appellees. Gill v. Whitford, 585 U.S. ___ (2018) (no. 16-1161). Retrieved from https://gking.harvard.edu/publications/brief-heather-k-gerken-jonathan-n-katz-gary-king-larry-j-sabato-and-samuel-s-h.

[26] Nicholas O. Stephanopoulos and Eric M. McGhee. Partisan gerrymandering and the efficiency gap. *University of Chicago Law Review*, 82:831, 2015.

[27] Moon Duchin. Gerrymandering metrics: How to measure? what's the baseline? *American Academy of Arts and Sciences*, Retrieved from https://www.amacad.org/news/redistricting-and-representation.

[28] Kristopher Tapp. Measuring political gerrymandering. *The American Mathematical Monthly*, 126(7):593–609, 2019.

[29] Mira Bernstein and Moon Duchin. A formula goes to court: Partisan gerrymandering and the efficiency gap. *Notices of the AMS*, 64(9):1020–1024, 2017.

[30] Ellen Veomett. Efficiency gap, voter turnout, and the efficiency principle. *Election Law Journal: Rules, Politics, and Policy*, 17(4):249–263, 2018.

[31] Benjamin Plener Cover. Quantifying partisan gerrymandering: An evaluation of the efficiency gap proposal. *Stanford Law Review*, 70:1131, 2018.

[32] Christopher P. Chambers, Alan D. Miller, and Joel Sobel. Flaws in the efficiency gap. *Journal of Law and Politics*, 33:1, 2017.

[33] Jan E. Leighley and Jonathan Nagler. *Who votes now?: Demographics, issues, inequality, and turnout in the United States.* Princeton University Press, 2013.

[34] Rucho v. Common Cause. 588 U.S.___ (2019).

[35] Common Cause v. Rucho. 318 F. Supp. 3d 777, 807–808 (MDNC 2018).

[36] Gregory Herschlag, Han Sung Kang, Justin Luo, Christy Vaughn Graves, Sachet Bangia, Robert Ravier, and Jonathan C. Mattingly. Quantifying Gerrymandering in North Carolina. *Statistics and Public Policy*, 7(1):30–38, 2020.

[37] Jowei Chen. Unintentional Gerrymandering: Political Geography and Electoral Bias in Legislatures. *Quarterly Journal of Political Science*, 8(3):239–269, June 2013.

[38] Daryl DeFord and Moon Duchin. Redistricting reform in Virginia: Districting criteria in context. *Virginia Policy Review*, 12(2):120–146, 2019.

[39] Rucho v. Common Cause. 588 U.S. ___ (2019) (Kagan E. dissenting opinion).

[40] Arizona Const. Art. IV, pt II, §1, cl. 14F.

[41] Ohio Const. Art. XI, §06.

3 *Interviews: Concepts of representation*

MOON DUCHIN AND OLIVIA WALCH

This chapter features interviews with four political thinkers.

Q. What is your specialty?

RYAN MULDOON PHILOSOPHY, UNIVERSITY AT BUFFALO

I work on Social Contract Theory and what's called "Public Reason," particularly with respect to diverse societies. There's been an interesting move since the 1970s where people became increasingly aware of how much diversity matters for ideal political philosophy, what the rules should be, social justice, coercion by the state, and so on.

BRIAN KOGELMANN PHILOSOPHY, POLITICS, AND ECONOMICS, UNIVERSITY OF MARYLAND

I'm interested in normative questions about what our institutions should look like, what our voting rules should be, how we should structure our systems.

LIZ ANDERSON PHILOSOPHY & WOMEN'S STUDIES, UNIVERSITY OF MICHIGAN

I work on egalitarianism, democratic theory, social epistemology, and pragmatism as a methodology of moral inquiry. I am a leading proponent of what is known as a "relational egalitarian" framework, which focuses on creating the conditions for people to relate to each other as equals, rather than in relations of domination/subordination, or honor/stigmatization, or counting/not counting in deliberation.

CLAUDINE GAY GOVERNMENT & AFRICAN AND AFRICAN AMERICAN STUDIES, HARVARD

I'm trained as a political scientist, with a focus on American politics. In that work, I look at political behavior, public opinion, and the politics of ordinary people. I'm especially interested in how all of this intersects with, and how it's inflected by, race.

M. Duchin, O. Walch (eds.), *Political Geometry*, https://doi.org/10.1007/978-3-319-69161-9_3

Q: The United States was set up as a democratic republic—what does that mean? What are the relevant design principles in the U.S. case?

First of all, "democracy" is a system where individual preferences determine political decisions, such that (in the ideal) each person counts the same.

I define a democracy as involving 3 levels (obviously this is an ideal and all democracies fall short):

1. As a membership organization, it offers all permanent residents an easy pathway to citizenship. There is only one class of citizens; all adult citizens must be equal under the law.
2. As a formal mode of government, democracy is either direct (the citizens directly vote on laws) or representative (citizens vote for legislative and executive offices). There must be a universal franchise of all adult citizens, periodic elections, free speech, assembly, and rights to petition government, a free press, an open right to compete for office, and provisions for fair competitive elections. All voters count equally.
3. As an informal way of life, democracy is a mode of relating to fellow citizens as equals, communicating public concerns with them in a spirit of open and sincere exchange, working together toward just compromises.

"Republic" just means *not a monarchy*—it's a contrastive term. Democratic republics are usually for big nation-states; setting up multiple levels lets you pump the brakes a bit, to mediate straight public will. The U.S. Constitution is designed around the states having significant authority and interacting with the federal government. Cities, towns, and districts don't have this kind of special authority.

In part, the institutional design was intended to promote the election of representatives that are a little more elite than the average citizen—which was supposed to mean more educated, more enlightened, and so on. The Electoral College in particular is a weird hybrid compromise between those who wanted a popularly-elected president and those (like James Madison) who wanted the president selected by Congress.

The U.S. is a republic and not a full democracy because there are arbitrary exclusions (D.C. citizens can't vote for Senators and only have a nonvoting member of the House) and because the Senate is a grossly unequal form of representation, with California and Wyoming each getting two senators despite the vast population difference. The U.S. only became an approximate democracy with the 1965 Voting Rights Act, empowering Black citizens to vote.

Q: What would it mean to have a *more* representative democracy?

One important aspect is the feedback loop between representative government and citizens' own orientation. When you have an elected body that looks more like constituents—what is sometimes called *descriptive representation*—that may create greater trust, so that the decisions that emanate have more legitimacy. As individual political actors, voters might feel a greater sense of agency and efficacy, manifested in more participation. Those are all empirical questions and hypotheses, but they were borne out in my studies.

I think inputs and processes matter for assessing whether a system is democratic and representative, and not just outputs. The system should be responsive to citizen inputs. A clairvoyant dictatorship that just happened to deliver what the people want is not democratic. This is partly because "what the people want" has itself to be a public thing shaped by public discourse about what the problems are, and not just an aggregate of private opinions formed without considering what other citizens think.

I would tend to try to cash this out in some notion of civic equality, equalizing the voice of the people. "Voice" works through two mechanisms: a formal or procedural voice (i.e., votes), or a civic "exercise of voice" (like protests, or letters of complaint or support).

The key justification for representative rather than direct democracy is that a smaller elected body, especially if it is full-time with serious investigative powers, is better able to gather the information needed to construct policies likely to deliver good outcomes. So this is an epistemic justification: representatives can study the issues, consult experts, get testimony, etc., so that they can legislate with an informed view. At the same time the perspective one brings to the table must be sensitive to the concerns of one's constituents, which vary depending on who is in the district. (Voters are more certain about their concerns than about what policies would effectively address them.)

Q: What is the role of parties in a representative democracy? (What could it be, and what is it actually in today's United States?)

Political parties are needed to set agendas—that is, to lay out and organize concerns to be addressed in the legislature with a broad sense of how to address those concerns. Otherwise representatives would come to the legislature with such disparate concerns from diverse constituents that there would be no working majority for anything.

Political scientists used to say that parties would be good at ensuring the "trustee model" of representation: parties would restrict who would be allowed to run, so that we would be more confident of getting wise representatives whom we should trust to operate with autonomy. (This is in contrast to the "delegate model" where the representative is merely a mouthpiece.) Primaries undermine this screening role! Trump, in particular, is unimaginable without primaries.

In the U.S., things changed greatly with post-1968 reforms. Before that, primaries were primarily a tool to give the party bosses insight into who would be more popular—they were not binding. The parties reformed themselves in the wake of the 1968 Democratic convention riots and other upheaval, which was partly driven by public frustration with backroom decisions. Primaries became binding in the 1970s.

One of the problems with our current party system is that the parties are too weak. They should be able to perform a screening function to screen out demagogues and irresponsible actors. Obviously, the GOP has failed at that and now the U.S. is paying the price.

Q: What's the best case for proportionality? And is race different from party?

I think that a representative body that demographically reflects the population is not a necessary condition to produce substantive policy representation, but it has a value in and of itself.

This idea of the body looking like the polity is something like proportionality. But aspects of proportionality worry me. The spectrum of political views is quite wide, and there is a critical mass in support of some fringe views. We don't necessarily want those institutionalized in an elected body—this is the risk of going all-in for proportionality.

Look, for instance, at the Voting Rights Act, where rough proportionality has been used as a guide to locating shortfalls in minority representation. Its role is remedial, not a statement of ideal theory. Proportionality is used as a rule of thumb for what would happen in a normal, well-functioning democracy where everyone has equal voice. We wouldn't need to articulate a positive right to proportionality in order to impose heightened scrutiny when we deviate too far.

When it comes to party, the current U.S. system is not in great shape. One party is multi-ethnic while the other is not at all, and geographic factionalism helps to lock the parties in place. But if we posit nimble, responsive parties, then over a long enough time-span, consistent disproportionality would signal that something is broken.

The problem with proportionality as a barometer of fairness is that not every cognizable group can have proportional representation. Being represented in proportion to your numbers, or at least getting enough representation to be effective, is only critically important if a group and its concerns are marginalized.

But everyone has cross-cutting identities and must be free to define for themselves which ones matter for representation. Race comes to the fore because in the U.S., groups that suffer continuing racial discrimination and marginalization have an overwhelming common concern in overcoming that.

The best argument for party proportionality is that it signals a more equitable distribution of representatives, in the sense of ideological equity. The knock on this view is that it leads to incoherent policy agendas. Perhaps the one good thing to say about a first-past-the-post voting system is that, if you couple it with strong parties, you get coherent policy.

Q: How can any of this help us to think about gerrymandering? What is wrong with gerrymandering from the point of view of democracy?

I don't know! It's quite difficult to articulate. Regardless of how gerrymandered your district is, you still have a 1/N say in who's elected. And quantifying something like influence on who's elected is elusive. If we make sense of this in terms of influence on policy decisions, the harm would have to be really long-lasting to be recognizable.

It may be useful to think in terms of "communities of interest." This is a very legitimate concept—there do truly exist cognizable COIs, such as along lines of race and class and urbanicity, reflecting different life chances, different infra-structure needs, and so on. Part of what we want our political process to reflect, what we want our policies to be shaped by, are the preferences and values of these communities.

What are the mechanisms that allow those shared values to be articulated and to be introduced meaningfully into the process? We can understand gerrymandering as disruptive of communities of interest, ensuring that some interests never make their way into the policymaking process.

Other than by ensuring that everyone has effective access to the polls (time, transportation, no legal hassles), it is hard to think about this at the individual level. One has to analyze this at the level of salient interest and identity groups that need representation, including political party as one kind of identity.

To best meet the goals of representative democracy, a system must be responsive to changes in public opinion. Officeholders can't be allowed to entrench themselves by choosing who gets to vote for them, as in the current gerrymandering system. Without a real risk of being unseated in an election, there is no accountability of representatives to the people. Michigan voters, on a bipartisan basis, passed a referendum to establish an independent citizens' redistricting commission. Voters were acting against politicians being able to entrench their power by drawing districts they will always win. Entrenchment gives politicians a kind of electoral security that enables them to ignore their voters. I think this is a compelling argument!

ECLECTIC READING RECOMMENDATIONS

Classical democracy

- Bernard Manin, *The Principles of Representative Government*—Athenian government described as a mix of councils of the enfranchised and some "lottocracy" (random selection).
- Danielle Allen, *The Origins of Political Philosophy* (in the Oxford Handbook of the History of Political Philosophy) and *Talking about revolution: on political change in fourth-century Athens.*

18th–19th centuries; French and American Revolutions

- David Hume *Of Parties in General* (1742)—Hume distinguishes between parties (which is used broadly here, like factions) that are based on shared interests as opposed to religions or personalities. He is writing just as Whigs and Tories are starting to emerge in Britain; he notes that alignments based on shared interests make it more feasible to compromise on policy.
- Jeremy Bentham—consequences of democracy; are people too selfish? *A fragment on government* (1776), *Short Review of the Declaration* (1776), *An Introduction to the Principles of Morals and Legislation* (1780), *Deontology or, The science of morality* (1834).
- *The Federalist Papers*—a series of essays where Hamilton, Madison, and friends are working out their defense of the new U.S. Constitution. See for instance Federalist 47 (Madison on factionalism).
- Benjamin Constant—How did the French revolution channel ancient democratic theory? *Réflexions sur les constitutions, la distribution des pouvoirs et les garanties dans une monarchie constitutionnelle* (1814).
- Alexis de Tocqueville, *Democracy in America* (1835–1840)—thoughts on the young American democracy; he observes democratic culture and small-scale communities of interest giving people practice with democracy.

20th century

- Hanna Pitkin, *The Concept of Representation* (1967). A classic book-length essay about what representation, and good representation, might mean.
- John Rawls was an important abstracter who only gradually realized how diversity is inescapable. In *Theory of Justice* (1971) he thinks he's got it figured out abstractly; by his last major book (*Political Liberalism,* 1993) he's really grappling with what he calls the "burdens of judgment"—we can have different ontological commitments, and considered moral views, that we treat as irresolvable disputes.
- Amartya Sen's works are generally essential reading—especially *Collective Choice and Social Welfare* (1970) and *Inequality Reexamined* (1992).
- Arend Lijphart is one of the first people to do what you might call econometrics of democracy. Notable reads include *Dimensions of democracies* and

Power-sharing and group autonomy in the 1990s and the 21st century.

- Lani Guinier, *The Tyranny of the Majority* (1994)—essential reading to shake us up and remind us that other systems are possible.
- Robert Dahl, *On Democracy* (1998)—a good overview at the turn of the century.

Up to the minute

- Claudine Gay, *Spirals of Trust: The Effect of Descriptive Representation on the Relationship Between Citizens and Their Government* (2002)—I use survey data to test intuitions about the effects of racial identification of voters with their representatives, in terms of individual and institutional perceptions as well as the likelihood of contacting one's representative.
- Elizabeth Anderson, *The Imperative of Integration* (2010)—I make a case for racially integrated districts in order to maximize accountability.
- For the history and dynamics of U.S. political parties, see for instance Marty Cohen et al, *The Party Decides* (2008); John Aldrich, *Why Parties?* (1995, 2011); Levitsky and Ziblatt, *How Democracies Die* (2018); or Rosenbluth and Shapiro, *Responsible Parties* (2018).
- Sam Issacharoff, Rick Pildes, Pam Karlan are all modern legal thinkers who have a lot to say on representation.
- Alex Guerrero, *Against Elections: The Lottocratic Alternative* (2014)—a normative case for random selection in government.
- Gerald (Jerry) Gaus pioneered a New Diversity Theory. See especially *The Tyranny of the Ideal: Justice in a Diverse Society* (2016). You can read work of Gaus (on Property) and Anderson (on Equality) in the Oxford Handbook of Political Philosophy.
- Ryan Muldoon, *Social Contract Theory for a Diverse World*—I develop a post-Rawlsian framework for handling the challenges posed by very diverse societies.
- Brian Kogelmann, *Secret Government: The Pathologies of Publicity*—my new book is on transparency and its relationship to democracy. Sometimes transparency doesn't serve our democratic goals!
- Hélène Landemore, *Open Democracy*—can modern representative government recover some of the anti-elite openness of ancient democracies?

Chapter 4

Is the redistricting problem math, systems, or people?

KEITH GÅDDIE

CHAPTER SUMMARY

Keith Gåddie will give us the political scientist's view of the redistricting landscape from 30,000 feet, helping us to taxonomize the problems and start to think about strategies for the way forward.

1 INTRODUCTION

"Tell Mike it was only business." Salvatore Tessio, *The Godfather*

Gerrymandering is the act of drawing legislative district boundaries for political advantage. And, in American politics, gerrymandering is a profanity. There is no positive conversation where gerrymandering is deemed to be virtuous; it is only a problem, and, in American politics the source of all that ails democracy.

Concerns about gerrymandering revolve around the role of gerrymandering in distorting popular democracy. Free and democratic elections translate votes into government. Good democratic systems translate majorities of votes into majorities of delegates, thereby fulfilling one of the fundamental assumptions of democracy, that majorities should govern. In the representative assemblies of the Anglosphere[1] the desire to reflect majority will while still representing minority constituencies

[1] The Anglosphere is a recent term, defined as countries of the world where the English language and cultural values derived from Great Britain predominate. To understand those linguistic traditions and cultural values in the American case, I recommend Fisher [17].

M. Duchin, O. Walch (eds.), *Political Geometry*, https://doi.org/10.1007/978-3-319-69161-9_4

has led to the use of geographic constituencies—districts—to choose delegates to serve and make law.

When existing political subdivisions are used to allocate power, the strength of representation is dictated in part by where someone lives, which creates distortions in the translation of votes into government and deviations from overall proportionality. District systems are in part meant to alleviate these malapportionment problems, but they potentially introduce new distortions through the act of redistricting. These distortions can deny access by popular majorities to legislative majorities; or they can enhance or temper their size. In this chapter, I ask the reader to consider the extent to which redistricting is a **people problem**, a **systems problem**, or a **math problem**.

People. Are the problems we ascribe to gerrymandering really just artifacts of where people live or how we count them?

Systems. The single-member district system (SMD) is dominant in America—one bounded geographical area sends one representative to government. Are single-member districts fundamentally to blame for the paradoxes and distortions of gerrymandering? And are our correctives (like the Voting Rights Act) doomed to backfire?

Math. Is the shortcoming in our metrics of fairness? Have we all been distracted by meaningless quantification?

Systems again. How does technology compound the problems, and could it relieve them?

People again. All of these elements interlock, but can't be teased apart without contending with the most central of all. The incentives of the people who currently control the process of redistricting do not align with the aims of the fullest and most responsive representative democracy.

2 THE REDISTRICTING PROBLEM IS A PEOPLE PROBLEM, VERSION 1

How is redistricting a people problem? It has to do with where people live; how many people live there; how eligibility to participate is determined; and how geography is used to allocate representation and voting power. The earliest 'people problems' in Anglo America had two origins: first, the question of eligibility; and then the abuses of apportionment known as 'rotten boroughs.'

ELIGIBILITY AND APPORTIONMENT

An early example of the eligibility problem appears in an exchange between James Madison and Thomas Jefferson regarding voter eligibility for the Virginia general assembly, with Madison advancing a property basis for eligibility to vote (acreage or town lots), while Jefferson argued for a broader definition allowing any free man who had borne arms in the Revolution to be eligible to vote. The former plan gave more political power to the counties of the Tidewater, dominated by planters and those with large slave populations. Jefferson's proposal gave more power to the frontiersmen who lived west of the Piedmont in the Blue Ridge, Appalachia, and in the Kentucky counties. This gives us our first glimpse of a recurring issue: the entanglement of principle with regional or political interests.

The rotten borough problem finds its origins in Great Britain, where a political subdivision called a *borough* was used as a vehicle to elect parliamentarians. Many boroughs were possessed of tiny populations, often in previously prosperous areas where population loss had ensued—for instance, Newtown, on the Isle of Wight, had fourteen residences and twenty-three voters in the 18th century [26, 31]. The tiny populations nonetheless commanded a significant share of political power, so they could be easily controlled by wealthy landowners who thereby magnified their voice in government. These came to be called "rotten boroughs."

Taken together, restrictive participation standards and the use of fixed political subdivisions for apportionment (usually counties, parishes, or townships) could significantly distort political power.

Before 1962, the allocation of political power by county was common in the U.S., with every county in a state often guaranteed one representative in the state legislature, no matter how small the population. The example I am most familiar with is the state of Georgia, which was the primary defendant in two different one person, one-vote lawsuits, including one of the 1960s cases that coined the term.[2] In Georgia, all political power was deeply devolved down to the counties. Each of Georgia's 159 counties was guaranteed at least one state representative, and no county had more than three representatives. In state Democratic Party primaries, the legislative apportionment system was used to allocate votes in the primary. For each state representative a county had, it got two votes in congressional or statewide primaries. The consequence of this system was to make the value of a vote in Georgia's smallest counties, such as Echols or Taliaferro, worth several hundred times more than a vote in the major urban counties like Fulton or DeKalb (which both contained the city of Atlanta).[3]

The consequence was that half of all political power in Georgia was invested in counties containing less than a quarter of the state's population—and far less of the functionally eligible electorate, due to the broad-based disfranchisement of Black voters [9, 22]. And, given the predisposition of those who have political power to keep political power, it took decades of litigation culminating in the one-person, one-vote cases to disrupt the broad-based use of rotten boroughs, thereby establishing the standard that the value of one's vote should not depend on where you live—that rural votes (and especially rural White votes) should be neither practically nor morally superior to the votes of any other Americans.

[2] *Gray v. Sanders*, 372 U.S. 368 (1963); *Wesberry v. Sanders*, 366 U.S. 1 (1964).

[3] The political consequences of Georgia's reapportionment problem are described in Bullock et al., Bullock and Gaddie, Buchanan, Cornelius, and Saye [7, 8, 11, 15, 30].

WHERE THE PEOPLE ARE

Even with equal-sized districts, there is still a potentially distorting impact of where people live in modern redistricting. When we form modern legislative districts, residential patterns can interact with the line-drawing to dilute or enhance the political power of some types of voters over others.

Elections turn votes into government. And, in single-member districts (which are the most common representative district in the United States) this means that the candidate with the most votes wins the district, and the party that wins the majority of districts organizes the legislative chamber. Not all people in a district are voters. Minors do not vote. Unregistered adults do not vote. In many states, people with felony convictions do not vote. And, noncitizens do not vote. But minors, unregistered adults, and noncitizens are not equally distributed across all districts. Even when all districts in a state are apportioned to exactly equalize overall population, some districts in a state have far fewer people who are potentially eligible to vote than other districts, among other factors driving differential turnout. For example, in 2008, the 20th District of New York cast 325,706 votes for the two major presidential candidates while New York's 16th District cast just 167,108. Even more stark differences are observed in Florida and California and other states with geographically concentrated noncitizen populations. Political scientist Jim Campbell refers to this as the 'cheap seats' problem in American legislative elections—and the cheap seats problem actually creates a Democratic Party advantage in the elections he analyzed [12]. A multivariate analysis for voter participation rates in 21st century congressional districts indicates that larger non-White populations, large noncitizen populations, being more rural areas, and having fewer districts (smaller states) is associated with a greater variation in turnout from district to district within a state [19, 23, 29]. (This distortion problem is effectively eliminated in democracies that use proportional representation. See Sidebar 2.3 and also Amy [3].)

Then, there is a related issue: Is representation about voters, citizens, or people? A conservative argument advanced for the past decade has held that we should consider apportioning political power and crafting district populations based on citizenship data. This can be an intellectually seductive argument—elections are constrained to voters, who are in turn always supposed to be citizens. They have the sovereignty, which allows for the American government and the state governments. But voters are not the only citizens, let alone the only people, who are represented by elected officials. Representative needs extend to residents in general—as do representative responsibilities of the members of government.[4] And there are data issues that can attach to trying to measure citizenship, as legal scholar Nathaniel Persily ably notes [29].

One of the questions of the modern redistricting conversation asks if gerrymandering drives political polarization. Polarization has grown in legislatures and among voters for three decades. The crafting of districts that cobble together people who vote of a like mind is often pointed to as a cause.

But, why is it so easy to form ideologically homogeneous districts? One theory is

[4]See Kozinski's partial concurrence and dissent in *Garza v. County of Los Angeles*, 918 F.2d 763 (1990).

that of journalist Bill Bishop, that people are making choices of where to live based on lifestyle and culture. And, lifestyle and culture ripple through politics. Termed *the Big Sort*, the half-century of American political separation into homogeneous communities based on race and ethnicity, the retreat of conservative Whites to their suburbs, and the gentrification wave of progressive Whites into heterogeneous communities like trendy urban centers and college towns have made it easy for mapmakers to lasso voters and create competitive or safe districts, whatever their goals [2, 5, 25]. Sophisticated community profiles by demographers, marketers, and social scientists tie political and other cultural and consumer data together to craft typologies of the communities in what Chinni and Gimpel term 'the patchwork nation' [14]. These granular data are available to mapmakers because of the detail of data developed by the Census, the American Community Survey, and also an expansive market research industry (which relies on Census Bureau products to support their work).

The ways that sorting amplifies the power to gerrymander are compounded by the political behavior and motivations of homogeneous groups. In part, this is just political partisanship. American politics has grown a more racially and ethnically diverse electorate, and non-White groups have trended heavily Democratic in their voting behavior. Meanwhile, White voters have trended more Republican. This means that tuning the racial balance of a district often gives a predictable party outcome. The divergence of White and non-White political preferences has a very specific definition under voting rights law: *racially polarized voting*. When blocs of Whites and blocs of minorities have consistently opposing preferences, and there is also evidence of racial appeals in campaigns (the famous 'dog whistles' of race, ethnicity, and immigration), then the foundation has been laid for potential legal challenge under the Voting Rights Act, which is the operational form of the 14th Amendment [10].

3 THE REDISTRICTING PROBLEM IS A SYSTEMS PROBLEM, VERSION 1

We have seen a variety of people problems that insinuate themselves into redistricting, including the forces that control clustering and segregation, the differing political preferences of homogeneous groups, and the larger political and legal context that maps group preferences onto party politics. But are our representative systems at the heart of the democratic problems we ascribe to redistricting and gerrymandering?

In order to explore this part of the puzzle, let's recall the predominant forms of gerrymandering and what each seeks to accomplish: the population gerrymander, the geographic gerrymander, the partisan gerrymander, the bipartisan gerrymander (also called the incumbent gerrymander), and the racial gerrymander.

MANY WAYS TO GERRYMANDER

The population gerrymander, as we've seen, gives more or less power to different districts by malapportionment, or skewing the population from district to district. This practice is largely nullified by case law requiring that population across districts be "as equal as practicable," which has practically meant a population range of 10 points (effectively ±5%) for state legislative districts and much tighter balance for Congressional districts.[5] High population deviations have to serve some general state policy or goal, and cannot advantage one party or geographic region over others.[6]

Even in district schemes that balance population, it is possible to advantage geography or regions. Rural districts that are losing population often have their populations bolstered by extending "fingers" into suburban or even urban areas. Those precise extensions of the district allow the rural areas to dominate the overall vote in the district. Sometimes termed the "rural reach" or the "rural stretch," such practices have been found to be permissible so long as equal population and voting rights protections are not violated.[7]

Partisan gerrymanders are the best known form, where "packing" and "cracking" allow one party to amplify its representation [6, 35]. A related version of the partisan gerrymander is the bipartisan gerrymander, which seeks to protect as many incumbents as possible for reelection, regardless of party. This gerrymander most often occurs under circumstances of low political polarization, or when there is split control of the legislature and the executive in a state, thereby necessitating compromise.[8] Bipartisan gerrymanders work by carefully constructing districts to put incumbents in front of familiar or friendly voters.

Then, there is the racial gerrymander. This is the ultimate legal, empirical, and rhetorical double-edged sword of the redistricting and apportionment process. Racial gerrymanders use race as a criterion when crafting districts, often by creating constituencies with majorities of homogeneous racial or ethnic minority groups. Race-conscious redistricting is 'good' when the practice creates a district that is a remedy to racially polarized voting that disadvantages the minority electorate. Race-conscious redistricting is 'bad' when it is designed to limit minority voter influence by cracking and packing. For example, if African American voters can readily control the outcome in a district that is 55% African American, increasing the African American vote share to 80% 'wastes' Black votes and limits broader African American representation. However, when minority voters are geographically concentrated and nearly homogeneous, a different problem occurs: it might

[5] *Voinovich v. Quilter*, 507 U.S. 146 (1993); *Brown v. Thomson*, 462 U.S. 835 (1983).

[6] This last point was reiterated by the courts in Georgia in *Cox v. Larios*, 542 U.S. 947 (2004). In that case, rural south Georgia and the urban core of Atlanta—predominantly Democratic areas—had been given an advantage in getting state legislative seats through the manipulation of district population deviations. Put simply, Democratic districts were typically underpopulated by 4-5%, Republican areas were similarly overpopulated. Regionally specific biases to advantage one region over others are not permissible [18].

[7] Beyond "reaches" and "stretches," the various metaphors for strange shapes in redistricting are ably described in Owen and Grofman [28].

[8] Some authors have argued that there is less partisan bias in district schemes when the process goes through commissions, courts, or split party control [1, 13].

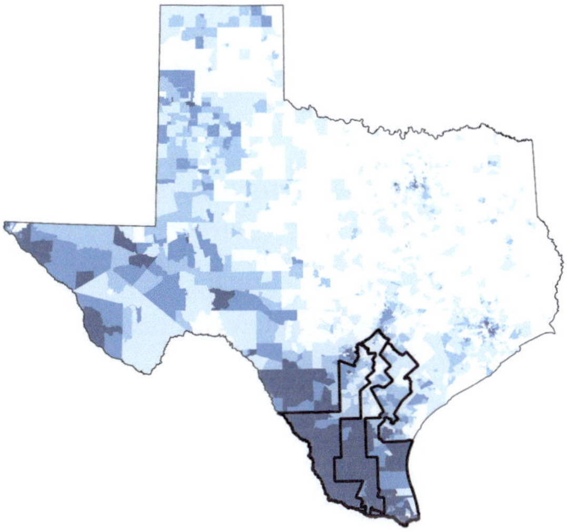

Figure 1: Districts in the Rio Grande Valley of Texas. The blue coloring indicates a higher concentration of Hispanic voters.

be necessary to break up clustered minority communities in order to avoid wasting votes and to secure greater impact for community preferences. This is highly visible in the Rio Grande Valley of Texas, which is approaching Hispanic homogeneity (see Figure 1). Drawing districts in the Valley can easily pack Hispanic voters in and similarly limit Hispanic impact in the overall Texas electoral map.[9] A similar problem can occur in parts of the Mississippi, Alabama, and Georgia "Black Belts."

IS THE SYSTEM TO BLAME?

We have all these challenges, and all of these opportunities for arbitrage, because we demand representativeness, responsiveness, diversity, and sometimes even proportionality from single-member districts (SMDs). And this system is ideally suited to allow for certain forms of community or geographic representation, but not to guarantee the balance of party or ethnicity in a representative system. Responsiveness is not built in. Yet we demand through legislation and litigation that these SMD systems do exactly these things.

There is a significant body of literature detailing the consequences of the SMD system itself. If a districting scheme combines a mixture of competitive and safe districts, the advantages to the favored party will be nonlinear. Therefore, the relationship between votes cast for a party and seats won will look curvilinear; many have described the votes to seats relationship with reference to a "seat bonus" or "winner's bonus" for the majority party, so that winning 51% of the vote might result in, perhaps, 54% or 55% of the seats.[10]

[9] An excellent take on Texas redistricting, the issues of Hispanic representation, and redistricting in general is found in Bickerstaff [4].

[10] Ideas about seats–votes curves are found in classic political science literature like Taagepera and

Districts also lend themselves to incumbency advantage. By electing singular representatives and allowing them to stand for reelection with representative advantages of office, we enable incumbents to defy shifts in the popular vote [20]. And, because incumbents draw the districts, they take care of themselves and selected colleagues.

Historically, regulated single-member districts were thought to be a cure for racially exclusive voting practices. But racial fairness in representation is complicated, even with rules for monitoring district composition. Historical patterns of differential voter turnout and population dispersal can make it difficult to draw districts of sufficient number and composition to allow minorities to secure representation proportional to population.[11]

Majoritarian values and racial fairness—two special cases of the goal of proportionality in representation—are implicit in the American ethos that has been refined through Constitutional amendments and case law over almost 250 years of U.S. history. The repeated use and interpretation of Article I and the 14th Amendment throughout the redistricting battles of the last half-century have been to protect equal access to the vote and minimize discriminatory distortions. Perhaps aggressive monitoring and regulation are necessary but not sufficient to cure the quirks, pathologies, and invitations to abuse in the system of single-member districts.

4 THE REDISTRICTING PROBLEM IS A MATH PROBLEM!

Political scientists and mathematicians enjoy redistricting for the same reason they enjoy baseball: it is a game filled with things that can be measured and correlated, and there are consequences to the play of the game.

Elections are iterative and dynamic. So is baseball. Nuanced changes in the playing field, playing conditions, and the composition of teams can alter the game. And, the home team has an advantage.

To think of redistricting as a math problem, one has to consider all the things that can be measured, and the uncertainty that attaches to those measures. We already know that there is a mathematical basis for the distortions from straight proportionality of votes-translated-to-seats in single-member district systems. And, there is an ongoing debate over how to measure bias and responsiveness of SMD systems based on changes in votes and the 'wasting' of votes. How does one disentangle the inherent advantages associated with SMDs, from the perception that those natural advantages are somehow reprehensible bias introduced by mapmakers? Some of the metrics and their logic were reviewed in Chapter 2. So we've seen that the

Tufte [32, 33]. See also Chapter 2.

[11]The level of minority voter concentration within an SMD required to allow minority voters an equal opportunity to elect a representative of choice is variable within a state. So, for example, a lower proportion of Hispanic voters in a district is required to elect a representative in the South Valley of Texas than in the Big Bend country. The same might be said of Black Belt constituencies in Georgia or North Carolina when compared to Louisiana.

sheer volume of measures available to evaluate representative maps is impressive. There is not always a clear cardinal hierarchy to the rules of redistricting, or a clear measure of their importance and influence in the process. And there is not always a clear metric for each qualitative rule. But, even when the criteria come with metrics and a clear cardinal hierarchy, they are still stubbornly interactive and interdependent. And, nearly all of the measures we use suffer from internal and external problems of validity and consistency.

Let's briefly review a few of the rules and their measurements to see how much we can convince ourselves that gerrymandering can be cured with math.

Population equality is deemed to be a background assumption, one that must be achieved over and above all others. But even though the Census gives us the legal fiction of complete accuracy, it is just as shaky as most attempts at large-scale quantification. It is usually two years old by the time districts are drawn and used. It suffers from coverage problems and measurement errors that are sources of enduring controversy. These potential problems exist in the general enumeration; in the measurement of race; and in the identification of citizen populations.

The only other national imperative is the Voting Rights Act and its mandate for racial fairness, whose enforcement depends on the circumstances in individual states and the necessity to remedy historic discrimination. States must consider race just the right amount—not too much and not too little—in the process of drawing districts. The numbers used in VRA relate to two empirical questions: (1) is there racially polarized voting, and (2) is there vote dilution?

To answer these questions, one must measure minority voter participation rates. These are variable from state to state with regard to their precision, and must often be estimated, often from relatively imprecise data. The legal demands have led political scientists to develop a huge secondary literature of statistical techniques—homogeneous precinct analysis, ecological regression techniques, and ecological inference—each with their virtues and limitations [16, 21, 22, 24, 34].

Justice Sandra Day O'Connor observed in the landmark case *Shaw v. Reno,* which discarded a North Carolina congressional plan as a racial gerrymander, "We believe that reapportionment is one area in which appearances do matter." She was talking about shape and "compactness," the geometry of districts, which has also eluded clean measurement (see Chapter 1).

Political scientists have been players in the game of quantifying compactness. Niemi et al sorted compactness scores into what you might call "measures of circlitude" and "measures of fillitude" [27]. Or, put another way, to what extent does a district look like a circle, or fill in the space within a circle? But there are other dimensions of the geometry to consider when talking about compactness. The shapes of districts are laid on top of complex, growing, and interacting communities. While it is possible to draw nice square districts in some parts of the country, the development of population corridors, the habits of movement and commuting, and even the concept of community might not fit in those squares. Appalachia and the coast of Florida form different challenges for making districts look like brownies in a pan. And even the outer outlines of a state like West Virginia or Texas will have a great impact on the numerical compactness of the districts within the state, when compared to a state cut out by cleaner lines like New Mexico or Colorado. Shape requires context to fully understand its impact.

There are myriad other measures that might inform us about a mapmaker's impact on representation. Judges are tasked to craft maps to resolve litigation without putting a needless thumb on the political scale. To assess partisan impact requires still more metrics, which in turn require having data on election outcomes. To compare across all districts, some sort of counterfactual measurements will be needed for uncontested races, such as by crafting "reconstituted elections" or other imputation techniques (see Sidebar 2.2).

We might have to consider the differential burden borne by the parties in the changes to a map. For example, Florida's redistricting criteria (which are part of the state constitution) hold that no plan or individual district be drawn with the intent to favor or disfavor a political party or incumbent.[12] To consider the impact on an incumbent or a party requires understanding not only of the partisan balance of districts, but also the extent to which incumbents confront new voters. This means measuring the retention of the "district core" from the old map to the new map. The more newer constituents an incumbent encounters, the presumptively tougher their reelection.

The measurement uncertainties can start to pile up, leading to a great deal of difficulty in identifying the best strategies. This may well lead us to doubt that fair redistricting can be achieved with math alone.

5 THE REDISTRICTING PROBLEM IS A SYSTEMS PROBLEM, V.2: TECHNOLOGY

The last problem for redistricting is one that is still being defined and written, that redistricting is a systems problem, and specifically a technology problem. In the last 35 years, we have gone from crayons to computers in the shaping of districts. And, now, with the incredible gains in computing speed, data storage, software, and online interfaces, most anyone can make districts.

As you'll read throughout this book, computers can now sample thousands or millions of alternative districting plans. This does allow for the crafting of a huge assortment of scenarios. But, this may be of limited use in a world where the maps are usually created and ultimately approved by legislatures, which only need one map that satisfies the rules and advances their goals. And, incumbents will create real constraints on the scope of potential maps that might be in play.

Where technology helps is in acting as a diagnostic and political forensic tool. The ability to run scenarios and view a range of possible plans allows one to also measure those scenarios in terms of every measure of maps available, including partisanship. If we can measure it, we can determine what map is an outlier or rare event, and what maps are typical of a process more guided by the rules and norms. This lets us ask questions of motive behind outlier maps, and just as importantly,

[12] Florida Constitution Article III, §20(a), 21(a).

the process of creation and assessment will hopefully give us insight that improves not just our *measures*, but our *systems* for deciding how maps will be crafted.

6 THE REDISTRICTING PROBLEM IS A PEOPLE PROBLEM, V. 2: POLITICS

In the end, the problems of redistricting are not created, nor can they be solved, by technology or by mathematics. Certainly the system of single-member districts and the landscape of human geography create challenges for fairness and opportunities for abuse. The main concern with single-member districts, however, is that the system encourages gerrymandering for political advantage because the process is infused with legislative politics. Those who benefit from the process also control the process in most every instance. Unified party majorities have little incentive or reprimand beyond conscience and a sense of fair play to not seek political advantage.

This makes redistricting problems, at heart, people problems after all. The reality of human existence and especially politics is that people want to win power, and this is especially the case for politicians. Political parties want to control government, and they want predictability. Status quo majorities that are afraid of the prospect for change, whether political or demographic. Democrats held on to political power in the Georgia legislature and the Texas congressional delegation long after demographics demanded partisan change[13] simply by virtue of favorable legislative maps and strategically placed incumbents. And now Republicans in those same places will need aggressive gerrymandering to resist the demographic pull the other way, brought on by rising immigration and a growing urban vote. In the 21st century, efforts to craft maps to a party's advantage entails identifying and minimizing the impact of demographic trends that swing the pendulum away from the party in power. The use of primary political data and secondary demographic data can inform the evolution of politics in a state or community, and therefore

[13]The Democratic "Solid South" that had held up from the 1880s to the 1960s had rapidly transformed by the 1980s into the red states we know today. But Republicans did not take a majority in the Texas congressional delegation, or in the Georgia House of Representatives, until the election of 2004.

how to draw districts. And, the politicians in power have the computers, the data, and the legislative votes.

If the courts won't act, and legislatures are compromised by either cynicism or the effects of gerrymandering, only direct democracy through referenda or constitutional revision remain as the means to effect change. Such initiatives focus on either changing the standards for drawing maps; changing the rules by which maps are drawn; or changing the venue, by instituting commissions. The failure of other institutions to act in a manner consistent with substantive democratic theory requires new solutions to either instill those values, or to create new institutions.

ACKNOWLEDGMENTS

The genesis for this chapter comes from a public lecture to the Voting Rights Data Institute at Tufts University, on June 27, 2018 (`https://sites.tufts.edu/vrdi/`). My thanks to all of the Institute participants and convener Prof. Moon Duchin for their feedback and encouragement in developing this essay.

REFERENCES

[1] Alan I. Abramowitz (1983). Partisan Redistricting and the 1982 Congressional Elections. The Journal of Politics 45 (3): 767–770.

[2] Alan Abramowitz, Brad Alexander and Matthew Gunning (2006). *Don't Blame Redistricting for Uncompetitive Elections.* PS: Political Science and Politics 39 (1): 87–90.

[3] Douglas J. Amy (2002). Real Choices/New Voices: How Proportional Representation Elections Could Revitalize American Democracy 2d. New York: Columbia University Press.

[4] Steve Bickerstaff (2007). *Lines in the Sand: Congressional Redistricting in Texas and the Downfall of Tom DeLay.* Austin: University of Texas Press.

[5] Bill Bishop (2004, 2008). *The Big Sort: Why the Clustering of Like-Minded America is Tearing Us Apart.* New York: Houghton Mifflin.

[6] Charles S. Bullock III (2010). Redistricting: The Most Political Activity in America. Lanham, MD: Rowman and Littlefield.

[7] Charles S. Bullock III, Scott E. Buchanan, and Ronald Keith Gåddie (2015) *The Three Governors Controversy: Skullduggery, Machinations, and the Decline of Georgia's Progressive Politics.* Athens, GA: The University of Georgia Press.

[8] Charles S. Bullock III and Ronald Keith Gåddie (2009). *The Triumph of Voting Rights in the South.* Norman, OK: The University of Oklahoma Press.

[9] Charles S. Bullock III and Ronald Keith Gåddie (2018). Georgia Politics in a State of Change, 3rd. New York: Pearson/Longman.

[10] Charles S. Bullock III, Ronald Keith Gåddie, and Justin J. Wert (2016). *The Rise and Fall of the Voting Rights Act.* Norman, OK: The University of Oklahoma Press.

[11] Scott E. Buchanan (1997). *The Effects of the Abolition of the Georgia County-Unit System on the 1962 Gubernatorial Election.* Southeastern Political Review 25 (4): 687–704.

[12] James Campbell (1996). *Cheap Seats: The Democratic Party's Advantage in U.S. House Elections.* Columbus, Ohio: Ohio State University Press.

[13] Jamie L. Carson and Michael H. Crespin (2004). The Effect of State Redistricting Methods on Electoral Competition in United States House of Representatives Races. State Politics & Policy Quarterly 4 (4): 455–469.

[14] Dante Chinni and James Gimpel (2011). *Our Patchwork Nation: The Surprising Truth About the "Real" America.* Paperback. New York: Avery.

[15] William Cornelius (1961). *The County Unit System of Georgia: Facts and Prospects.* The Western Political Quarterly 14 (4): 942–960.

[16] Richard L. Engstrom and Michael D. McDonald (1988). Definitions, Measurements, and Statistics: Weeding Wildgren's Thicket. Urban Lawyer 20: 195–194.

[17] David Hackett Fisher, *Albion's Seed: Four British Folkways in America.* New York: Oxford, 1989.

[18] Ronald Keith Gåddie and Charles S. Bullock, III, *From Ashcroft To Larios: Recent Redistricting Lessons from Georgia.* Fordham Urban Law Journal 34 (April 2007): 997–1048.

[19] Ronald Keith Gåddie, Justin J. Wert, & Charles S. Bullock, III (2012). Seats, Votes, Citizens, and the One-Person, One-Vote Problem. Stanford Law & Policy Review 23: 431–454.

[20] Andrew Gelman and Gary King (1990). Estimating Incumbency Advantage without Bias. American Journal of Political Science 34 (4): 1142–1164.

[21] Gary King. 1997. A Solution to the Ecological Inference Problem: Reconstructing Individual Behavior from Aggregate Data. Princeton: Princeton University Press.

[22] J. Morgan Kousser (1973). Ecological Regression and the Analysis of Past Politics. The Journal of Interdisciplinary History 4 (2): 237–262.

[23] Justin Levitt (2011). Essay: Weighing the Potential of Citizen Redistricting. Loyola of Los Angeles Law Review 44: 513–543.

[24] Allan J. Lichtman and Laura Irwin Langbein (1978). Ecological Regression Versus Homogeneous Units: A Specification Analysis. Social Science History 2 (2): 172–193.

[25] Seth E. Masket, Jonathan Winburn and Gerald C. Wright (2012). *The Gerrymanderers Are Coming! Legislative Redistricting Won't Affect Competition or*

Polarization Much, No Matter Who Does It. PS: Political Science and Politics 45 (1): 39–43.

[26] D.C. Moore, *Concession or Cure: the Sociological Premises of the First Reform Act.* The Historical Journal 9 (1) 1966: 39–59.

[27] Richard G. Niemi, Bernard Grofman, Carl Carlucci, and Thomas Hofeller (1990). Measuring Compactness and the Role of a Compactness Standard in a Test for Partisan and Racial Gerrymandering. Journal of Politics 52 (4): 1156–1181.

[28] Guillermo Owen and Bernard Grofman (1988). Optimal Partisan Gerrymandering. Political Geography Quarterly 7 (1): 5–22.

[29] Nathaniel Persily (2017). Who Counts for One Person, One Vote? UC Davis Law Review 50 (4): 1397–1421.

[30] Albert B. Saye, (1950) Georgia's County Unit System of Election, Journal of Politics 12: 93–106.

[31] Rosemary Sweet, *Freemen and Independence in English Borough Politics c. 1770–1830.* Past & Present 161 (Nov 1998): 84–115.

[32] Rein Taagepera (1973). Seats and Votes: A Generalization of the Cube Law of Elections. Social Science Research 2(3): 257–275.

[33] Edward R. Tufte (1973). The Relationship between Seats and Votes in Two-Party Systems. American Political Science Review 67(2): 540–554.

[34] John K. Wildgen (1988) Adding Thornburg to the Thicket: The Ecological Fallacy and Parameter Control in Vote Dilution Cases. Urban Lawyer 20 155–174.

[35] Jonathan Winburn (2009). The Realities of Redistricting: Following the Rules and Limiting Gerrymandering in State Legislative Redistricting. Lanham, MD: Lexington Book.

Chapter 5

Political geography and representation: A case study of districting in Pennsylvania

JONATHAN RODDEN AND THOMAS WEIGHILL

CHAPTER SUMMARY

So geography matters... but how? This chapter offers a detailed look, both qualitative and quantitative, at districting with respect to recent voting patterns in one state: Pennsylvania. Rodden and Weighill investigate how much the partisan playing field is tilted by political geography. In particular, they closely examine the role of *scale*. They find that partisan-neutral maps rarely give seats proportional to votes, and that making the district size smaller tends to make it even harder to find a proportional map.

1 INTRODUCTION

The impressive success of recent ballot initiatives in Michigan, Missouri, Colorado, Ohio and Utah demonstrate that redistricting reform is broadly popular with voters. But there is little agreement about what type of reform is optimal, and little knowledge about what is at stake when choosing between alternatives. One type of reform, referred to by political and legal theorists as *process-oriented* reform, focuses on creating a redistricting process that is fair and transparent, giving authority to either independent commissioners or equal numbers of Democrats and Republicans, and encouraging them to pay little attention to the potential political outcomes associated with alternative maps, or perhaps even forbidding the

M. Duchin, O. Walch (eds.), *Political Geometry*, https://doi.org/10.1007/978-3-319-69161-9_5

analysis of electoral data altogether. An alternative approach to reform is to focus explicitly on partisan outcomes—encouraging or requiring commissioners to draw maps that are fair to both parties according to some agreed criteria about how votes should translate into seats.

Advocates of an outcome-based approach point out that while a pure process-oriented approach may be easy to explain and implement, it does not necessarily satisfy all the definitions of fairness stakeholders may have in mind. For example, many citizen observers prefer outcomes where the seat share for the parties matches the vote share (so-called "proportional" outcomes), but neutral redistricting tends to lead to maps that result in representation that is far from proportional. This phenomenon is the result of the particular way in which votes are distributed within a state (what we will refer to as *political geography* in this book). This distribution has a signature form in the United States: Democrats are often highly concentrated in city centers and educated suburbs and Republicans are more dispersed in exurbs and rural areas [21].

The partisan tendencies of neutral redistricting, and hence the stakes of debates about redistricting reform, are driven by each state's political geography, and perhaps especially by its urban political geography. An influential paper calling attention to quantifying the partisan tendencies of neutral redistricting was Chen and Rodden [2], which dubbed the phenomenon "unintentional gerrymandering." The size, structure, and geographic arrangement of cities is extremely important for political representation in the United States [21]. Moreover, it has been argued that the impact of urban geography on representation is conditioned by the geographic scale at which districts are drawn, as well as the overall level of support for the two parties [11].

Adding apparently neutral criteria like "competitiveness" [8] or the procedure for defining adjacency across bodies of water [1] can have unforeseen and sometimes dramatic effects on the partisan statistics of maps drawn under a neutral process. See also DeFord et al. [6] for an extensive survey of the impact of redistricting criteria for Virginia, including compactness, population balance, racial balance, and locality splits. Also included in that paper are redistricting criteria that explicitly depend on vote data such as efficiency gap and mean–median scores. It is very clear that a neutral redistricting process that focuses only on creating compact, contiguous districts, and minimizing county or municipal splits, for instance, can lead to what many would deem to be a "fair" outcome in some states but not in others, particularly if the definition of "fair" under consideration is based on the ability to translate overall vote share into seat share. Moreover, the fairness or lack thereof can depend in unexpected ways on which criteria are chosen and how they are measured.

Within a given state, inferences about the fairness of the maps created through a neutral process might change as the scale of districts varies from massive 700,000-person Congressional districts to, for example, 3,000-person New Hampshire State House districts. Reformers often point out that U.S. Congressional districts are extremely large relative to districts in other countries that use winner-take-all districts, and a popular reform proposal is to make the U.S. Congress considerably larger by reducing the size of districts.[1] Part of the logic of this type of reform is the hope that disproportionalities in the transformation of votes to seats would be reduced if districts were drawn at a smaller scale. In this chapter, however, our finding will be that the Democratic disadvantage in turning votes into seats in Pennsylvania persists at every hypothetical district size, from 4 million people down to just 55,000.

Social scientists and mathematicians are in the early stages of understanding the complex interplay of political geography, spatial scale, and statewide partisanship that determine patterns of political representation when districts are drawn without regard for partisanship. This chapter makes progress by presenting a detailed case study of Pennsylvania. We choose Pennsylvania in part because in the wake of a recent state court decision, Pennsylvania reformers are in the midst of serious debates about process-oriented versus outcome-oriented reform [18]. We make use of modern statistical sampling methods, discussed in more depth in the Introduction and Chapter 17, to generate large neutral ensembles of possible districting plans in order to study the baseline of representation for each party at a wide range of feasible spatial scales.

Our central conclusion is that given current patterns of political geography in Pennsylvania, purely process-oriented reforms would typically result in the Democratic Party falling significantly short of proportional representation, even when it has a majority of votes, showing that spatial effects overcome the usual “winner’s bonus.”[2] We are able to draw inferences about this not only by observing outcomes of very close elections, like the 2016 presidential election, but also by examining statewide elections, where Democratic candidates won significant victories, as well as elections in which Republican candidates were victorious. We are also able to learn subtle lessons about the importance of spatial voting patterns by observing surprisingly different anticipated seat shares associated with elections held on the same day, and with very similar overall partisan vote shares, but with different underlying spatial support patterns.

Second, we demonstrate that while the scale of districts does affect the baseline for representation, the effect is largely to decrease the variance and not to reduce the gap between expected seat share and the statewide vote share. In other words, the Democrats’ geography problem does not simply go away if districts become sufficiently small. In closely contested elections, at no plausible scale of redistricting do our neutral ensembles produce Democratic seat shares that match their vote shares.

Third, the main reason for choosing Pennsylvania as our case study is that by dividing the state in half and treating Eastern and Western Pennsylvania as two separate states, we are able to gain a better understanding of what exactly lies behind the Republican advantage. That is, we are able to gain insights by “modularizing” the problem into two smaller problems. Pennsylvania gives us the opportunity to examine two very different, and perhaps somewhat representative, patterns of political geography. Eastern Pennsylvania contains not only a large, extremely

[1] See, for instance, “America Needs a Bigger House,” New York Times, 9 November, 2018.

[2] The “winner’s bonus” is the established idea in the literature that parties that win the statewide vote should generally have a seat share that exceeds their statewide vote share—see Chapter 2 for more.

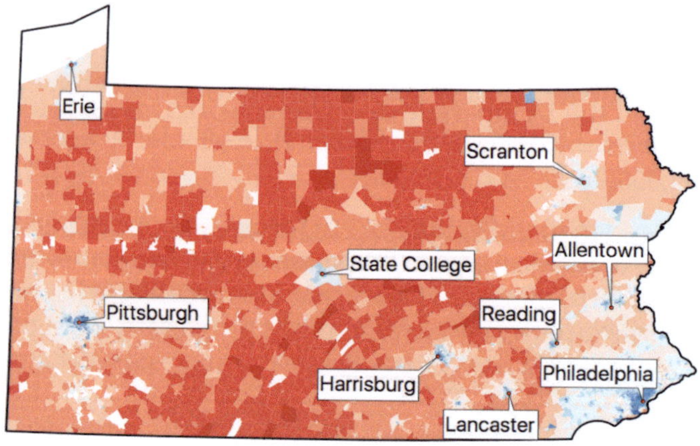

Figure 1: Map of Pennsylvania with results from the 2016 Presidential election.

Democratic "primate" city, but also, due to the geography of coal and the associated 19th-century process of rail-based city formation, a series of smaller Democratic urban agglomerations located in close proximity to one another (Figure 1). This pattern of smaller, geographically proximate corridors of post-industrial cities that grew up along rail lines in the periphery of larger regional primate cities resembles other early-industrializing states along the Eastern Seaboard. In Pennsylvania, this pattern is associated with an unambiguous but relatively modest pattern of Republican advantage in our ensembles of nonpartisan redistricting plans.

Western Pennsylvania, in contrast, contains a single large "primate" city that is overwhelmingly Democratic, while smaller Democratic enclaves are few in number and quite isolated from one another. This pattern of political geography is also found in other states on the Western and Southern fringes of the early 20th century manufacturing core, like Missouri, Tennessee, and Louisiana, which contain relatively large, extremely Democratic cities, but lack a network of smaller, proximate rail-based agglomerations on the order of Allentown, Easton, and Reading.

Our analysis demonstrates that relative to the Eastern Pennsylvania pattern of dense industrialized corridors, this Western Pennsylvania structure featuring a single isolated 19th century industrial outpost is associated with a much greater under-representation of 21st century Democrats. Nonpartisan redistricting plans grant Republicans substantially more seats than proportional representation would suggest in Western Pennsylvania, with this phenomenon driving a large part of the overall story of Democratic under-representation in our ensembles of statewide maps.

We begin with a brief discussion of American political geography and the normative challenge that it creates for a scheme of representation that relies on dividing the states up into winner-take-all districts. Next, we describe our empirical strategy for generating samples of nonpartisan redistricting plans at various spatial scales. We then explore the key characteristics of our sampled statewide plans, paying special attention to issues of 1) spatial scale and 2) the heterogeneity in statewide partisanship and political geography associated with different statewide Pennsylvania elections.

2 URBAN GEOGRAPHY AND THE PARTISAN TILT OF NEUTRAL REDISTRICTING

Until recently, the debate about redistricting reform in the United States pitted those who believe that redistricting should remain in the hands of legislative majorities against those who believe it should be delegated to either nonpartisan or bipartisan commissions. The prevailing model among the latter group was that of the nonpartisan commissions employed in Great Britain, Canada, and Australia, or in the U.S. context, the Iowa process. All of these prevent those drawing the maps from having access to data on partisanship. More recently, among those who favor redistricting reform, a new debate has emerged. Should reformers attempt to stick with some form of party-blind process, or include some measure of anticipated partisan symmetry in the marching orders of commissions?

This debate has been spurred by a body of literature that builds on observations of classic British and Australian political geographers [12, 13, 14]. Ever since the rise of modern parties of the left in the late 19th and early 20th centuries in the era of labor mobilization in industrialized societies, voters for these parties have been highly concentrated in city centers. This relative concentration is widely believed to underlie their difficulty in transforming votes into seats.

In the United States, Democrats today are still quite concentrated in the urban core of cities—large and small—that emerged in the era of rapid industrialization, railroad construction, and labor mobilization. Much has changed since the Democrats emerged as an urban party in the New Deal era, however, and as new issues have been politicized, from civil rights to abortion to guns and now immigration and globalization, the correlation between population density and Democratic voting has increased substantially, and it has spread from the early industrializing states to the entire country, including the deep South.

The Democratic Party today often suffers from the same difficulty in transforming votes to seats as that faced by Labor parties in the Commonwealth countries in the early postwar era, and so one may wonder if this is also an effect of urban concentration. The reasons behind the under-performance of the Democrats have been hard to nail down, however, because the era of intense urban-rural polarization coincided with highly visible attempts at partisan gerrymandering. For good reasons, Americans came to see stark disjunctures between votes and seats as phenomena that could be explained purely by partisan gerrymandering. However, by drawing a series of alternative neutral maps through a simple automated redistricting algorithm, Chen and Rodden [2] showed that in a number of states, substantial Republican advantage would have emerged even in their samples of nonpartisan maps, which the authors attributed to the concentration of Democrats in city centers. This technique was then used to generate a set of comparison maps that was used in court as part of a lawsuit that led to the invalidation of Florida's

Congressional redistricting plan in 2014 [3]. The Republican advantage in neutral redistricting is not universal, we should be careful to note. For some elections in Massachusetts, for example, no map (and hence in particular no process, neutral or otherwise) could have garnered the Republican party a congressional seat despite statewide vote shares of above 30% [10].

Subsequently, a number of scholars have adopted a series of alternative approaches to sampling from the distribution of possible nonpartisan plans, often with the goal of contrasting that distribution with the plan drawn by a state legislature in order to challenge it in court [4, 5, 7, 9, 15, 16, 17, 19].

This technique has now been used to invalidate redistricting plans in state court in Florida, North Carolina, Michigan, and Pennsylvania.

As the body of research relying on computer-generated redistricting samples evolves and matures, and as the conversation shifts from court challenges to redistricting reform proposals, it is useful to bring these tools back to the original questions about political geography. The key question remains: what partisan tendencies should we expect in the absence of gerrymandering intent? In other words, what is the impact of the unavoidable consequences of districts on proportionality and other fairness measures, and how does this impact change from election to election and between scales of redistricting?

Although U.S. political geography is always changing, in the current moment, when population density and Democratic voting are correlated at unprecedented levels, the answer to this question appears to lie largely in the size, structure, and arrangement of cities and suburbs relative to their rural surroundings. A basic problem is that large cities like Philadelphia and Pittsburgh are overwhelmingly Democratic, and neutral redistricting plans will tend to produce overwhelmingly Democratic districts. On the other hand, "rural" districts often encompass not only a large number of Republicans but also nontrivial clusters of far-flung Democrats in agglomerations like Erie and State College, Pennsylvania, that are too small to produce Democratic majorities. Moreover, districts that are largely exurban will often contain fragments of heavily Democratic parts of inner- and middle-ring suburbs. As a result, even if their statewide vote shares are similar, Republican candidates tend to win victories by smaller margins than do Democrats, whose votes are inefficiently concentrated in the districts they win.

The effect of urban concentration on representation is highly dependent on the size, arrangement, and structure of cities, as well as the scale at which districts are drawn. When cities are very large relative to the size of districts, e.g., Philadelphia and Pittsburgh, a nonpartisan process would likely create extremely Democratic urban districts. When cities are too small relative to the size of districts, as with Erie and State College relative to Congressional districts, Democrats are unable to form majorities. But sometimes the size of a city is better for the representation of Democrats. For example, some of Pennsylvania's mid-sized cities, like Reading and Bethlehem, are close to the ideal size for producing comfortable but not overwhelming Democratic state Senate Seats. But at the much smaller scale of State House districts, these cities can produce overwhelming Democratic majorities akin to Philadelphia. What is far from clear, however, is what the effects of chang-

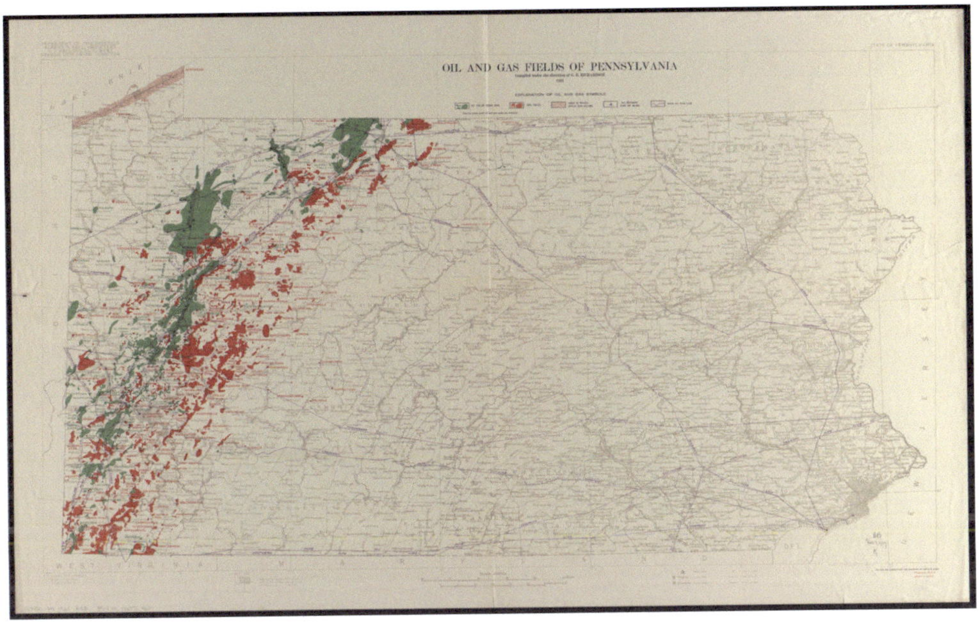

Oil and gas fields of Pennsylvania, from 1921. Taken from the Norman B. Leventhal Map Center Collection.

ing redistricting scales will be on aggregate: smaller districts may lead to small Democratic towns electing Democratic representatives, but is this effect enough to change the overall seat share in the chamber in question?

As we demonstrate, it also matters a great deal how these cities are arranged in space. In addition to Philadelphia, the cities of Eastern Pennsylvania grew up in the late 19th century along a dense web of railroads that were built around the economic geography of coal mining and heavy industry. As a result, Eastern Pennsylvania ended up with a series of small, proximate rail-based industrial towns. Scranton and Wilkes-Barre blend together along a seam of coal to the North. Further South, Easton, Bethlehem, and Allentown blend together into a Democratic corridor. Continuing in a ring around Philadelphia is a series of smaller, extremely Democratic railroad cities, including Reading, Lancaster, Harrisburg, and Chester. At the scale of Congressional districts, these cities are sufficiently proximate to one another, and to some Democratic suburbs of Philadelphia, that they can string together to produce Democratic majorities.

Another consideration is urban form. Some 19th-century cities, like Pittsburgh, have a dense and Democratic urban core, and as one moves to the suburbs, the Republican vote share increases rapidly, which generates a highly concentrated Democratic population. In contrast, the growth of Republican vote shares is slower as one moves from the core to the outer-ring suburbs in Philadelphia, in part because of the locations of high-technology employers, colleges, and universities, whose employees have become important parts of the Democratic coalition.

Even more distinct are cities like Orlando or Phoenix, where there is no 19th century

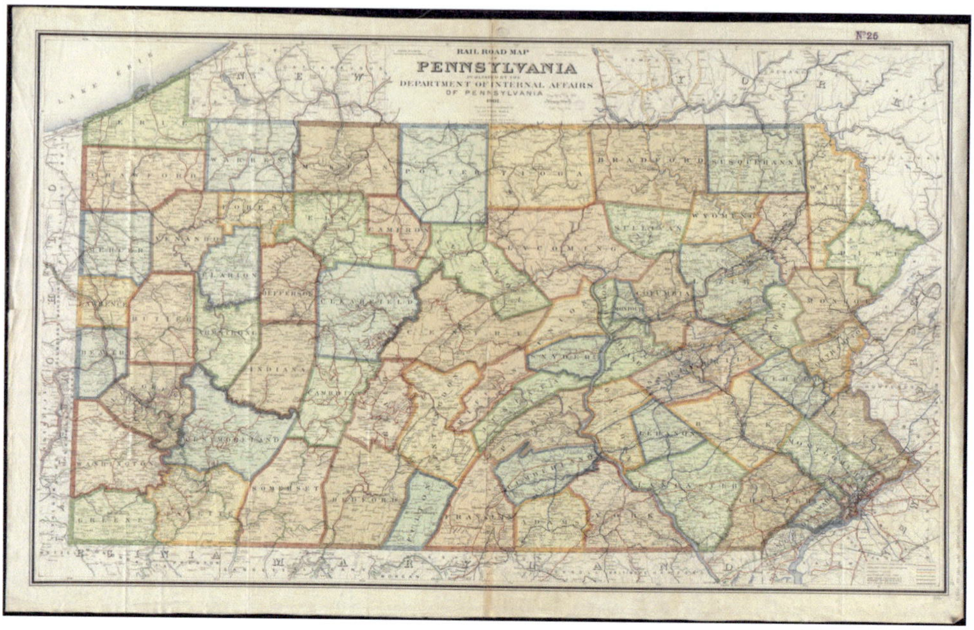

Railroad map of Pennsylvania from 1901. Taken from the Norman B. Leventhal Map Center Collection.

core, and Democrats and Republicans are interspersed in a sprawling poly-centric metropolis. Moreover, there are parts of the United States with important pockets of rural support for Democrats, including African American communities in the South, tribal lands, and communities with a history of mining.

With the continued focus on the possible detrimental effects of urban concentration on representation, we should be careful not to exclude the possibility that in some cases this concentration may *help* a party in the transformation of votes to seats, particularly in states where that party typically expects to lose the overall popular vote. Indeed, the worst case scenario for a losing party is to have its votes perfectly evenly distributed in space, a scenario that would cause it to lose every single district. This sounds far-fetched, but the situation is Massachusetts is not so far off [10]: the Republicans are unable to gain even one seat in some cases precisely because they are too diluted, not because they are too concentrated.

In short, the location of Democrats in cities is not a sufficient condition to produce a Republican advantage in neutral redistricting plans, and the extent of that advantage, when it exists, is potentially a function of political geography, the scale at which districts are drawn, and the overall level of support for the party. Our goal in this chapter is to take a close look at Pennsylvania, varying the spatial scale of which districts are drawn, exploring variation in overall vote shares by drawing on a diverse set of recent statewide elections, and exploring the role of heterogeneous urban structure by contrasting the Eastern and Western parts of the state.

3 SAMPLING PENNSYLVANIA REDISTRICTING PLANS AT DIFFERENT SPATIAL SCALES

METHOD

We seek to understand the general properties of the universe of all redistricting plans for Pennsylvania. No computer could ever enumerate every possible redistricting plan, but we will use sampling methods that allow us to effectively sample from this vast space in a mathematically rigorous way. The algorithm we will use is the so-called recombination (or ReCom) Markov chain algorithm, which is implemented in the freely available GerryChain software developed at MGGG and which is discussed elsewhere in this book (Chapter 17). Beginning with a randomly generated "seed" plan, the algorithm merges and redivides two adjacent districts at every step, resulting in a large ensemble of randomly generated plans, all drawn without the use of any partisan data. In each case, districts are built out of fixed geographic units: precincts for low numbers of districts and census blocks for higher numbers (in order to make population-balanced plans feasible).[3]

Typically, this kind of algorithm is used to generate a large ensemble of districting plans with a fixed number of districts. This is because one is usually interested in studying the districts for a particular level of government (for example, U.S. Congressional districts, of which there are eighteen in Pennsylvania). For our purposes, we want to study redistricting plans with a variety of different numbers of districts. We thus run the algorithm multiple times (once for each number of districts) to generate many ensembles of 50,000 plans each. For a plan in any one of these ensembles, we can choose a past election and use the vote data to determine how many districts the Democrats would have won using that election and that redistricting plan;[4] dividing this number by the total number of districts in the plan gives us the *Democratic seat share*. We chose nine elections to base our analysis on: the presidential elections from 2012 and 2016 (PRES12 and PRES16), the U.S. Senate elections from 2010, 2012, and 2016 (SEN10, SEN12, and SEN16), the Attorney General elections from 2012 and 2016 (ATG12 and ATG16), and the Gubernatorial elections from 2010 and 2014 (GOV10 and GOV14). In each case we just treated the election as a head-to-head election between Democrats and Republicans, ignoring any third-party candidates.

RESULTS

Figures 3, 4, and 5 summarize the results of this multi-ensemble analysis, with each plot indicating a different choice of election. Our goal with these plots is to indicate the frequency of different seats outcomes at every scale. In order to make

[3]Some intermediate scales of redistricting were studied with both precincts and blocks, with the two methods producing almost identical results.

[4]Since election data are only reported at the precinct level we prorate it down to blocks based on voting age population when needed.

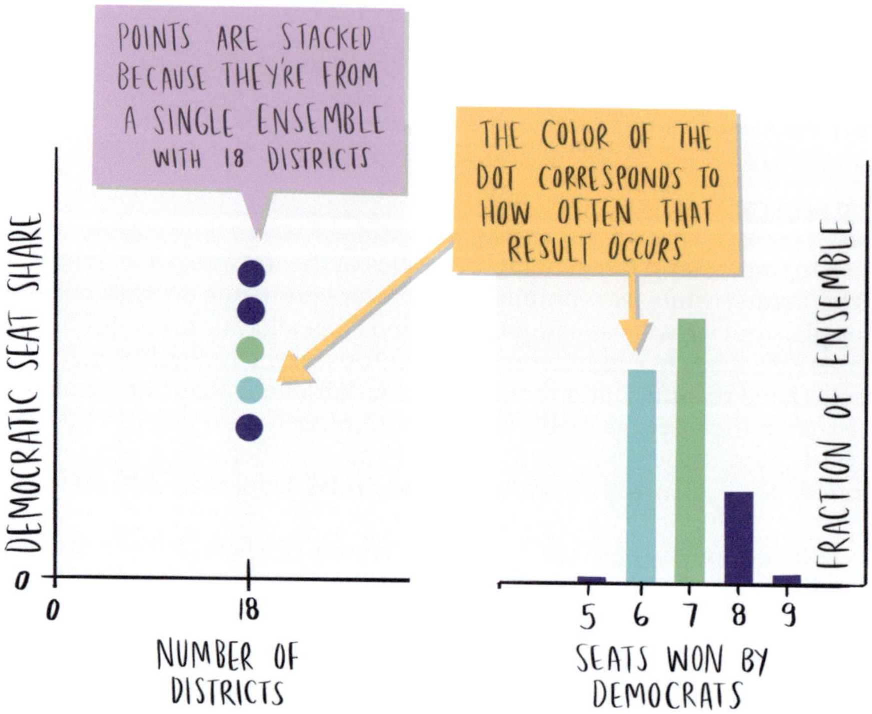

Figure 2: An illustration of how the plots in this section are constructed. On the right is a histogram of seats outcomes for an ensemble of 18-district plans, using PRES16 vote data. Each bar of this histogram is converted to a dot in the left-hand plot, with the brightness of the color indicating the bar height, the x value denoting the number of districts (in this case, 18) and the y axis indicating the fraction of seats won.

the mode of representation clear, first consider the plots in Figure 2, where we have isolated just one scale, 18 districts, and where we use PRES16 vote data. On the right is a histogram indicating the seats outcomes for Democrats in the ensemble. This histogram is represented vertically by the colored dots on the left-hand plot: dark blue represents a low fraction, with the fraction increasing through light blue (and later to yellow and to red). Notice that we have also switched from number of seats won to fraction of seats won so that we can later compare multiple scales.

In Figures 3, 4, and 5, we employ this scheme to show the seats outcome for all scales at once. Each dot represents a choice of number of districts and a Democratic seat share value. The dot is colored based on what fraction of the plans with that many districts had that Democratic seat share, just as in Figure 2. Lighter colors therefore represent more common Democratic seat share outcomes for that particular number of districts. The green line represents the statewide vote share achieved by the Democrats in the specified election. Finally, even though we are including hypothetical plans with a large range of numbers of districts, we also indicate three real-world districting scales by dotted lines: U.S. Congress (18 districts), PA state Senate (50 districts), and PA state House (203 districts).

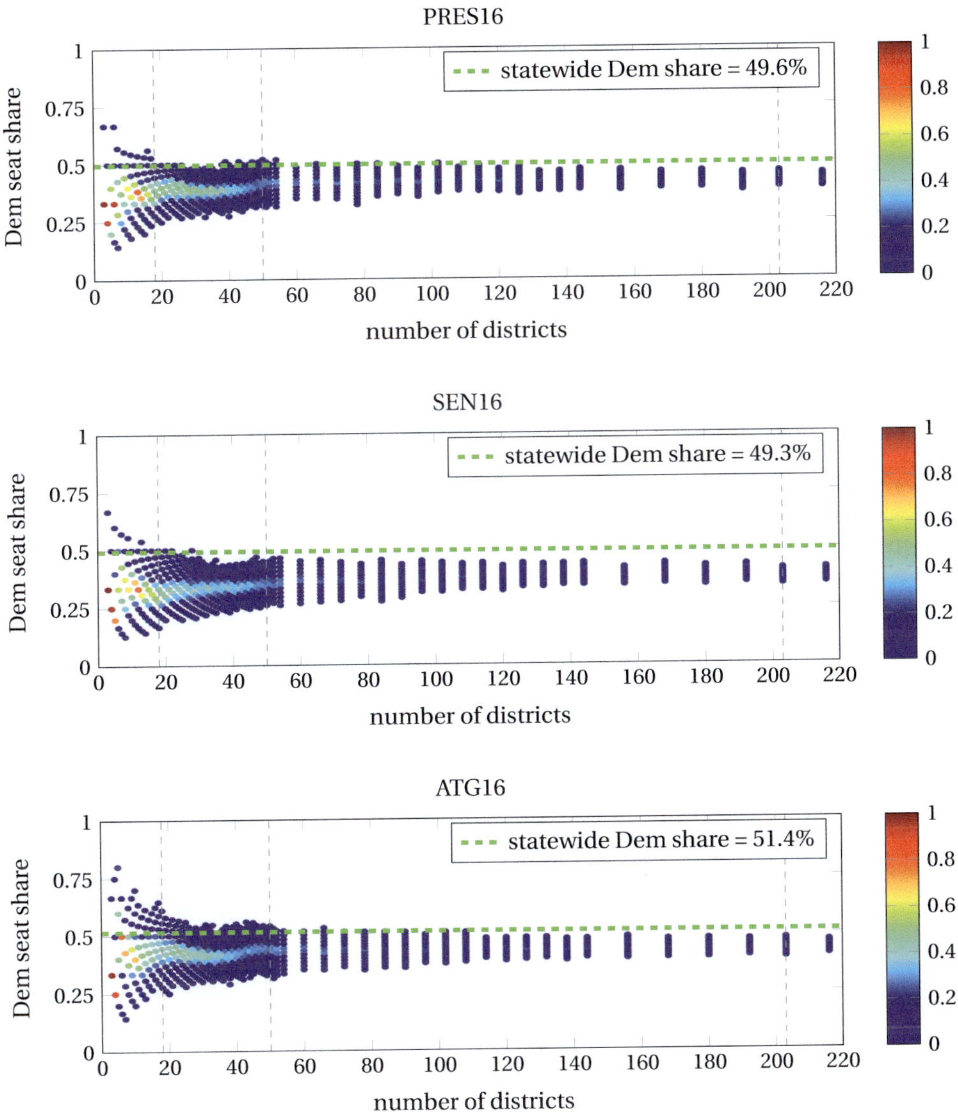

Figure 3: Democratic seat shares in neutral ensembles at various redistricting scales, 2016 elections

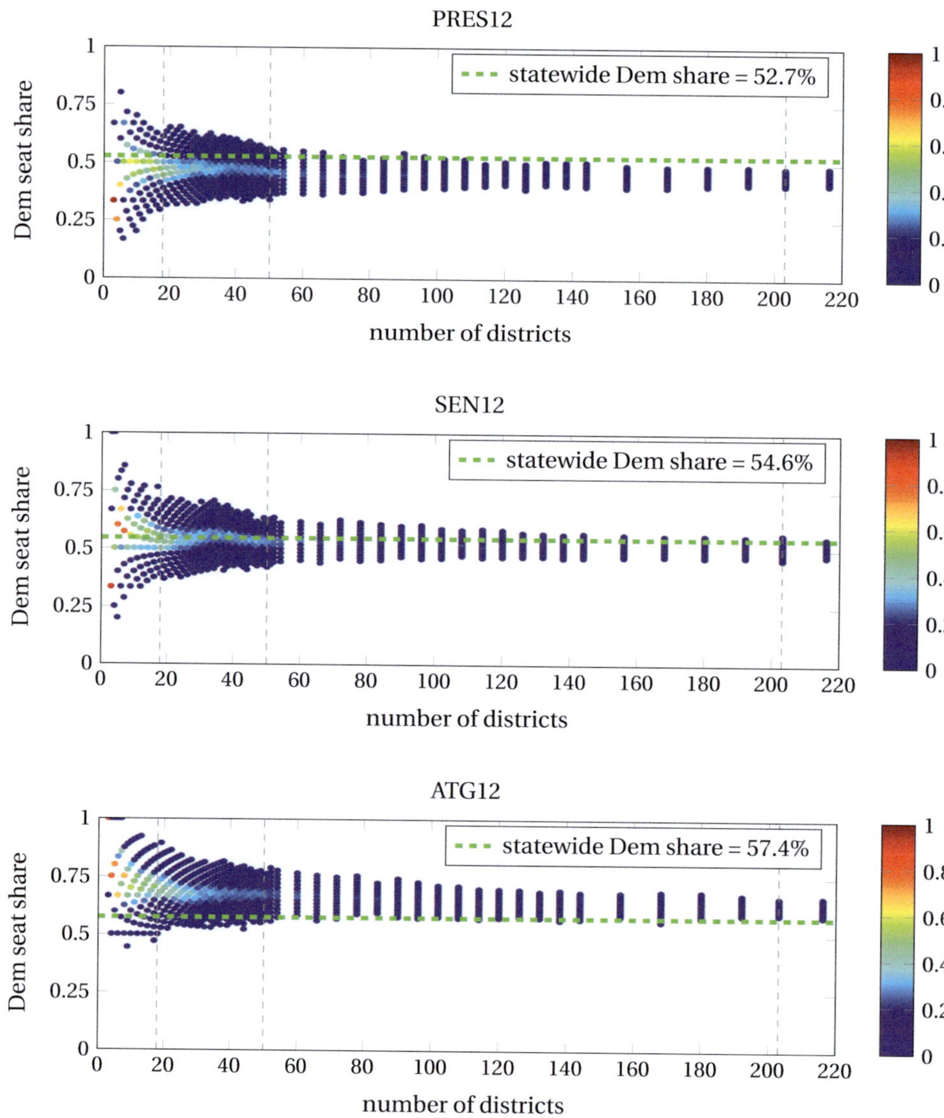

Figure 4: Democratic seat shares in neutral ensembles at various redistricting scales, 2012 elections.

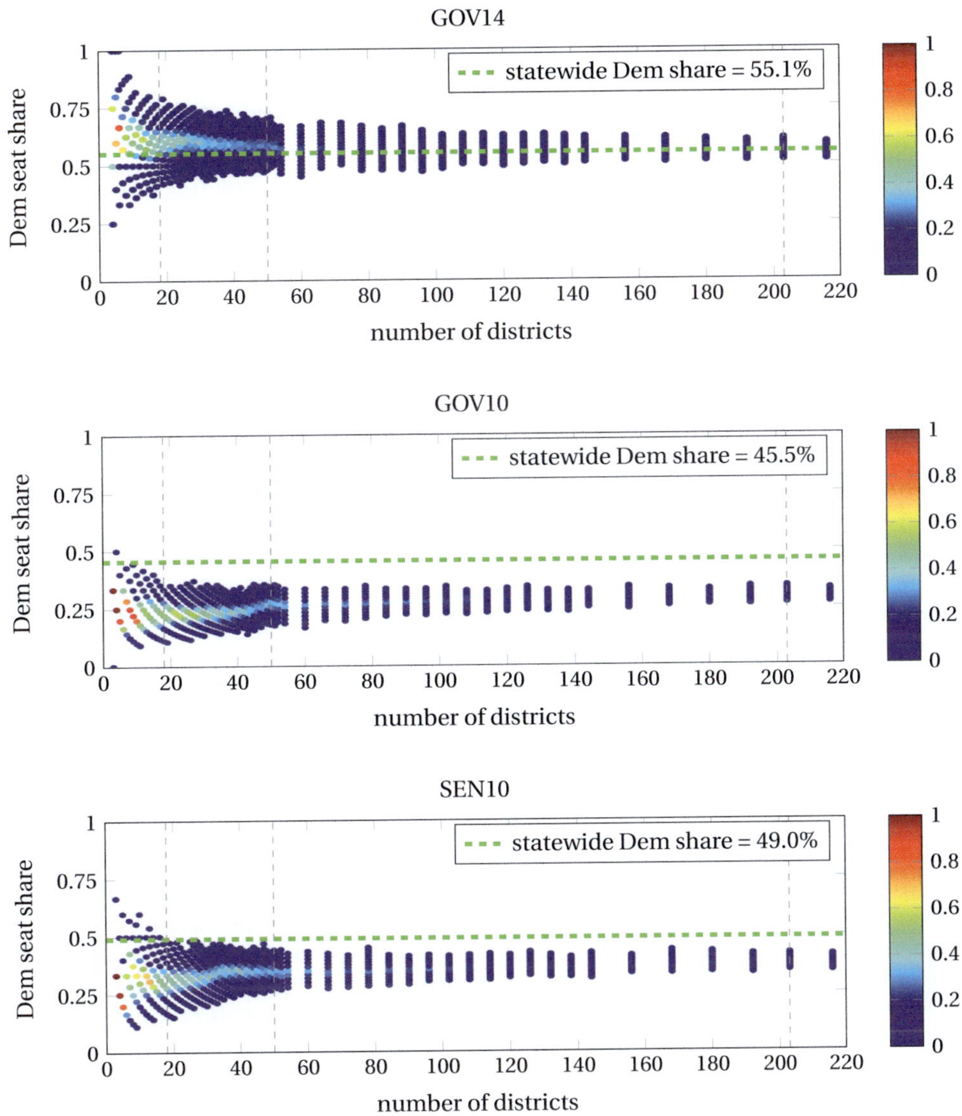

Figure 5: Democratic seat shares in neutral ensembles at various redistricting scales, other elections.

DISCUSSION

The most striking conclusion to be drawn from the plots in this section is that the Democrats underperform proportional representation in seven of the nine elections considered, with the only exceptions being the ATG12 and GOV14 elections. In those two exceptional elections, Democratic candidates performed unusually well, with 57% and 55% of the statewide vote respectively. For elections with relatively even statewide splits between the two parties, like those in 2016, the neutral ensembles showed substantial tendencies to award more seats to the Republican Party.

Moreover, the Democratic under-performance was more or less unaltered by changing district scales. If anything, when the Democratic statewide vote share was relatively low, as in GOV10 and SEN10, the Democratic seat share increased very slightly as the scale of districts became smaller. But when Democrats performed well, as in SEN12 and ATG12, their seat share declined as the scale of districts became smaller. This suggests that a general mismatch between smaller Democratic urban centers and particular district sizes (for example, Congressional districts) cannot be the only reason for the Democrats' disadvantage, if it plays a role at all. Indeed, these experiments suggest the absence of any significant scale effects.

The only clear pattern related to the scale of districts in these graphs is the much wider range of seat shares produced by the neutral ensembles when districts are larger (on the left-hand side of the graphs). The range of outcomes produced in the neutral ensembles narrows considerably as the scale of districts becomes increasingly fine-grained. Let us focus on the very hotly contested 2016 races, all of which were very close to an even split between the two parties, and where one might expect that a "fair" districting plan would produce a roughly similar number of Democratic and Republican seats. Imagine that a redistricting commission or special master was tasked with the job of randomly selecting a plan from the ensembles. This would likely lead to a rather large Republican advantage of roughly similar size, whether the plan was for Congress or either state legislative chamber.

However, imagine an alternative rule in which a commissioner or special master was told to choose from among the relevant neutral ensemble a plan for which the anticipated seat share of each party was 50 percent when the vote share was 50 percent. At the scale of Congressional districts or state senate districts, the range of outcomes in the ensemble is sufficiently large that this could be achieved by selecting one of the most pro-Democratic plans. However, this becomes impossible as districts become smaller and more numerous. The range of outcomes is much narrower at the scale of Pennsylvania House districts, where even the most Democratic plan falls short of proportionality. To be clear, the lesson is not that a "fair" plan with 203 districts cannot be drawn in Pennsylvania. Rather, such a plan does not emerge from the neutral ensembles, and it might take a conscious effort to consider partisanship in order to produce one.

Some interesting inferences—and questions for further analysis—emerge from comparisons of the graphs for different elections. One lesson, explored further below, is that the statewide vote share is important. The Democrats' seat share is especially far from proportionality when their vote share is low (e.g., GOV10), and they are still quite far from proportionality even in elections that are very close to 50 percent. In fact, even in an election with 55 percent of the vote (SEN12), they do not achieve proportionality. Only when they received 57 percent of the vote (ATG12) did they significantly surpass proportionality.

This latter comparison suggests that perhaps there are differences between these two races that go beyond the difference in vote shares between SEN12 and ATG12. In the Attorney General election, the Democratic candidate, Kathleen Kane, outperformed the Democratic Senate candidate, Bob Casey, who was on the same ballot on the same day, by 2.86 percentage points. But the difference in seats was substantially larger. At the scale of Congressional districts and state senate seats, Casey came out ahead, on average, in less than 55 percent of the districts, while Kane came out ahead in well over 65 percent. This indicates that it matters not only that Kane received more votes than Casey, but also *where* she outperformed him. That is to say, she had stronger support than Casey in some geographic areas where, in the ensembles, Casey fell below 50 percent. For instance, Kane outperformed Casey in many of the counties surrounding her home town of Scranton, as well as in the counties along the Western border of the state.

Another election pair that stands out as a place where subtle geographic factors play a big role is PRES16 and SEN16. These two elections were on the same ballot and their statewide shares differed by a mere 0.37 percentage points, yet the Democrats' ability to turn votes into seats in a neutral redistricting process is substantially worse in SEN16 than in PRES16. In other words, as with Kathleen Kane vis-à-vis Bob Casey, Hillary Clinton was stronger than the Democratic Senate candidate, Katie McGinty, in parts of the state that leaned Republican—as it turns out, parts of suburban Philadelphia—in the Senate race.

We should take this as a warning that subtle changes in voting patterns can result in significant swings in representation that elude simple explanation. It is true that in the era of polarization and nationalized politics, results of various statewide races are highly correlated. Nevertheless, split-ticket voting is still alive and well, and the spatial distribution of votes varies across races in ways that are consequential for inferences about representation.

4 SEATS–VOTES PLOTS

In the previous section, we broke down the data by election. In this section, we will plot all elections together for each of three districting scales. We organize the elections by their statewide vote share to produce a seats–votes plot. The values on the horizontal axis correspond to the observed statewide vote share in each of the nine statewide elections examined above. This is meant to parallel the traditional seats–votes curves used in partisan symmetry analysis (see Chapter 2). However, our plots contain far more information than curves since they cover the full range of possibilities encountered in a neutral ensemble associated with each election.

In each of the images in Figure 6, we have selected points from the plots in the previous section that correspond to a particular scale of redistricting. Each dot therefore represents a set of districting plans. The colors are the same; lighter colors represent more frequent seat share outcomes. What has changed is the x-axis, which now represents the statewide vote share of the election used to calculate the seat share. The dotted lines indicate two different doctrines of "fairness" one might adopt that are not based on ensembles. The gray dotted line indicates proportionality (that is, seat share equals vote share). The green line indicates outcomes which correspond to an efficiency gap of zero.[5] The efficiency gap is a measure of fairness found in the literature based on the concept of "wasted votes" [22].

The Republican advantage we observed in the previous section is strikingly visible in these plots as well. Where the dots are below the gray line, Democrats are underperforming relative to proportionality: this is the case for all but the two most Democratic elections considered (ATG12 and GOV14). A good way to appreciate the asymmetry present in these graphs is to contrast elections in which Democrats receive around 55 percent of the vote (SEN12 and GOV14) with an election in which the Republican candidate received around 55 percent of the vote (GOV10). At the scale of Congressional districts, on average, the neutral plans produce an expected seat share of around 53 and 60 percent respectively for Democrats, but around 77 percent for Republicans.

While proportionality is considered by some as the mark of a fair districting process, others recognize that a "winner's bonus" is a reasonable property to expect in a districted system. That is, while 50% of the vote should win you half the seats, 70% of the vote—an overwhelming victory—could easily result in far more than 70% of the seats, depending on how the extra 20% advantage is spatially arranged. If it is uniform, of course, then all seats go to the majority party. The efficiency gap boils down to a specific recommendation for this bonus: $50 + x$ percent of the vote should roughly translate into $50 + 2x$ of the seats. This is where the green line comes from. All this discussion is to say that where the dots are below the green line for Democratic vote shares less than 0.5, the Democrats are not only failing to achieve proportionality, but are not even able to achieve the representation predicted by a common standard in the literature that takes into account the Republican winner's bonus.

An interesting but subtle scale phenomenon is visible on these plots for the most Democratic election of all—ATG12. With 57 percent of the statewide vote, Democrats exceed the efficiency gap standard for 18 and 50 districts for many plans, but far more rarely for the plans with 203 districts. In other words, the Democratic winner's bonus diminishes as the scale of redistricting grows finer for this particular election.

And again, these graphs demonstrate the much tighter range of outcomes produced in the neutral ensembles as districts become smaller. As mentioned above, some reformers anticipate that smaller districts on the scale of Canadian or British parliamentary constituencies, or Pennsylvania State House districts, might reduce the level of Republican advantage observed in recent Congressional elections. However, these graphs suggest that in Pennsylvania, neutral districting at a smaller scale might not produce any maps at all that meet a rather uncontroversial standard of

[5] This is the simplified EG formula, assuming equal turnout in each district.

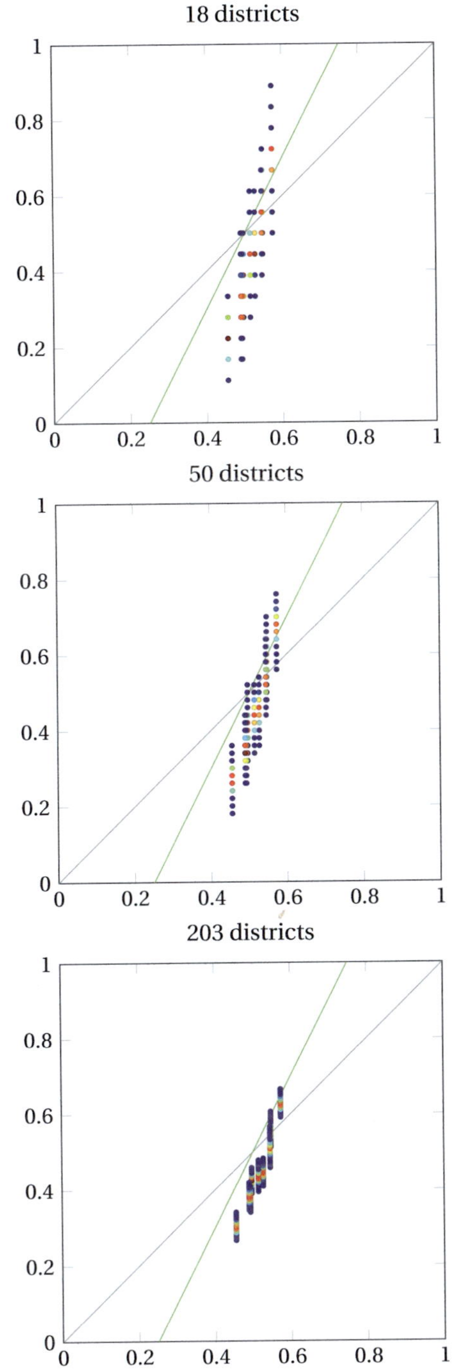

Figure 6: Seats–votes plots for Pennsylvania. The x-axis indicates statewide Democratic vote share and the y-axis indicates Democratic seat share in each case. Gray lines indicate proportionality, while green lines indicate an efficiency gap of zero.

partisan fairness.

5 EAST VERSUS WEST

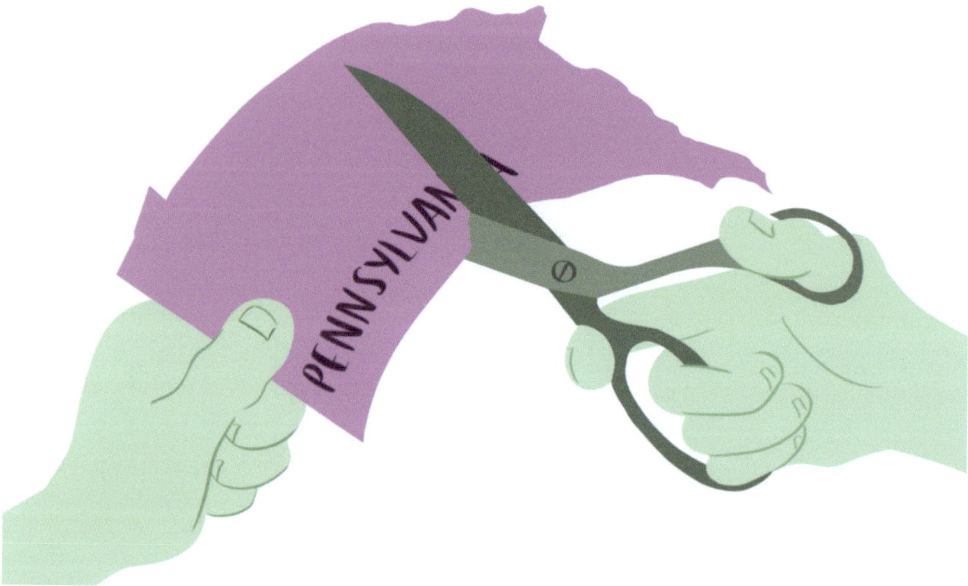

METHOD

In this section we examine the difference in political geography between the western and eastern parts of Pennsylvania at the level of Congressional and Pennsylvania state Senate redistricting (we omit the state House level for the sake of brevity). We choose a subdivision of Pennsylvania along county boundaries which results in two pieces, West and East, as shown in Figure 7. Up to an error of just over 3000 people, the West has half the population of the East. Since Pennsylvania has 18 congressional districts, it thus makes sense to consider plans of six districts for the West and plans of twelve districts for the East. To best approximate state Senate plans, we consider plans of seventeen districts for the West and plans of thirty-four districts for the East, adding up to a total of fifty-one, one over the correct number of fifty. For ease of reference, the overall vote shares for each party in each piece are shown in Tables 5.1 and 5.2, along with the mean seat shares coming from the ensemble analyses.

RESULTS

Figures 8, 9, and 10 in this section each have four histograms showing the Democratic seats outcomes for four different ensembles based on the specified election data. The "West" ensemble is an ensemble of 50,000 plans with a third of the

	West		East		Full	
	Seat %	Vote %	Seat %	Vote %	Seat %	Vote %
PRES16	20.76%	41.66%	46.83%	53.60%	37.83%	49.65%
SEN16	22.08%	43.36%	36.32%	52.18%	31.62%	49.28%
ATG16	27.22%	46.04%	51.18%	54.11%	43.61%	51.43%
PRES12	23.49%	46.03%	57.47%	56.04%	46.46%	52.71%
SEN12	33.64%	48.83%	62.79%	57.45%	53.05%	54.56%
ATG12	67.55%	54.00%	69.74%	59.12%	69.23%	57.42%
GOV14	48.61%	50.63%	64.89%	57.30%	59.81%	55.12%
GOV10	10.59%	40.49%	29.07%	48.08%	22.69%	45.48%
SEN10	20.72%	45.14%	36.26%	50.92%	31.23%	48.95%

Table 5.1: Vote shares and mean seat shares for 18 districts (6 West, 12 East).

	West		East		Full	
	Seat %	Vote %	Seat %	Vote %	Seat %	Vote %
PRES16	24.85%	41.66%	49.64%	53.60%	41.90%	49.65%
SEN16	20.59%	43.36%	42.40%	52.18%	35.63%	49.28%
ATG16	28.60%	46.04%	51.30%	54.11%	43.96%	51.43%
PRES12	25.70%	46.03%	55.74%	56.04%	46.23%	52.71%
SEN12	35.02%	48.83%	61.57%	57.45%	52.91%	54.56%
ATG12	63.25%	54.00%	69.84%	59.12%	67.30%	57.42%
GOV14	47.63%	50.63%	65.14%	57.30%	58.32%	55.12%
GOV10	17.78%	40.49%	32.03%	48.08%	27.16%	45.48%
SEN10	26.00%	45.14%	38.89%	50.92%	34.96%	48.95%

Table 5.2: Vote shares and mean seat shares for 51 districts (17 West, 34 East).

targeted number of districts for the West piece of Pennsylvania only. The "East" ensemble is an ensemble of 50,000 plans with two-thirds of the targeted number of districts for the East piece. The "Full state" ensemble is an ensemble of 50,000 plans for the entire state with the targeted number of districts. Finally, the "E–W pairs" ensemble consists of every possible plan that can be created by putting together a plan from the "West" ensemble and a plan from the "East" ensemble. This last ensemble should be thought of as an ensemble of plans that respect the East–West subdivision of the state we chose. As mentioned above, we chose two districting scales: 18 districts (for Congressional) and 51 districts (as the closest multiple of three to the state Senate number of 50).

DISCUSSION

One observation we should immediately make is that the seats outcomes for the unsplit state plans and the East–West combination plans are in all cases extremely similar. In other words, forcing plans to respect our arbitrary East–West division does not have a substantial impact on the baseline for redistricting in Pennsylvania. This gives us the confidence to examine the impacts of the East and West on baseline representation separately, since combining them pairwise reproduces

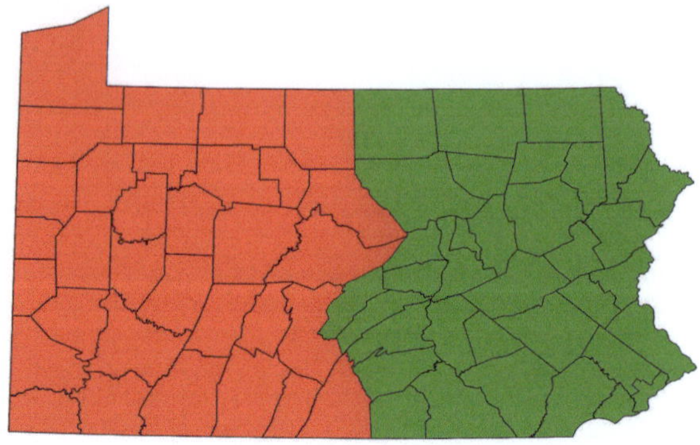

Figure 7: Dividing Pennsylvania into West and East.

the redistricting phenomena we are trying to study for the whole state.

The plots reveal that the general Democratic under-performance is more pronounced in the West than in the East. In the West, in both PRES16 and SEN16, the Democrats were able to secure only one Western Congressional seat in a majority of the plans (in Pittsburgh), despite the Western vote share being well above 40% in both cases. Even when the Democrats receive 49 percent of the votes in the West, as they did in SEN12, they only received 34 percent of the Congressional seats. There is some contrast between the two elections where the Democrats achieve a higher mean seat share than vote share. For ATG12, when Kane received a statewide vote share of 57 percent, both the West and East mean seat share exceed the vote share (the West by a greater margin than the East in fact). For GOV14, when the Democratic candidate received 55 percent statewide, the mean seat share falls short of the vote share in the West but not in the East, and the two combine to result in a statewide seat share that is above the statewide vote share.

The political geography of Western Pennsylvania seems to make it quite difficult for the Democrats to transform votes to seats. At the scale of Congressional districts, in a typical election, the ensembles tend to produce a single Democratic Pittsburgh seat. Perhaps there is a hint of a scale effect here, since the Democratic seat share is somewhat higher at the scale of state Senate than Congressional districts in the West for 6 of the 9 elections. This may have to do with the nature of the partitioning of Pittsburgh, and the greater likelihood of Democratic victories occurring in Erie at the smaller scale of state Senate districts. To be sure, the Democrats' political geography is still quite inefficient in the East, but the Democrats' difficulty in the West is especially striking at both spatial scales analyzed here.

The East–West comparison is also useful for shedding light on the puzzling gap, described above, between SEN16 and PRES16. The right-hand columns of Tables 1 and 2 illuminate that in the state as a whole, Clinton's presidential vote share was more efficiently distributed than that of McGinty in the Senate race. With very similar vote shares, on average, neutral ensembles produced a seat share about 6 percentage points higher for Clinton at both spatial scales considered here. We can now see that McGinty outperformed Clinton in the West, and Clinton

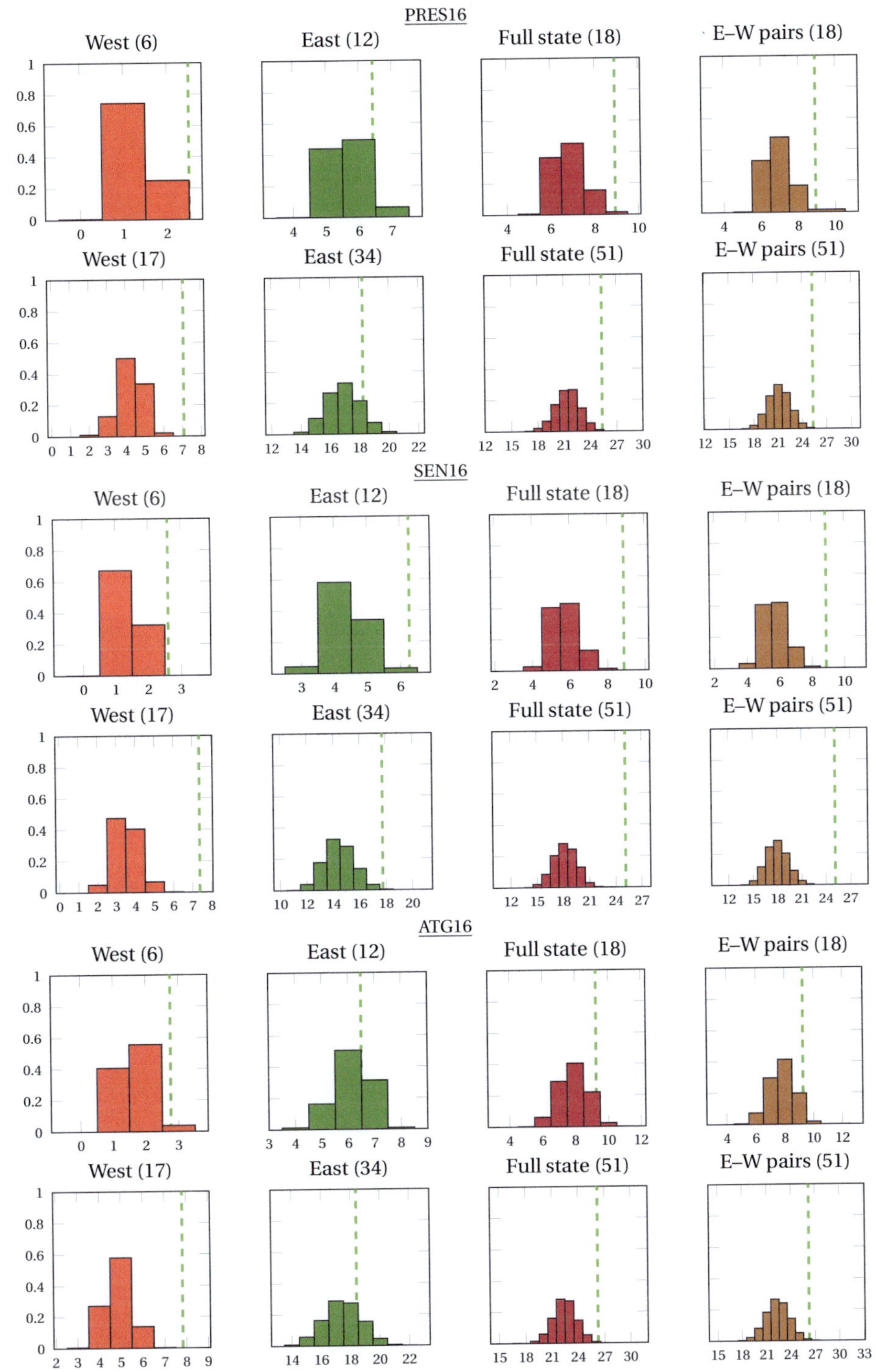

Figure 8: East-West comparison for the 2016 elections. The x-axis and y-axis in each plot represent Democratic seats won and fraction of the ensemble respectively. Numbers in parentheses indicate the number of districts in each plan. Dotted green lines indicate proportionality.

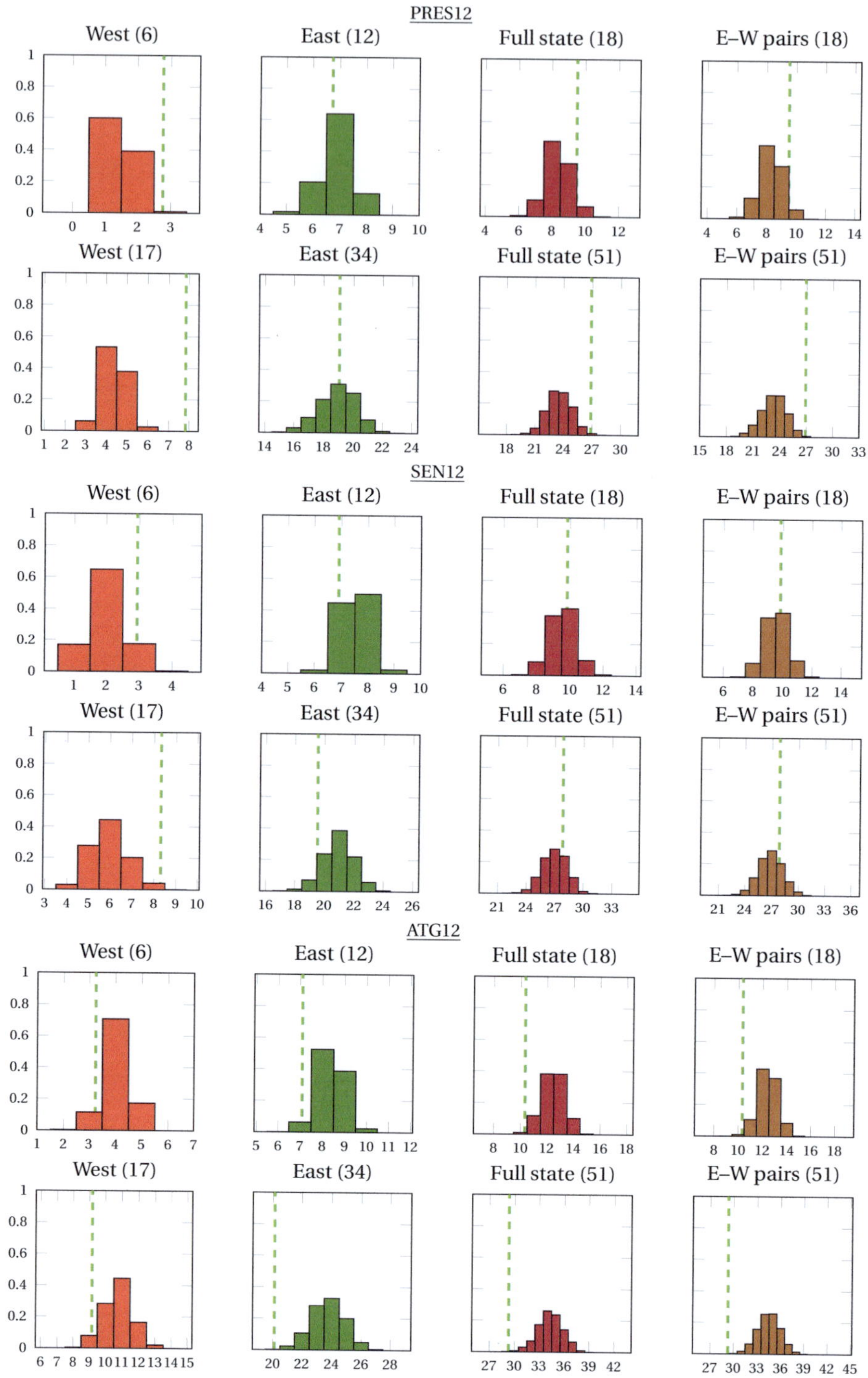

Figure 9: East-West comparison for the 2012 elections. The x-axis and y-axis in each plot represent Democratic seats won and fraction of the ensemble respectively. Numbers in parentheses indicate the number of districts in each plan. Dotted green lines indicate proportionality.

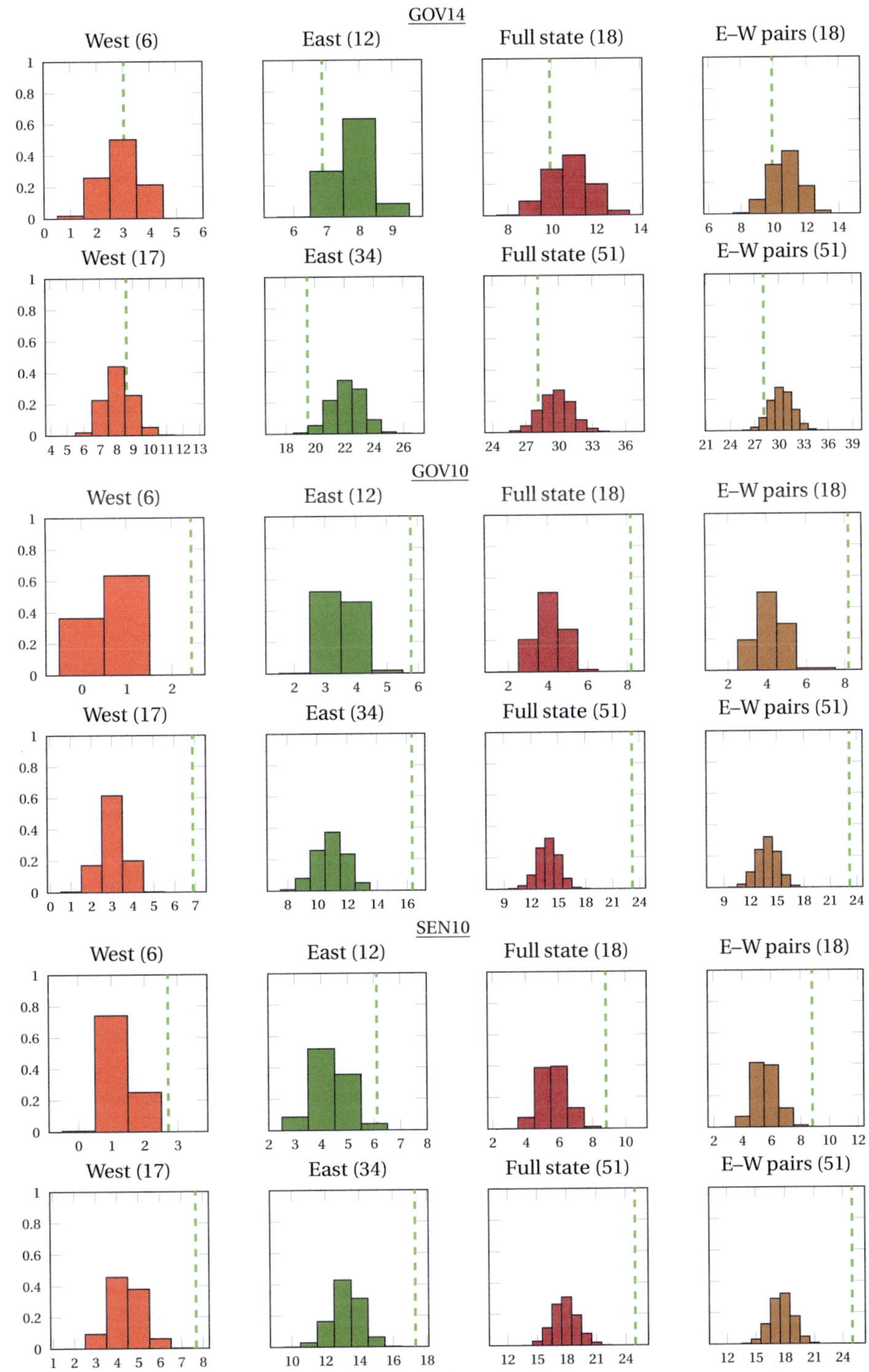

Figure 10: East-West comparison for other elections. The x-axis and y-axis in each plot represent Democratic seats won and fraction of the ensemble respectively. Numbers in parentheses indicate the number of districts in each plan. Dotted green lines indicate proportionality.

outperformed McGinty in the East. Inspection of precinct-level maps reveals that split-ticket voters favoring the Democratic Senate candidate while favoring Donald Trump in the presidential election in the West were located in nonurban working-class areas, especially in the Southwest. And ticket-splitting in the East, where those voting Republican in the Senate race chose Clinton in the presidential race, were located largely in educated suburbs of Philadelphia.

Clinton's better overall performance than that of McGinty in transforming votes to seats is driven primarily by the East. This is clearest at the scale of Congressional districts, where McGinty's higher vote share corresponded to a higher seat share. In the East, on the other hand, where Clinton outpolled McGinty by 1.42 percentage points, she received a seat share that was more then 10 percentage points higher than that of McGinty. This phenomenon persists to a lesser degree at 51 districts: both the West and East have higher mean seat shares for PRES16 than SEN16, but the difference is greater for the East (around 7 percentage points) than for the West (around 4 percentage points). It appears that Clinton's spatial pattern of support was more efficient at winning seats than McGinty's because she out-polled McGinty in marginal areas of greater Philadelphia that produced districts with small majorities for Clinton in the presidential race but small majorities for Toomey (the Republican candidate) in the Senate race.

6 CONCLUSION

This chapter has focused on a single state, but we have been able to exploit useful variation of several kinds: different vote shares and spatial patterns in different elections, different spatial scales for drawing districts, and the very different political geography of Eastern and Western Pennsylvania.

Perhaps the most basic conclusion of this study is that because of the spatial distribution of partisanship, a neutral approach to redistricting would likely lead to the under-representation of the Democratic Party relative to its statewide strength. In the vast majority of neutral redistricting ensembles, the Republican Party would be able to win a very comfortable majority of seats with a little less than half of the votes. Democrats cannot expect to win a majority of seats until they win somewhere around 54 percent of the votes. They do not benefit from a disproportionate "winner's bonus" until they obtain well over 56 percent of the statewide vote. In contrast, the Republican Party can receive a massive winner's bonus even with very slightly more than 50 percent of the statewide vote. This pattern can be seen both in Eastern and Western Pennsylvania, but it is more pronounced in the Western part of the state, where a large share of the Democrats are concentrated in Pittsburgh.

It was useful to examine a wide variety of elections, not only in order to assess the implications of neutral ensembles at different statewide vote shares, but also to explore differences in the spatial support for candidates even when the overall vote shares were similar. For instance, we discovered that in 2016, Hillary Clinton's support distribution led to a significantly better seat share than that of Katie McGinty in the Senate race, even though their statewide vote shares were quite similar. This appears to be driven above all by Clinton's relative success in marginal suburban areas in Eastern Pennsylvania.

This observation suggests that a state's political geography is not static, but constantly changing with time and between elections (even on the same ballot!). Geographic realignments can and do take place. Neutral redistricting ensembles might produce important differences in seat shares for the parties, even without large differences in statewide vote shares, if enough geographically proximate voters in marginal areas shift from one party to the other. In many U.S. states, large swaths of suburbia have been marginally Republican for a period of time, but recent shifts in favor of the Democratic Party among educated voters in those areas—even if offset by losses in more rural areas—could lead to changes in seat shares. This is an important topic for further research.

We have also explored the proposition that as the scale of districts becomes smaller, seat shares should come closer to mirroring statewide proportionality. We explored the range from two districts to 220 districts for Pennsylvania, and found no consistent relationship between geographic scale and Republican advantage across elections. It is entirely plausible, however, that scale effects might exist in other states over a similar range of district sizes. In fact, we see a hint of a scale effect in Western Pennsylvania that we do not see in the East.

We also note that the range of seat shares produced in the neutral ensembles narrows considerably as the state is divided up into more and more districts. This leads to an interesting observation. When the state is carved up into a relatively small number of districts, the range of outcomes is sufficiently wide that, if one draws from the most pro-Democratic tail in the distribution of plans in the ensemble, it is possible to select a plan in which 50 percent of the votes corresponds with 50 percent of the seats. However, as the state is partitioned into smaller and smaller districts, even the most pro-Democratic plan still demonstrates substantial disadvantage for Democratic candidates.

These findings have implications for debates about reform of redistricting in Pennsylvania and beyond. All the ensembles used here were generated by an algorithm that is independent of partisan data, and yet substantial deviations from proportionality occurred. This suggests that even a neutral process involving commissioners or demographers without access to partisan data might result in maps that lead to disproportionate results such as awarding a majority of the seats to a party that loses the statewide vote. To be clear, our results do not show that political geography is so constraining that fair plans (as defined by measures like proportionality or a minimal efficiency gap) are impossible to draw. Rather, some volition, based on analysis of partisan data, would be required.

ACKNOWLEDGMENTS

The authors would like to thank the editors for their valuable input; in particular, Olivia Walch for the illustrations and Moon Duchin for suggestions on how to design the experiments. They would also like to thank Ruth Buck, former MGGG geographer, for her help with the data. Thomas Weighill acknowledges the support of the NSF Convergence Accelerator Grant No. OIA-1937095.

REFERENCES

[1] Caldera, Sophia, Daryl DeFord, Moon Duchin, Samuel C. Gutekunst & Cara Nix. 2020. "Mathematics of nested districts: The case of Alaska." *Statistics and Public Policy* 7(1):39–51.

[2] Chen, Jowei & Jonathan Rodden. 2013. "Unintentional Gerrymandering: Political Geography and Electoral Bias in Legislatures." *Quarterly Journal of Political Science* 8(3):239–269.

[3] Chen, Jowei & Jonathan Rodden. 2015. "Cutting Through the Thicket: Redistricting Simulations and the Detection of Partisan Gerrymanders." *Election Law Journal* 14(4):331–345.

[4] Chikina, Maria, Alan Frieze & Wesley Pegden. 2017. "Assessing Significance in a Markov Chain Without Mixing." *Proceedings of the National Academy of Science* 114:2860.

[5] Cho, Wendy K. Tam & Yan Y. Liu. 2016. "Toward a Talismanic Redistricting Tool: A Computational Method for Identifying Extreme Redistricting Plans." *Election Law Journal* 15:351.

[6] DeFord, Daryl & Moon Duchin. 2019. "Redistricting Reform in Virginia: Districting Criteria in Context." *Virginia Policy Review* 12(2):120–146.

[7] DeFord, Daryl, Moon Duchin & Justin Solomon. 2018. Comparison of districting plans for the Virginia House of Delegates. Technical report MGGG. `https://mggg.org/VA-report.pdf`.

[8] DeFord, Daryl, Moon Duchin & Justin Solomon. 2020. "A Computational Approach to Measuring Vote Elasticity and Competitiveness." *Statistics and Public Policy* 7(1):69–86 .

[9] Duchin, Moon. 2018. Outlier analysis for Pennsylvania congressional redistricting. Technical report. `https://mggg.org/uploads/md-report.pdf`.

[10] Duchin, Moon, Taissa Gladkova, Eugene Henninger-Voss, Ben Klingensmith, Heather Newman & Hannah Wheelen. 2019. "Locating the representational baseline: Republicans in Massachusetts." *Election Law Journal: Rules, Politics, and Policy* 18(4):388–401.

[11] Eubank, Nicholas & Jonathan Rodden. 2020. "Who is my Neighbor? The Spatial Efficiency of Partisanship." *Statistics and Public Policy* 7(1):87–100.

[12] Gudgin, Graham & P.J. Taylor. 1979. *Seats, Votes, and the Spatial Organisation of Elections.* London: Pion Limited.

[13] Johnston, Ronald. 1977. "Spatial Structure, Plurality Systems and Electoral Bias." *The Canadian Geographer* 20:310–328.

[14] Johnston, Ronald, Charles Pattie, Dany Dorling & David Rossiter. 2001. *From Votes to Seats: The Operation of the UK Electoral System since 1945.* Manchester and New York: Manchester University Press.

[15] Magleby, Daniel & Daniel Mosesson. 2018. "A New Approach for Developing Neutral Redistricting Plans." *Political Analysis* 26(2):147–167.

[16] Mathematicians' Amicus Brief, Rucho v. Common Cause. 2018. "Amicus Brief of Mathematicians, Law Professors, and Students in Support of Appelleees and Affirmance." Amicus Brief, Supreme Court of the United States, Rucho et al. v. Common Cause et al.

[17] Mattingly, Jonathan C. & Christy Vaughn. 2014. "Redistricting and the Will of the People." `https://arxiv.org/abs/1410.8796`.

[18] Nagle, John F. 2019. "What Criteria Should be Used for Redistricting Reform?" *Election Law Journal* 18(1):63–77.

[19] Pegden, Wesley. 2017. "Pennsylvania's Congressional Districting is an Outlier: Expert Report." Expert report submitted in League of Women Voters of Pennsylvania v. Commonwealth of Pennsylvania.

[20] Pegden, Wesley, Jonathan Rodden & Samuel Wang. 2018. "Brief of Amici Curiae Professors Wesley Pegden, Jonathan Rodden, and Samuel Wang in Support of Appellees." Supreme Court of the United States.

[21] Rodden, Jonathan. 2019. *Why Cities Lose: The Deep Roots of the Urban-Rural Divide.* Basic Books.

[22] Stephanopoulos, Nicholas O & Eric M McGhee. 2015. "Partisan gerrymandering and the efficiency gap." *University of Chicago Law Review* 82:831.

II

Law

6 *Explainer: A brief introduction to the Voting Rights Act*

ARUSHA GORDON AND DOUGLAS M. SPENCER

HISTORY

With the end of the Civil War in 1865, equal rights for African Americans were formally recognized in the U.S. through the passage of the 14th and 15th Amendments to the Constitution. These events initially led to a surge of African American voter registration and to the election of Black representatives, but the gains were quickly rolled back. By the late 1870s, a series of Supreme Court decisions, political deals, and legislative actions spelled the end of the Reconstruction Era and the dawn of Jim Crow, the long period of legal repression of Black civil rights coupled with violent intimidation campaigns. Historians typically bracket the Jim Crow Era from 1877, when the federal troops who were enforcing anti-discrimination laws were withdrawn from the South, to 1965, with the passage of the Voting Rights Act.

The Voting Rights Act of 1965 (VRA) reflects "Congress' firm intention to rid the country of racial discrimination in voting" (in the words of Justice Earl Warren) and was one of the most important pieces of legislation passed during the Civil Rights era. Before the passage of the VRA, less than one-third of Black adults were registered to vote in Southern states, while White voter registration was closer to 75 percent. The decades following the VRA's passage coincided with a 30-fold expansion in the number of Black elected officials, from about 300 in 1964 to 9,430 in 2002 (see Figure 1 for a visualization).[1] The number of elected Hispanic officials saw similar growth in the decades since the VRA was passed.

The VRA was passed in the wake of a methodical, courageous, and at times bloody campaign led by John Lewis, Martin Luther King, Jr., Ella Baker, and other civil rights leaders. After World War II, the campaign against Jim Crow and voter suppression

[1] The Census Bureau cites the latter figure in `perma.cc/Q8UT-DP3C`.

M. Duchin, O. Walch (eds.), *Political Geometry*, https://doi.org/10.1007/978-3-319-69161-9_6

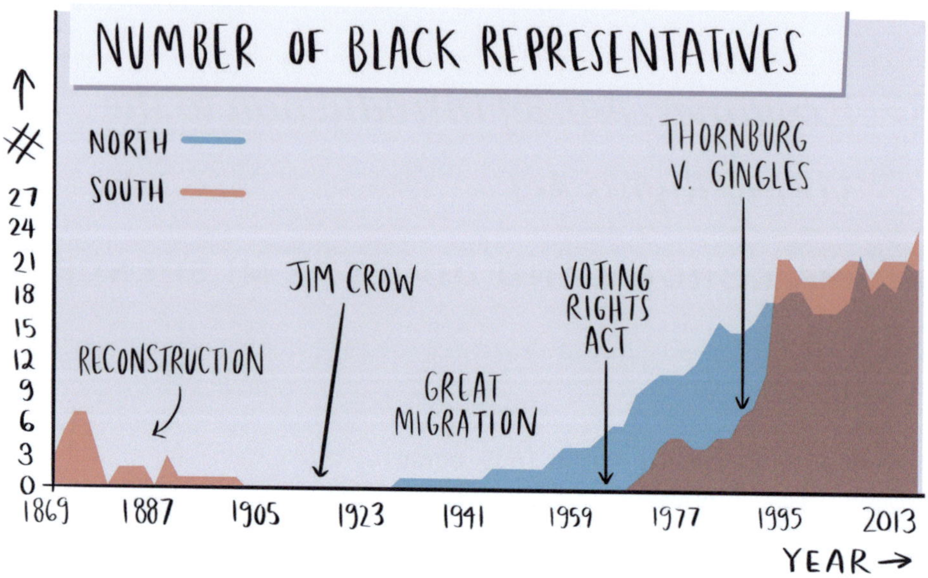

Figure 1: Number of Black representatives in the U.S. Congress by year, split by North and South. Adapted from a figure created by Mira Bernstein.

picked up momentum. Organizations like the Student Nonviolent Coordinating Committee sent young people to the South to help to register and educate Black residents; civil rights leaders adeptly used the media to draw public attention to discrimination in the South; and large events—like the march across the Edmund Pettus bridge in Selma, Alabama, in March of 1965, in which Representative John Lewis and others were badly beaten by local police—helped to force the federal government to act.

The violent attacks in Selma created an urgency that propelled Congress and President Johnson to push for new legislation. Just days after the Selma attacks, President Johnson addressed the nation on television, echoing the words used in the civil rights movement by calling on southern jurisdictions to "[o]pen your polling places to all your people," and to "[a]llow men and women to register and vote whatever the color of their skin." Five months later, Johnson signed the Voting Rights Act of 1965 into law. The VRA was amended and reauthorized by Congress five times—1970, 1975, 1982, 1992, and 2006. It was in the 1975 reauthorization that "language minorities" were added, opening the door to claims on behalf of Latino, Asian, and Native American plaintiffs. Over the years, the core provisions of the Act remained largely the same, but these important clarifications and expansions of scope—passed with strong bipartisan majorities—helped keep the law in sync with shifting American racial realities.

KEY PROVISIONS

The 1965 Act included a number of provisions that drastically expanded the ability of the federal government, and the executive branch specifically, to address discrimination in voting rights. The Act has five sections: Section 1 is just the name of the Act; Section 2 is a powerful and detailed restatement of the promises of the 15th Amendment to provide equal access to voting; and Sections 3-5 collectively describe a stronger set of rules called "preclearance" under which jurisdictions with a history of discrimination would face sustained scrutiny. In particular, Section 3 explains how some jurisdictions might "bail in" to covered status, Section 4 details "bail out" and also lays out the "formula" or system for deciding which states and localities would be covered. Section 5 requires covered jurisdictions to submit any proposed changes to their election procedures to the Attorney General or U.S. District Court of D.C. for approval, so as to prevent any election changes that might have a discriminatory impact or be based on discriminatory intent. Let's review some of this in more detail.

SECTION 2

The strong provisions of Section 2 prohibit any "voting qualification or prerequisite to voting or standard, practice, or procedure" that is "imposed or applied by any State or political subdivision in a manner which results in a denial or abridgement of the right of any citizen of the United States to vote on account of race or color..." Part b of Section 2 further states that a violation "is established if, based on the **totality of circumstances**, it is shown that the political processes leading to nomination or election in the State or political subdivision are not equally open to participation by... citizens protected by [Section 2] in that [they] have less opportunity than other members of the electorate to participate in the political process and to elect representatives of their choice" (emphasis added). Taken as a whole, Section 2 is a tool to prevent not only vote denial, but, much more broadly, structures and practices that *dilute* voting strength.

In its early years, it was unclear whether Section 2 prohibited just intentional discrimination or whether it could also be read to prohibit practices and procedures just on the basis of discriminatory effect. In 1980, in *Mobile v. Bolden*, a case challenging the practice of a municipality electing its city council members at large, the Supreme Court held that a successful Section 2 claim required a finding of intentional discrimination, and that establishing a practice's discriminatory effect on minority voters was not enough. The finding of the Supreme Court dealt a major blow to the ability of advocates to use the VRA to attack and rout out discrimination in electoral practices. However, just two years later, Congress responded to the decision in *Mobile* by amending the VRA to expressly allow for "effects" or "results" claims – i.e., to allow plaintiffs bringing claims under Section 2 to succeed without establishing any intent to discriminate, but just on the basis of outcomes.

In amending the VRA, Congress used its investigatory powers to hold hearings, then ultimately drafted what would become known as the "Senate Report." It "elaborates on the nature of Section 2 violations," essentially codifying the totality-of-the-circumstances standard from the text of the Act. The Senate Report listed out numerous elements that courts should consider in assessing a claim under Section 2. These factors include: the history of voting-related discrimination in the state or political subdivision; the extent to which voting in the elections of the state or political subdivision is racially polarized; the extent to which the state or political subdivision has used voting practices or procedures that tend to enhance the opportunity for discrimination against the minority group, such as unusually large election districts, majority vote requirements, and prohibitions against bullet voting; the exclusion of members of the minority group from candidate slating processes; the extent to which minority group members bear the effects of past discrimination in areas such as education, employment, and health, which hinder their ability to participate effectively in the political process; the use of overt or subtle racial appeals in political campaigns; and the extent to which members of the minority group have been elected to public office in the jurisdiction.

Reviewing these today, we might be surprised at some of these inclusions—for instance, bullet voting (or the practice of listing just one name on a ballot designed for choosing multiple candidates) sounds race-neutral on its face, but it had become an organizing tactic for gaining Black voting power and so was expressly prohibited in certain White-controlled jurisdictions. The Report stresses, however, that this list of factors is not comprehensive and that courts may also consider additional factors.

These Senate Factors are broad and interdisciplinary, and today there will often be historians brought in as expert witnesses to speak to some of these issues in a particular locality. But the second Senate Factor, the presence of racial polarization in voting, would soon be elevated to a quantitatively specific threshold test, bringing new clarity to VRA litigation. A few years after *Mobile* and the 1982 amendments, another key case interpreting the VRA was decided by the Supreme Court. In *Thornburg v. Gingles*, the Supreme Court delineated a short checklist—now known as the "Gingles factors" or "Gingles preconditions"—which plaintiffs must complete in order to advance a claim of vote dilution under Section 2. These are:

Gingles 1. the minority group is "sufficiently large and geographically compact to constitute a majority in a single-member district";

Gingles 2. the minority group is "politically cohesive"; and

Gingles 3. "the white majority votes sufficiently as a bloc to enable it... usually to defeat the minority's preferred candidate."

The first condition is essentially established by drawing a suitable demonstration plan, that is, a plan with an additional majority-minority district that still adheres to traditional principles.[2] The second and third require a showing that members of different racial groups vote differently in a way that thwarts the minority from electing candidates of choice.

[2] Importantly, this is the *only* role for majority-minority districts in current VRA case law; you don't have to draw one at the end of the day, but just have to show that one *could have been* drawn in order to launch your lawsuit.

The *Thornburg* decision holds that, if plaintiffs satisfy these three preconditions, then the court must conduct "a searching practical evaluation of the 'past and present reality,'"—that is, the totality of the circumstances—to learn "whether the political process is equally open to minority voters."

Over the years, the statistical inference machinery to establish Gingles 2-3 has become sufficiently standardized that, taken together, the Gingles factors have introduced some sense of routine to the holistic endeavor of proving vote dilution, making it more legally manageable. This helps explain the big uptick in Black representation that coincides with the *Thornburg* decision (Figure 1).

SECTION 5

As we heard above, Section 5 of the VRA requires covered jurisdictions to submit any proposed changes to their election procedures to the Attorney General or U.S. District Court of D.C. for preapproval.

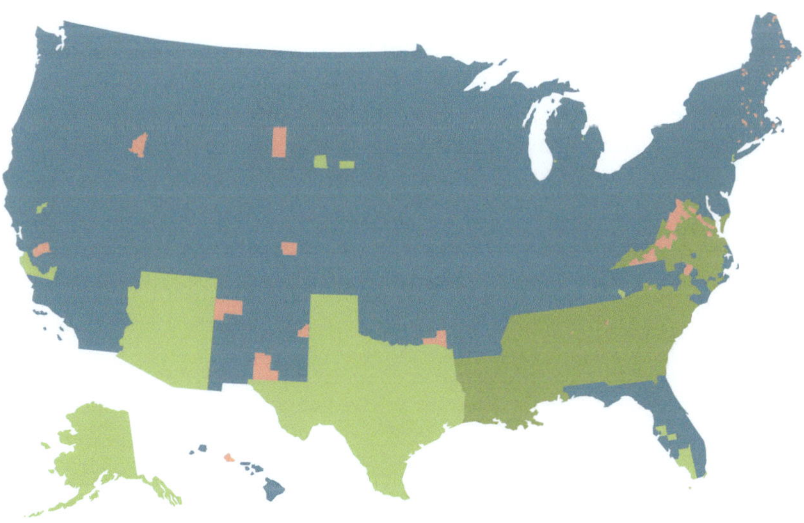

Figure 2: Map of preclearance regions, adapted from the New York Times. Dark green areas were covered from 1965, light green areas were added in 1970 or 1975, and orange areas were released from coverage by a court ("bailed out").

The map in Figure 2 shows areas formerly covered by Section 5. The original list was built through a "formula" involving measures of low Black voter registration and turnout disparities between Black and White voters. In some cases, entire states were covered, as in Alabama, Alaska, Arizona, Georgia, Louisiana, Mississippi, South Carolina, Texas, and Virginia. Other states were only covered in part, such as California, Florida, New York, North Carolina, South Dakota, and Michigan. Certain jurisdictions were "bailed out" under Section 4(a) of the VRA after convincing the courts that preclearance was no longer needed.

In the decades after the VRA was passed, Section 5 proved one of the most effective tools for preventing voter discrimination, especially under the emerging standard prohibiting "retrogression"—in other words, minority electoral opportunity in covered jurisdictions should not get worse over time. The U.S. Department of Justice (DOJ) denied more than 3,000 voting changes between 1965 and 2013, including over 500 proposed redistricting plans, due to the discriminatory or retrogressive effect of those changes.[3] Besides a districting plan, a preclearance review could be triggered by many other kinds of rule changes, such as the size of an elected body, the system of election, the availability of early voting, and so on.

However, on June 25, 2013, the Supreme Court issued its opinion in *Shelby v. Holder*, finding Section 4(b) unconstitutional because the formula determining coverage was said to be outdated. As a result, while Section 5 technically remains in place, it now applies to an emptied list of locations.[4] Restoring the promise of the VRA will require new Congressional action to develop an updated model for coverage. But the protection of voting rights has now become divisive and gridlocked, so that a change to Senate rules might be needed to even squeak voting legislation through Congress today—in marked contrast to the strong bipartisan support the VRA enjoyed throughout its first 50 years.

Today, the VRA is under new threat. As of February 2022, the Supreme Court has stepped in to hit the brakes on a routine voting rights case in Alabama and seems poised to declare that racial fairness in elections must be secured in a race-blind way. We are all watching to see the next act of the Voting Rights Act.

[3] See Posner, Mark A. "The real story behind the Justice Department's implementation of Section 5 of the VRA: Vigorous enforcement, as intended by congress." *Duke J. Const. L. Pub. Pol'y* 1 (2006): 79.

[4] After the list was emptied, Pasadena, TX got "bailed in," so it is now a lonely member of the preclearance list.

Chapter 7

Race and redistricting: The legal framework

ELLEN D. KATZ

CHAPTER SUMMARY

Legal scholar Ellen Katz gives a 60-year history of American jurisprudence around race and redistricting, from the cases that set the stage for the Voting Rights Act to its current precarity in the Roberts Court.

1 INTRODUCTION

This chapter examines the federal legal framework governing questions of race and redistricting in the United States. Organized chronologically, it examines the foundational laws and cases that define the ways in which race *may not* be used in the redistricting process, as well as the ways in which race *must* be used in that process. It explores the tension between the prohibited and required uses of race in redistricting.

Stage I: Into the Thicket tracks the Supreme Court's development of a constitutional framework to address issues of race and redistricting. It begins with the Court's 1960 decision, *Gomillion v. Lightfoot,*[1] which struck down an Alabama gerrymander that redefined the borders of the City of Tuskegee so as to exclude almost every African American resident from the municipality. This Part discusses how *Gomillion* led the Court to enter a realm Justice Felix Frankfurter once described

[1] 364 U.S. 339 (1960).

M. Duchin, O. Walch (eds.), *Political Geometry*, https://doi.org/10.1007/978-3-319-69161-9_7

as the "political thicket"[2] and subsequently to develop the concept of racial vote dilution in *Whitcomb v. Chavis*,[3] *White v. Regester*,[4] and *Mobile v. Bolden*.[5]

Stage II: Elaboration turns to the 1982 amendments Congress made to the Voting Rights Act (VRA). Those amendments were, in part, a response to the Supreme Court's ruling that the Constitution prohibits racial vote dilution only when policymakers intentionally draw district lines to burden minority voters. The 1982 amendments to Section 2 prohibit electoral practices that "result" in discriminatory burdens, regardless of the intent underlying their enactment.[6] This Part examines the contours of the new statutory prohibition, the so-called "Senate Factors" that Congress indicated should guide interpretation of the provision, and the substantial gloss that the Supreme Court placed on this statutory claim in *Thornburg v. Gingles*.[7]

Stage III: Uneasiness examines the decisions of the Rehnquist Court during the 1990s that show the Court's increasing discomfort with what has long been the preferred remedy for racial vote dilution, namely, the majority-minority district. This Part describes the Court's development of a new constitutional injury in *Shaw v. Reno*[8] and its progeny,[9] one that limited the ways in which jurisdictions may rely on race when drawing electoral districts.

Stage IV: Hostility shows how judicial uneasiness about the role of race in redistricting evolved into outright hostility in the Roberts Court. The Part traces the development of this hostility from Chief Justice Roberts' early description of efforts to comply with the VRA as "a sordid business, this divvying us by race,"[10] to sweeping decisions such as *Bartlett v. Strickland*[11] and *Shelby County v. Holder*[12] that substantially reduced the reach of the VRA. This Part closes by examining the Roberts Court development of the *Shaw* doctrine in a series of recent cases.[13]

A brief conclusion considers how federal law addressing race and redistricting might develop in the coming years.

---

[2]See *Reynolds v. Sims*, 377 U.S. 533 (1964); *Baker v. Carr*, 369 U.S. 186 (1962); see also *Colegrove v. Green*, 328 U.S. 549 (1946).

[3]403 U.S. 124 (1971).

[4]412 U.S. 755 (1973).

[5]446 U.S. 55 (1980)

[6]See 52 U.S.C. §10301 (formerly 42 U.S.C. §1973).

[7]478 U.S. 30 (1986).

[8]509 U.S. 630 (1993).

[9]See *Bush v. Vera*, 517 U.S. 952 (1996), *Shaw v. Hunt*, 517 U.S. 899 (1996); *Miller v. Johnson*, 515 U.S. 900 (1995).

[10]*League of United Latin Am. Citizens v. Perry*, 548 U.S. 399, 511 (2006) (Roberts, C.J., concurring in part and dissenting in part).

[11]556 U.S. 1 (2009).

[12]570 U.S. 529 (2013).

[13]See *Cooper v. Harris* 137 S.Ct. 1455 (2017); *Bethune-Hill v. Va. State Bd. of Elections*, 137 S.Ct. 788 (2017); *Alabama Legislative Black Caucus v. Alabama*, 135 S.Ct. 1257 (2015)

7.1 GOMILLION V LIGHTFOOT, 1960

Can the city of Tuskegee legally redraw its own boundaries, shrinking itself to less than half of its former size, in a manner that exploits racial segregation to change the city population from 80% Black to 100% White with the stroke of a pen?
The Court says no, 9-0.

Calling the map "uncouth" and "irregular," Justice Frankfurter wrote that,

> "[T]antamount for all practical purposes to a **mathematical demonstration**... that the legislation is solely concerned with segregating White and colored voters by fencing Negro citizens out of town so as to deprive them of their **pre-existing** municipal vote."

He also sought to make a distinction between this case and *Colegrove* (of "political thicket" fame):

> "The appellants in Colegrove complained only of a **dilution** of the strength of their votes... The petitioners here complain that affirmative legislative action **deprives** them of their votes and the consequent advantages that the ballot affords."

2 INTO THE THICKET: THE CONSTITUTIONAL FRAMEWORK

In July, 1957, Alabama's legislature voted unanimously to redraw the boundaries of the City of Tuskegee. The City had been home to a highly educated African American population ever since Booker T. Washington set up his renowned institute there in 1881. By 1957, Tuskegee had long been a majority-Black city. The new law changed that by redefining the city limits from a square into what Justice Felix Frankfurter would describe as an "uncouth twenty-eight-sided figure."[14] This action removed "all save four or five" of the Tuskegee's African American voters "while not removing a single White voter or resident."[15]

Charles Gomillion, a sociology professor at the Tuskegee Institute and president of the Tuskegee Civic Association, was one of twelve Black voters who challenged the new boundaries as unconstitutional. The lower courts tossed out the claim, but a unanimous Supreme Court agreed with the *Gomillion* plaintiffs. Writing for the Court, Justice Frankfurter held that Alabama's action was "not an ordinary exercise in redistricting even within familiar abuses of gerrymandering." Recognizing that states normally have wide latitude to draw municipal boundaries as they see fit, Frankfurter nevertheless held Alabama's plan unconstitutional, finding that the State's purpose was to segregate White and Black voters by "fencing Negro citizens out of town so as to deprive them of their pre-existing municipal vote." The Court held that the new city limits were an illegitimate racial gerrymander barred by the Fifteenth Amendment.[16]

The illegality of Alabama's action seems patent today. Whatever leeway States enjoy to draw district lines, excising a racially defined population from a city falls well outside the realm of the permissible. But what seems clear to the contemporary observer was far from self-evident in 1960 when the Court decided *Gomillion*. At the time, the non-justiciability of districting lines was firmly established. The Court, in an earlier opinion by none other than Justice Frankfurter, had held unequivocally that it lacked "competence to grant" relief from discriminatory electoral lines, and, that it "ought not enter this political thicket."[17] In *Gomillion* itself, Justice Whittaker thought that Alabama's action amounted to "unlawful segregation of races of citizens" but nevertheless did not deny anyone the right to vote "inasmuch as no one has the right to vote in a political division, or in a local election concerning only an area in which he does not reside."[18]

Justice Frankfurter insisted his opinion in *Gomillion* did nothing to unsettle existing precedent. The ruling, he wrote, carved a very narrow exception to the non-justiciability of district lines. For Frankfurter, it was Alabama's affirmative decision to withdraw what was a "pre-existing" vote that critically distinguished

[14] *Gomillion v. Lightfoot*, 364 U.S. 339, 340 (1960)
[15] Id. at 341
[16] Id.
[17] *Colegrove v. Green*, 328 U.S. 549, 552, 557 (1946).
[18] Gomillion, at 349 (Whittaker, J., concurring).

the gerrymander from what he viewed to be non-justiciable electoral disputes.[19] Cases like *Colegrove v. Green* involved a complaint "only of a dilution of the strength of ... votes as a result of legislative inaction over a course of many years."[20] The *Gomillion* plaintiffs, by contrast, challenged "affirmative legislative action" that gave state "approval ... to unequivocal withdrawal of the vote solely from colored citizens."[21]

Justice Frankfurter's effort to maintain immunity for legislative inaction of this sort failed. Shortly after *Gomillion*, the Court held that malapportionment resulting from such inaction was non-justiciable, with *Gomillion*'s holding that electoral lines are not untouchable paving the way. *Baker v. Carr*[22] rejected the idea that challenges to malapportionment among electoral districts were justiciable political questions, at least under the newly crafted standard for assessing political questions that the Court adopted in and for the dispute.[23] Justice Brennan's opinion for the Court held that a state apportionment scheme that yielded electoral districts with vastly different populations was subject to constitutional challenge in federal court and that the Equal Protection Clause offered "well developed and familiar" standards for judicial assessment of the claim.[24]

Justice Frankfurter dissented, complaining that *Colegrove v. Green* (1946) had rejected the same claim as being beyond judicial competence and should control the result in this case as well. The dissent added that the majority's approach in *Baker* was incoherent. Justice Frankfurter wrote that "[t]alk of 'debasement' or 'dilution' is circular talk" because "[o]ne cannot speak of 'debasement' or 'dilution' of the value of a vote until there is first defined a standard of reference as to what a vote should be worth."[25] For Frankfurter, opening the door to legal challenges to malapportionment meant that the Court would need "to choose among competing bases of representation—ultimately, really, among competing theories of political philosophy—in order to establish an appropriate frame of [state] government." Justice Frankfurter was convinced that the Court was ill-equipped to engage in this inquiry.[26]

Two years after *Baker, Reynolds v. Sims*[27] located within the Equal Protection Clause the principle of one person, one vote. Chief Justice Earl Warren wrote that "an individual's right to vote for state legislators is unconstitutionally impaired when its weight is in a substantial fashion diluted when compared with votes of citizens living in other parts of the State."[28] Deeming the Alabama districting plans challenged in the litigation incompatible with this principle and hence "irrational," *Reynolds* mandated that legislatures, including many that had not altered district

[19]Id. at 346–347.
[20]Id. at 346.
[21]Id. at 347 (emphasis added).
[22]369 U.S. 186 (1962).
[23]Id. at 226–227.
[24]Id. at 226.
[25]Id. at 300 (Frankfurter, J., dissenting).
[26]Id.
[27]377 U.S. 533 (1964).
[28]Id. at 568. See also *Wesberry v. Sanders*, 376 U.S. 1 (1964) (locating the same principle in Article I, section 2 of the U.S. Constitution as governing congressional districts).

boundaries for decades, equalize the population among electoral districts.

Justice Harlan dissented alone. (Justice Frankfurter had retired shortly after *Baker.*) His dissent emphasized "the cold truth that cases of this type are not amenable to the development of judicial standards."[29] Echoing earlier voiced concerns about "political philosophy," Justice Harlan objected to the Court's selection of population equality as the controlling principle, pointing out that "people are not ciphers and that legislators can represent their electors only by speaking for their interests—economic, social, political—many of which do reflect the place where the electors live."[30]

The Court's foray into the political thicket soon moved beyond malapportionment. Plaintiffs alleging racial vote dilution challenge electoral rules that they claim minimize or "dilute" the voting strength of a specific, racially defined group with which they identify. In 1971, African American voters in Indianapolis pressed such a claim, arguing that a countywide multi-member districting plan allowed them "'almost no political force or control over legislators because the effect of their vote is cancelled out by other contrary interest groups.'"[31] In *Whitcomb v. Chavis,* plaintiffs argued that replacing the multi-member structure with single-member districts would allow them to elect legislators who were more responsive to their interests.[32]

Whitcomb held that federal courts are available to entertain claims of this sort. The Court thus rejected Justice Harlan's objection, raised in his dissent, that claims asserting minority vote dilution should not be cognizable in a majoritarian system.[33] Justice White's opinion for the Court nevertheless concluded that the *Whitcomb* plaintiffs had failed to establish a constitutional violation on the facts they presented. Justice White noted the absence of evidence indicating either that the multi-member districting plan "was conceived or operated as purposeful devices to further racial . . . discrimination" or that the plaintiffs confronted meaningful obstacles when registering to vote, joining the political party of their choice, participating in party affairs, and selecting party candidates responsive to their needs.[34] Finding no "built-in" bias against the plaintiffs, Justice White dismissed the allegation that Black voting power was "cancelled out" as a "mere euphemism for defeat at the polls."[35] Justice White made clear that the absence of proportional representation was not alone sufficient to support a viable vote dilution challenge, and rejected the idea that "any group with distinctive interests must be represented in legislative halls if it is numerous enough to command at least one seat and represents a majority living in an area sufficiently compact to constitute a single member district."[36]

Two years later in 1973, a group of African American and Mexican American plain-

[29] Reynolds, at 621 (Harlan, J., dissenting).

[30] Id. at 623-624.

[31] 403 U.S. 124, 129 (1971).

[32] Id. at 129.

[33] Id. at 166 (Harlan, J., dissenting).

[34] Id. at 149–150.

[35] Id. at 153.

[36] Id. at 156.

tiffs in Texas succeeded where the *Whitcomb* plaintiffs had failed. In *White v. Regester*,[37] the Supreme Court held unanimously that Texas violated the Equal Protection Clause by relying on multi-member legislative districts in Dallas and Bexar counties. Justice White explained that "multimember districts are not per se unconstitutional" and that "it is not enough that the racial group allegedly discriminated against has not had legislative seats in proportion to its voting potential."[38] Instead, plaintiffs challenging multi-member districts (and district lines generally) must "produce evidence to support findings that the political processes leading to nomination and election were not equally open to participation by the group in question—that its members had less opportunity than did other residents in the district to participate in the political processes and to elect legislators of their choice."[39]

Black plaintiffs from Dallas made the requisite showing by establishing a number of crucial factors, specifically, the state's history of racial discrimination in voting; its reliance on majority vote requirements and a "place" system not tied to residency; the existence of a White-controlling slating organization that ignored the African American interests; the prevalence of racial appeals in elections; the absence of Black elected officials; and the insufficient responsiveness of the officials elected.[40] The Mexican American plaintiffs from Bexar County established unconstitutional dilution by showing socioeconomic discrimination in education, employment, and housing; cultural and language barriers that inhibited political participation; the lack of Mexican American elected representatives; and the insufficient responsiveness of the elected county delegation to their interests.[41]

In both communities, the trial court had found that the multi-member district left the plaintiffs with less opportunity than other residents to participate in the process and elect representatives of choice. In affirming that holding and the order to replace the multi-member structure with single-member districts, *White v. Regester* emphasized that findings of unconstitutional racial dilution depended not on a single factor but instead on the "totality of circumstances" evaluated through an "intensely local appraisal."[42] Lower courts developed this framework in dilution cases in the years that followed.[43]

The Court, however, jettisoned *White v. Regester*'s approach to racial vote dilution in its 1980 decision, *Mobile v. Bolden*.[44] African American residents in Mobile challenged the city's longstanding reliance on at-large elections to select members of a three-person city commission. At the time the case was brought, African Americans comprised approximately one-third of the city's population, White and Black voters consistently supported different candidates, and no African American candidate had ever won a seat on the commission. Housing remained segregated, Black city employees were concentrated in the lowest city salary classification, and

[37] 412 U.S. 755 (1973).
[38] Id. at 765–766.
[39] Id. at 766.
[40] Id. at 766–767.
[41] Id. at 767–769.
[42] Id. at 769.
[43] See *Zimmer v. McKeithen*, 485 F.2d 1297 (5th Cir. 1973).
[44] 446 U.S. 55 (1980).

"a significant difference and sluggishness" characterized the city's provision of city services to Black residents when compared to those provided to Whites.[45]

Both the federal district and appellate courts used the framework set forth in *White v. Regester* to find that Mobile's at-large system unconstitutionally diluted Black voting strength.[46] A divided Supreme Court reversed, holding that Mobile's at-large system was permissible so long as the plaintiffs were unable to demonstrate that the city adopted it intentionally to dilute Black voting strength. Justice Stewart's controlling plurality opinion dismissed as inconsequential evidence that no African American commissioner had ever been elected to the city commission; that the commission itself was not only unresponsive to African American interests but affirmatively discriminated against Black residents in city employment and the provision of public services; that Alabama had a long history of employing racially discriminatory practices in voting; and that the city's at-large system relied also on majority vote requirements and number posts, devices long recognized to limit minority influence. Unless the plaintiffs could show the city adopted the at-large system for the purpose of discriminating against Black residents, the plaintiffs had no claim.[47]

All the while, Justice Stewart insisted that *Mobile v. Bolden* was charting no new ground. *White v. Regester*, he argued, both required and found evidence of discriminatory intent,[48] while the evidence in *Mobile* simply mandated a different result on intent. For his part, Justice White, who authored *White v. Regester*, agreed that the earlier decision required and found discriminatory intent; Justice White thought, however, that the plaintiff's evidence in *Mobile* made the requisite showing as well.[49] Justice Marshall, joined by Justice Brennan, argued that *White v. Regester* had applied a "discriminatory-effect standard," and that electoral practices may be unconstitutionally dilutive notwithstanding the absence of evidence showing that the practice was adopted with discriminatory intent.[50] He wrote, "Whatever may be the merits of applying motivational analysis to the allocation of constitutionally gratuitous benefits, that approach is completely misplaced where, as here, it is applied to the distribution of a constitutionally protected interest."[51] Justice Marshall added that insofar as intent was to be required, "foreseeability" rather than "but-for" causation should satisfy the requirement.[52]

White v. Regester, like *Whitcomb v. Chavis*, was far more equivocal on the question of intent than the various opinions in *Mobile* suggested. Written before the Court explicitly limited the Equal Protection Clause's proscription to acts of intentional discrimination,[53] Justice White's opinions in both cases include language

[45] *Bolden v. City of Mobile*, 423 F. Supp. 384, 391 (S.D. Ala. 1976).

[46] 446 U.S. at 58.

[47] On remand, the district court struck down Mobile's at-large system based on its conclusion that the City had indeed adopted the contested at-large system with the invidious purpose of diluting Black voting strength. *Bolden v. Mobile*, 542 F. Supp. 1050 (S.D. Ala. 1982).

[48] *Mobile v. Bolden*, 446 U.S. at 69.

[49] Id. at 101 (White, J., dissenting).

[50] Id. at 112 (Marshall, J., dissenting).

[51] Id. at 121.

[52] Id. at 136–137.

[53] *Washington v. Davis*, 426 U.S. 229 (1977).

suggesting that evidence of intent was required, but also some suggesting that a discriminatory effect sufficed. In *Whitcomb*, Justice White noted that the plaintiffs did not allege that the challenged districts "were conceived or operated as purposeful devices" to dilute, but the opinion nevertheless proceeded to analyze and reject their claim.[54] *White v. Regester*, meanwhile, queried whether the challenged multi-member districts were being used "invidiously" but also focused on various factors, sounding in effect, that diminished opportunities for the plaintiffs to participate and elect representatives of choice.[55]

Mobile v. Bolden nevertheless held that an electoral practice is not unconstitutionally dilutive unless a jurisdiction specifically adopted it in order to dilute minority voting strength. The ruling was both controversial and consequential. Attorney Armand Derfner called *Mobile* "devastating" and wrote that it brought "[d]ilution cases . . . to a virtual standstill; existing cases were overturned and dismissed, while plans for new cases were abandoned."[56] Congress noticed, and soon took action in response.[57]

3 ELABORATION: THE STATUTORY FRAMEWORK

Prior to the Voting Rights Act of 1965, state and local officials, primarily in the South, had relied on various mechanisms or "devices" to exclude African Americans from the franchise. Tactics ranging from outright violence to explicit race-based exclusions to "grandfather clauses," literacy tests, and redistricting practices successfully prevented African American voters from participating in local, state, and federal elections.[58] Lawsuits brought to displace these mechanisms were both expensive and time-consuming. They produced some victories but little progress, as state and local officials simply replaced invalidated electoral practices with new discriminatory measures that would require more litigation to displace.

In 1965, Congress enacted the VRA to address this entrenched opposition by targeting the most recalcitrant and discriminatory jurisdictions and subjecting them to intrusive requirements designed to secure African American access to the ballot. Among the VRA's most notable features was the manner in which it targeted jurisdictions in which disenfranchisement was most widespread. Section 4(b) designated jurisdictions "covered" if they presently used a "test or device" to limit registration or voting, and less than half the jurisdiction's eligible citizens were either registered to vote on November 1, 1964 or actually cast ballots in the presidential election that year.[59] Section 4(a) prohibited jurisdictions covered under 4(b) from denying the

[54] 403 U.S. 124, 149 (1971)

[55] 412 U.S. 755, 756, 765-770 (1973).

[56] See Armand Derfner, Vote Dilution and the Voting Rights Act Amendments of 1982, in Minority Vote Dilution 161 (Chandler Davidson, ed. 1989).

[57] See infra notes and accompanying text.

[58] See Quiet Revolution in the South: The Impact of the Voting Rights Act, 1965–1990, at 3 (Chandler Davidson & Bernard Grofman eds., 1994).

[59] 52 U.S.C. §10303 (b).

right to vote to any person who failed to comply with a test device.[60] Section 5 required that covered jurisdictions obtain federal approval, known as "preclearance," before changing any aspect of their voting rules, and specifically, demonstrating that the changes they proposed were not discriminatory in purpose or effect.[61]

As originally crafted and construed, the VRA targeted literacy tests and other barriers to accessing to the ballot. But the statute also took aim at districting practices that limited minority influence. Such practices, which came to be known as "second-generation" devices and more prominent objects for concern beginning in the 1970s, predated the VRA by decades.[62] Such practices stand with the racially exclusive White primary, the literacy test, the poll tax, and other tactics that were used concurrently in the Jim Crow South to ensure that African American citizens lacked meaningful opportunities to participate in the electoral process. As such, the practices grouped as "second generation" were, in fact, part and parcel of the practices the original statute targeted.[63]

The Supreme Court recognized as much in 1969 when it held that the decision to replace single-member electoral districts with an at-large system was a change with respect to voting for which preclearance was required.[64] Chief Justice Warren observed that "[t]he right to vote can be affected by a dilution of voting power as well as by an absolute prohibition on casting a ballot."[65] Justice Harlan objected, claiming that Congress meant for the preclearance obligation to apply only to "tests and devices" and not to districting structures.[66] The Court, however, identified congressional intent to give the Act "the broadest possible scope," and that the statute encompassed a move from districted elections to an at-large system given that it might "nullify" the ability of members of a racial minority "to elect the candidate of their choice just as would prohibiting some of them from voting."[67] Subsequent cases applied the VRA to a host of districting decisions. Congress made no effort to displace these rulings[68] and, indeed, expressly affirmed these rulings when it amended and extended the statute in 1970 and 1975.[69]

By contrast, the Court's 1980 ruling in *Mobile v. Bolden* generated considerable opposition in Congress. *Mobile's* conclusion that racial vote dilution is illegal only when intentionally imposed prompted Congress to amend Section 2 of the VRA to create an explicit "results" based test for discrimination in voting. Section 2 as amended made clear that plaintiffs need not establish discriminatory intent to establish dilution.[70] Instead, Congress codified the more expansive standard from

[60] 52 U.S.C. §10303 (a).

[61] 52 U.S.C. §10304 (a).

[62] See generally Quiet Revolution in the South (Chandler Davidson and Bernard Grofman, eds. 1994).

[63] See Ellen D. Katz, What was Wrong With the Record?, 12 Elec. L. J. 329–331 (September 2013).

[64] *Allen v. Board of Elections*, 393 U.S. 554, 567 (1969).

[65] Id. at 569.

[66] Id. at 585 (Harlan, J., concurring in part and dissenting in part).

[67] Id. at 567, 569.

[68] See, e.g., *Georgia v. United States*, 411 U.S. 526 (1973); *Perkins v. Matthews*, 400 U.S. 379 (1971).

[69] See *Shelby County v. Holder*, 570 U.S. 529, 564 (2013) (Ginsburg, J., dissenting).

[70] 52 U.S.C. §10301(a) (formerly cited as 42 U.S.C. §1973(a)) (providing that "No voting qualification or prerequisite to voting or standard, practice, or procedure shall be imposed or applied by any State or political subdivision in a manner which results in a denial or abridgement of the right of any citizen of the United States to vote on account of race or color, or in contravention of the guarantees set forth in

White v. Regester, and provided that, to prevail under Section 2, plaintiffs must demonstrate that, "based on the totality of circumstances . . . the political processes leading to nomination or election in the State or political subdivision are not equally open to participation by members of a [racial or language minority]." Plaintiffs need to show that members of these protected classes "have less opportunity than other members of the electorate to participate in the political process and to elect representatives of their choice." Relevant to the inquiry is "the extent to which members of a protected class have been elected to office in the State or political subdivision," although the statute is explicit in that it creates no right to proportional representation.[71]

A report of the Senate Judiciary Committee accompanying the 1982 amendments identified several factors to guide courts when assessing whether a challenged practice violates Section 2. Derived from *White v. Regester* and a subsequent appellate decision,[72] these so-called "Senate Factors" include the extent to which the jurisdiction (1) has a history of discrimination in voting; (2) has elections that are marked by racially polarized voting; (3) relies on majoritarian electoral devices that enhance opportunities for discrimination against minority groups; (4) denies minority voters access to candidate slating; (5) includes minority group members who suffer from the effects of discrimination in education, employment, and health in ways that hinder their ability to participate in the political process; (6) has experienced racial appeals during campaigns; (7) elected members of the minority group to office. The Senate Report added that courts might also assess any lack of responsiveness to minority interests by elected officials and the extent to which the policy supporting the challenged practice is tenuous.[73]

With the 1982 amendments, Congress rejected a bright-line rule for Section 2 liability, opting instead for what *White v. Regester* labelled as an "intensely local appraisal" of the challenged "in the light of past and present reality, political and otherwise."[74] The result is an intentionally complex inquiry, and one that quickly led to disagreements among federal courts called upon to adjudicate whether a particular electoral rule results in a denial or abridgement of the right to vote under Section 2.

The Supreme Court attempted to simplify the inquiry in *Thornburg v. Gingles.*[75] Addressing "a claim of vote dilution through submergence in multimember districts," Justice Brennan's controlling opinion acknowledged that while "many or all of the factors listed in the Senate Report may be relevant" to a Section 2 injury, a multimember district typically will not violate Section 2 unless three "preconditions" are met. Specifically, plaintiffs must be able to demonstrate that the minority group is "sufficiently large and geographically compact to constitute a majority in a single member district" and that both minority and White voters vote cohesively and in opposition to one another—i.e., evidence of racially polarized voting.[76]

section 10303(f)(2) of this title, as provided in subsection (b).").

[71] 52 U.S.C. §10301(b)(formerly cited as 42 U.S.C. §1973(b)).

[72] See *Zimmer v. McKeithen,* 485 F2d 1297, 1304 (5th Cir. 1973).

[73] S. Rep. No. 417, 97th Cong., 2d Sess. 1, 28 (1982).

[74] See *White v. Regester,* 412 U.S. 755, 769–770 (1973).

[75] 478 U.S. 30 (1986).

[76] Id. at 48–49.

Justice O'Connor objected both to the distillation of "preconditions" and to the substance of preconditions selected. She noted "an inherent tension between what Congress wished to do and what it wished to avoid" when it amended Section 2; namely, it wanted to prohibit racial vote dilution without creating a right to proportional representation.[77] The problem, Justice O'Connor wrote, is that "any theory of vote dilution must necessarily rely. . . on a measure of minority voting strength that makes some reference to the proportion between the minority group and the electorate at large."[78] In Justice O'Connor's view, the Court's preconditions exacerbated this tension by measuring minority voting strength "solely in terms of the minority group's ability to elect candidates it prefers" when concentrated as a voting majority in a single-member district. This measure, Justice O'Connor argued, assumes that undiluted minority voting strength "means the maximum feasible minority strength." As such, she argued, it necessarily mandates liability whenever a challenged electoral rule fails to yield proportional representation.[79]

A majority of the Court, however, was not persuaded. It concluded that the *Gingles factors*, as they quickly became known, both comported with congressional intent and provided useful guidance for the evaluation of Section 2 claims. In short order, lower courts were applying the Gingles factors to both single-member and multi-member districts challenged under Section 2. In both contexts, plaintiffs who successfully established the Gingles factors typically prevailed, while those unable to satisfy one or more of them did not.[80]

In the years following *Gingles*, districting plans began including a greater proportion of districts in which members of a racial minority constituted a majority of a district's electorate. *Gingles* itself did not mandate the formation of these "majority-minority" districts,[81] but the framework it established nevertheless encouraged their creation both to remedy and to avoid Section 2 violations.[82] Such districts provide a meaningful remedy for dilution insofar as they allow for minority influence when voting is racially polarized, and, as several scholars have argued, may help to erode racial polarization among voters.[83] Voters in these districts largely, albeit not exclusively, elected minority candidates to office. By the mid-1990s, more minority representatives were serving on school boards, city councils, state legislatures, and

[77] Id. at 84 (O'Connor, J., concurring in the judgment).

[78] Id. at 89–91.

[79] Id.

[80] See Ellen Katz et al., Documenting Discrimination in Voting: Judicial Findings Under Section 2 of the Voting Rights Act Since 1982. Final Report of the Voting Rights Initiative, University of Michigan Law School, 39 U. Mich. J.L. Reform 643, 660 (2006).

[81] See *Bone Shirt v. Hazeltine*, 461 F.3d 1011, 1019 (8th Cir. 2006); see also Ellen D. Katz, Reviving the Right to Vote, 68 Ohio St. L.J. 1163, 1165, 1178–1179 (2007). *Gingles* also prompted jurisdictions subject to the VRA's preclearance requirement to include more majority-minority districts in proposed districting plans than they had done previously, in part because compliance with the preclearance requirement was understood to require compliance with Section 2 (at least until the Court ruled otherwise in 1997). See *Reno v. Bossier Parish Sch. Bd.*, 520 U.S. 471, 476 (1997); see also 28 C.F.R. §51.55(b)(2) (1996).

[82] See generally Alexander Keyssar, The Right to Vote: The Contested History of Democracy in the United States 238–239 (2009); Reviving the Right to Vote, supra note, at 1178–1179.

[83] See, e.g., Dale Ho, Minority Vote Dilution in the Age of Obama, 47 U. Rich. L. Rev. 1041, 1070–75 (2013); Christopher S. Elmendorf, Making Sense of Section 2: Of Biased Votes, Unconstitutional Elections, and Common Law Statutes, 160 U. Pa. L. Rev. 377, 395–400 (2012); Michael S. Kang, Race and Democratic Contestation, 117 Yale L.J. 734, 773–88 (2008); David T. Canon, Race, Redistricting, and Representation: The Unintended Consequences of Black Majority Districts 204–205, 261 (1999).

in the U.S. House of Representatives than at any time since Reconstruction.[84]

4 UNEASINESS: RECASTING FRAMEWORKS IN THE 1990S

As the number of majority-minority districts proliferated in the 1990s, critics of the VRA grew increasingly uneasy about the race consciousness inherent in their creation and both the type of political participation and quality of representation critics said the districts fostered. These concerns shaped a number of Supreme Court decisions that restricted reliance on the majority-minority district.[85]

Most directly, the Court read the VRA narrowly to limit the instances in which liability might arise and a new majority-minority district might be required or adopted. In the 1994 case *Holder v. Hall*,[86] for instance, African American voters in Bleckley County, Georgia, had challenged the County's reliance on a single-member county commission, arguing that a five-member commission elected from single-member districts—a structure widely used throughout the State—would provide Black voters with the ability to elect a preferred candidate to the Commission.[87] Justice Kennedy's controlling opinion in *Holder* rejected this claim, holding that the size of a legislative body was not subject to challenge under Section 2 of the VRA.[88] Justice Thomas, joined by Justice Scalia, agreed, but would have gone much further and scrapped Section 2's application to vote dilution claims entirely. Justice Thomas equated the creation of majority-minority districts, fostered by Section 2 and the Court's construction of it, to "an enterprise of segregating the races into political homelands that amounts, in truth, to nothing short of a system of 'political apartheid.'"[89]

Decided the same day as *Holder*, *Johnson v. DeGrandy*[90] adopted a more sympathetic stance toward the majority-minority district, but still urged jurisdictions to resist drawing them unless "necessary." Justice Souter's opinion for the Court rejected the plaintiffs' argument that a Florida districting plan needed to include more majority-minority districts given that the challenged plan contained "majority-minority districts in substantial proportion to the minority's share of voting-age population,"[91] Justice Souter distinguished such proportionality from proportional representation, which linked the success of minority candidates (rather than majority-minority districts) to the population, and stated, "One may suspect vote dilution from political famine, but one is not entitled to suspect (much less

[84] See Richard H. Pildes, The Politics of Race: Quiet Revolution in the South, 108 Harv. L. Rev. 1359, 1364–1365 & n.31 (1995) (book review).
[85] See Ellen D. Katz, Enforcing the Fifteenth Amendment, in Oxford Handbook on the U.S. Constitution (Tushnet, Levinson & Graber eds., 2015).
[86] 512 U.S. 874 (1994).
[87] Id. at 878 (plurality opinion of Kennedy, J.).
[88] Id. at 885.
[89] Id. at 90 (Thomas, J., concurring in the judgment) (quoting *Shaw v. Reno*, 509 U.S. 630, 647 (1994)).
[90] 512 U.S. 997 (1994).
[91] Id. at 1013.

infer) dilution from mere failure to guarantee a political feast."[92]

Justice Souter went on to warn against drawing majority-minority districts in circumstances that do not absolutely require them. Such districts, he explained, have "virtues ... as remedial devices," and are something necessary, given "society's racial and ethnic cleavages, . . . to ensure equal political and electoral opportunity," but they nevertheless "rely on a quintessentially race-conscious calculus aptly described as the 'politics of second best.'"[93] As such, majority-minority districts should not be used in communities where meaningful cross-racial coalitions are possible, even when such coalitions elect candidates who "may not represent perfection to every minority voter." No voter is "immune from the obligation to pull, haul, and trade to find common political ground."[94]

Decisions like *Holder* and *DeGrandy* limited the creation of new majority-minority districts by construing Section 2 of the VRA narrowly. The Supreme Court further curbed the proliferation of such districts with a series of constitutional rulings that significantly limited the circumstances in which such districts could be drawn. Specifically, in *Shaw v. Reno*[95] and its progeny,[96] the Court recognized a new "analytically distinct" injury under the Equal Protection Clause that arose when jurisdictions created oddly shaped majority-minority districts that were not absolutely required by the VRA.[97] By design, this new constitutional injury discouraged jurisdictions from drawing majority-minority districts prophylactically to avoid liability under the VRA.[98]

Shaw itself arose in North Carolina, where, at the time, African American residents comprised twenty percent of the State's population but a majority in only five of its 100 counties. A covered jurisdiction subject to the VRA's preclearance obligation, North Carolina needed federal approval in order to implement a new redistricting plan after the 1990 census. The Attorney General objected to the first plan North Carolina proposed, noting that only one of its 12 proposed congressional districts would be majority-minority. North Carolina's second plan created an additional majority-minority district—District 12—which, in contrast to the one the Justice Department recommended, was highly irregular in shape, crossing multiple town and county lines as it wound along the I-85 corridor. White voters challenged the plan as an unconstitutional racial gerrymander.

Shaw held that these voters stated a cognizable claim. Emphasizing that reapportionment is "one area in which appearances do matter,"[99] Justice O'Connor's majority opinion stated that, "a reapportionment plan may be so highly irregular

[92] Id. at 1014 & n. 11.

[93] Id. at 1020 (quoting B. Grofman, L. Handley, & R. Niemi, Minority Representation and the Quest for Voting Equality 136 (1992)).

[94] Id. at 1020.

[95] 509 U.S. 630 (1993).

[96] See *Easley v. Cromartie*, 532 U.S. 234 (2001); *Bush v. Vera*, 517 U.S. 952 (1996), *Shaw v. Hunt*, 517 U.S. 899 (1996); *Miller v. Johnson*, 515 U.S. 900 (1995).

[97] *Shaw v. Reno*, 509 U.S. at 653.

[98] See, e.g., *Holder v. Hall*, at 905 (Thomas, J., concurring in the judgment) (characterizing *Shaw* and the cases that would follow as an "attempt[] to undo, or at least to minimize, the damage wrought by the system we created.") (citing, inter alia, *Shaw v. Reno*, 509 U.S. 630 (1993)).

[99] *Shaw v. Reno*, 509 U.S. at 647.

Figure 1: Maps from the *Bush v. Vera* decisions. On the top, two unconstitutional districts from the majority opinion. On the bottom, two constitutional districts from Justice Stevens' dissent.

that, on its face, it rationally cannot be understood as anything other than an effort to 'segregat[e] ... voters' on the basis of race."[100] Justice O'Connor thus likened what North Carolina did to create District 12 to Alabama's expulsion of African American residents from Tuskegee four decades earlier. The opinion states that aggregating individuals "who belong to the same race, but who are otherwise widely separated by geographical and political boundaries, and who may have little in common with one another but the color of their skin, bears an uncomfortable resemblance to political apartheid."[101]

In the cases that followed *Shaw*, the Court came to understand the injury at issue as an "expressive" one, i.e., linked to the message a State sends to its citizens when it uses race as the predominant factor to draw district lines without a sufficiently

[100] Id. at 646–647 (quoting *Gomillion v. Lightfoot*, 364 U.S. at 341)).

[101] Id. at 647.

compelling reason for doing so.[102] *Shaw*'s progeny made clear that "bizarre" shape was not a prerequisite to a *Shaw* claim, but instead "circumstantial evidence that race for its own sake, and not other districting principles, was the legislature's dominant and controlling rationale in drawing its district lines."[103] As distilled, the *Shaw* test posited that race may permissibly predominate in the drawing of district lines only when such predominance is necessary and narrowly tailored to achieve a compelling governmental interest.[104] *Shaw*'s progeny assumed that compliance with the VRA was a compelling interest but held that the VRA did not mandate the race based moves evident in the plans challenged in those cases.[105]

The *Shaw* cases presented a serious concern during the round of redistricting following the 2000 census as jurisdictions confronted the confounding challenge of crafting districting plans that met both their obligations under Sections 2 and 5 of the VRA and the constitutional restrictions the *Shaw* cases imposed.[106] The Supreme Court's 2001 decision in *Easley v. Cromartie*[107] initially seemed to make the task even more complex. Ostensibly applying clear error review, the Court reversed a lower court ruling that race predominated over traditional districting principles when North Carolina redrew its congressional districts to comply with *Shaw*'s requirements. Justice Breyer's majority opinion examined the trial record in excruciating detail and found insufficient evidence to support the conclusion that "race, rather than politics, predominantly accounts" for the challenged districting lines.[108]

Cromartie was widely read to compound the difficulties jurisdictions faced in complying with *Shaw* and its progeny. The decision portended that amorphous, uncertain, fact-intensive judicial scrutiny would be the norm going forward. Surveying precedent in 2002, Pam Karlan observed that *Cromartie* itself "cannot be explained in any sort of principled terms that provide guidance for future cases."[109] And yet, the widely predicted flood of *Shaw* litigation after 2001 never materialized, due, in part, to *Cromartie* itself. The decision implied, and accordingly, advised, that saying "party" instead of "race" would disprove a racial motivation,[110] as discussed in the previous chapter. Articulating the goal of protecting incumbents performed a similar function. Officials involved in the districting process understood this as advice, and their resulting professions of partisan motivation immunized plans that might otherwise have been subjected to scrutiny under the *Shaw* doctrine.

At the same time, jurisdictions controlled by Democrats began a concerted effort to disperse rather than concentrate minority voters in majority-minority districts.

[102] See Richard H. Pildes and Richard G. Niemi, Expressive Harms, "Bizarre Districts," and Voting Rights: Evaluating Election–District Appearances after *Shaw v. Reno*, 92 Mich. L. Rev. 483, 506–507 (1993).

[103] *Miller v. Johnson*, 515 U.S. 900, 913 (1995).

[104] *Bush v. Vera*, 517 U.S. 952, 957 (1996); *Shaw v. Hunt*, 517 U.S. 899, 902 (1996).

[105] *Bush v. Vera*, 517 U.S. at 982; *Shaw v. Hunt*, 517 U.S. at 917.

[106] See Pamela S. Karlan, Exit Strategies in Constitutional Law: Lessons for Getting the Least Dangerous Branch Out of the Political Thicket, 82 B.U. L. Rev. 667 (2002).

[107] 532 U.S. 234 (2001).

[108] Id. at 257.

[109] Karlan, supra note, at 677.

[110] Richard H. Pildes, The Supreme Court, 2003 Term-Foreword: The Constitutionalization of Democratic Politics, 118 Harv. L. Rev. 28, 67–69 (2004).

Declines in racial bloc voting in certain parts of the country allowed for the formation of viable cross-racial coalitions, a development that meant minority voters need not always comprise a majority of a district's electorate to elect representatives of choice.[111] Democratic legislators, moreover, saw how the concentration of minority voters, particularly of African American voters in the South, helped make legislatures more Republican overall.[112] As a result, when Democrats found themselves in control of the districting process after 2001, they sought systematically to reduce the percentage of minority voters in majority-minority districts, with the hope that strategically dispersing their most reliable voters would enable Democrats to preserve and even gain legislative seats.[113]

In 2001, for instance, Democratic control led Georgia to adopt a redistricting plan for its State Senate that substantially reduced the African American voting-age population, or BVAP, in a number of majority-minority districts. A highly politicized dispute followed, in which Georgia Republicans, the U.S. Department of Justice, and the ACLU argued that the plan discriminated against African American voters in Georgia by making it more difficult for them to elect representatives of choice. Democratic members of the Georgia legislature, many of whom were African American, countered that the plan was valid under the VRA, because, they claimed, it provided Black voters more influence in the political process than would have a plan that concentrated these voters in fewer districts.[114]

In *Georgia v. Ashcroft*,[115] the Supreme Court sided with Georgia's Democrats, holding that the challenged plan satisfied Section 5's mandate to avoid changes that worsened participatory opportunities for minority voters. Justice O'Connor's majority opinion emphasized that, under Section 5, Georgia had discretion to choose what type of districting plan best serves minority voters. Georgia, she explained that, chose not to concentrate African American voters in a few "safe" majority-minority districts, but instead to disperse Black voters among so-called "coalition" districts "in which it is likely—although perhaps not quite as likely" that they will elect candidates of their choice. Justice O'Connor noted that the plan also placed large numbers of minority voters in districts in which they would be unable to elect representatives of choice but where "'candidates elected without decisive minority support would be willing to take the minority's interests into account.'"[116] Justice O'Connor's opinion recognized Georgia's power to disperse minority voters in this way. "The State may choose, consistent with §5, that it is better to risk having fewer minority representatives in order to achieve greater overall representation of a minority group by increasing the number of representatives sympathetic to the interests of minority voters." [117]

Justice Souter's dissent pointed out that Georgia never substantiated its claim

[111] See, e.g., *Page v. Bartels*, 144 F. Supp. 2d 346 (D.N.J. 2001).

[112] See, e.g., Grofman, Handley & Lublin, Drawing Effective Minority Districts: A Conceptual Framework and Some Empirical Evidence, 79 N.C.L. Rev. 1383 (2001).

[113] David Lublin, The Paradox of Representation 99 (1997).

[114] Samuel Issacharoff, Is Section 5 of the Voting Rights Act a Victim of its Own Success? 104 Colum. L. Rev. 1710, 1716–1717 (2004).

[115] 539 U.S. 461 (2003).

[116] Id. at 481 (quoting *Thornburg v. Gingles*, 478 U.S. at 100 (O'Connor, J., concurring).

[117] Id. at 483.

that Black voters would be able to elect representatives of choice in the identified coalition districts, and that, in fact, the uncontested record indicated that they would not.[118] Justice Souter also questioned whether African American voters enjoyed meaningful or even measurable influence in districts in which the elected representative won without their support, and how various suggested measures of influence could be assessed or meaningfully compared.[119] More broadly, the dissent argued that the majority's ruling gutted Section 5, substituting what had been a critical review of state action that diminishes a minority group's ability to elect representatives of choice with an approach that simply deferred to state judgments as to optimal arrangements. If the Court "allows the State's burden to be satisfied on the pretense that unquantifiable influence can be equated with majority-minority power, §5 will simply drop out as a safeguard against the 'unremitting and ingenious defiance of the Constitution' that required the procedure of preclearance in the first place."[120]

Congress responded to *Georgia v. Ashcroft* in the Voting Rights Act Reauthorization and Amendments Act of 2006. It amended the preclearance standard to prohibit a covered jurisdiction from adopting an electoral change that "has the purpose of or will have the effect of diminishing the ability of [the minority group] to elect their preferred candidates of choice."[121] In so doing, Congress embraced the position advanced by the dissent in *Georgia v. Ashcroft* that the Section 5 retrogression standard should bar electoral changes that deprive or diminish the ability of minority voters to elect candidates of choice.[122] Controversial from its enactment, this amendment had far less impact than intended given other dramatic changes the Court would soon make to the VRA.

5 HOSTILITY: RACE, REDISTRICTING, AND THE ROBERTS COURT

The rules governing race and redistricting shifted dramatically after John Roberts became Chief Justice in 2005 and Justice Samuel Alito joined the Court in 2006. Within a just a few years, the Roberts Court markedly narrowed the reach of Section 2 of the VRA, and effectively eliminated Section 5 as a constraint. More recently, it has reinvigorated the *Shaw* doctrine as a foundational restriction on the use of race in the redistricting process and thereby limited yet further permissible applications of the VRA.

The Roberts Court decided its first redistricting case on June 28, 2006. *LULAC v. Perry*[123] involved a multi-party challenge to the congressional redistricting plan Texas adopted in 2003. That plan supplanted a court-drawn plan adopted two years

[118] Id. at 492–493 (Souter, J., dissenting)

[119] Id. at 494–496.

[120] Id. at 497 (quoting *South Carolina v. Katzenbach*, 383 U.S. 301, 309 (1966)).

[121] 52 U.S.C. §10304(b); see also §10304(d) (the "purpose of subsection (b) ... is to protect the ability of such citizens to elect their preferred candidates of choice").

[122] *Alabama Legislative Black Caucus v. Alabama*, 135 S.Ct. 1257, 1273 (2015).

[123] 548 U.S. 399 (2006).

earlier after the Texas Legislature had been unable to agree to one. Republicans gained control of the Texas House of Representatives in 2002 and with that control, acquired the ability to adopt a new districting plan without Democratic support. Designed to maximize Republican influence, the 2003 plan promised to (and, in fact, did) replace six Democratic members of Texas's congressional delegation with Republican representatives. Texas Democrats challenged the plan as the blatant partisan gerrymander it was.[124]

The Supreme Court was unable to identify a constitutional problem with either the partisanship that propelled the Texas plan or the partisan effects it yielded.[125] A majority of the Court nevertheless concluded that a portion of that gerrymander diluted minority voting strength in the southwest portion of the State. More specifically, the Court held that Texas violated Section 2 of the VRA when it displaced 100,000 Latino residents from a congressional district in Laredo to protect the Republican incumbent they did not support.[126] Justice Kennedy's majority opinion observed that, "the State chose to break apart a Latino opportunity district to protect the incumbent congressman from the growing dissatisfaction of the cohesive and politically active Latino community in the district."[127] As Justice Kennedy explained, Texas "took away the Latinos' opportunity because Latinos were about to exercise it," and thereby violated Section 2 of the VRA.[128]

Chief Justice Roberts, joined by Justice Alito, argued in dissent that the State had provided adequate representation to Latino voters in southwest Texas.[129] The new Justices did not join a separate dissent by Justice Scalia and Justice Thomas, which adhered to their view that Section 2 does not prohibit racial vote dilution and hence could not constrain actions of the sort Texas took in Laredo.[130] More limited in scope, the Chief Justice's dissent nevertheless voiced a deep aversion to the VRA and the type of race-based decision-making it was understood to mandate. Chief Justice Roberts wrote, "It is a sordid business, this divvying us up by race."[131]

This sentiment propelled subsequent decisions that dramatically limited the VRA. Decided in 2009, *Bartlett v. Strickland*[132] rejected the claim that Section 2 protects the ability of minority voters to form cross-racial coalitions to elect representatives of their choice. Resolving a question that had been left open since *Gingles*,[133] *Bartlett* held that Section 2 offers no protection to minority voters who are too few in number to comprise the majority of a single-member district. As such, Section 2 neither required district lines that allowed for the formation of cross-

[124] Id. at 412–413.
[125] Id. at 416–423.
[126] Id. at 442. See generally Ellen D. Katz, Reviving the Right to Vote, 68 Ohio St. L.J. 1163 (2007).
[127] LULAC, 548 U.S. at 441.
[128] Id. at 440.
[129] Id. at 502–503 (Roberts, C.J, concurring in part and dissenting in part).
[130] Id. at 512 (concurring in the judgment and dissenting in part). See also supra notes and accompanying text.
[131] 548 U.S. at 511 (2006) (Roberts, C.J., concurring in part and dissenting in part).
[132] 556 U.S. 1 (2009).
[133] See *Thornburg v. Gingles*, 478 U.S. 30, 46 n. 12 (1986). Cf. *LULAC v. Perry*, 548 U.S. at 443–447 (holding that Section 2 does not prohibit Texas's dismantling of a district in Fort Worth in which African American voters comprised less than half the district's electorate because the pervasive lack of competition meant that meaningful voter preferences could neither be expressed nor ascertained).

racial coalitions, nor did it preclude lines that actively stymied them.

Bartlett rejected a broad reading of Section 2 that would have made more conduct subject to challenge under the statute and thus would have increased the junctures in which districting officials needed to consider race to comply with the regime. The narrow reading, however, cut off an application of the VRA that promised to encourage the type of political participation the Court has long claimed it wants to promote—namely, the type that yields cross-racial coalitions. Justice Souter highlighted this point in dissent, arguing that Section 2 should be read to promote drawing a "crossover district," which he described as "superior to a majority-minority district precisely because it requires polarized factions to break out of the mold and form the coalitions that discourage racial divisions."[134] Invoking *DeGrandy's* observation that minority-opportunity districts implicate a "quintessentially race-conscious calculus," the dissent argued that Section 2's application in this context would "moderat[e] reliance on race as an exclusive determinant in districting decisions."[135] The *Bartlett* majority, however, was unpersuaded, and unwilling to read Section 2 in a manner that would expand its reach.

This aversion to the VRA found even stronger expression when the Court confronted a constitutional challenge to the VRA's preclearance regime in *Shelby County v. Holder*. Congress enacted Sections 4 and 5 of the VRA as time-limited measures in 1965, but repeatedly extended them. The last reauthorization was in 2006, when Congress extended preclearance for twenty-five more years. The 2006 amendments reversed the VRA rulings adopted in *Georgia v. Ashcroft* and another decision,[136] but otherwise left the statute to operate as it long had. In particular, Congress neither added nor removed jurisdictions from coverage and instead maintained the statute's pre-existing geographic reach by preserving the Section 4(b) formula it last altered in 1975.[137]

A constitutional challenge to the 2006 amendments was filed within days of its enactment.[138] The Supreme Court had previously upheld broad congressional power to enact the preclearance regime,[139] but a series of decisions in the 1990s adopted a more restrictive view of Congress's power and thus raised significant questions about the validity of the preclearance regime going forward.[140] Supporters of reauthorization argued that the 15,000-page record Congress produced after 21 hearings provided ample evidence showing both that preclearance continued to play a critical role in covered jurisdictions, and that the discrimination that

---

[134] 556 U.S. at 35 (Souter, J, dissenting). See also Richard Pildes, Is Voting Rights Law Now At War With Itself? Social Science and Voting Rights in the 2000s, 80 N.C. L. Rev. 1517, 1547–1548 ("Coalitional districts would seem to encourage and require a kind of integrative, cross-racial political alliance that might be thought consistent with, even the very ideal of, both the VRA and the U.S. Constitution").

[135] Bartlett, at 34 (quoting *Johnson v. DeGrandy*, 512 U.S. at 1020).

[136] See Fannie Lou Hamer, Rosa Parks, and Coretta Scott King Voting Rights Act Reauthorization and Amendments Act of 2006, Pub. L. No. 109-246, §2(b)(6), 120 Stat. 577, 578 (2006) (overturning *Georgia v. Ashcroft* and *Reno v. Bossier Parish Sch. Bd*).

[137] *Shelby County, Ala. v. Holder*, 570 U.S. 529, 539 (2013).

[138] Id.

[139] See supra notes and accompanying text.

[140] See *City of Boerne v. Flores*, 521 U.S. 507 (1997). See also Ellen D. Katz, Justice Ginsburg's Umbrella, in A Nation of Widening Opportunities? The Civil Rights Act at Fifty (Samuel Bagenstos and Ellen D. Katz, eds., 2015), at 268.

persisted in these regions remained distinct. Opponents countered that minority voters no longer faced distinct obstacles in places subject to Section 5 and thus that the preclearance requirement was no longer justified.[141]

In *Shelby County*[142], a divided Court voted to scrap the statute's coverage formula and thereby render the preclearance regime inoperative. Writing for the majority, Chief Justice Roberts posited that the Section 4(b) formula once "made sense" but was no longer justified by "current conditions."[143] Observing that "things have changed dramatically," the Chief Justice wrote that the tests and devices that had triggered coverage had long been outlawed, voter turnout and registration rates in covered jurisdictions had risen dramatically, minority officials had been elected to office, and overt discrimination was no longer pervasive. These improved conditions meant that a formula that had been "rational in both practice and theory" in 1965 was no longer responsive to "current political conditions."[144]

Justice Ginsburg's dissent chided the majority for making "no genuine attempt to engage with the massive legislative record Congress assembled" to support reauthorization.[145] The dissent argued that the Court focused too narrowly on registration and turnout data, and ignored critical record evidence, including the scores of documented instances of intentional racial discrimination in voting; hundreds of Department of Justice (DOJ) objections interposed to proposed electoral changes; hundreds more that were withdrawn after the DOJ began investigating; the scale of racially polarized voting in covered jurisdictions; and comparative evidence that showed "that the coverage formula continues to identify the jurisdictions of greatest concern."[146] This record, Justice Ginsburg wrote, amply supported Congress's determination that preclearance was effective and should be retained. The dissent stated: "Throwing out preclearance when it has worked and is continuing to work to stop discriminatory changes is like throwing away your umbrella in a rainstorm because you are not getting wet."[147] Justice Ginsburg closed: "[h]ubris is a fit word for [the majority's] demolition of the VRA."[148]

Disdain is another word Justice Ginsburg might have invoked. Animating *Shelby County*'s willingness to discard the preclearance regime was the majority's belief that the disputed provisions were not only obsolete but also the source of affirmative harm. True, the Justices comprising the *Shelby County* majority likely thought that the conditions on the ground were less dire than Justice Ginsburg described and the record documented.[149] And yet, the *Shelby County* ruling suggests that had these Justices shared the dissent's view of "current conditions," they would have still scrapped the preclearance regime because they thought that it made matters worse. Justice Scalia made this point when, at oral argument, he characterized

[141] See Justice Ginsburg's Umbrella, supra, at 266.

[142] *Shelby County v. Holder*, 570 U.S. 529 (2013).

[143] Id. at 553, 554.

[144] Id. at 546–548, 552 (quoting Katzenbach, 383 U.S. at 308; NAMUDNO, 557 U.S. at 204).

[145] Id. at 580 (Ginsburg, J., dissenting).

[146] Id. at 570–579. See also Ellen D. Katz, Mission Accomplished? 117 Yale L.J. Pocket Part 142, 145 (2007).

[147] 570 U.S.. at 590 (Ginsburg, J., dissenting).

[148] Id. at 587.

[149] Justice Ginsburg's Umbrella, supra note, at 267.

preclearance and the VRA more generally as "a racial entitlement." Similarly, Chief Justice Roberts' majority opinion described the 2006 amendments as requiring measures that "favored" minority voters through the creation of majority-minority districts.[150] These arguments dismissed the primary mechanism used to elect minority-preferred candidates as objectionable "favored" treatment rather than as a means of implementing the VRA's longstanding commitment to equality of political opportunity.[151] Put differently, *Shelby County* bluntly resolved the Court's longstanding discomfort with the creation of majority-minority districts under the VRA by eliminating a key mechanism that fostered their creation.[152]

Previously covered jurisdictions responded quickly to *Shelby County*, and imposed new electoral rules that preclearance had, or would have, blocked.[153] These new rules included districting plans that reduced or restructured majority-minority districts, and a deluge of measures wholly unrelated to districting, including new stringent voter identification requirements,[154] which will be discussed further in Chapter 23. Legal challenges followed, raising claims under Section 2 of the VRA and the Constitution itself. Plaintiffs prevailed in some of these actions, but lost a good deal more.

Meanwhile, plaintiffs challenged several post-2010 redistricting plans as racial gerrymanders. Invoking *Shaw* and its progeny (which had been largely dormant for more than decade), these plaintiffs objected to what they claimed was an excessive concentration of minority voters in specific districts. State defendants countered that the challenged redistricting lines either were not race-based or were necessary to comply with Sections 2 and 5 of the VRA. These cases resembled the *Shaw* claims of the 1990s, only now the state officials were Republicans, and the plaintiffs were primarily African American Democrats.[155]

The Supreme Court upheld one challenged district in Virginia as necessary to comply with the VRA,[156] but otherwise held that the statute did not mandate the race-based population percentages used in the challenged plans.[157] The Court also refined the *Shaw* doctrine to make racial gerrymanders easier to establish. It held unanimously, for instance, that a plan's compliance with traditional districting principles does not neutralize a racially predominant motive.[158] It also expanded the ways in which plaintiffs may establish that race, rather than partisanship, predominantly shaped district lines.[159] Both moves allowed *Shaw* plaintiffs to

[150] 570 U.S. at 549.

[151] Ellen D. Katz, A Cure Worse Than the Disease?, 123 Yale L.J. Online 117 (2013), http://www.yalelawjournal.org/forum/a-cure-worse-than-the-disease.

[152] Charles & Luis Fuentes-Rohwer, The Voting Rights Act in Winter: The Death of a Superstatute, 100 Iowa L. Rev. 1389, 1393 (2015).

[153] See, e.g., Michael Cooper, After Ruling, States Rush to Enact Voting Laws, N.Y. Times (July 5, 2013), at http://www.nytimes.com/2013/07/06/us/politics/after-Supreme-Court-ruling-states-rush-to-enact-voting-laws.html.

[154] See generally Ellen D. Katz, Section 2 After Section 5: Voting's Race to the Bottom, 51 Wm & Mary L. Rev. 1961 (2018).

[155] See *Cooper v. Harris*, 137 S. Ct. 1455 (2017); *Bethune-Hill v. Va. State Bd. of Elections*, 137 S.Ct. 788 (2017); *Alabama Legislative Black Caucus v. Alabama*, 135 S.Ct. 1257 (2015).

[156] See *Bethune-Hill*, 137 S.Ct. at 800–802.

[157] See *Cooper*, 137 S. Ct. at 1460; *Alabama Legislative Black Caucus*, 135 S.Ct. at 1273.

[158] See *Bethune-Hill*, 137 S.Ct. at 796–797.

[159] See *Cooper*, 137 U.S. at 1481 (disavowing language in *Cromartie* that suggested plaintiffs must

prevail more easily.

This resurrection of *Shaw* is a notable development, but whether it is a productive one remains to be seen. Consider *Cooper v. Harris*, which held that a disputed majority-minority district in North Carolina was a racial gerrymander under *Shaw*. Justice Kagan's majority opinion observed that a sufficient number of White voters supported Black-preferred candidates so as to make a majority-minority district unnecessary. *Cooper* accordingly found that Section 2 of the VRA did not require the district to be majority-minority and hence could not be used to justify the race-based moves evident in the district's creation.[160]

The editors of a leading casebook predict that *Cooper* will help to foster cross-racial political coalitions by barring the "mechanical creation" of unnecessary majority-minority districts. The decision, they write, "opens up more space for the creation of coalitional or cross-over districts in which minority and White political coalitions unite behind the same candidates."[161] In a narrow sense, this is correct. *Cooper* and the other recent *Shaw* cases provide plaintiffs with a mechanism to displace the undue packing of minority voters and thus provide jurisdictions that are so inclined with "space" to craft electoral districts that foster cross-racial coalitions.

These recent *Shaw* cases, however, do not require jurisdictions to do so. They invalidated or called into question "packed" districts that the Court thought relied unduly on race. New districting lines that do not predominantly rely on race will wholly remedy that identified injury, regardless of whether the new lines aggregate previously packed minority voters in numbers sufficient to exert meaningful influence. Put differently, the recent *Shaw* cases do nothing to prevent States from dispersing "excess" minority voters among numerous districts in which they will be too few in number to influence, much less control, electoral outcomes. The cases make clear that minority voters may be so dispersed either because the VRA does not independently require their aggregation in a majority-minority district, or because the voters at issue are not needed to create such a required district.

Back in the 1990s, *Shaw* and its progeny self-consciously articulated an analytically distinct constitutional injury that differed both from dilution and conventional intentional discrimination. The harm came to be understood as an expressive one, inflicted on and experienced by individuals when state action relies too heavily on race without sufficient justification. This injury never hinged on electoral outcomes, and arose regardless of whether elected representatives behaved as though beholden or indifferent to particular district residents.[162] The new *Shaw* cases continue to understand the injury in these terms. They do so, however, at a time when the VRA no longer provides a vigorous check on redistricting practices. Section 5 is inoperative and Section 2 inapplicable to minority communities that are either too small to constitute a majority in a single-member district or enjoy minimal support from White neighbors.[163]

present alternate maps in order to establish that race, rather than partisanship, motivated a challenged plan).

[160] See id. at 1471–1472.

[161] See Samuel Issacharoff et al., 2018 Supplement to the Law of Democracy 96 (5th ed. 2018).

[162] See supra notes and accompanying text.

[163] See supra notes and accompanying text.

Indeed, the new *Shaw* cases narrow the VRA even further by subjecting the deliberate creation of majority-minority districts, including ones that otherwise comport with traditional districting principles, to rigorous judicial scrutiny. *Cooper* found that race "furnished the predominant rationale" for a disputed district given that "[u]ncontested evidence . . . shows that the State's mapmakers . . . purposefully established a racial target: African–Americans should make up no less than a majority of the voting-age population."[164] The Court thus dismissed efforts to comply with the VRA as involving the pursuit of "a racial target." Whether or not this characterization and the resurrection of the *Shaw* doctrine more generally portend the demise of the VRA in its entirety,[165] the new *Shaw* cases expose just how minimally the VRA presently limits the redistricting process. Whether it will provide any meaningful limits in the years ahead remains to be seen.

6 CONCLUSION: FUTURE OF THE VRA

As a new round of redistricting got underway following the 2020 Census, the Supreme Court further narrowed the reach of the VRA's Section 2. First, *Brnovich v. Democratic National Committee* involved a Section 2 challenge to Arizona's practice of discarding rather than partially counting ballots cast by voters outside their assigned precincts, and its refusal to allow third parties to collect absentee ballots.[166] Sitting *en banc*, the Court of Appeals for the Ninth Circuit held that both practices violated Section 2, emphasizing how each imposed significant and disproportionate burdens on minority voters in the State. The *en banc* court further held that Arizona acted with racially discriminatory intent when it adopted the ballot collection ban.[167] The Supreme Court, voting 6-3, reversed.[168]

Writing for the Court, Justice Alito's opinion offered a template for addressing Section 2 claims that arose outside of the redistricting context. The opinion identified "several important circumstances" that, while not exhaustive, critically inform such claims. For instance, an electoral practice is more likely to violate Section 2 the larger the burden it imposes and the greater the racial disparity it produces. Neither a "[m]ere inconvenience" nor the "mere fact there is some disparity in impact" is itself sufficient. Practices that were "in widespread use" at the time of the 1982 amendments are less likely to run afoul of the statute, as are those that become less burdensome when evaluated in light of "the opportunities provided by a State's entire system of voting." Finally, practices supported by strong state interests are likely to comport with Section 2, even if the challenged practice is not the only or even the best means by which the interest might be advanced. Turning to the case at hand, the Court held that consideration of the identified "circumstances" established that the challenged Arizona practices passed muster under Section 2.[169]

[164] See *Cooper*, 137 S.Ct. at 1468

[165] See Guy-Uriel E. Charles & Luis Fuentes-Rohwer, Race and Representation Revisited: The New Racial Gerrymandering Cases and Section 2 of the VRA, 59 William & Mary L. Rev. 1559, 1567 (2018) (arguing that the new *Shaw* cases are on a "collision course" with the mandates of Section 2).

[166] *Brnovich v. Democratic National Committee*, 141 S.Ct. 2321 (2021).

[167] *Democratic National Committee v. Hobbs*, 948 F.3d 989 (9th Cir. 2020).

[168] *Brnovich*, 141 S.Ct. 2330.

[169] *Id.* at 2338-2340.

Joined by Justices Breyer and Sotomayor, Justice Kagan's dissenting opinion charged the majority with "creat[ing] a set of extra-textual exceptions and considerations to sap the [Voting Rights] Act's strength... This Court has no right to remake Section 2." On Justice Kagan's reading, the appellate court got it right when it concluded that the challenged practices denied minority voters in Arizona the equal opportunity guaranteed by Section 2.[170]

The limits *Brnovich* read into Section 2 are not directly applicable in the redistricting context. Still, the decision confirmed the present Court's inclination to narrow the reach of Section 2 and suggests that it remains deeply unhappy with the race-based considerations that statute mandates.

Second, an emergency ruling issued as this volume went to press suggests that Section 2's application in the redistricting contest is likely to be reduced yet further, and perhaps drastically so. In *Merrill v. Milligan*, the Court noted probable jurisdiction and stayed the ruling of a three-judge District Court that had held that Alabama's new congressional redistricting plan violated Section 2.[171] The lower court concluded that Alabama ran afoul of the statute by failing to include a second majority-Black district when the *Gingles* factors and totality of circumstances review dictated that an additional district was required. Chief Justice Roberts, who dissented from the order issuing the stay, stated that "the District Court properly applied existing law in an extensive opinion with no apparent errors for our correction."[172] Justices Kagan, joined by Justices Breyer and Sotomayor, likewise objected to the stay, noting that the lower court "did everything right under the law existing today" and that "[s]taying its decision forces Black Alabamians to suffer what under that law is clear vote dilution."[173]

Merrill promises to restructure dramatically Section 2's application to the redistricting process. The Chief Justice indicated as much, stating that while he thought the stay was improper, he nevertheless agreed that the Court should review the decision on its merits. His opinion posited that "*Gingles* and its progeny have engendered considerable disagreement and uncertainty regarding the nature and contours of a vote dilution claim" and that the Court should "resolve the wide range of uncertainties" that arise under existing doctrine.[174]

Together, *Brnovich* and *Merrill* signal that Section 2 is likely to occupy, at best, a diminished role in future electoral disputes involving racial and language-based discrimination. For now, the redistricting plans produced following the 2020 Census—the first in more than a half century enacted absent the constraints of the VRA's Section 5—remain subject to Section 2. But the impact of that provision on districting practices is likely to be limited.

Decades ago, Congress crafted Section 2, and the VRA more broadly, to promote equality of opportunity in the electoral process. The deluge of restrictive electoral restrictions enacted in the years since Shelby County make clear that equality of

[170] Id. at 2373 (Kagan, J., dissenting).
[171] Merrill v. Milligan, 2022 WL 354467 (Feb. 7, 2022).
[172] *Id.* at *4 (Roberts, C.J., dissenting).
[173] *Id.* at *4 (Kagan, J., dissenting).
[174] *Id.* at *4 (Roberts, C.J., dissenting).

opportunity remains unrealized. Section 2 might yet be read as it was intended. It might still be applied widely to developing problems and resurgent practices and implemented to ensure that all voters have an equal opportunity to participate in the electoral process. Recent judicial rulings and the sentiments they manifest nevertheless suggest that this prospect is most unlikely.

Chapter 8

Law, computing and redistricting in the 1960s

ALMA STEINGART

CHAPTER SUMMARY

Computers have been used to draw maps for decades. This chapter takes a look at how computing and redistricting intersected in the 1960s, along with the critical role "one person, one vote" played in opening the door for them.

1 AGAINST COMPUTERS

THE CASE AGAINST ONE PERSON, ONE VOTE

In 1963, political scientist Alfred de Grazia published *Apportionment and Representative Government,* a polemical attack on what he called the "equal-populations doctrine" [9]. An expert in the history and theory of political representation, de Grazia was a professor of government and social theory at New York University and a founder of *The American Behavioral Scientist.*[1] Over his career, de Grazia wrote monographs on numerous topics such as the American government, the welfare system, and international politics, but he found the problem of representation uniquely pressing: "no political problem is more fascinating than that of arranging the representation of the many local and general interests that compose modern society, and none is more frustrating to the scientist and the man of good will" [8].

[1] The journal was founded in 1957 under the title *PROD: Political Research: Organization and Design* and was renamed in 1961.

M. Duchin, O. Walch (eds.), *Political Geometry*, https://doi.org/10.1007/978-3-319-69161-9_8

If de Grazia found the apportionment problem to be frustrating at the best of times, in 1962, when the Supreme Court ruled in *Baker v. Carr* that malapportionment cases are justiciable, de Grazia was livid. The case originated when citizens sued the state of Tennessee, claiming that the state legislatures violated their Fourteenth Amendment guarantee of equal protection under the law by refraining from apportioning the state's General Assembly since 1901. The Court's ruling—that the appellants had standing and that their claim could be heard in the District Court—overturned the Court's earlier position that districting and apportionment were legislative rather than judicial matters.[2] De Grazia believed that the new decision set a dangerous precedent, and he directed his critique at legal and political scholars, believing that they were skewing the conversation by presenting a consensus when none existed.[3]

It is important to note that in *Baker v. Carr*, the Supreme Court did not specify a standard by which lower courts could determine the validity of a given apportionment scheme. Only in 1964 (*Reynolds v. Sims*), after legal challenges to legislative malapportionment mounted in the states, did the Court rule that districts must be comprised of roughly equal population. Even though de Grazia published his work before the Court clearly articulated the now famous "One Person, One Vote" doctrine, he focused much of his critique on that ideal.[4] In the nine months following the Supreme Court ruling in *Baker v. Carr*, litigation on apportionment occurred in twenty-one states, and in almost all of these cases the lower courts based their decision on the equal population principle.

De Grazia argued that if the goal of apportionment and districting was to ensure better representation, then protecting community and interest groups must be more important than numerical equality. Geography, local, ethnic, and religious interests had all been taken into account in forming constituencies in the past—properly, in his view.[5] And while the courts never explicitly suggested that these other factors were less crucial for effective representation, de Grazia worried that an emphasis on equal population would give rise to a "number obsession," which not only would lead to degraded representation but would also give rise to "a mechanical view of man and human relations" [9]. The ideal of equal population, he believed, risked treating citizens as nothing more than numbers that can be interchanged for one another.

[2] *Colegrove v. Green*, 328 U.S. 549 (1946)

[3] "With few exceptions, the legal briefs and numerous research reports that I have since read have been deplorably shallow and unobjective. Our national history has been rummaged largely with an eye toward illustrating the feelings for numerical equality to be found among the people. But the case on apportionment is far from closed. What stands now cannot be permitted to represent all that political science has to say about apportionment and representative government" [9].

[4] In fact, the principle was initially called "One Man, One Vote," but the language has since evolved.

[5] For instance, a system of randomized district assignments could achieve numerical equality, but each district's residents would have nothing in common besides the very fact of belonging to the same district.

FROM PARITY TO PARODY

To illustrate the absurdity of the principle in his opinion, de Grazia concluded his introductory chapter with a section entitled "Drawing of State Apportionment Boundaries by Computer," in which he offered an algorithm to accomplish just that. The "instructions to machine" included steps such as assigning a set of numbers to all the people in a given district and transferring one individual at a time from one district to an adjunct one until all possible permutations were exhausted and millions of maps were produced. The algorithm was a theoretical one. His goal was not to implement it but to document the futility of such an endeavor: "which of these millions should be chosen?" By adding voters' affiliations, the algorithm could be further adjusted to either favor one party or even "group people around existing communities," but it would not solve the question, which map to choose? De Grazia was not inherently opposed to the use of computers. There could be a time in the future when computers would be able to be used for determining constituencies, but de Grazia insisted that this possibility was still in the future. "Until a machine can do human work," he concluded:

> "It is best to limit its use strictly and so also limit the use of machine-like theories that try to organize society. American society is not a collection of faceless particles. It is composed of highly diverse and yet interconnected sets of people. A political theory suggesting that people are interchangeable like nuts and bolts is likely to be both fallacious and detrimental to the personal happiness of the citizenry."

Using computers in the work of redistricting exposed the absurdity of the principle of equal population. It also threatened to promote the "number obsession" into an organizational democratic principle.

In the years following the publication of his book, de Grazia's parody became a reality, as computers paved the way to strict numerical equality. But the impact was perhaps even more extensive than de Grazia first anticipated. Malapportionment cases arrived in the courts because the votes of rural citizens had more weight than those of urban dwellers. It is this problem that the principle of equality addressed. Yet as strict equality started to overshadow other factors such as counties and communities of interest, it also became a shield protecting legislators from potential challenges. As the history of racial and partisan redistricting in the second half of the century demonstrates, with time equality became a proxy for equity.

2 THE APPORTIONMENT REVOLUTION MEETS THE COMPUTER REVOLUTION

Throughout the 1960s, political scientists and allies began turning to computers to aid them in redistricting. In response to *Baker v. Carr* and the consequent court decision in *Reynolds v. Sims*, computer redistricting moved from a theoretical musing to an active research agenda, bringing in people from industry and the academy to enter what Justice Frankfurter had famously termed the "political thicket."

The principle of One Person, One Vote opened up apportionment to computational remedies. Before *Baker v. Carr*, many states had districts of widely differing sizes. For example, when Tennessee's apportionment plan came before the Supreme Court, Moore County (population 2,340) and Rutherford County (population 25,316) each sent two representatives to the Tennessee General Assembly. This was a way for lawmakers to maintain power for rural areas despite growing urbanization. Computational methods for ensuring mathematical equality would have not been appealing in such cases because the balance being struck was not between any two abstracted citizens, but between the interests of an urban and a rural citizenry. For that matter, there might not even be an opening for a calculative approach because states like Tennessee had simply not districted in decades. In other states, factors such as existing coherent geographical units (counties, cities, etc.) and historical antecedent (such as incumbency) guided most apportionment efforts, with little to no regard for numerical equality. Computers only became useful to the problem because both a legal demand and a numerical standard had been established. Furthermore, computers not only became tools for implementing One Person, One Vote—their use even helped to establish a definition of what One Person, One Vote might mean. During the 1960s, courts and legislatures around the country were still unclear about the scope and strictness of the equal population principle.[6] Without any formal criteria by which to address this ambiguity, computers artificially foreclosed the problem. As political scientists, computer scientists, and state legislators began advocating the use of computers in redistricting, a computable definition of perfect equality as a rough proxy for political equity cemented the principle of One Person, One Vote as the governing criterion for apportionment and redistricting.[7]

The "apportionment revolution," as the mushrooming of legal challenges to districting plans around the country were referred to in the 1960s, began just as computers were becoming more ubiquitous in academic and industrial research. In the 1950s, the market for digital computers was still limited. By the 1960s, as computers became smaller, cheaper, and easier to operate, a growing number of Americans had the opportunity to interact with the new technology and imagine new uses for it. As historian of computing Paul E. Ceruzzi puts it, in the 1960s "the computer was reinvented yet again."[8] The transformation in the computing hardware industry was matched by growth of a new software industry [4]. New companies offering programming services for interested parties were growing at a fast rate, and some of these new startups became involved in computerized districting efforts.[9] The

[6]Public Law 94-171 was passed in 1975 calling for the Census Bureau to create official population tables for redistricting. See Chapter 13.

[7]Today, the terms apportionment and redistricting mainly designate two distinct activities. Now, *apportionment* names the process by which the number of seats in a governing body is designated for a particular territory, while *redistricting* is the division of the geography of the state into this number of districts. However, in the 1960s the two were often thought of together—because allowable deviation was a major live question—and were not as clearly separated as they are today. As a vestige of the earlier way of using the terms, population imbalance in districts is still called *malapportionment*.

[8]At first, digital computers were understood to be giant calculators. It was only in the 1950s that the computer's potential for data processing and scientific machines was established [5].

[9]Both Computer Applications Inc. and the Council for Economic and Industrial Research (C-E-I-R)

1960s also witnessed a growing interest by political scientists in digital computers as a new research tool for mathematical modeling, simulation, and large-scale survey analysis.[10] It is across this background that some political scientists turned to computers to study apportionment and redistricting. Finally, early efforts at computerized redistricting were spearheaded by postwar engineers, scientists, and mathematicians trained in operations research, systems theory, dynamic processing, and game theory. Often working in industry, these researchers had access to computing equipment and were eager to put into practical use their academic training. For them redistricting was just another problem that proved to be amenable to their new theoretical tools.

In what follows I survey three of the earliest efforts to tackle the problem of apportionment and districting in light of *Baker v. Carr* using digital computers. As will become clear, the individuals who turned to computer districting in the 1960s employed a wide range of methods and sought to forge differing political alliances. Yet, they were united by the belief that computers were apt for the task—not for their ability to crunch the numbers much faster than any human could, but for potentially offering an *objective* solution to a deeply political problem.[11] This is not to say that these historical figures were naïve or believed that computerized redistricting would bring an end to gerrymandering. On the contrary, they were acutely aware that their products would be used in a highly partisan environment. Their efforts, as such, can be seen as attempts to isolate a technical aspect of redistricting that is independent of its political implementation. The belief that this is possible persists in many venues today.

IN SEARCH OF A MECHANICAL FORMULA

One of the earliest papers offering a computer solution to apportionment was published by James B. Weaver and Sidney W. Hess in 1963 [17]. Weaver was a member of The Committee of 39, a civic group in Delaware that gathered historical data and information to inform the redistricting battles in the state. Weaver recalled that in 1963 the urban district he lived in had thirty-five times as many people as the smallest farming district in the state. "It hit me like the electric lights over people's heads on the comic pages," Weaver recalled years later, "that the computer could create districts which were blind to politics. Not only do they process numbers better than humans, they can't introduce politics if no party affiliation is in their memories" [16].

Weaver enlisted Hess, who was his younger colleague in Atlas Chemical Industries and a recent PhD in operations research from Case Western University, and the two began testing potential programs. What Hess and Weaver proposed as an objective function was a new measurement for compactness that took into account not only the shape of a district but also the population density in given areas. The goal of the computer program was to draw maps that optimize this measurement of

got involved in computerized districting in the 1960s.

[10] In his book, de Grazia points to some of the earliest work on computer simulation of voting behavior when discussing computers.

[11] For analyses from 2005 and 2010 of the various computing approaches to redistricting, see Altman et al. [2, 1]. A current overview is found in this volume, in Chapter 16.

compactness overall. The main breakthrough for Hess came when he realized that this optimization problem was structurally similar to a well-known class of problems in operations research, known as warehouse allocation or optimal transport: How should you direct customers' orders to specific warehouses so as to optimize freight costs? Hess was thus able to apply existing ideas and algorithms to offer new methods for redistricting.

In the introduction to their paper, Weaver and Hess explain that the goal of such computerized redistricting was not to replace legislative efforts, but rather to intervene in cases in which the legislator failed to provide a map. In the wake of *Baker v. Carr*, the authors explained, courts may strike down representation schemes as unconstitutional, but there was no clear guidance as to how "to administer relief" [17]. More often than not, the courts would choose to give the legislature another opportunity to draw a districting map, but in cases when this additional attempt failed, the court might have to offer a direct solution. It is at this remedial stage that the court might wish to appeal to computers to produce a map: "since redistricting usually affects the political balance of a legislature, a court undertaking affirmative apportionment and districting is likely to become the subject of highly partisan appeal and criticism... To avoid this 'political thicket,' a court may desire to limit its own discretion in creating a new legislative district."

In other words, the appeal of computer programming was that it removed human discretion from the mapping process. "One means of accomplishing this end," Weaver and Hess wrote, "could be to adopt a mechanical formula which makes the actual drafting of district lines non-discretionary once general principles of representation have been determined." To accomplish this goal, Weaver and Hess came up with a new definition of compactness, which took into account both geography and population. Intuitively, their goal was to keep centers of population intact. They therefore instructed their program to generate a series of maps through an iterative process and kept only those maps whose "compactness score" was highest and whose population deviation among the different districts was smallest. Whereas the former measure was unique to their program, the latter would become a standard for computerized districting efforts. In other words, Weaver and Hess subscribed to a mode of mechanical objectivity that reduced the nuanced notion of democratic "fairness" to a computational problem that could be studied and (like freight costs) optimized.[12,13]

In his concurring opinion in *Baker v. Carr*, Justice Tom C. Clark described the Tennessee apportionment plan as "a crazy quilt without rational basis."[14] No consistent scheme or standard, he explained, could account for the representational disparity among the various Tennessee counties. "Certainly there must be some

[12] "Mechanical objectivity" is a epistemic style articulated by Lorraine J. Daston and Peter Galison, who place its emergence in the mid-nineteenth century: "to be objective is to aspire to knowledge that bears no trace of the knower—knowledge unmarked by prejudice or skill, fantasy of judgment, wishing or striving" [7].

[13] The first call for automated redistricting came from economist William Vickrey in 1961. Vickrey did not call upon the computer directly, but instead described a theoretical algorithm for achieving automatic districting. The "elimination of gerrymandering would seem to require the establishment of an automatic and impersonal procedure for carrying out a redistricting" [15].

[14] *Baker v. Carr*, 369 U.S. 186, 254 (1962) (Clark concurring opinion)

rational design to a state's districting," Justice Clark added.[15] Early attempts to use digital computers to draw congressional and state legislative districts could be viewed as a response to Justice Clark's criticism. Numerous computer programs could be written and each one could produce thousands of maps, but an interested observer could always discern an operating rationale behind each map, one which had to be literally written into computer code.[16]

Weaver and Hess were able to put their program to use because the Committee of 39 had already gathered much of the necessary data. Still, when it came to running the algorithm on a computer, they quickly realized that the one purchased by Atlas Chemical Industries did not have the necessary computational power. They described the problem to operations researchers at DuPont, who got interested in it and offered the two access to the company's computer. In the following years, Weaver, Hess, and their DuPont colleagues founded Computer Research on Nonpartisan Redistricting, or CROND, Inc. For a while they continued to use DuPont's computer, but as their research expanded, they received a major ($96,000) grant from the Ford Foundation through the National Municipal League. As they improved their method and computational capacity, the group also began contracting directly with legislatures. They first drew districting maps for the Delaware state legislature and in the following years consulted with other states, including Pennsylvania, New York, and Nevada.

"THE ART OF DECISION MAKING"

Another early approach came from Chandler Harrison Stevens, who was supported through CROND's grant. Stevens earned a Ph.D. in economics from MIT in 1967 and served in the Massachusetts House of Representatives from 1965–1968.[17] He was impressed by Weaver and Hess's computational approach, but he did not believe that the process should be fully automated. In 1966, while working at MIT's Center for International Studies, Stevens developed a new computer districting application, building upon Weaver and Hess's algorithm, but he added a graphic component. The addition of a graphic display was conceived by Stevens as a way of construing computer districting as not fully automated, but instead a hybrid "man-machine" endeavor. A cathode-ray tube (TV screen) could be connected to the computer, enabling the user to see in real time how the new districts mapped onto the geography of the state. A user could visually inspect a districting map the program produced and then instruct the computer to move a given town from one district to an adjunct one. The program would then automatically evaluate the new

[15] Justice Clark did not require numerical equality nor did he believe that mathematical exactness is required to produce a "rational" plan. *Baker v. Carr*, 369 U.S. 186, 258 (1962) (Clark concurring opinion)

[16] Historians of science have argued that a novel brand of rationality, one they termed "Cold War Rationality," dominated much academic and policy thinking at the time. Algorithmic and rule-bound, this way of thinking stood in opposition to judgment and experience. Cold War Rationality was a child of the atomic age, a world in which computerized algorithms might be a safeguard against the human fallibility that could lead to total destruction [10].

[17] In fact, he was the first Independent elected to the Massachusetts state legislature. Stevens was at various points a programmer at the Pentagon, a science advisor to the governor of Puerto Rico, and a retirement-aged Peace Corps volunteer in Ukraine. In his obituary, his family describes him as a futurist committed to social justice (https://perma.cc/46G3-6HVR).

map according to a series of criteria such as population deviation, compactness, and contiguity.

When he presented his program at the 73rd National Conference on Government as part of a CROND-organized session, Stevens explained that the motivation for the work came from desire to couple "human judgment with the computer, which I felt the districting problem required" [13]. Stevens was quick to acknowledge that the insistence on human judgment might seem antithetical to those who researched computerized districting, but he insisted that the approach is both more practical and sounder. Certain districting criteria, Stevens explained, simply cannot be quantified or computerized and thus require human intervention. Moreover, "human judgment such as contained in the subjective weights applied to conflicting criteria can never be completely eliminated."

Drawing on his experience as an elected official, Stevens maintained that additional criteria besides population equality, compactness, and contiguity should enter into the decision process of the legislature. "I personally feel that we need much better correspondence between election districts and other districts used for regional planning, mass transportation, mental health, pollution control, welfare and employment services and a myriad of other programs." Using the visual display, Stevens envisions a system in which the user could project on top of a given map slides of additional information such as "newspaper reading patterns" or "commuting patterns." It was man–machine interaction that Stevens believed showed promise, not full automation. Unlike Hess and Weaver, Stevens believed that not only was there room for subjective judgment, but that it was in fact necessary.[18] He advanced this view when he declared that "there is an *art* as well as a *science* to drawing election districts." Stevens did not mean "art" in a derogatory sense to refer to gerrymandering. Rather, what he had in mind was "the art of decision-making, for the qualitative as opposed to the quantitative" [14].

Still, Stevens recognized that it was now possible to minimize population deviation using computers, a point he made sure to impress upon his colleagues in the legislature. When he appeared before the districting committee with some preliminary results, he emphasized two points above all. First was that it was possible to achieve districting plans approaching population equality, with as little as 1% deviation. Second was that, even within such limits, great variation in the plans themselves was possible. Foreshadowing what would become a pattern in decades to come, Stevens reported that the committee "seemed to understand the first point, for every districting plan which they subsequently considered had population deviation of 1% or less. But they ignored the second point." The districting maps the committee ended up considering were the least compact. Even within the court-imposed criterion of population equality, he concluded, "traditional Gerrymandering had plenty of room in which to operate."

This realization made Stevens decide to expand the potential utility of his program. As Stevens was developing his program, the United States District Court for the

[18]Thus, Stevens did not subscribe to the same "mechanical objectivity" that Hess and Weaver did, but to what Daston and Galison have named "trained judgment." This regime of objectivity, which emerged in the twentieth century, did not seek to completely erase the subjective. Rather, practitioners believed that intuition, gained through experience, plays a critical role in the production of true knowledge.

District of Massachusetts ruled that the state's current Congressional plan was unconstitutional and instructed the legislature to produce a new map. Stevens had a chance to put his program to the test sooner than he had planned. The sole Independent lawmaker in the Assembly, Stevens tried to convince his colleagues in the Joint Legislative Committee, which was tasked with producing a new plan to use his computer program. His goal, fundamentally, was not to suggest a new plan, but rather to use the computer as a tool with which to evaluate the existing proposals. Stevens renamed Weaver and Hess's measurement of compactness the "Gerry Index," and in a public demonstration at MIT, Stevens sought to expose the Joint Legislative Committee's plan as the least compact. "This type of public exposure," he wrote later, "should help hasten the day when state constitutions will be revised and/or courts will be armed to force fairer districting" [14]. Rather than simply serving as an aid to mathematical or calculative equality, Stevens also put computing to work as a tool to oversee, evaluate, and even expose unfair apportionment and districting.

POLITICAL REALISM

A third early paper calling for the use of computers in redistricting came from political scientist Stuart S. Nagel. Nagel received his Ph.D. from Northwestern in 1961 and directly joined the faculty of the University of Illinois Urbana-Champaign. Over the years Nagel published on numerous topics, authoring dozens of articles and books, with a long-term focus on technological applications in public policy. As one Illinois colleague put it: "like early Progressives, Nagel faithfully believed that the application of scientific rigor to public tribulations would virtually guarantee human progress" [18].

In 1964–1965, Nagel was a fellow at Stanford's Center for Advanced Study in the Behavioral Sciences, where he tackled the districting problem using the university's IBM 7090 computer. In 1965, he published his results, "Simplified Bipartisan Computer Redistricting" in the Stanford Law Review [12]. Unlike Weaver and Hess, Nagel promoted the use of computers in districting as a bipartisan rather than nonpartisan tool. Nagel believed computers could help solve legislative deadlocks. Instead of helping the court, Nagel wanted to help elected officials.

In Illinois, the inability of Republicans and Democrats to agree on new districts resulted in an at-large election; at the same time, the parties were coming to similar dead ends around the country. Nagel's Stanford Law Review piece envisioned a computer program that would save time and money by letting politicians examine a set of possible maps and arrive at a compromise. He explained that his program was "designed with realistic politics in mind." It therefore included past voting records for each district and could in principle produce a plan favoring one party over another, or remain neutral. "Such a parameter," Nagel explained, "might be needed to convert a compromise between the political parties into a redistricting pattern." Democrats and Republicans, Nagel reasoned, would need to be able to predict how each party will fair in future elections before they would be willing to sign onto a new districting map.

In emphasizing the importance of political information for the districting map,

Nagel called into question early efforts to present computer redistricting as objective and an antidote to gerrymandering. There was nothing inherent in the computational approach to redistricting that prevented politicians from seeking to maximize their position. In particular, Nagel was pushing back against a 1964 paper by Edward Forrest describing a computer program for districting as not only generally useful for avoiding legislative deadlock but even capable of producing "truly neutral and unbiased" maps. "Since the computer doesn't know how to gerrymander—because two plus two always equals four—the electronically generated map can't be anything but unbiased" [11]. An academic, Nagel provided his program for free to anyone who was interested and he was eager for others to use it. In contrast, because Forrest was working for Computer Applications, Inc., one of the earliest software service companies, the details of his program remained classified.

Nagel's belief that the computerized districting should be used by politicians shaped the algorithmic approach he proposed. Unlike Hess and Weaver, the program was aimed at improving upon existing districts rather than overhauling the entire districting plan. Instead of starting from scratch, in each step the program created new maps by trading a geographical unit from one district for another and then checking if equality and compactness were improved. Nagel decided to adhere as closely as possible to a state's existing map because it aligned with his "realistic" approach. "It is naïve to think that incumbent politicians are likely to want to upset the status quo any more than the minimum extent required by the state constitution and courts or by the federal constitution and courts" [12]. Nagel never forgot that it was politicians who approved new districting plans, even if a computer drew the lines. Fairness, for him, was a pragmatic rather than a theoretical concept, a compromise aided by computers.

Nagel first put his program to the test in Illinois, but he was unable to impact the deliberation of the state legislature. Then, in 1965, an analyst named William Below adapted the program in order to help the California Assembly to produce a new districting plan. Below was on staff for the Assembly Committee on elections and reapportionment. During a November committee meeting, Below explained to committee members that the program he worked on "does not try to go to any unique plan for redistricting but one which is rational and which conforms to population and political registration. This is a fairly accurate method of maintaining the political status quo in California legislative district" [6]. When asked about the criteria used in the program, he dodged the question, asserting that the "mathematical plan is the best plan... it is the plan that fits best." As Below's remarks make clear, he believed that Nagel's program carved out a technical aspect of the problem that could be solved independent of the political context of its applicability. As long as a set of some criteria that were legal (population equality, contiguity) or political (maintaining the status quo) could be agreed upon and programmed into a computer, the creation of the maps themselves was better left to a computer than to humans.

8.1 WHAT'S ACTUALLY IN THESE ALGORITHMS?

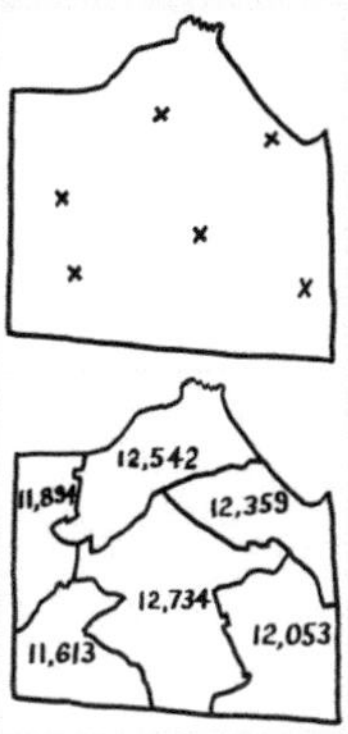

Weaver and Hess (1963) introduced a district compactness metric closely related to the moment of inertia of a physical object (sum the squared distances from the units to district centers). To create compact districting plans, they used what is now called a *naive k-means* approach—the term was not coined until a few years later, but they recognized this as an already fairly standard technique. Starting with an initial guess of good locations for population centers of k districts, they assigned units to districts to minimize the moment of inertia. When a plan was complete, they recalculated the population centers of the just-created districts, which would likely differ from their initial guesses. Using these new centers as their next guess, they then iterated until the procedure stabilized.

The most obvious problem is that this optimizes for compactness alone, and might sacrifice population balance or even contiguity, let alone the more human criteria. Another problem is that there's no practical way to be sure that you iterate toward a global optimum. In the meantime, they imagined that the program could be used to produce a small number of acceptable plans, leaving a situation where "some discretion may be necessary to choose."

Stevens' `Geodata` was not really an algorithm at all, but what today we would call a GUI, or graphical user interface. Researchers used punch cards to feed data into an IBM 360 computer, which could then display a map of the districting plan on a TV screen. Users could type in instructions—e.g., "MOVE LEXINGTON FROM DISTRICT 5 TO DISTRICT 3"—and see the map change accordingly. The evaluation of plans by a pre-programmed score could be done quickly, and Stevens revived the moment of inertia score, now branded the Gerry Index. He mused about compactness with respect to travel time or social difference, but that was a more sophisticated version planned for the future. As a good reminder of the state of the hardware in 1965, Stevens used plastic overlays on the screen for fixed boundaries so that the "computer-controlled beam of light" could be saved for dynamic information. He also lamented that "we do not have a means as yet for making permanent the maps which show on our TV screen" and suggested using a camera to take a picture of the screen.

Nagel (1965) proposed a program that modified a given districting plan into something "better," working in a language called ALGOL (and he actually mocked Weaver and Hess for working in hard-to-understand FORTRAN!). To define what would be better, Nagel built a compound score by essentially multiplying a population deviation penalty by a compactness (normalized moment of inertia) penalty. The program looped over the units (say, the counties) in a state and proposed to "Move" a unit from one district to another or to "Trade" two units from adjacent districts. Proposals had to pass validity checks and had to improve the score to be adopted. Because this was essentially just hill-climbing with respect to a score, he could build in a partisan objective as part of the score function—this could be a deviation from proportionality penalty if you want to seek partisan balance or a partisan-favoring score if you seek a gerrymander! To gerrymander for a party, he proposed using the average margin of victory in a winning district. Making this lower would translate to more wins by efficient margins.

3 THE WIDENING GAP BETWEEN EQUALITY AND FAIRNESS

The computational approach pursued in each of the cases outlined above was highly determined by the authors' understanding of the nature of the problem. Weaver and Hess, operating outside the political process, believed that districting can (and should) be reduced to a purely technical optimization problem. As a state elected official, Stevens held that human judgment had a place in the districting process and hence emphasized the human–machine interaction. Finally, Nagel, as a behavioralist political scientist, chose pragmatism over optimization. Computers were a tool for negotiation, *not* full automation. Despite their differences, all of these early researchers wished to eliminate direct human intervention from some part of the districting process. Computers, they believed, were apt for the job because they fundamentally subsumed human intentionality to programmed randomness.

As these three examples make clear, *Baker v. Carr* had served as the impetus for a swell of early research studying computerized redistricting. As legal cases mounted around the country, the idea that newly available digital computers might offer a tool to counter the practice of gross malapportionment grew in popularity.

On the whole, their attempts to promote algorithmic plan generation were unsuccessful. As the Advisory Council on Reapportionment to the Legislature of the State of New York learned in 1965, existing computational approaches were still far from being practically useful. Having contracted Forrest to help the planning process, the Council concluded that "it ultimately became apparent that, in the present state of computer technology, there are some problems that can be handled more effectively by human judgment than by machines."[19] The impact of early attempts at computerized districting has nonetheless been profound. As several commentators noted, by the end of the 1960s, far from removing human intentionality from the process, computers only made the problem more acute by allowing increased and unnecessary emphasis on the principle of One Person, One Vote. With the help of computers, the legal requirement got subsumed by its mathematical proxy.

Even though *Reynolds* had set population balance as the ideal in 1964, it was still not clear what deviation from perfect equality the court would tolerate. In the wake of One Person, One Vote, legislatures understood that they should strive to produce districts of equal population, but no one, not even the justices, believed that perfect equality could be demanded. As Chief Justice Earl Warren wrote for the majority, "we realize that it is a practical impossibility to arrange legislative districts so that each one has an identical number of residents, or citizens, or voters. Mathematical exactness or precision is hardly a workable constitutional requirement."[20]

In the following years, numerous cases arrived in state and federal courts testing the limits of, and the justifications for, permissible population deviation. It is in

[19] "Report of the Advisory Council on Reapportionment to the Legislature of the State of New York," December 23, 1965, 14.
[20] *Reynolds v. Sims* 377 U.S. 577

the gray zone between "mathematical exactness" and practical considerations that researchers had first hailed computers as a possible solution. Legislatures found computers attractive exactly because they could quickly compare how different districting plans deviated from total equality.[21] The preservation of counties' historical and natural boundaries gave way to statistical equality.[22]

By the end of the decade, the Supreme Court began to tighten the expectations when it struck down a Missouri congressional districting plan (*Kirkpatrick v. Preisler*, 1969) and a New York congressional districting plan (*Wells v. Rockefeller*, 1969) for excessive population deviation. Following these court decisions, though they did not establish precise numerical bounds, top-to-bottom deviation of congressional districts is typically no more than 1% of ideal district size, and state legislative districts deviate no more than 10%. Taken together, the two decisions were seen to elevate numerical equality to a guiding principle. Population equality became an end in itself rather than an indicator of fairness.

In his dissenting opinion in *Wells v. Rockefeller*, Justice Harlan wrote that "the Court's exclusive concentration upon arithmetic blinds it to the realities of the political process, as the Rockefeller case makes so clear. The fact of the matter is that the rule of absolute equality is perfectly compatible with 'gerrymandering' of the worst sort."[23] He then noted that a "computer may grind out district lines which can totally frustrate the popular will on an overwhelming number of critical issues. The legislature must do more than satisfy one man, one vote." Justice Harlan recognized that mathematical exactness was a byproduct of computerization and he foresaw a future where "a computer can produce countless plans for absolute population equality, one differing very little from another, but each having its own very different political ramifications." If the dominant criterion for districting was population equality, then computers, so the logic went, could easily (and more efficiently than before) produce numerous configurations that would satisfy the legal requirement while allowing for political manipulations of any kind.[24]

By the early 1970s, Justice Harlan's critique was taken by some political scientists as fully just. Gordon E. Baker, who had been directly attacked by de Grazia in 1963 for promoting numerical equality, later wrote that "the singleminded quest for mathematical equality of districts at the expense of some adherence to local governmental subunits carries with it the potential for extensive gerrymandering" [3]. Baker celebrated the Supreme Court's decision allowing a less strict numerical standard for state-level as opposed to congressional districting. "This development could help minimize some of the computerized equal-population gerrymandering that ignored local governmental subunits as well as communities of interest." The

[21]As Gordon E. Baker reflected, in the years following *Baker v. Carr*, "lawmakers increasingly sought to make redistricting statutes invulnerable to legal challenge by aiming for statistical precision at the expense of geographic boundary constraints."

[22]In their early paper, Weaver and Hess argue that state geographical units such as counties must be broken down to prioritize a map that was both compact and adhered to population equality.

[23]*Wells v. Rockefeller*, 394 U.S. 542, 551 (1969) (Justice Harlan dissenting)

[24]In 1983, Weaver was quick to distinguish his and his colleagues' early efforts from those that came later. CROND, Inc. spent the grant money by first holding a seminar in Washington for interested national lawmakers, and then focusing on research for publication. "Then the political parties took over—both have now developed ways to gerrymander districts to their heart's content, using the computer" [16].

computer, seen at first as a solution for gerrymandering, had within ten years come to seem like a source. If numerical equality was intended as a constraint on lawmakers' ability to construct district maps to favor their own interests, in the end the implementation of population equality through computer technology increased their power to do so—and all within the bounds of the law.

REFERENCES

[1] Micah Altman and Michael McDonald, "The Promise and Perils of Computers in Redistricting," *Duke Journal of Constitutional Law & Public Policy* 5, no. 1 (2010): 69–111.

[2] Micah Altman, Karin MacDonald, and Michael McDonald, "From Crayons to Computers: The Evolution of Computer Use in Redistricting," Social Science Computer Review 23, no. 3 (August 1, 2005): 334–46, doi.org/10.1177/0894439305275855

[3] Gordon E. Baker, "Quantitative and Descriptive Guidelines to Minimize Gerrymandering," *Annals New York Academy of Science* 219, no. 1 (1973).

[4] Martin Campbell-Kelly, *From Airline Reservations to Sonic the Hedgehog: A History of the Software Industry*. Cambridge, MA: MIT Press, 2004.

[5] Paul E. Ceruzzi, *A History of Modern Computing*. Cambridge, MA: MIT Press, 2003.

[6] Committee on Election and Reapportionment, "Transcript of organizational meeting—election and reapportionment," Vacation Village, San Diego, California, November 29, 1965.

[7] Lorraine J. Daston and Peter Galison, *Objectivity*. (Cambridge, MA: Zone Books, 2010).

[8] Alfred De Grazia, *Public and Republic: Political Representation in America*. New York: Alfred A. Knope, 1951.

[9] Alfred De Grazia, *Apportionment and Representative Government*. New York, NY: Frederick A. Praeger, Inc., 1963.

[10] Paul Erickson et al., *How Reason Almost Lost Its Mind: The Strange Career of Cold War Rationality*. (Chicago: University Of Chicago Press, 2013).

[11] Edward Forrest, "Apportionment by Computer," The American Behavioral Scientist 8, no. 4 (1964).

[12] Stuart S. Nagel, "Simplified Bipartisan Computer Redistricting," Stanford Law Review 17, no. 5 (May 1965): 863–99.

[13] Chandler Harrison Stevens, "On the Screen Man-Machine Interaction," Paper presented on November 13, 1967. Available at: dspace.mit.edu/bitstream/handle/1721.1/85123/826708272.pdf?sequence=1

[14] Chandler Harrison Stevens, "A Report on the Feasibility of GEO-DATA: A Man-Machine System for Political Districting." Available at: dspace.mit.edu/handle/1721.1/85124

[15] William Vickrey, "On the Prevention of Gerrymandering," Political Science Quarterly 76, no. 1 (1961): 105–110.

[16] James B. Weaver, "Commentary," *Business & Professional Ethics Journal* 4, no. 3/4 (1985): 141.

[17] James B. Weaver and Sidney W. Hess, "A Procedure for Nonpartisan Districting: Development of Computer Techniques," The Yale Law Journal 73, no. 2 (December 1963): 288–308.

[18] Robert Weissberg, "Technology Evolution and Citizen Activism: The Net and the Rebirth of Limited Government," The Policy Studies Journal 31, no. 3 (2003): 385.

Chapter 9

The law of gerrymandering

GUY-URIEL CHARLES AND DOUGLAS M. SPENCER

CHAPTER SUMMARY

How do judges think about partisan gerrymandering? This chapter, by two law professors, is an answer to that question. The authors highlight both parallels between racial and partisan gerrymandering and divergences in the legal logic.

1 BACKDROP

The challenges of apportionment and redistricting are as old as representative government itself, and certainly predate the founding of America. When the U.S. Supreme Court established its "one person one vote" doctrine and required congressional districts to have equal population, the majority pointed to the "rotten boroughs" of pre-colonial Britain "under which one man could send two members of Parliament to represent the borough of Old Sarum while London's million people sent but four."[1] The Founders grappled with these same concerns during the constitutional convention in 1787. In fact, George Washington's sole substantive proposal at the convention was that each member of Congress represent no more than 30,000 people to guard against such disparities and to ensure that representatives retained a familiarity with the local circumstances of their constituents. Had Washington's proposal been ratified, our House of Representatives would currently comprise more than 10,200 members! As it turns out, the Constitution enshrined a rule in the opposite direction, specifying that each member of Congress represent no fewer than 30,000 people. (As of 2020 each member of Congress represents a district of approximately 750,000 people).

[1] *Wesberry v. Sanders*, 376 U.S. 1, 14 (1964).

M. Duchin, O. Walch (eds.), *Political Geometry*, https://doi.org/10.1007/978-3-319-69161-9_9

While much of this book is devoted to shape metrics and vote metrics, our goal in this chapter is to outline the basic framework for how judges think about the challenges of gerrymandering. Many readers will find the jurisprudence of gerrymandering to be misguided or inadequate to the task. However, because judges often have the final say on whether a gerrymander violates constitutional principles, and if so, what kind of remedies are available to those who are wronged, it is imperative to have a productive understanding of the legal underpinnings of gerrymandering cases that have come before the courts and a sense for how judges may evaluate future redistricting challenges.

1.1 WHY ARE COURTS INVOLVED AT ALL?

At the root of all redistricting cases is a simple, but important question: why are federal courts involved in the redistricting process at all? Courts are organized at both the state and the federal level. Federal courts have limited jurisdiction to hear cases about federal law, but the guiding document—the U.S. Constitution—does not explicitly address the problem of gerrymandering. In fact, the Constitution places immense responsibility on the states when it comes to the design and composition of the electoral structures of representative democracy. Furthermore, when it comes to partisan gerrymandering, courts not only recognize that they are on uncertain constitutional footing, but they also recognize that redistricting is primarily a political task that involves a pull and haul between different theories of democracy and political representation. All of these factors raise the question whether gerrymandering ought to be "justiciable" in the first place—that is, whether this is an issue that the federal courts ought to resolve or whether this is an issue that is best left to the political process.

On June 27, 2019 the U.S. Supreme Court, by a 5-4 margin, held that federal courts should not intervene in partisan gerrymandering cases. In its opinion in *Rucho v. Common Cause*, the Court overturned two lower court rulings that had been consolidated on appeal. The first ruling had invalidated a partisan gerrymander in North Carolina that awarded 10 of the state's 13 congressional districts to Republicans despite Republican candidates winning just 53.2% of the overall vote. The second ruling had invalidated a partisan gerrymander in Maryland that awarded 7 of the state's 8 congressional districts to Democrats despite Democratic candidates winning less than two-thirds of the overall vote. While the Supreme Court acknowledged that these gerrymanders were "incompatible with democratic principles" and "lead to results that reasonably seem unjust," the majority concluded that "partisan gerrymandering claims present political questions beyond the reach of the federal courts."[2] The decision in *Rucho v. Common Cause* returns the Court to its posture prior to 1962 when it consistently avoided "political questions" related to electoral design, apportionment, and the drawing of districts.

[2] *Rucho v. Common Cause*, No. 18-422, slip op. at 30.

9.1 STRUCTURE OF THE COURTS

Federal cases normally start in the lower courts (district/trial court) and can proceed "up" from there to appeals and then sometimes to SCOTUS, the Supreme Court of the United States. Often several cases will be fused into one as they work their way up. The Supreme Court can sometimes *remand* cases back to lower courts, so a single case can sometimes travel up and down the ladder repeatedly.

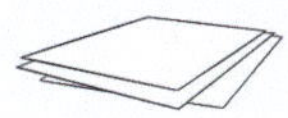

Supreme Court SCOTUS has discretion over which cases they hear. To have your case heard, you have to file a "writ of certiorari" or "cert petition." Of the 3,888 cert petitions filed in 2019–2020, they agreed to hear **74** cases. Cases are heard *en banc*, with all nine Justices together.

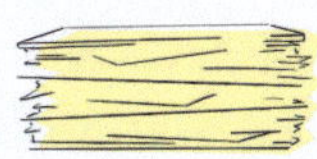

Circuit/Appeals Court Whereas district court cases deal with law and facts, appeals typically just focus on questions of law. There are 12 regional circuit courts and one federal circuit court of appeals; each appeal is heard by a panel of three judges. **51,693** cases were appealed to the level of appellate/circuit court in 2019–2020.

District/Trial Court Facts of the case are established at this stage, sometimes with the use of experts who provide reports and testimony. There's a single judge at trial who is usually randomly assigned to each case, and if you lose you can appeal. **425,925** district/trial court cases were held in 2019–2020.

The circuit courts have various reputations, making judge-shopping possible. This is why challenges to Obamacare often started in Texas (5th circuit) and challenges to Trump's immigration policies often started in Hawaii (9th circuit).

Redistricting cases (and a very small number of other kinds of cases) do **not** proceed through the regular process of working their way up this chain. In 28 U.S.C. § 2284, Congress specified that redistricting cases in particular should start with a three-judge hearing at the district court level—in addition to the usual district court judge, the chief judge of their circuit appoints two additional judges, one of whom must be a circuit court judge. Decisions of these three-judge panels are then appealed directly to the Supreme Court, which has mandatory jurisdiction (see 28 U.S.C. § 1253), meaning they have to do *something* with the case.

In a landmark 1962 case *Baker v. Carr*, the Court abandoned its hands-off approach to apportionment and entered the political thicket by holding that questions about how lines are drawn implicate the Equal Protection Clause of the 14th Amendment to the U.S. Constitution.[3] After *Baker v. Carr*, courts had relatively little difficulty formulating a principle for individual equality—"one person, one vote"—yet a benchmark for evaluating *group harms* proved more difficult to articulate. Thus, even though the Court had decided in 1962 that it had a constitutional role to play in resolving disputes about the design of districts, there was a lot of uncertainty about whether the Court would also resolve political gerrymandering disputes.

In 1973 the Supreme Court seemed willing to apply the Constitution to partisan gerrymandering claims in the case *Gaffney v. Cummings*. *Gaffney* was a bipartisan gerrymandering case where both parties agreed to protect their incumbents. (This arrangement is sometimes referred to as a "sweetheart" gerrymander). The Supreme Court was asked to scrutinize the gerrymandered plan and determine whether it violated the Equal Protection Clause of the U.S. Constitution's 14th Amendment. The Court obliged, but ultimately upheld the plan "because it attempted to reflect the relative strength of the parties in locating and defining election districts."[4]

Gaffney was the last time the Court would unanimously agree that partisan gerrymandering claims are justiciable (though not the last time the Court would uphold a gerrymandered map). In the two major cases after *Gaffney*, *Davis v. Bandemer* and *Vieth v. Jubelirer*, the Court badly split on the question of justiciability and in both cases the Court upheld the gerrymandered plans in the face of strong dissenting opinions.

Thirteen years after *Gaffney*, in the 1986 case of *Davis v. Bandemer*, the Court heard a case about a partisan gerrymander. The Democrats in Indiana filed suit against the Republicans, claiming that the Republicans had violated their constitutional rights by gerrymandering the state's electoral districts to minimize Democratic political power. *Bandemer* showed that the Court had become very divided when it came to the issue of partisan gerrymandering.

Three Justices said that the judicial enterprise of regulating partisan gerrymandering was "flawed from its inception."[5] These Justices would have dismissed the case altogether as nonjusticiable.[6] From their perspective, the federal courts had no business resolving partisan gerrymandering claims. Six Justices, however, reiterated the importance of judicial intervention, but were split on how the courts should proceed. Four of the Justices criticized the trial court's reliance on election outcomes from a single election, writing that "the power to influence the political process is not limited to winning elections"[7] and that "[r]elying on a single elec-

[3]"We conclude that the complaint's allegations [that the state of Tennessee's refusal to redistrict for 60 years] present[s] a justiciable constitutional cause of action upon which appellants are entitled to a trial and a decision," *Baker v. Carr*, 369 U.S. 186, 237 (1962).

[4]*Gaffney v. Cummings*, 412 U.S. 735, 752 (1973).

[5]*Davis v. Bandemer*, 478 U.S. 109, 147 (1986).

[6]"The Equal Protection Clause does not supply judicially manageable standards for resolving purely political gerrymandering claims, and no group right to an equal share of political power was ever intended by the Framers of the Fourteenth Amendment,"478 U.S. at 147.

[7]478 U.S. at 132.

tion to prove unconstitutional discrimination is unsatisfactory."[8] These Justices voted to reverse the lower court, which had struck down the Indiana map, but they failed to offer an evaluative standard for the lower court to apply on "remand" (the term used when a higher court sends a case back to a lower court and asks it to try the case again with new instructions). Finally, two Justices articulated what they believed to be a judicially manageable standard by importing a set of criteria for fair redistricting to guide the lower court:

> "The most important of these factors are the shapes of voting districts and adherence to established political subdivisions. Other relevant considerations include the nature of the legislative procedures by which the apportionment law was adopted and legislative history reflecting contemporaneous legislative goals. To make out a case of unconstitutional partisan gerrymandering, the plaintiff should be required to offer proof concerning these factors, which bear directly on the fairness of a redistricting plan, as well as evidence concerning population disparities and statistics tending to show vote dilution. No one factor should be dispositive."[9]

Overall, a majority of the Court concluded that partisan gerrymandering claims were justiciable, but reversed the decision of the lower court, which had erred in its reliance on the principle that "any apportionment scheme that purposely prevents proportional representation is unconstitutional."[10] The upshot of the Court's 3-4-2 decision was that the partisan gerrymander in Indiana remained in place for the rest of the decade. The split decision was also a harbinger of the courts' inability to develop a consensus on the issue of partisan gerrymandering.

The Supreme Court did not hear another partisan gerrymandering case until 2004, when it split badly again in *Vieth v. Jubelirer*. The question in *Vieth* was "whether [the Court's] decision in *Bandemer* was in error, and, if not, what the standard [for adjudicating partisan gerrymanders] should be."[11] In the eighteen years between *Bandemer* and *Vieth*, the lower courts had struggled to coalesce around a single standard. In his *Vieth* concurrence, Justice Scalia lamented that the *Bandemer* decision "has almost invariably produced the same result (except for incurring of attorney's fees) as would have obtained if the question were nonjusticiable: Judicial intervention has been refused."[12] Scalia was joined by three other Justices in holding that there are no judicially manageable standards to distinguish benign partisan gerrymanders from gerrymanders that go too far. Five Justices disagreed. Pointing to the forty-year history of judicial involvement in highly political cases since *Baker v. Carr*[13] and the Court's duty to protect the fundamental right to

[8] Id. at 135.

[9] Id. at 173.

[10] Id. at 129-130.

[11] *Vieth v. Jubelirer*, 541 U.S. 267, 272 (2004).

[12] Id. at 279.

[13] See, e.g., *Wesberry v. Sanders*, 376 U.S. 1 (1964); *Reynolds v. Sims*, 377 U.S. 533 (1964); *Avery v. Midland County*, 390 U.S. 474 (1968); *Kirkpatrick v. Preisler*, 394 U.S. 542 (1969); *Karcher v. Daggett*, 462 U.S. 725 (1983); *Davis v. Bandemer*, 478 U.S. 109 (1986); *Board of Estimate of City of New York v. Morris*, 489 U.S. 688 (1989).

vote,[14] these Justices emphasized the importance of allowing courts to intervene in gerrymandering cases. However, like the Justices in *Bandemer*, they could not agree on a standard for identifying an invidious gerrymander. Two of the Justices argued that gerrymanders are unconstitutional when excessively partisan; either when partisanship is the *sole* motivation in their design (Stevens) or when partisanship is used in an unjustified way (Breyer). Justices Ginsburg and Souter argued that if a plaintiff could show (1) that she belonged to a cohesive political group, (2) that her group was intentionally cracked or packed into a district with borders that violated traditional districting principles (e.g., contiguity, compactness, and respect for existing political and geographic boundaries), and (3) if she could produce a hypothetical district that increased the political power of her political group with fewer deviations from these traditional principles, then the burden of proof would shift to the state legislature to rebut the plaintiff's evidence.

Justice Kennedy cast a split vote. While he agreed that partisan gerrymandering was a justiciable issue, he rejected the standards proposed by Justices Stevens, Breyer, Souter, and Ginsburg and wrote that he could not think of a single judicially manageable standard. Kennedy concluded: "That no such standard has emerged in this case should not be taken to prove that none will emerge in the future."[15] The resulting 4-1-4 plurality opinion either perfectly illustrates why the Court should not be involved in partisan gerrymandering cases in the first place or reflects just how close the Court is to reining in an abuse of power (just one vote!).

By 2019, Justice Kennedy had retired. In his absence the Supreme Court (again by a single vote) decided that federal courts should extract themselves from the political thicket of partisan gerrymandering. The majority in *Rucho v. Common Cause* wrote

> "Sometimes...the law is that the judicial department has no business entertaining the claim of unlawfulness—because the question is entrusted to one of the political branches or involves no judicially enforceable rights. In such a case the claim is said to present a 'political question' and to be nonjusticiable—outside the courts' competence and therefore beyond the courts' jurisdiction."[16]

The Court then cited to *Baker v. Carr* to signal that the jurisprudence on partisan gerrymandering had come full circle. To fully understand this development, we need to unpack what the Court means when it says a claim presents a "political question." In *Rucho* the Court explains that not all cases dealing with political issues are political questions. Rather, a case poses a political question when there is "a lack of judicially discoverable and manageable standards for resolving it."[17] This definition begs two questions: (1) how does a court distinguish between standards and rules? and (2) what does it mean for a standard to be judicially discoverable and manageable?

[14] "Where important rights are involved, the impossibility of full analytical satisfaction is reason to err on the side of caution," 541 U.S. at 311.

[15] Id.

[16] *Rucho v. Common Cause*, slip op. at 7

[17] Id.

1.2 LEGAL STANDARDS

Simply put, a legal standard is a guideline for judges. A standard typically identifies factors for judges to consider and might provide direction for evaluating the relative weight of each factor, without any single factor being dispositive. Legal standards are less strict than legal rules, which provide bright-line definitions and thresholds, or that otherwise compel a particular outcome. In other words, a legal standard provides judges with more discretion than a legal rule. For example, in the gerrymandering context, a legal rule might dictate that any district with a Reock score less than .1 is unconstitutional or that any districting plan where the efficiency gap is greater than .07 in three or more consecutive elections is unconstitutional. A legal standard, on the other hand, may direct judges to reject districts with "irregular shapes" or to invalidate plans with "historically anomalous" partisan asymmetry. Legal standards may also require judges to balance various interests simultaneously. For example, judges may be asked to balance the rights of individuals to be treated equally against the rights of elected officials to draw new districts after each Census. Or a judge may be asked to balance the interest of government responsiveness (which might suggest more competitive districts) with government representativeness (which might suggest more homogeneous districts). Legal standards provide flexibility to individual judges, which means there is no guarantee of fairness *ex ante*. However, legal standards allow judges to tailor their considerations to each individual case, thus protecting against unintended consequences and ensuring fair outcomes *ex post*.

Importantly, standards can be relevant to both the procedure and substance of a case. For example, in *Rucho* the plaintiffs presented the Court with a procedural standard: a districting plan should be ruled unconstitutional if the evidence shows (1) that districts were drawn with partisan intent, (2) that a districting plan discriminates against voters based on their partisanship, and (3) that there is a nexus between the intent (1) and the effect (2). Baked into this legal standard are empirical questions—how much partisan discrimination is too much?—that can be addressed in ways that look like a "standard" (e.g., outliers among an ensemble of possible districting plans) or that look like a "rule" (e.g., two standard deviations from the median plan among an ensemble of districting plans).

The holding in *Rucho v. Common Cause* is complicated by the fact that the Justices conflated these points in their consideration of the case. At oral argument, the attorney for Common Cause began his testimony with reference to an expert's ensemble of districting plans that showed why North Carolina's gerrymander was an outlier; the 10-3 split in favor of Republicans did not appear a single time in the ensemble of 1,000 plans. Justices Kavanaugh and Roberts interrupted each other to immediately push back on this *empirical* finding with questions about what the appropriate *legal* standard should be: how relevant is the inquiry into the legislature's intent in the first place? When the attorney suggested that the Court adopt the sliding-scale balancing-test legal standard[18] used in other election law cases

[18]According to the "Anderson Burdick" standard referred to by the attorney for Common Cause, courts pragmatically balance the burden or injury imposed by a given election law against the justifications for this burden offered by the state. As the burden becomes more severe, courts require more persuasive and compelling justifications from the state.

(see *Anderson v. Celebrezze* and *Burdick v. Takushi*), Justices Alito and Gorsuch replied by asking the attorney what empirical standard the court should adopt for distinguishing benign gerrymanders from unconstitutional gerrymanders. Further complicating the discussion, Justice Gorsuch proposed non-standard-like numerical cutoffs:

> "So aren't we just back in the business of deciding what degree of tolerance we're willing to put up with from proportional representation? We might pluck a number out of the air or see that, you know, maybe two-thirds is used for veto overrides, so we like that. Where are we going to get the number on the business end of this?"[19]

When the lawyer representing the League of Women Voters stepped up to the lectern, she tried to redirect the Justices back to the three-prong test described above (that courts should look for evidence of (1) discriminatory intent, (2) discriminatory effect, and (3) a nexus between the two). Justice Gorsuch continued his push for a numerical cutoff:

> "...we talked a lot about last year the efficiency gap, which is how far a deviation from proportional representation. And we were told, I think, six or seven percent of deviation would be okay, and that would not be an untoward effect. But anything above six or seven percent. Today we're talking about two-thirds is an effect. We need to have a number or some formula to determine what effect is enough to state a claim and what isn't, otherwise every case is going to come to this Court. And I'm—I'm—I'm still waiting to hear what that might–what that number, what that formula might be..."[20]

The Supreme Court ultimately concluded that "partisan gerrymandering claims present political questions beyond the reach of the federal courts." However, it remains unclear whether the Court reached this conclusion because of unsatisfactory answers related to legal standards, to empirical standards, or to something else. What is clear is that whatever the standards may have been, the majority believed they were not judicially discoverable and manageable, which cuts right to the heart of the political question doctrine.

1.3 JUDICIALLY DISCOVERABLE AND MANAGEABLE

The phrase "judicially discoverable and manageable" is a term of art that was first articulated in *Baker v. Carr*. In *Baker* the Supreme Court decided to assert itself in the debate about apportionment and redistricting just 16 years after explicitly avoiding what it had called the political thicket. In *Baker* the Court distinguished between cases that posed "political questions" (and were thus not justiciable) and cases where judicial intervention would be appropriate. The Court was not seeking to sidestep cases that were political in nature. Instead, the Court was asking whether the federal courts have the competency to adjudicate a particular category

[19] *Rucho v. Common Cause*, Transcript of Oral Argument at 43–44.
[20] Id. at 60–61.

of cases and, if not, to leave the resolution of these cases to the political process. The competency of the courts is measured in two ways.

First, courts must be able to look to the U.S. Constitution for a theory of harm and provide a remedy that flows from a constitutional framework, past practice, and precedent. In other words, for a standard to be "judicially discoverable and manageable" the courts should be focused primarily on legal questions (e.g., has the plaintiff suffered a cognizable harm? Has the government violated the plaintiff's constitutional rights?) as opposed to questions about political theory, public policy, and/or empirical data.

Second, courts as an institution must be properly situated to adequately address the central conflict of a case. Are courts able to adequately compile all of the relevant information to resolve the dispute? Are courts an effective venue for the contest of ideas related to the dispute? Do judges have the capacity to digest the implications of the dispute? Are judicial opinions likely to be viewed as legitimate and binding? Are other branches of government and the general public likely to adhere to judicial conclusions? Because these considerations are largely pragmatic, the Court's invocation of the political question doctrine can appear idiosyncratic and inconsistent over time. Thus, just as the Court invoked the political question doctrine to shield itself from apportionment cases in 1946 only to drop that shield sixteen years later, the Court could easily reassert its voice with respect to partisan gerrymandering in the not-too-distant future. If and when the federal courts reopen their doors to these claims, what kinds of legal arguments would they likely consider? Below we provide an overview of the legal landscape of partisan gerrymandering. We begin by distinguishing the legally thorny and open question of partisan gerrymandering from the equally thorny but more settled caselaw dealing with racial gerrymandering.

2 PARTISAN VS. RACIAL GERRYMANDERING

One argument that opponents of partisan gerrymandering sometimes make is that partisan gerrymandering is just like racial gerrymandering. This is because, in general, judges agree that racial gerrymanders—redistricting plans that dilute the voting power of racial minorities—raise constitutional questions that can be heard in court. One reason that racial gerrymandering has been deemed justiciable is that race is a protected class under the Constitution, meaning laws that treat people differently based on their race can only be justified if they are necessary for achieving a compelling governmental interest (for example, race-based school funding programs in pursuit of educational equality). Courts have assumed the responsibility of refereeing disputes about whether laws that draw distinctions based on race are narrowly tailored and in pursuit of compelling interests. In addition to this constitutional protection, Congress created additional statutory protections for racial minorities in the Voting Rights Act (VRA) of 1965, discussed in detail in Chapter 6–Chapter 7. The VRA explicitly prohibits any election practice or procedure that abridges the right of racial minorities to vote and to elect candidates of their choice. As a result, courts have been open to legal challenges of racial gerrymanders. In sum: over the course of many cases, the Supreme Court has articulated a set of judicially manageable standards to enforce both the Equal Protection Clause of the 14th Amendment and the Voting Rights Act. In the context of the 14th Amendment, courts require plaintiffs to provide evidence that race was the "predominant factor" in deciding where to draw district lines (*Miller v. Johnson*).[21] Courts sometimes point to "bizarrely" or "irregularly" shaped districts as evidence that race was central to their design, but the shape of a district is insufficient by itself to overturn a racial gerrymander. For cases alleging vote dilution under the VRA, courts do not explicitly require plaintiffs to provide evidence of intentional discrimination (though that evidence is always very powerful). Instead they require plaintiffs to show how, in the "totality of circumstances," racially drawn districts "interact with social and historical conditions to cause an inequality in the opportunities enjoyed by black and white voters to elect their preferred representatives" (*Thornburg v. Gingles*).[22] Both the "predominant factor" standard and the "totality of circumstances" test have proven to be judicially manageable, and on their face these standards could easily be imported to cases dealing with partisan gerrymandering. Indeed, one vital skill for lawyers is the ability to draw analogies to relevant, settled precedent and racial gerrymanders present a tempting analogy. It is not surprising, then, that the standards suggested by Justices Stevens and Breyer in *Vieth* to address partisan gerrymandering parallel the "predominant factor" test from the race-based case of Miller v. Johnson. Or that the various factors proposed by the dissenting Justices in *Bandemer*, as well as the multi-pronged test proposed by Justices Souter and Ginsburg in *Vieth*, are easily analogized to the "totality of circumstances" test in VRA cases.

On the other hand, racial gerrymandering differs from partisan gerrymandering in important ways that may limit the power of these analogies. Perhaps most importantly, the original purpose of the 14th Amendment's Equal Protection Clause was to provide legal protection from political oppression directed at former slaves. As an original matter, the equal protection clause was not intended to protect the legal rights of other groups such as women, gays and lesbians, and certainly not Democrats or Republicans as such.[23] Writing for the plurality in *Vieth*, Justice Scalia argued that "the Constitution clearly contemplates districting by political entities" whereas "the purpose of segregating voters on the basis of race is not a lawful one." Thus, "to the extent that our racial gerrymandering cases represent a model of discernible and manageable standards, they provide no comfort here [in the partisan context]."[24] Justice Kennedy was silent in *Vieth* on the question of importing the judicial standards from racial gerrymandering cases. Although he presumably agreed with Justice Scalia that racial gerrymandering cases start on stronger constitutional footing given America's long history of race discrimination, Kennedy's particular concern in *Vieth* was not about constitutional footing; he argued that courts *should* hear partisan gerrymandering challenges. The open question to Kennedy was whether there were judicially discoverable and manage-

[21] 515 U.S. 900 (1995).

[22] 478 U.S. 30, 47 (1986).

[23] As a practical matter, the Supreme Court has interpreted the Equal Protection Clause to guard against all race-based, ethnicity-based, and gender-based discrimination. The Court has also cited to the Equal Protection Clause to protect nonracial groups such as minorities defined by a shared language. See, e.g., *Meyer v. Nebraska*, 262 U.S. 390 (1923).

[24] *Vieth v. Jubelirer*, 541 U.S. at 286.

able standards to evaluate a gerrymander. The racial gerrymandering cases proved that there were, yet Kennedy chose not to adopt them, with very little commentary about his thinking. In *Rucho v. Common Cause* the Supreme Court definitively dismissed the parallels between racial and partisan gerrymandering:

> "Nor do our racial gerrymandering cases provide an appropriate standard for assessing partisan gerrymandering. Nothing in our case law compels the conclusion that racial and political gerrymanders are subject to precisely the same constitutional scrutiny. In fact, our country's long and persistent history of racial discrimination in voting—as well as our Fourteenth Amendment jurisprudence, which always has reserved the strictest scrutiny for discrimination on the basis of race—would seem to compel the opposite conclusion. Unlike partisan gerrymandering claims, a racial gerrymandering claim does not ask for a fair share of political power and influence, with all the justiciability conundrums that entails. It asks instead for the elimination of a racial classification. A partisan gerrymandering claim cannot ask for the elimination of partisanship."[25]

Because the Court limited its comparison to just racial gerrymandering cases, it is possible that a future Court could find important parallels and lessons from racial vote dilution cases, which *do* ask for a fair share of political power and influence, which *do* provide a group-based remedy for an individualized harm, and which *do* involve an inquiry into both the discriminatory intent and effect of the law. Notably, however, the appellees (Common Cause et al.) in *Rucho* relied heavily on the logic of vote dilution in their briefs and testimony, with almost no references to racial gerrymandering. The parties and the Court were talking past each other. In any case, future attempts to articulate a new standard will likely need to distinguish themselves from, not analogize themselves to, the standards used by courts in the race cases.

3 CONSTITUTIONAL PROVISIONS REGULATING PARTISANSHIP

3.1 GUARANTEE CLAUSE

One of the earliest proposed legal standards for evaluating partisan gerrymanders was based on the argument that these gerrymanders deprive individuals of a truly representative government. Article IV of the U.S. Constitution reads, "The United States shall guarantee to every State in this Union a Republican Form of Government." According to one possible legal standard based on this Guarantee Clause, a partisan gerrymander would violate the constitution when districts skew electoral outcomes so much that members of the state legislature do not properly represent a state's population. Relevant data for this inquiry might include

[25] *Rucho v. Common Cause*, slip op. at 21.

a comparison of policy outcomes to the preferences of the electorate,[26] or more simply a comparison of electoral outcomes to the distribution of registered voters as evidence that elected officials and their constituents are not ideologically aligned.[27] The Supreme Court has repeatedly rejected the Guarantee Clause as an appropriate basis for any legal challenge, let alone partisan gerrymandering cases. In 1849 the Supreme Court declared that legal challenges based on the Guarantee Clause are nonjusticiable. The case, *Luther v. Borden*, held that "no one, we believe, has ever doubted [that] the sovereignty in every State resides in the people of the State, and that they may alter and change their form of government at their own pleasure."[28] Later challenges to the ballot initiative process and other direct democracy procedures that allegedly violate the Guarantee Clause (for excluding elected officials) were dismissed by the Court.[29] In 1946, the Court summarily dismissed a challenge that Illinois's congressional districts violated the Guarantee Clause by writing a "violation of the great guaranty of a republican form of government in States cannot be challenged in the courts." (*Colegrove v. Green*).[30] As we outlined above, the Supreme Court changed its position about the justiciability of apportionment and districting procedures in 1962, but its decision was not rooted in a new understanding of the Guarantee Clause. Instead the Court held that malapportionment and gerrymandering implicate rights protected under the Equal Protection Clause of the 14th Amendment. Thus, future challenges to partisan gerrymandering based on the Guarantee Clause are likely dead on arrival, barring a reversal of long-standing precedent.

3.2 EQUAL PROTECTION CLAUSE

The Supreme Court's holding in *Baker v. Carr* that an apportionment scheme might deprive voters of equal protection of the laws not only introduced judicial review into the districting process, but also helped to clarify what harm(s) the Court was worried about. As a result, nearly every challenge to partisan gerrymandering since 1962 has relied on the Equal Protection Clause. The 14th Amendment to the U.S. Constitution states, in part, that "No State shall...deny to any person within its jurisdiction the equal protection of the laws." A legal standard based on equal protection requires a court to examine whether a partisan gerrymander is rationally related to a legitimate state interest.[31] This inquiry has two factors: (1) are partisan considerations illegitimate? and (2) is the gerrymander itself so invidious that it lacks any rational justification? With respect to the first factor, the Supreme Court has acknowledged that as long as redistricting is overseen by legislatures, partisan considerations will naturally play a role in the process. Thus, a holding that any gerrymander created with partisan intent violates the Equal Protection Clause will open the floodgates of litigation since every single district in the United States is

[26] See, e.g., Lax, Jeffrey R. and Justin H. Phillips. 2012. "The Democratic Deficit in the States," American Journal of Political Science, Vol. 56, No. 1, pp. 148–166.

[27] See, e.g., Stephanopoulos, Nicholas O. 2014. "Elections and Alignment," Columbia Law Review, vol. 114, pp. 283–365.

[28] 48 U.S. 43 (1849).

[29] See, e.g., *Pacific States Telephone & Telegraph Co. v. Oregon*, 223 U.S. 118 (1912).

[30] 328 U.S. 549, 556 (1946).

[31] Compare this with the stricter standard for racial gerrymanders discussed above, which must be narrowly tailored to achieve a compelling state interest (a standard referred to as "strict scrutiny").

infected with some level of partisanship. To avoid this floodgate, the Court could evaluate partisan consideration along a continuum and articulate a threshold beyond which partisan motives become illegitimate.[32] As a reminder, two Justices in the 2004 case *Vieth v. Jubelirer* adopted this approach and argued that partisan motivation is okay, but becomes illegitimate when "the minority's hold on power is purely the result of partisan manipulation and not other factors"[33] or "when any pretense of neutrality is forsaken unabashedly and all traditional districting criteria are subverted for partisan advantage."[34] However, this approach did not garner a majority of votes in *Vieth* and one of the two Justices that espoused this view (Stevens) has since retired.

The remaining factor for assessing whether a gerrymander violates the Equal Protection Clause asks courts to evaluate whether the redistricting plan is so extreme that it lacks any rational relationship to representative government. This inquiry has split the Supreme Court as well, with several Justices presenting their own view about when a gerrymander becomes too extreme. Some Justices have focused on the districting *process*: were both parties involved in the plan's design? Did the legislature hold public hearings or meetings? Were traditional districting criteria considered and/or followed? Others have focused on the real and predicted effects of the redistricting plan. This latter assessment of a plan's *effects* has garnered the most attention by the news media, academics, and others. Much like the judicial inquiry into the motivations behind a gerrymander, the analysis of a districting plan's *outcomes* looks at a continuum of partisan differences and tries to identify a threshold beyond which the unequal political opportunities of the minority party goes too far. Courts have measured this unequal opportunity in various ways, using many of the metrics featured in other chapters of this volume. One theme that runs through many lower court opinions is partisan symmetry, which the Supreme Court has described as "a helpful (though not talismanic) tool."[35] The principle of partisan symmetry states that the two parties should be treated equally in structural or systematic terms. One way to operationalize this principle is to compare counterfactual voting data: for instance, if Democrats win 48% of the vote but earn 55% of the seats in a given election, partisan symmetry demands that in other circumstances, if Republicans had won 48% of the vote they should have earned 55% of the seats. The high-level logic of partisan symmetry has guided the courts in their evaluations of various gerrymanders, though the courts have not coalesced around any particular measure, and in a 2006 challenge to a partisan gerrymander in Texas Justice Kennedy wrote that "asymmetry alone is not a reliable measure of unconstitutional partisanship."[36] Metrics inspired by the logic of partisan symmetry include numerical differences in district-based voter distributions such as the efficiency gap and the mean–median difference. After

[32] Justin Levitt has argued that partisanship should be evaluated along a spectrum that distinguishes partisan considerations by type (e.g., coincidental, ideological, responsive, tribal) rather than by degree. See Levitt, Justin. 2014. "The Partisanship Spectrum," William & Mary Law Review, vol. 55, pp. 1787–1868. See also Levitt, Justin. 2018. "Intent is Enough: Invidious Partisanship in Redistricting," William & Mary Law Review, vol. 59, pp. 1993–2051.

[33] 541 U.S. at 360.

[34] Id at 318.

[35] *League of United Latin American Citizens v. Perry*, 548 U.S. 399, 468 (2006) (fn. 9).

[36] *LULAC v. Perry*, 548 U.S. 399, 420 (2006)

Davis v. Bandemer, all measurements are additionally expected to demonstrate the durability of a partisan effect (e.g., predicted losses over multiple election cycles). In *Rucho* the plaintiffs relied on both symmetry measures and on ensembles of districting plans—hundreds of thousands of plans sampled from the universe of all possible plans—that incorporated traditional districting criteria such as compactness and contiguity. In this approach, the challenged plan was flagged as an extreme gerrymander because it was an outlier among the ensemble's distribution.

In short, judges have a plethora of metrics available to them for deciding when, how, and why a redistricting plan potentially violates the Equal Protection Clause. This surplus of options has proven to be a double-edged sword as individual judges have gravitated toward different measures in different cases. However, all four of the dissenting Justices in *Rucho* signed on to the opinion that outlier analysis of ensembles is the most promising approach for determining a partisan gerrymander. No other single approach has ever been endorsed by as many Justices, which should serve as a signal to lower court judges if the federal courts reopen their doors to partisan gerrymandering claims. Whether or not this happens, equality-based arguments at the state level (more on this below) are likely to be buttressed by ensembles and outlier analysis.

3.3 FIRST AMENDMENT

Whereas the Equal Protection Clause protects individuals from being treated differently in general, the First Amendment protects individuals from viewpoint discrimination by the government and has been used to protect the associational rights of political organizations. The First Amendment states that "Congress shall make no law...abridging the freedom of speech." This Free Speech Clause prevents the government from treating people differently *based on their political beliefs*. Justice Kennedy highlighted this possible constitutional hook in his *Vieth* opinion when he wrote:

> "The First Amendment may be the more relevant constitutional provision in future cases that allege unconstitutional partisan gerrymandering. After all, these allegations involve the First Amendment interest of not burdening or penalizing citizens because of their participation in the electoral process, their voting history, their association with a political party, or their expression of political views."[37]

Pointing to various First Amendment cases, Kennedy concluded that "In the context of partisan gerrymandering, First Amendment concerns arise where an apportionment has the purpose and effect of burdening a group of voters' representational rights." Justice Kagan picked up on this idea in her concurring opinion in *Gill v. Whitford* in 2018, citing to Kennedy six times and arguing that "partisan gerrymandering no doubt burdens individual votes, but it also causes other harms," specifically associational harms under the First Amendment.[38] Unlike the Fourteenth Amendment, which has generated many metrics measuring deviations from equality, there is no developed caselaw or clearly articulated standard for

[37] *Vieth v. Jubelirer*, 541 U.S. 267, 314 (2004) (J. Kennedy, concurring).
[38] Id.

courts to consider in the face of a First Amendment challenge to partisan gerrymandering. In *Rucho v. Common Cause* the Court considered a First Amendment challenge to Maryland's 6th congressional district, which was represented by a Democrat for twenty-two years between 1971 and 1993, and then by a Republican for twenty years from 1993 to 2013. (This case was consolidated with the challenge to North Carolina's districts.) During the 2011 redistricting cycle, the 6th district was redrawn to heavily favor Democrats who controlled both legislative chambers and the governor's mansion. The result was immediate and drastic. The Republican incumbent who was reelected in 2010 by 28 points lost his bid in 2012 by more than 20 points. A group of Republican voters filed suit, alleging that the 6th district was reconfigured in 2011 as retaliation for supporting Republican candidates—in other words, retaliation based on how they voted—in violation of the First Amendment. At trial, the lower court held that to prevail the plaintiffs must provide evidence that (1) the district was intentionally drawn to burden voters based on how they voted, (2) that the burden resulted in a "tangible and concrete adverse effect," and finally that (2) was caused by (1). Upon a showing of all three, the burden would shift to the state to prove there was some lawful alternative to explain the district's design. Notice that the trial court's inquiry is not very different from the framework of the Equal Protection Clause. In both cases, courts look for evidence of partisan intent and partisan effects. The difference is that under the First Amendment the effects are framed in terms of their burden on voters instead of the relationship of voting power between different groups.

While the lower court concluded that the redistricting plan in Maryland violated the First Amendment associational rights of Republican voters, the Supreme Court in *Rucho* vacated the ruling on the grounds that there was "no 'clear' and 'manageable' way of distinguishing permissible from impermissible partisan motivation."[39] The lower court found Republicans in Maryland's 6th district had suffered in their attempts to fundraise, attract volunteers, and generate interest in voting after having their voting power diluted. The majority in *Rucho* was unconvinced, asking "how many door knocks must go unanswered? How many petitions unsigned? How many calls for volunteers unheeded?"[40] Appellees did not have a satisfactory answer for these questions.

To date, relatively scant attention has been paid to the First Amendment in the gerrymandering context, and there is room for important contributions by mathematicians, social scientists, computer scientists, and others defining the scope of adverse impacts on the associational rights of voters and parties, measuring the severity and extent of this "burden," and conceptualizing the operation of "but-for causation" as part of the inquiry.

3.4 ELECTIONS CLAUSE

Legal scholars have also pointed to the Elections Clause of the Constitution as a way to walk the fine line of maintaining a republican form of government while

[39] *Rucho v. Common Cause*, slip op. at 27.
[40] Id. at 26.

limiting the judicial role as much as possible.[41] The Elections Clause is found in Article 1§4 of the Constitution, which reads, “The times, places and manner of holding elections for Senators and Representatives, shall be prescribed in each state by the legislature thereof; but the Congress may at any time by law make or alter such regulations.” Richard Pildes notes that under the Elections Clause, partisan gerrymandering is an explicit yet limited power. Because the Supreme Court has accepted since *Marbury v. Madison* in 1803 that one of its central roles is to police the limits of powers that are enumerated in the Constitution, the Elections Clause provides support to the claim that partisan gerrymandering is a justiciable issue.[42] The majority in *Rucho v. Common Cause* curtly dismissed this argument in a single sentence: “we are unconvinced by that novel approach.”[43]

In a different line of attack, Ned Foley has argued that because the Elections Clause grants ultimate authority over the “places and manner” of elections to Congress, Congress could immediately supersede any federal court ruling it disagreed with. This power-sharing arrangement should presumably clip the wings of those who worry that the Court is overstepping its properly prescribed judicial role, such as Justice Kennedy and Chief Justice Roberts. According to Foley, when a court nullifies a state's congressional map under the Elections Clause, it acknowledges that Congress is free, even welcome, to intervene. The Court in *Rucho* recognized this power-sharing arrangement with Congress, but held that Congress must be the first mover, no matter how small the action. In fact, the Court acknowledged that a resolution from Congress merely stipulating that no districting plan shall be drawn to unduly favor any person or party would provide the green light for courts to reassert themselves as gerrymandering referees.

3.5 STATE CONSTITUTIONS

All nine Justices in *Rucho v. Common Cause* encourage states to take up the task of policing gerrymanders, whether through litigation or the political process. The majority concludes its opinion with the following consolation:

> “Our conclusion does not condone excessive partisan gerrymandering. Nor does our conclusion condemn complaints about districting to echo into a void. The States, for example, are actively addressing the issue on a number of fronts.”[44]

Most state constitutions explicitly protect the right to vote and provide for free and fair elections. For example, Article I §10 of the North Carolina state constitution declares that “all elections shall be free.” In September 2019, a panel of three state judges pointed to this clause in the state's constitution to invalidate the same gerrymandered plan that the U.S. Supreme Court had side-stepped in *Rucho*. The judges argued that the “Free Elections Clause is one of the clauses that makes the North

---

[41] Foley, Edward B. 2018. “Constitutional Preservation and the Judicial Review of Partisan Gerrymanders,” University of Georgia Law Review (forthcoming 2018).

[42] Pildes, Richard H. 2018. “The Elections Clause as a Structural Constraint on Partisan Gerrymandering of Congress,” SCOTUSblog Symposium, 19 June, 2018.

[43] Slip op. at 29.

[44] Slip op. at 31.

Carolina Constitution more detailed and specific than the federal Constitution in the protection of the rights of its citizens"[45] and held that "extreme partisan gerrymandering. . . is contrary to the fundamental right of North Carolina citizens to have elections conducted freely and honestly."[46] The Pennsylvania state constitution includes a similar provision that "Elections shall be free and equal; and no power, civil or military, shall at any time interfere to prevent the free exercise of the right of suffrage." Eighteen plaintiffs, one from each congressional district in Pennsylvania, recently challenged the 2011 congressional maps that were drawn by the Republican-majority legislature. The plaintiffs, all Democrats, alleged that the maps were unconstitutional under the state constitution. In February 2018 the state supreme court sided with the plaintiffs and struck down the plan as an unconstitutional gerrymander under Article 1§5 of the state constitution. The court wrote that "while federal courts have, to date, been unable to settle on a workable standard by which to assess such claims under the federal Constitution, we find no such barriers under our great Pennsylvania charter...We conclude that in this matter [the state constitution] provides a constitutional standard, and remedy, even if the federal charter does not."

The standard adopted by the court looked similar to the approach taken in federal courts with respect to the Equal Protection Clause. The court noted that traditional redistricting criteria were subordinated to partisan motivations. Experts at trial also provided evidence that the congressional map was an outlier compared to an ensemble of thousands of alternative plans with respect to measures of compactness and partisanship. The efficiency gap was also used to show that Republicans experienced significant partisan advantage under the challenged plan. Based on this evidence, the court held that because the 2011 congressional map "aimed at achieving unfair partisan gain," it "undermines voters' ability to exercise their right to vote in free and 'equal' elections if the term is to be interpreted in any credible way." When the legislature could not generate a satisfactory new map, the court hired an outside expert called a "special master" to draw a new map that was hastily installed in time for the 2018 midterm election, which saw the delegation swing from 13-5 to 9-9 under the new plan.

The experience in Pennsylvania is not without controversy. The Republican-majority legislature was upset by the court's ruling and the president of the state senate filed ethics complaints against the judges. A dozen state lawmakers later threatened to impeach the judges who voted to strike down the map, an act that threatened the separation of powers and independent judiciary in Pennsylvania, but also plays into the fears of Chief Justice Roberts who has repeatedly worried that when the U.S. Supreme Court inserts itself into political matters the entire judiciary risks being perceived as partisan and biased. Nevertheless, the Pennsylvania state supreme court illustrates that state constitutions and state courts are relevant and important to the inquiry into partisan gerrymanders. In other words, there are fifty legal frontiers waiting to be explored.

[45] *Common Cause v. Lewis*, 18 CVS 014001 at p. 299.

[46] Id. at p. 302.

4 ALTERNATIVE APPROACHES

Although nearly every challenge to partisan gerrymandering relies on constitutional language, two alternative approaches are worth noting.

4.1 RACE AS PARTY

First, because the distinction between partisan and racial gerrymanders is not always very clear, one possible strategy is to focus on the racial effects of a partisan gerrymander. In other words, instead of analogizing to the legal standards used in racial cases, the idea is to coopt the racial gerrymandering framework altogether. Race and party are currently correlated quite closely in most states, meaning a partisan gerrymander is likely to look like a racial gerrymander.[47] To the extent that courts are receptive to challenges based on race, not party, plaintiffs are more likely to succeed in striking down a partisan map if they focus on the racial effects. In 2016, the 4th Circuit invalidated a partisan-motivated voter suppression bill in North Carolina (that did not include a redistricting plan) in part because the state legislature was shown to have used race as a proxy for partisanship. See North Carolina State Conference of *NAACP v. McCrory* (2016). In the gerrymandering context, however, the correlation between race and party has not yet doomed any partisan gerrymanders on racial grounds. On the contrary, states have defended racial gerrymanders by arguing that the true motivation was partisanship, and in one case (see below) a racial gerrymander was replaced with an openly partisan one.

In Texas, a three-judge panel struck down the state's 2011 congressional redistricting plan because it was an impermissible racial gerrymander. The state had argued that there was no proof their plan was "enacted for the purpose of diluting minority voting strength rather than protecting incumbents and preserving Republican political strength won in the 2010 elections." Nevertheless, the state updated its redistricting plan in 2013. This new plan met a similar fate at the lower court, which found that Texas had used "race as a tool for partisan goals" with the goal of "intentionally destroy[ing] an existing district with significant minority population that consistently elected a Democrat." However, in *Abbott v. Perez* (2018) the U.S. Supreme Court reversed the lower court and upheld the 2013 districting plan.

North Carolina's 2011 congressional redistricting plan was also struck down by a three-judge panel because the court held that two districts were unconstitutional racial gerrymanders. The Supreme Court upheld this ruling in *Cooper v. Harris* (2017) while applauding the "formidable task" of lower courts who must make a "sensitive inquiry into all the circumstantial and direct evidence of intent to assess whether the plaintiffs have managed to disentangle race from politics and prove that the former drove a district's lines." Recognizing the legal implications of racial vs. partisan gerrymandering, the North Carolina state legislature responded by enacting a bold and extreme partisan gerrymander. Representative David Lewis,

[47] See, e.g., Hasen, Richard L. 2018. "Race or Party, Race as Party, or Party All the Time: Three Uneasy Approaches to Conjoined Polarization in Redistricting and Voting Cases," William & Mary Law Review, Vol. 59, No. 5, pp. 1837–1886.

who co-chaired the legislature's Joint Select Committee on Redistricting, openly acknowledged his desire to maximize Republican seats. At one hearing he argued that the goal was to "draw the maps to give a partisan advantage to ten Republicans and three Democrats because I do not believe it's possible to draw a map with 11 Republicans and two Democrats." When confronted with the observation that this was the very definition of a partisan gerrymander, Lewis responded that "a political gerrymander is not against the law." Representative Lewis was right, at least for the time being. Despite his brazen statements, the Supreme Court in *Rucho v. Common Cause* (2018) declined to prevent the districts from being used during the 2018 election, and finally found for the defendants in 2019, with a sweeping new determination that partisan gerrymandering is nonjusticiable.

4.2 STATE AND FEDERAL STATUTES

Finally, legal standards and remedies are available through political channels. Much like the Voting Rights Act has proven especially powerful in cases challenging racial gerrymanders, Congress and state legislatures can enact statutes to complement (or substitute for) constitutional protections. Remember that the Supreme Court has explicitly rejected numerous times the argument that the Constitution guarantees a right to proportional representation. For example, when the lower court in *Bandemer* invalidated Indiana's 1981 partisan gerrymander because it "purposely prevented proportional representation," the Supreme Court reversed the decision and wrote that "our cases clearly foreclose any claim that the Constitution requires proportional representation or that legislatures in reapportioning must draw district lines to come as near as possible to allocating seats to the contending parties in proportion to what their anticipated statewide vote will be," 478 U.S. 109, 130 (1986). Although the Constitution doesn't *require* proportional representation, this does not mean that proportional representation is *forbidden.* Any state legislature is free to enact a benchmark of proportional representation into its redistricting process (or partisan symmetry, or a mean–median threshold, or some other metric of fairness). And Congress is also free to create a benchmark of proportional representation (or other criteria) for states to follow when drawing congressional districts. Although this outcome is unlikely in many states, the point is that partisan gerrymandering is as much a political issue as a legal one. In other words, states are not constrained by the limited number of provisions in the U.S. Constitution that speak to fairness and representation. While proportional representation may be a pipe dream, states are also free to adopt redistricting standards that incorporate any new ideas at all: ensembles, curvature, the efficiency gap, Reock scores, you name it.

5 A CALL TO ACTION

The purpose of this book is to introduce readers from diverse backgrounds to the challenges of defining and evaluating all kinds of gerrymanders. Our goal in this chapter is to press these constituencies to understand the multiple levels of why political gerrymandering is a difficult legal problem to solve. Gerrymandering is a sickness whose cause(s) and cure(s) are hotly contested. Imagine you are sick and a series of doctors says "you are going to die, but we have no clue why and no idea how to cure you. All we know is that we are unanimous in the fact that you are going to die." Understanding the cause(s) of gerrymandering is just as important as, and perhaps necessary to, understanding its cure(s). This understanding can be evidence-based, driven by logic and theory, or even based on intuition. But our understanding needs to be articulated clearly, widely disseminated, and ultimately popularized.

Empirical data may or may not provide a magic bullet that resolves this issue, but quality data will no doubt play an important role. For example, to the extent that the Court is concerned about the durational impact of a gerrymandering plan (see, e.g., in *Bandemer v. Davis*) data on population shifts, changing sentiments, and campaign strategy may prove useful. To the extent that the Court's preferred standard is based on the idea of partisan symmetry, metrics like the efficiency gap and the mean–median difference will be more relevant. Hand-drawn demonstration plans and computer-generated alternative plans will presumably continue to be used as persuasive evidence. And whether courts evaluate electoral harms under the Equal Protection Clause or the First Amendment, debates about the best way to combine data from past elections, to project future elections, or to harness other predictive analytics are likely to come into play.[48]

Perhaps most importantly, despite our disproportionate focus on the U.S. Supreme Court, there are real opportunities for action at the state and local level. Justice Brandeis famously wrote that "it is one of the happy incidents of the federal system that a single courageous State may, if its citizens choose, serve as a laboratory; and try novel social and economic experiments without risk to the rest of the country."[49]

Regardless of the outcome of the various gerrymandering cases the Court will hear in the next few years, there is room for readers to exercise their preferred solutions; first at the local and state level and then perhaps at the federal level. School boards, city councils, and state legislatures can be laboratories for establishing new frameworks that will change the way people think about gerrymandering. These laboratories are also venues for testing out new theories, experimenting with different methodologies, and observing the effectiveness of various remedies (e.g., single-member districts vs. transferable votes). More testing will lead to more understanding, and more understanding will improve the quality of challenges raised in the courts in the future.

Justice Felix Frankfurter served on the Supreme Court from 1939 to 1962. He was skeptical that courts were equipped to address the problems of political gerrymandering. Whether or not you agree with his view on the justiciability of these issues, Frankfurter gently reminded readers that the source of all government power "ultimately lies with the people...the vigilance of the people in exercising their political rights."[50] We are far from a settled legal equilibrium, meaning all hands on deck.

[48] For a more detailed framework about how empirical data and mathematical tools can most effectively be utilized in litigation, see Wang, Samuel S.-H. 2018. "An Antidote for Gobbledygook: Organizing the Judge's Partisan Gerrymandering Toolkit into a Two-Part Framework," Harvard Law Review Blog, 11 April at https:// blog.harvardlawreview.org/an-antidote-for-gobbledygook-organizing-the-judges-partisan-gerrymandering-toolkit- into-a-two-part-framework/.

[49] *New State Ice Co. v. Liebmann*, 285 U.S. 262, 388 (1932)

[50] *Colegrove v. Green*, 328 U.S. 549, 554–555 (1946).

III

Geography

Chapter 10

Race, space, and the geography of representation

CHRISTOPHER S. FOWLER

CHAPTER SUMMARY

This chapter examines gerrymandering through the lens of race and geography. Fowler combines specific case studies with high-level concepts in geography, such as the Modifiable Areal Unit Problem.

1 INTRODUCTION

From its concrete links to the Voting Rights Act to broader concerns of social justice and representation, the geography of race is critical to any discussion of districting. Drawing districts that offer fair representation for minority populations requires that we take into account not only how and to what extent populations are currently clustered or dispersed, but also that we understand and account for the spatial processes that have led to clustering or dispersion.

In this chapter, I look at the geography of race as it applies to districting. In §2, I argue that contemporary population distributions may strongly impact prospects for district representation, but that they should not be taken for granted or considered neutral, natural, or fixed. I turn to a case study of Philadelphia to examine the ways that historical policies of the local and federal government segregated the Black population. The resultant distribution of Philadelphia's Black communities structures the possibilities for district-based representation, simultaneously producing opportunities for representation and vulnerability to gerrymandering. Considering these and other effects of segregation, I make the case that we must be

M. Duchin, O. Walch (eds.), *Political Geometry*, https://doi.org/10.1007/978-3-319-69161-9_10

cautious in treating the current spatial distribution of the Black population as the basis for an unbiased map. In §3, I survey ways that geographers grapple with the effects of units and scale, introducing the *Modifiable Areal Unit Problem* (MAUP). I suggest that multiscale measures can illuminate questions of community and neighborhood and can give us traction on fair practices in redistricting. By examining the underlying geographic processes that produce what we ultimately call "the data" used for districting, we can do more to adapt the districting process to treat marginalized groups fairly.

2 POPULATION DISTRIBUTION: CAUSE AND EFFECT

Geographers examine both the distribution of things in space and the processes that produced those distributions. This includes outcomes like segregation or intermixing of populations as well as processes like the expression of housing preferences or the impact of lending institutions on home buying outcomes. Disentangling proximity and causality requires careful attention to boundaries, spatial outliers, edge effects, and scale. In this chapter, I want to look at some factors relevant to the spatial arrangements of people and try to bring all of these concepts to bear.

2.1 CONCENTRATION AND SEGREGATION IN PENNSYLVANIA

How do various population-level attributes come to spread out or concentrate in the ways that they do? And what are the impacts of these human distributions? We have already seen in Chapter 0, Chapter 2, and Chapter 5 (and really throughout the book) that the distributions of people can cause various effects in redistricting, but the distributions themselves are caused by other effects. Walking back along this chain of causality can help us to better understand what constitutes fair practices in drawing districts.

In collaboration with political scientist Linda Fowler, I examined the distribution of various socio-economic variables across tens of thousands of randomly generated 18-district Congressional plans for the Commonwealth of Pennsylvania. Our research asks how much the district-level properties can look different from the state-level properties, as we sample districting plans at random. Our research asks how much difference there can be (a) among districts in the same plan and (b) between each district and the state as a whole, as the lines vary. Based on data from the 2010 decennial Census and the 2008–2012 American Community Survey we find that for characteristics like **percent over 65** and **percent in Manufacturing** there is very little variation between districts and between plans. In other words, no matter where the boundaries between electoral districts are placed, the population composition in each district remains close to that of the Commonwealth as a whole.

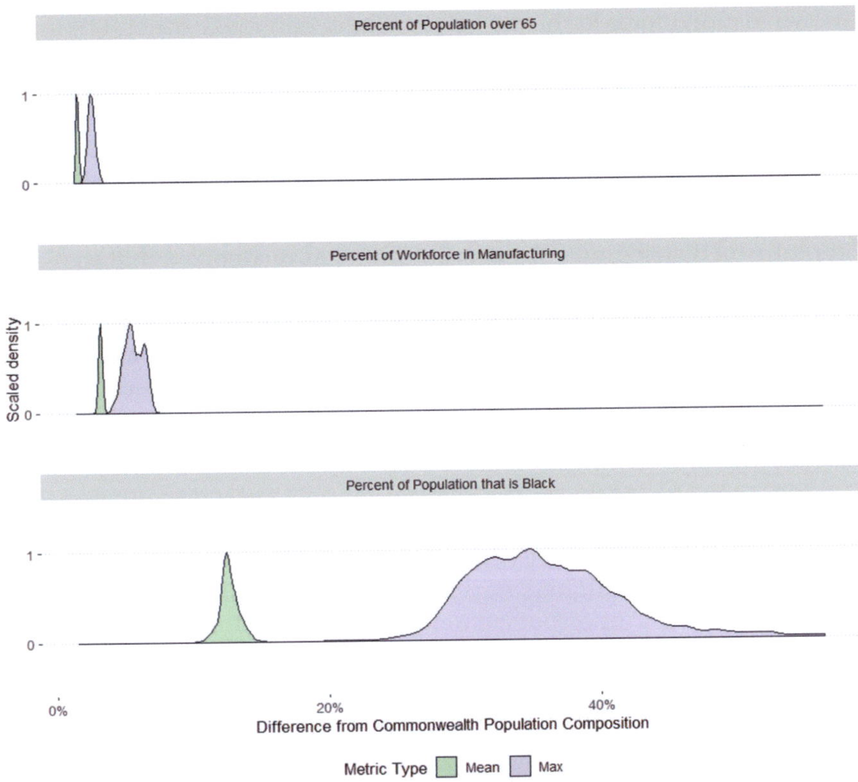

Figure 1: For Pennsylvania's 18 Congressional districts, it is far easier to control the composition of the Black population than it is for other socio-economic characteristics that are more evenly distributed. In these density plots, districting plans appearing farther to the right vary more from the state average on the measured variable. Likewise, a wider curve means that boundary shifts can change how evenly the variable is distributed within a plan. For the share of seniors, most plans show an average of 1–2% variation from the state average of 15.42%. In these plans the district with the maximum difference from the state is usually only about 3–4% different. Manufacturing employment varies a bit more, with an average of 3–5% and maximally different districts as high as 7–8% more or less than the State average of 11.83%. In comparison, for percent Black the average plan has a mean difference of 12–14% from the state average with maximally different districts frequently more than 40% above the state average.

In contrast, the **percent Black** variable varies widely between districts and plans. This is because the Black population is concentrated primarily in a few parts of the state. Thus, the variation of Black population between districts is itself subject to significant change depending on how districts are drawn. This contrast is illustrated by Figure 1, which shows the distribution of both mean district distance and maximum district distance from the state average for selected population measures across a set of randomly drawn plans. In applied terms, this analysis demonstrates that it is relatively difficult to draw districts that concentrate or disperse seniors or the manufacturing workforce, but relatively easy to manipulate the Black population share.

However, this analysis of manipulability has to be viewed alongside representability in the first place. Seniors make up 15.42% of Pennsylvania's population in the 2010

Census, while individuals in the manufacturing sector make up 11.83% and those identifying as Black make up a nearly identical 11.87%. It is never the case that seniors or those in the manufacturing sector can constitute the majority in a district, while random plans found up to *three* majority-Black districts out of 18.

REPRESENTATIONAL CONSEQUENCES

The literature on districting and representation has recognized that spatially concentrated populations can be at an advantage or a disadvantage in the districting process. On one hand, concentrated populations are more easily 'cracked' or 'packed' than their dispersed counterparts, but on the other hand, dispersion makes representation impossible (Chen and Rodden 2013; Rodden 2019; Webster 2013; but see Jiahua Chen et al. 2019).

Even aside from how distribution drives representational possibilities in spatial terms alone, it may have other consequences for political representation. Concentration is closely associated with what political theorists call "descriptive representation"—electing representatives who share identity characteristics with their voting base—and some researchers have argued that district-based elections have negative consequences for dispersed minorities, but not for concentrated minorities (Bowler, Donovan, and Brockington 2003). And sociologist Robert Vargas (2016) notes other significant impacts from cracking that go beyond electoral representation. Latino communities in Chicago that were kept whole in the districting process could successfully request support from local government programs, whereas cracked communities went without.

In general, though, there is a risk that geographic concentration might be treated as a 'natural' situation with remedies that are outside the purview of 'fair' districting. This book is coming at a time when algorithmic sampling (ensemble) methods are on the rise, leveraging distributions of randomly generated plans to distinguish gerrymandering from neutral consequences of the spatial reality (Chen and Cottrell 2016; Chin, Herschlag, and Mattingly 2019; Duchin 2018; Ramachandran and Gold 2018). Finding fairness in this way risks cementing the distribution of each population as a given. However, contemporary population distributions of race, economics, and partisanship are far from natural, in the sense of indifferent or unbiased. To see this, let's consider a few of the historical factors leading to segregation and concentration, and investigate the ways that this has been and still is associated with significant costs to certain segregated communities. For our main example, we will look at the Black population in Philadelphia, Pennsylvania.

10.1 LEGACIES OF SEGREGATION

Neighborhood change in Philadelphia offers a lens into the mixed legacies of segregation and mobility. In the maps below (Figure 2) we present a series of maps of population by race. The first map shows White versus non-White population in 1940, the first year that Census tract-level data were available for Philadelphia (https://data2.nhgis.org). It is also the closest Decennial Census to the height of redlining. Historical sources—including redlining maps themselves—suggest that many, though far from all, people, in the "non-White" category in Philadelphia were Black. After this, the maps show relative Black and White populations for each Decennial Census. It is important to note here that the maps are all by Census tract. Tract boundaries are shown, and the maps should not be used to make inferences about segregation at the sub-tract level. Also, except for the first map, the series only shows Black and White populations, and does not include people of more than one race.

The maps thus offer only a limited set of snapshots. What these maps do show, however, is that Philadelphia has grown both more integrated and more segregated in the last eighty years. Major Black populations North, South, and West of downtown have indeed been present from 1940 to 2010. However, the more precise locations of some of these groups has varied. For example, the area west of downtown has long been considered West Philadelphia, an historically Black, working class neighborhood. The Easternmost part is where the University of Pennsylvania is located, and the University's presence has, especially more recently, caused gentrification and pushed Black communities further west. On an only somewhat similar note, Black populations have indeed been mobile, finding residence in more areas of the city. This is perhaps especially obvious starting in 1970, but overall the maps suggest that the problematic constraints of redlining and racial covenants had lingering effects, but that Black people were able to overcome some of them in terms of finding more places in the city to establish communities. A final point worth making is in terms of integration. Philadelphia is far from fully integrated, but some integration between Black and White people is present.

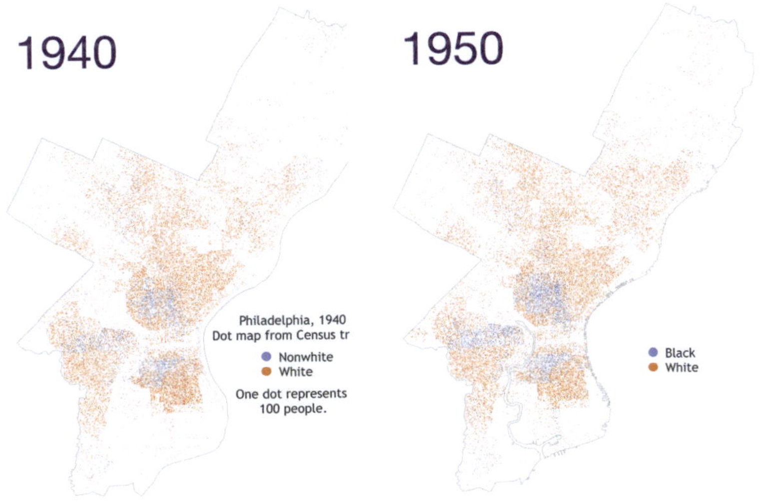

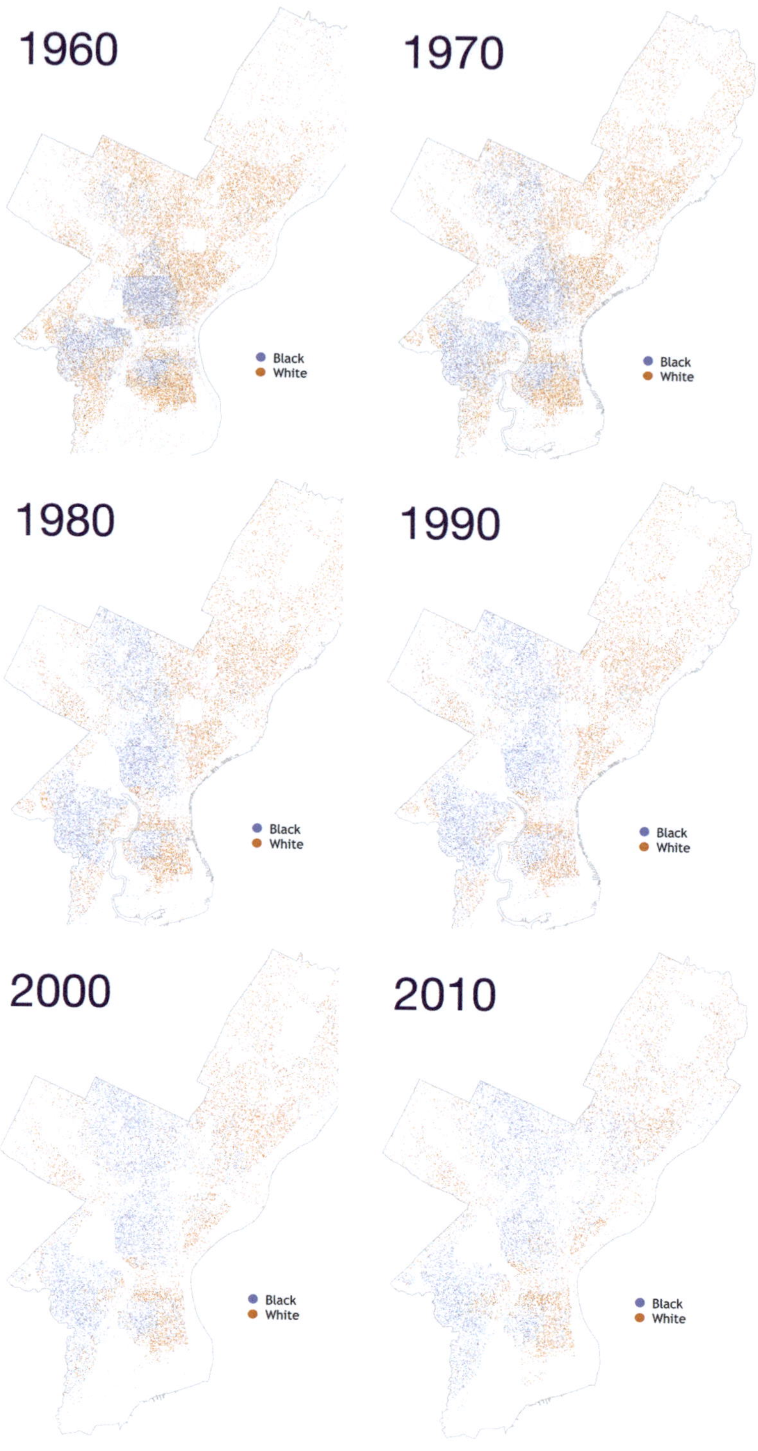

Figure 2: A history of integration and segregation in Philadelphia

2.2 MIGRATION AND STATE-SPONSORED SEGREGATION

An extensive body of literature has documented the ways that state actions and policies have constrained where members of minoritized groups can reside (Massey and Denton 1993; Wilson 1987). In Detroit (Thomas and Bekkering 2015), Chicago (Sampson 2012), Los Angeles (Davis 1992), and many other cities, the segregation and concentration of Black residents can be traced at least in part to the combined effects of state racism.[1] Individuals, agencies, and jurisdictions may 1) *create* laws that have intentionally disparate impacts on Black communities; 2) *interpret* the law in a way that marginalizes Black people; or 3) *enforce* the law selectively so that Black disadvantages are exacerbated. This section details selected policies that led to segregation patterns in Philadelphia.

The majority of Pennsylvania's Black population is concentrated in the Philadelphia metropolitan area (nearly two-thirds, according to the 2010 Census), though significant Black populations can be found in other cities like Pittsburgh and Harrisburg. Black people streamed to Philadelphia at the turn of the 20th century because of job opportunities and the presence of an established Black community dating back many years (Smucker and Hardy n.d.). Anthropologist Isabel Wilkerson has documented the acceleration of this pattern in the period known as the Great Migration, beginning around the time of the first World War, drawing people especially from the southeast states of Florida, Georgia, South Carolina, North Carolina, and Virginia (Wilkerson 2011). The growth in Philadelphia's longstanding Black community became a draw for other Black people around the country, resulting in what is known as linked migration. Once they arrived in Philadelphia, however, Black residents faced severe restrictions on their living choices, ranging from extreme income inequality to overtly racist practices in the housing market (Santucci 2019).

The situation of Black communities is very different than that of most European migrant groups, which were initially concentrated and exposed to significant constraints in housing choice in Philadelphia, but ultimately fully dispersed in the city (Pais, South, and Crowder 2012). The reasons for this difference are numerous. I address two of them here: restrictive racial covenants and redlining.

RESTRICTIVE COVENANTS

In response to the early migration waves, Northern cities and states rushed to enact laws designating restricted zones for Black residents, until the Supreme Court disallowed municipal racial zoning (Buchanan v. Warley, 1917). Once cities were barred from explicit racial zoning, authorities turned to individual contracts to do the same work.

[1] Following Kendi (2019:24), I understand racism to mean "the marriage of racist policies and racist ideas that produces and normalizes racial inequalities." Crucially for my purposes here with regard to districting, this includes not only overt beliefs in racial superiority but also ostensibly race-blind policies that allow established racial inequalities to persist uncontested.

Racial covenants significantly constrained where Black Philadelphians could live in the first part of the 20th century. Often drafted by property developers, racial covenants were statements in legal deeds that prevented homeowners from selling their property to non-White buyers (Figure 3). Recent work by Santucci and the Federal Reserve Bank of Philadelphia has assembled extensive documentation of this practice in Philadelphia specifically (Santucci 2019). These racial covenants were most prevalent in middle-income neighborhoods, which would have been the first step for Black families looking to move out of high-poverty neighborhoods (Santucci 2019, 3). In 1948, the U.S. Supreme Court heard the case of Shelley v. Kraemer and ruled racial covenants unconstitutional on 14th Amendment grounds. This put racial covenants in a legal gray area for decades: unenforceable, but still widespread and powerful. The practice was not conclusively ended until the Fair Housing Act in the 1968 Civil Rights Act (Brooks and Rose 2013; Santucci 2019, 10).

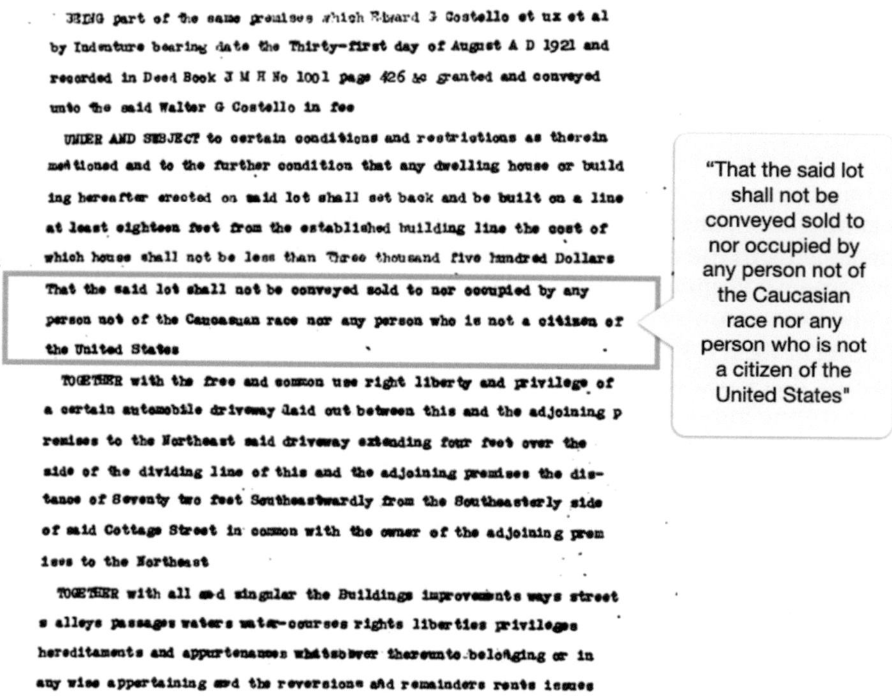
BEING part of the same premises which Edward J Costello et ux et al by Indenture bearing date the Thirty-first day of August A D 1921 and recorded in Deed Book J M H No 1001 page 426 &c granted and conveyed unto the said Walter G Costello in fee

UNDER AND SUBJECT to certain conditions and restrictions as therein mentioned and to the further condition that any dwelling house or build ing hereafter erected on said lot shall set back and be built on a line at least eighteen feet from the established building line the cost of which house shall not be less than Three thousand five hundred Dollars

That the said lot shall not be conveyed sold to nor occupied by any person not of the Caucasian race nor any person who is not a citizen of the United States

TOGETHER with the free and common use right liberty and privilege of a certain automobile driveway laid out between this and the adjoining p remises to the Northeast said driveway extending four feet over the side of the dividing line of this and the adjoining premises the dis- tance of Seventy two feet Southeastwardly from the Southeasterly side of said Cottage Street in common with the owner of the adjoining prem ises to the Northeast

TOGETHER with all and singular the Buildings improvements ways street s alleys passages waters water-courses rights liberties privileges hereditaments and appurtenances whatsoever thereunto belonging or in any wise appertaining and the reversions and remainders rents issues

Figure 3: Racially restrictive covenant forbidding sale to "any person not of the Caucasian race nor any person who is not a citizen of the United States." Philadelphia 1927 Source: Santucci 2019 and Philadelphia Department of Records.

Legal restrictions on home sales expressly blocked the Black community from dispersing in this and other cities. These practices were implicated in the exclusion of Black families from affluent neighborhoods in ways that constrained their access to jobs, education, and opportunity.

REDLINING

Another institutional mechanism dictating the locations of Black communities was *redlining*, which was at its peak in the 1930s. In brief, redlining was the practice of designating zones for mortgages and other home loans, and it entwined federal agencies, state and municipal planners, and banks in a system that locked in patterns of segregation and economic stratification. Areas designated as high risk—disproportionately made up of immigrant and Black communities—were often marked in red, leading to the term 'redlining.' For example, Amy Hillier (2003, 2005) has written extensively on the ways in which the Federal Home Owners Loan Corporation (HOLC) drew maps partitioning Philadelphia based on race. HOLC maps determined where federally backed mortgages would be available and where they would be withheld. Private mortgage lenders generally followed the HOLC's guidance in this regard, such that mortgages were extremely hard to access in places receiving the lowest HOLC rating: 'fourth grade' or 'high risk.' HOLC assessors were explicit in their reasoning for identifying some neighborhoods as 'fourth grade,' frequently citing the presence of Black communities (but also Jewish, Italian, and Irish people). A 1937 HOLC map for Philaldelphia is shown in Figure 4.

The high-risk designation constrained Black people's access to mortgage capital, which limited Black homeowners' capacity to upgrade their properties and decreased the desirability of these properties for new homebuyers (who could access credit for purchases in other parts of the city). Thus, redlining led to slow or negative growth in home values and held back wealth accumulation in Black neighborhoods. Even as policies like the Fair Housing Act eliminated legal and formal constraints on where Black Philadelphians could locate, the wealth disparities resulting from earlier policies and practices further entrenched barriers to movement because of the capital needed to afford homes elsewhere. Sociologist Rachel Woldoff (Woldoff 2011; Woldoff and Ovadia 2009) documents the long-term compounding of this wealth handicap for Black homeowners into the present day, noting how middle-class Black households have continually worked to "buy up" into more affluent neighborhoods, only to have those neighborhoods decline in value around them. This cycle leaves these families underwater on mortgages and thus with less capital to pass on to each successive generation than equivalent White households would gain through the housing market.

Wealth disparities further entrenched barriers to movement for Black Philadelphians because of the capital needed to afford homes elsewhere. Woldoff (Woldoff 2011; Woldoff and Ovadia 2009) documents the continued relevance of this wealth constraint for Black homeowners into the present day.

These were not the only government and individual actions that limited where Black people could live in Philadelphia and the rest of the United States. For instance, post-WWII restrictions in the GI Bill kept Black veterans from accessing mortgage support available to White veterans (Woods 2013). Planning decisions, particularly around the development of the Interstate Highway System, frequently split, emptied, or placed hard borders on Black neighborhoods (Flint 2011; Hanlon 2011; Karas 2015). These same policies facilitated the exodus of White residents, increasing racial disparities further. Discussing these and other policies and actions

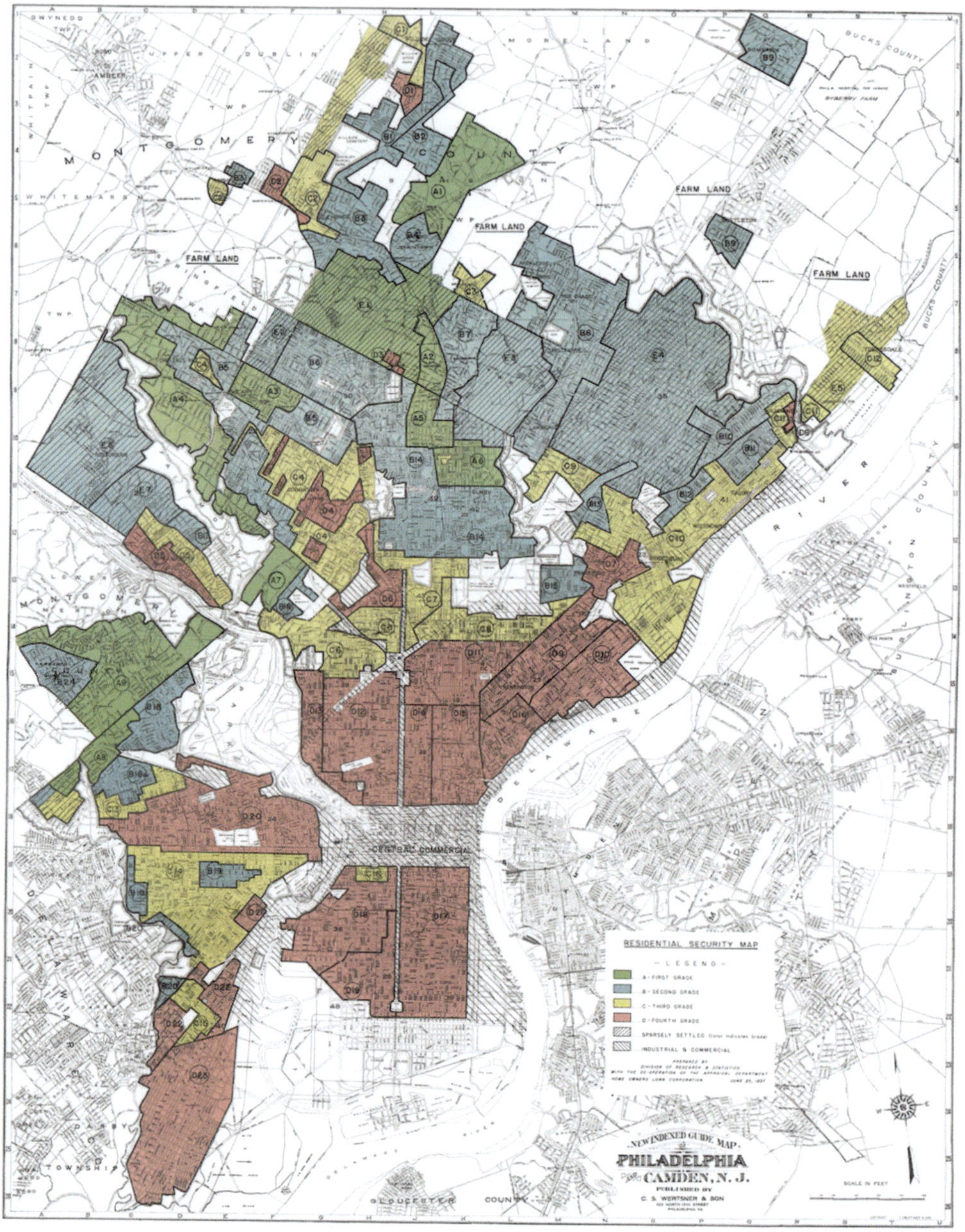

Figure 4: HOLC map for Philadelphia and Camden (Nelson et al. 2019)

in detail is beyond the scope of this chapter. The key takeaway is that the contemporary geography of race reflects many decades of layered public and private practices of exclusion.

2.3 DISTRICTS, SEGREGATION, AND FAIRNESS

The impact of spatial segregation has been high. In *American Apartheid*, sociologists Massey and Denton (1993) document the ways in which segregation dampens economic upturns and increases the impact of downturns for Black Americans. A broader body of literature on concentrated poverty and race has further established the significant negative consequences of segregation for political efficacy, wealth accumulation, education, and exposure to crime (Friedman, Tsao, and Chen 2013; Grengs 2007; Jargowsky 2014; Kozol 1991; Peterson, Krivo, and Browning 2006; Saporito and Sohoni 2007; Wilson 1987). More locally, Black respondents consistently report stronger preferences for living in mixed-race neighborhoods than White respondents (Bader and Krysan 2015). And research on both rental and home-buying markets indicates that minorities continue to face limits on their choices that go beyond constraints imposed by income (Ahmed and Hammarstedt 2008; Christensen and Timmins 2018; Korver-Glenn 2018).

Redistricting has a powerful tool meant to block discriminatory practices and promote electoral opportunity for racial minorities: the Voting Rights Act (VRA) of 1965. Despite the VRA—or sometimes under its cover—current districting practices continue to employ packing and cracking at the expense of Black communities (Curtis 2016). But the VRA only requires that we seek a districting plan that secures electoral opportunity; it does not tell us how to find such plans or whether they are possible.

Reformers should explore a range of remedies, both through districts and with alternative systems of election, for their effects on communities of color and other marginalized populations. Districts that are 'fair' to marginalized populations may fall outside the norms established by ensemble districting measures and may require careful manipulation (some would say gerrymandering) to achieve representation consistent with population share. Often, vulnerable communities that are composed of more recent immigrants or are spatially dispersed for other reasons would be better served by non-districted solutions. (see Chapter 23 and Chapter 21).

3 GEOGRAPHY ON MULTIPLE SCALES

Concentration is not the only population outcome that should concern parties seeking fair districting outcomes. If we take the lessons of the previous section to heart—that the spatial configuration of the population is not just a starting point for districting, but also the end point of a whole host of spatial processes that may or may not be just—then it is essential that we examine the data for other forms of spatial organization. Multiscale analysis, a set of exploratory tools for examining population distributions in space, can be an important step in this process.

3.1 THE MODIFIABLE AREAL UNIT PROBLEM

Population data are almost always aggregated into spatial units, such as zip codes, school districts, voting precincts, or census tracts. Spatial aggregation, or the process of making these units, turns out to be quite delicate when it comes to reporting statistics. Many common kinds of statistics are wildly sensitive to the choice of spatial subdivision. Even statistics that sound secure, such as "Percent Black" or "Democratic Vote Share in the 2016 Presidential Election," report only limited truths about the populations in a place, because they depend heavily on whether you report by county, by tract, or by something else. This feature of spatially aggregated data is called the Modifiable Areal Unit Problem (MAUP). This concept has a long history in geography, dating back at least to Gehlke and Biehl (1934), and has served as a thread uniting different eras of geographical inquiry (Fotheringham and Wong 1991; Openshaw and Taylor 1979, 1981; Wong 2004). To give one example: geographer Elizabeth Root found that the choice of units for designating the size of "neighborhoods" was significant for determining the magnitude of effects between class and certain kinds of birth defects (Root 2011). In Root's study, choosing one set of units could significantly mask the relationship that was unmistakable with another choice.

Geographers distinguish between two aspects of the MAUP: **scale** and **zoning**. Within the MAUP context, scale refers to the size of the units of aggregation. Sometimes a phenomenon can appear or disappear when you change the scale, as we will see below. Zoning is about the potential to change the observation by shifting the borders without necessarily changing scale. In this sense, all of gerrymandering can be considered one big zoning MAUP. Given a fixed number of electoral districts, boundary changes can produce plans with widely diverging election results, as we've seen. The wider the range of possible outcomes, the more zoning is an issue. For more viewpoints on MAUP, see the scale experiments in Chapter 5 and the discussion of segregation measures in Chapter 15.

3.2 MULTISCALE MEASURES

The population composition of an apartment building (a 'fine' or 'micro' scale) may shape the information its residents can access about job opportunities, but the composition of the labor market as a whole (a 'broad' or 'macro' scale) will also impact the likelihood that they get a job. Both contexts may be important for predicting whether an individual is employed and with what outcome. A growing body of literature within geography considers population sorting processes on multiple scales, encouraging us to look beyond a given set of observations to examine the spatial processes at work to produce them (Fowler 2016, 2018; Östh, Clark, and Malmberg 2014; Reardon et al. 2008). Each research contribution typically begins with clarifying why scale matters for the measurements at hand (in terms of health, education, political efficacy, and so on), and then goes on to argue that any analysis on a single fixed scale would fall short.

10.2 SCALE EFFECTS IN TWO CITIES

As a case study, we can consider two major American cities, Philadelphia and Chicago. Their non-White populations have very different demographic shares and distributions; we can consider the consequences for city council representation. The population of Philadelphia is around 12.29% Hispanic and 42.22% Black based on the 2010 Census. Chicago, on the other hand, is about 28.90% Hispanic and 32.36% Black. For this analysis, we sampled ten thousand possible city council districting plans for each city. As Chicago's city council consists of 50 members and Philadelphia's only 10, we considered both council sizes for each city. For each plan generated, we counted the number of districts with majority-Hispanic and majority-Black populations. The full results are shown in the histograms in Figure 5, and the table below records the expectations.

	Chicago	Philadelphia
Hispanic population	28.90 %	12.29 %
% Majority-Hispanic districts (out of 10)	15.37 %	.075 %
% Majority-Hispanic districts (out of 50)	22.95 %	5.395 %
Black population	32.36 %	42.22 %
% Majority-Black districts (out of 10)	28.9 %	41.28 %
% Majority-Black districts (out of 50)	28.51 %	38.45 %

One important observation we can make is that shifting to a finer scale of districting—50 districts instead of 10—boosts Hispanic representation in both cities, but actually slightly reduces the Black seat share.

For example, in research on diverse neighborhoods in Seattle, Lumley-Sapanski and Fowler (2017) found that diversity on the neighborhood scale was stable over time because of small, homogeneous communities within those neighborhoods. People interviewed for the project felt that they benefited from the ethnoracial mixing around them. Nevertheless, mixing was stable only because there were community anchors (a Hispanic community center, a Korean Baptist church) that kept very specific, ethnically homogeneous groups tied to the neighborhood. Thus, a positive contextual effect associated with (diverse) neighborhood ethnoracial composition was made possible by the opposite (homogeneous) composition on a smaller scale.

Sociospatial contexts can also have different meanings depending on the scale at which they appear in the population. Homogeneity measured on micro scales is consistent with positive effects on social capital (Merry 2013) and the reverse is also true: political scientist Robert Putnam has found evidence that in neighborhoods that score high in a certain kind of diversity, residents report fewer friends and less community cooperation (Putnam 2007). When homogeneity signifies a broadly segregated community, as when suburbs and city are segregated in a metropolitan area, the effect of the macro context may work in the opposite direction of the micro context. Significantly, both contexts may operate at the same time, with the negative effects of the macro context dampened by the positive micro effects and the positive micro effects masked by the negative macro context (Fowler 2016). An openness to processes taking place on multiple scales creates opportunities to

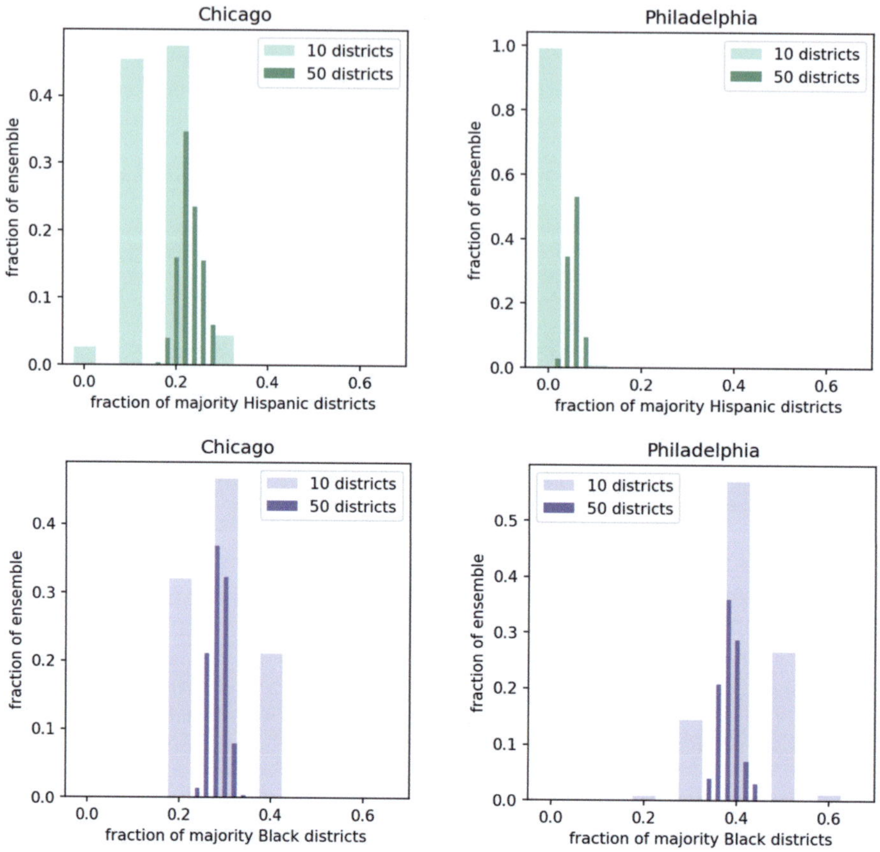

Figure 5: Sampling of city council districting plans for Chicago and Philadelphia. The choice of how many districts are in the plans being generated (10 or 50) has different implications for Black and Hispanic representation.

better understand why people are sorted as they are and how this sorting is likely to affect them. These examples suggest the enduring metaphor of a *checkerboard* configuration that looks very diverse (black and white neatly interspersed) when viewed from a certain distance, but looks completely homogeneous at both the smallest scale (zoomed in to a single square) or the largest (zoomed out to see a gray mix from far away). The analogy to neighborhood squares within a checkerboard city is one that more than a few geographers have used in their arguments about the importance of scale (Figure 6; see Morrill 1991, Wong 1993).

3.3 APPLICATIONS TO DISTRICTING

One area where multiscale analysis can support districting efforts is in helping to understand how communities inform the process. Through this, we can build a more comprehensive understanding of the human impact of different districting plans.

Figure 6: Zooming out from a checkerboard pattern changes the picture from homogeneous (all one color, left), to diverse (checkerboard, middle), to homogeneous again (gray blend of black and white, right).

To get an intuitive handle on this, consider that a competitive district (party affiliation mixed close to 50/50 at the district scale) where populations are clustered tightly within the district (e.g., segregation by ethnicity in neighborhoods within the district) may contain large groups with different priorities. If, on the other hand, population mixing takes place across multiple smaller scales within the district, then at least some of the key variables correlated with location are likely to be shared across party affiliation. In this context, a competitive district is more likely to function as intended, with candidates competing to represent shared interests rather than single-group interests.

A nuanced example is found in the work of Johnston and Pattie (2016), who conducted a multiscale analysis of ethnic groups in Sydney, Australia. One kind of scale they examine is not spatial but taxonomic: in their study of immigration, they break down broad categories such as "Asian" into cohorts that migrated for similar reasons and at similar times. They find that some cohorts are located all in one quadrant of the city but scattered evenly across that quadrant (micro-scale dispersal, meso-scale segregation). Other groups live in tight clusters, but those clusters are scattered across the city (micro-scale segregation, macro-scale dispersal). Their observation of these sorting patterns on different scales allowed them to better understand how different communities were shaped by changing views on race and immigration. And it would obviously carry strong implications for a districting analysis, because the ability of districts to secure representation depends on the relationship between the size of the district and the clusters of population. Furthermore, even being seen as a salient community might depend on these factors of scale and clustering. Meso-scale segregation becomes an asset in terms of making a community visible, whereas micro-scale segregation linked to macro-scale dispersal may be visible or not depending on how boundaries are drawn.

We should aspire to draw boundaries with a clearer understanding of which communities are being contained and which communities are being broken up by proposed boundaries. Garrett Nelson will take this up in the next chapter.

4 CONCLUDING THOUGHTS: PAY ATTENTION TO RACE

Districting and gerrymandering are fundamentally geographic problems because of their conjoining of spatial and social processes. Districting begins from the assumption that how we draw boundaries matters. This chapter has highlighted the uneven spatial processes that drive observed population distributions. This can help to see communities on different scales, identify shared versus conflicting interests in populations, and determine which groups it might be especially important to keep intact in a district. Overall, this can help us to reason about the best way to meet the fundamental goal of providing voters with good and fair representation.

A geographer's perspective on districting compels us to examine not only how our boundaries are drawn, but also who is made visible by our choices and what led to the configurations we observe. Race has played an outsized role in shaping America's human geography. Fair districting must therefore be attentive to the multiple landscapes of race.

REFERENCES

Ahmed, Ali M, and Mats Hammarstedt. 2008. "Discrimination in the Rental Housing Market: A Field Experiment on the Internet." Journal of Urban Economics 64(2): 362–72.

Bader, M. D. M., and M. Krysan. 2015. "Community Attraction and Avoidance in Chicago: What's Race Got to Do with It?" The Annals of the American Academy of Political and Social Science 660(1): 261–81. `http://ann.sagepub.com/cgi/doi/10.1177/0002716215577615` (18 June 2015).

Bowler, Shaun, Todd Donovan, and David Brockington. 2003. Electoral Reform and Minority Representation: Local Experiments with Alternative Elections . Ohio State University Press.

Brooks, Richard R.W., and Carol M. Rose. 2013. "Saving the Neighborhood: Racially Restrictive Covenants," Law, and Social Norms." 19.

Chen, Jiahua et al. 2019. "Minority Voter Distributions and Partisan Gerrymandering." `http://arxiv.org/abs/1911.09792`

Chen, Jowei, and David Cottrell. 2016. "Evaluating Partisan Gains from Congressional Gerrymandering: Using Computer Simulations to Estimate the Effect of Gerrymandering in the US House." Electoral Studies 44: 329–40.

Chen, Jowei, and Jonathan Rodden. 2013. "Unintentional Gerrymandering: Political Geography and Electoral Bias in Legislatures." Quarterly Journal of Political Science 8(3): 239–69.

Chin, Andrew, Gregory Herschlag, and Jonathan Mattingly. 2019. "The Signature of Gerrymandering in Rucho v. Common Cause." South Carolina Law Review 70(4).

Christensen, Peter, and Christopher Timmins. 2018. Sorting or Steering: Experimental Evidence on the Economic Effects of Housing Discrimination. National Bureau of Economic Research.

Curtis, Michael Kent. 2016. "Using the Voting Rights Act to Discriminate: North Carolina's Use of Racial Gerrymanders, Two Racial Quotas, Safe Harbors, Shields, and Inoculations to Undermine Multiracial Coalitions and Black Political Power." Wake Forest Law Review 51: 421.

Davis, Mike. 1992. "City of Quartz: Excavating the Future in Los Angeles. 1990." New York: Vintage.

Duchin, Moon. 2018. "Gerrymandering Metrics: How to Measure? What's the Baseline?" arXiv preprint arXiv:1801.02064 .

Flint, Anthony. 2011. Wrestling with Moses: How Jane Jacobs Took on New York's Master Builder and Transformed the American City . Random House Trade Paperbacks.

Fotheringham, A. Stewart, and David W.S. Wong. 1991. "The Modifiable Areal Unit Problem in Multivariate Statistical Analysis." Environment and planning A 23(7): 1025–44.

Fowler, Christopher S. 2016. "Segregation as a Multi-Scalar Phenomenon and Its Implications for Neighborhood-Scale Research: –2010." Urban Geography 37(1).

Fowler, Christopher S. 2018. "Key Assumptions in Multiscale Segregation Measures: How Zoning and Strength of Spatial Association Condition Outcomes." Environment and Planning B: Urban Analytics and City Science 45(6): 1055–72. `http://journals.sagepub.com/doi/10.1177/2399808318760570`

Friedman, Samantha, Hui shien Tsao, and Cheng Chen. 2013. "Housing Tenure and Residential Segregation in Metropolitan America." Demography 50(4).

Gehlke, Charles E, and Katherine Biehl. 1934. "Certain Effects of Grouping upon the Size of the Correlation Coefficient in Census Tract Material." Journal of the American Statistical Association 29(185): 169–70. `http://amstat.tandfonline.com/doi/pdf/10.1080/01621459.1934.10506247` (20 July 2017).

Grengs, Joe. 2007. "Reevaluating Poverty Concentration With Spatial Analysis: Detroit in the 1990s." Urban Geography 28(4): 340–60. `http://www.tandfonline.com/doi/abs/10.2747/0272-3638.28.4.340` (12 June 2017).

Hanlon, James. 2011. "Unsightly Urban Menaces and the Rescaling of Residential Segregation in the United States." Journal of Urban History 37(5): 732–56.

Hillier, Amy E. 2003. "Who Received Loans? Home Owners' Loan Corporation Lending and Discrimination in Philadelphia in the 1930s." Journal of Planning History 2(1): 3–24. `http://journals.sagepub.com/doi/10.1177/1538513202239694` (6 December 2019).

———. 2005. “Residential Security Maps and Neighborhood Appraisals.” Social Science History 29(2): 207–33. https://www.cambridge.org/core/product/identifier/S014555320001292X/type/journal_article (6 December 2019).

Jargowsky, Paul A. 2014. “Segregation, Neighborhoods, and Schools.” In Choosing Homes, Choosing Schools Choosing Homes, Choosing Schools , eds. A. Lareau and K.A. Goyette. New York: Russell Sage Foundation, 97–136.

Johnston, Ron J., and Charles Pattie. 2006. Putting Voters in Their Place. Oxford: Oxford University Press.

Karas, David. 2015. “Highway to Inequity: The Disparate Impact of the Interstate Highway System on Poor and Minority Communities in American Cities.” New Visions for Public Affairs 7: 9–21.

Kendi, Ibram X. 2019. How to be an Antiracist. New York: One World.

Korver-Glenn, Elizabeth. 2018. “Compounding Inequalities: How Racial Stereotypes and Discrimination Accumulate across the Stages of Housing Exchange.” American Sociological Review 83(4): 627–56.

Kozol, Jonathan. 1991. Savage Inequalities: Children in America’s Schools . New York: Crown.

Lumley-Sapanski, Audrey, and Christopher S. Fowler. 2017. “‘Planning Dissonance’ and the Bases for Stably Diverse Neighborhoods: The Case of South Seattle.” City & Community 16(1): 86–115. http://doi.wiley.com/10.1111/cico.12224

Massey, Douglas S., and Nancy A. Denton. 1993. American Apartheid. Cambridge, Massachusetts: Harvard University Press.

Merry, Michael S. 2013. Equality, Citizenship, and Segregation: A Defense of Separation . New York: Palgrave Macmillan. (29 September 2016).

Morrill, Richard L. (1991). On the measure of geographic segregation. In Geography research forum (Vol. 11, No. 1, pp. 25-36).

Nelson, Robert K., LaDale Winling, Richard Marciano, and Nathan Connolly. 2019. “Mapping Inequality.” In American Panorama , eds. Robert K. Nelson and Edward L. Ayers. https://s3.amazonaws.com/holc/tiles/PA/Philadelphia/1937/holc-scan.jpg

Openshaw, Stan, and Peter J. Taylor. 1979. “A Million or so Correlation Coefficients: Three Experiments on the Modifiable Areal Unit Problem.” In Statistical Applications in the Spatial Sciences , ed. Neil Wrigley. London: Pion, 127–44.

———. 1981. “The Modifiable Areal Unit Problem.” In Quantitative Geography: A British View , eds. Neil Wrigley and Robert John Bennett. London: Routledge, 60–69.

Östh, John, William A. V. Clark, and Bo Malmberg. 2014. “Measuring the Scale of Segregation Using k-Nearest Neighbor Aggregates.” Geographical Analysis 47(1): 34–49. http://doi.wiley.com/10.1111/gean.12053 (1 October 2014).

Pais, Jeremy, Scott J. South, and Kyle Crowder. 2012. “Metropolitan Heterogeneity and Minority Neighborhood Attainment: Spatial Assimilation or Place Stratification?” Social Problems 59(2): 258–81.

Peterson, Ruth D., Lauren J. Krivo, and Christopher R. Browning. 2006. “Segregation and Race/Ethnic Inequality in Crime: New Directions.” In Taking Stock: The Status of Criminological Theory , eds. F. T. Cullen, J. P. Wright, and K. R. Blevins. New Brunswick: Transaction, 169–90. (14 May 2014).

Putnam, Robert D. 2007. “E Pluribus Unum: Diversity and Community in the Twenty-First Century.” Scandinavian Political Studies 30(2): 137–74.

Ramachandran, Gowri, and Dara Gold. 2018. “Using Outlier Analysis to Detect Partisan Gerrymanders: A Survey of Current Approaches and Future Directions.” Election Law Journal: Rules, Politics, and Policy 17(4): 286–301.

Reardon, Sean F. et al. 2008. “The Geographic Scale of Metropolitan Racial Segregation.” Demography 45(3): 489–514.

Rodden, Jonathan. 2019. Why Cities Lose: The Deep Roots of the Urban-Rural Political Divide . New York: Basic Books.

Sampson, Robert J. 2012. Great American City: Chicago and the Enduring Neighborhood Effect . Chicago: University of Chicago Press.

Santucci, Larry. 2019. How Prevalent Were Racially Restrictive Covenants in 20th Century Philadelphia? A New Spatial Data Set Provides Answers. FRB of Philadelphia Payment Cards Center Discussion Paper. `https://perma.cc/64B7-G3R8`

Saporito, Salvatore, and D. Sohoni. 2007. “Mapping Educational Inequality: Concentrations of Poverty among Poor and Minority Students in Public Schools.” Social Forces 85(3): 1227–53. (29 December 2018).

Smucker, Janneken, and Charles Hardy. “Goin’ North.” `http://goinnorth.org/about`

Thomas, June Manning, and Hans Bekkering. 2015. Mapping Detroit: Land, Community, and Shaping a City . Detroit: Wayne State University Press.

U.S. Census Bureau. 2010. “Decennial Census of the Population.” `https://factfinder.census.gov` (12 December 2019).

U.S. Supreme Court. 1948. Shelley v. Kraemer.

Vargas, Robert. 2016. Wounded City: Violent Turf Wars in a Chicago Barrio. New York: Oxford University Press. (5 December 2019).

Webster, Gerald R. 2013. “Reflections on Current Criteria to Evaluate Redistricting Plans.” Political Geography 32: 3–14. `https://linkinghub.elsevier.com/retrieve/pii/S0962629812001151`

Wilkerson, Isabel. 2011. The Warmth of Other Suns: The Epic Story of America’s Great Migration. Vintage.

Wilson, William Julius. 1987. The Truly Disadvantaged: The Inner City, the Underclass, and Public Policy. Chicago: University of Chicago Press.

Woldoff, RA, and S. Ovadia. 2009. "Not Getting Their Money's Worth: African-American Disadvantages in Converting Income, Wealth, and Education into Residential Quality." Urban Affairs Review. (11 March 2013).

Woldoff, Rachael A. 2011. White Flight/Black Flight: The Dynamics of Racial Change in an American Neighborhood. Ithaca: Cornell University Press.

Wong, D.W., 1993. Spatial indices of segregation. Urban studies, 30(3), pp.559-572.

Wong, D.W.S. 2004. "The Modifiable Areal Unit Problem (MAUP)." Worldminds: Geographical Perspectives on 100 Problems. http://link.springer.com/chapter/10.1007/978-1-4020-2352-1_93 (2 February 2017).

Woods, Louis Lee. 2013. "Almost 'No Negro Veteran… Could Get A Loan': African Americans, the GI Bill, and the NAACP Campaign against Residential Segregation, 1917–1960." Journal of African American History 98(3): 392–417.

Chapter 11

The elusive geography of communities

GARRETT DASH NELSON

CHAPTER SUMMARY

This chapter is a philosophical and historical overview of geographers' concepts of community. Nelson highlights many of the challenges in defining "communities of interest" for redistricting. He also illustrates the idea of functional regionalization and discusses possible ties to redistricting.

1 THE GEOGRAPHIC COMMUNITY AS A PRINCIPLE OF REPRESENTATIVE GOVERNMENT

The justification for organizing votes and constituencies in a representative democracy according to spatially defined units is not merely convenience. Electoral systems where representatives are chosen to represent territorial areas are premised on a deeply rooted and ancient assumption that geography is a key structuring factor in political and social communities. But although there is an obvious relationship between geographical terms for electoral units (like *district, riding, precinct,* or *ward*) and the geographic terms for more substantively constituted spatial units (like *region, neighborhood, community,* or *polity*), defining the precise interplay between these terms—both conceptually and in terms of actual lines on the map—is far more elusive in practice. For this reason, the principle that electoral districts should represent something known as "communities of interest,"

M. Duchin, O. Walch (eds.), *Political Geometry*, https://doi.org/10.1007/978-3-319-69161-9_11

while widely recognized in common perceptions of good district-making and even established in some cases as a legal requirement, has exerted relatively little force in the campaign to salvage electoral boundary-drawing from partisan manipulation.

In this chapter, I provide a brief overview of the role of space and place in shaping social and political life, and push at the term "community" to investigate its complicated status vis-à-vis territorial definition. I then summarize how geographical methods have been brought to bear on the problem of defining and delineating functional areas for the purposes of administrative activities. Finally, I offer the concept of geographic "coherence" as one desideratum for drawing electoral maps that match as far as possible the underlying patterns of human interaction and codependence. Geographic coherence does not necessarily match exactly with other objectives of good districting schemes, such as competitiveness within a party system or racial proportionality. Yet it offers one potential principle for district-drawing that could mitigate against partisan chicanery while remaining flexible enough to accommodate the wide variety of spatial distributions of political interests found in the real world.

2 ARE COMMUNITIES PLACES OR NOT?

In common use, a kind of conceptual homology exists between *geographical places* and social and political *human groups*. Indeed, one of the broadest terms for both of these categories, *community*, may shift from indicating a place to indicating a social group within the course of a single sentence—we refer to communities in one breath as bordered locations inside which one might be spatially located and in the next breath as human groups linked together by some mutual common attribute. A register of other linguistic clues suggests the same conceptual affinity between places and groups: *region* comes from *regio*, the domain ruled by a monarch; *landscape* derives from the Germanic *Landschaft*, a unit of self-government; a *neighborhood* is both a spatial definition used by planners as well as a group of people who share the quality of neighborliness (Crary 1959; Jackson 1964; Minar and Greer 1969; Paasi 1991; Olwig 1996; Chaskin 1997; Looker 2015). And there is no better semantic proof of the relationship between politics and place than the fact that *politics* itself derives from *polis*, referring to the Greek form of social organization within a city's walls, a term "connecting a human community and a determinate territory." (Wolff 2014, 801)

Since it is easier to interact with people who are nearby than those who are far away, it stands to reason that space and social structure are closely linked in this way. A basic principle of geography is that people and things that are spatially proximate will be, *ceteris paribus*, more likely to form meaningful and durable patterns of mutual influence than people and things that are far apart. In fact, it is almost impossible to split apart the deeply engrained cognitive metaphor that links commonality and shared location: try to imagine a group of people "together" without automatically picturing them as physically proximate. And how do mathematicians visualize a set of objects with a common attribute? By clustering them together on a page and drawing a border around them, in a Venn diagram.

In various interpretations, the role of spatial proximity and spatial exclusion as a factor in social and political life is the basis for the entire enterprise of human geography. For geographers, space and territory are not neutral containers in which other phenomena take place; instead, they are treated as both the *producers of* and the *productions of* social, political, economic, and cultural formations. And although spatial proximity undoubtedly creates a condition for social interaction and interdependence, it does not necessarily follow that human groups will always coalesce neatly into internally homogenous and externally bounded geographic units. Such categories as ethnicity, race, religion, language, or position within an economic class system are all important dimensions of group affiliation, and, while these may be spatially correlated, they do not necessarily collapse into neatly spatially demarcated territorial objects.

This contingent relationship between territory and community becomes especially clear when considering the historical trajectory of modern social life. With the rise of modern forms of transportation and communication, the intensifying complexities of industrial integration, and the spread of diasporic populations, the "primary community" consisting of face-to-face relationships faded in importance. Thus, social scientists throughout the twentieth century increasingly dismissed the importance of spatial propinquity in producing meaningful and durable forms of community structure. The sociologist Louis Wirth, for example, argued in 1938 that "in the face of the disappearance of the territorial unit as a basis of social solidarity we create interest units," (Wirth 1938, 23) and the geographer George H. T. Kimble expressed a similar conclusion in 1951: "whatever the pattern of the new age may be, we can be sure that there will be no independent, discrete units in it—no 'worlds within worlds.'" (Kimble 1951, 173) To observers like these, the ability to read a daily newspaper published a continent away, the rapid migration of huge groups of people to new places, and the standardizing logic of industrial capitalism all offer examples of the de-spatializing forces that were rendering propinquity less and less important as a condition for binding individuals together into meaningful interest groups.

Not coincidentally, the rise of national political parties was one of the historical developments closely tied to the erosion of geographic solidarity. As Mac Donald and Cain (2013) note, the delegate theory of representation implicitly rests on the assumption that "constituents residing within the boundaries of a given district or territorial jurisdiction" will have "widely shared attributes and a greater sense of kinship." By contrast, in the virtual model of representation, "supra-territorial interests such as a political parties, classes, or organizations" are the basis for a representative's legitimacy. This duality—on the one hand, communities are geographic places, and, on the other hand, communities are nonspatial interest groups—therefore underlies a basic tension in representative theory. Such a tension is evident even in early disputes over electoral boundary drawing: in Massachusetts at the time of the original gerrymander in the state senate, for instance, the state's lower house still apportioned representatives based on the fundamental unit of the township—a geographic unit that seemed to exemplify the organic, bound-together form of place-based community (Nelson 2018a)—and thus retained an artifact of the geographic-community delegate model even as Elbridge Gerry's Democratic-Republican party redrew the map according to the logic of a virtual, party-based theory of political power. The scandal of this original gerrymander, then, was "not only the undue pursuit of political advantage, but also the disruption of organic geographic communities" (Stephanopolous 2012, 1408). Vermont retained a system in which townships each sent one representative to the lower house, and counties a representative each to the upper house, until 1965, representing the tenacious historical legacy of geographic-community maximalism.

3 THE FUNCTIONAL LOGIC OF REGIONAL DEFINITION

Even as the assumptions of social theory shifted away from the organic concept of place-bound community, the practical problems of how to measure, map, and govern spatially defined entities continued apace, often in the hands of planners, administrators, and statisticians—not to mention electoral commissions. Such questions can be grouped together as the *regionalization problem*: how do we divide space according to some sort of empirical justification rather than merely retaining the boundaries inherited from the past? Here it is worth pausing to note a semantic subtlety. While "region" often carries the connotation of a specific type of unit occupying a scalar size larger than a city but smaller than a nation-state, for regionalization studies in geography, it can refer generically to any spatial unit defined according to some organizational logic, and we can speak about regionalizing an area as small as a classroom or as large as the globe.

The regionalization problem has sharpened at moments where the physical transformation of spatiality has seemed to outpace the administrative functionality of older units. Indeed, the contradistinction of a "functional" area versus a "political" or "historical" area is a basic assumption of such work, one that is retained in the vocabulary of studies like those of "functional urban regions" (Coombes et al. 1982; Noronha and Goodchild 1992). The practical geographic exigencies of administration have long dictated how political units are drawn: to take just two older examples, the medieval parish was based on the layout of the manorial farming system and the limits of church congregation (Whyte 2007), and a circuit court's jurisdiction was once based on the area that could be traveled by a judge during an era of horse transportation (Glick 2003).

As nineteenth-century governments modernized and rationalized systems like census-taking and postal delivery, the relationship between administrative logic and place definition became even stronger. London, for instance, saw the introduction of a Metropolitan Board of Works in 1855, with a geographical boundary that allowed it to operate according to the drainage lines needed for sewer construction, dictated by topography and gravity, rather than the hodgepodge boundaries of ancient constituencies; and in 1857 a comparable London metropolitan postal district was established to rationalize mail routes. These two single-purpose geographies provided the basis for the wholesale reorganization of metropolitan London under the auspices of the London County Council in 1889, which exercised not only bureaucratic functions like its predecessors, but also gained the power of a representative body (Saint 1989). The London example suggests the reciprocal structuring process between the material imperatives of the administrative state (the need to run sewage downhill demanded a rupture of older municipal boundaries based on a vanished manorial system) and the geographical structure of a political body (a metropolitan public works district also needed a tax base, a representative body, and, ultimately, a defined constituency on which to base its legitimacy).

This period saw countless examples of industrialized metropolises experimenting with similar territorial reforms as their physical forms diverged from historic boundary lines. New York consolidated its five boroughs in 1898, a geographic fusion that seems obvious in retrospect, but that was fiercely opposed by many Brooklynites, who did not consider themselves part of the same political community as Manhattan (Henschel 1895; Coler 1899). Reformers in Boston sought a similar metropolitan consolidation in the 1890s, with one influential journalist calling the metropolitan area "the true Boston—geographical Boston, as distinguished from political Boston" (Baxter 1891). Figure 1 shows the multiple overlapping functional districts in the Boston metropolitan area in the 1930s, forming a ill-defined but nevertheless considerably integrated metropolitan community. So many cities across the industrialized world went through similar territorial explosions at this time that the influential Scottish planner-geographer Patrick Geddes coined the term "conurbation" to refer to the new type of urban form created by cities growing into one another (Geddes 1915).

Since the early twentieth century, geographers have taken considerable interest in the empirics and methodologies of this regionalization problem. Some of the earliest academic works in geography were attempts to classify the world into regions according to climatic, biological, and geomophological attributes. As the discipline formalized in the early twentieth century, the goal of identifying geographic objects remained paramount: when J. G. Granö sought to define geography's mandate in 1929, he called it "a science that forms entities" (Granö 1997). With the rise of urban geography and regional studies in the first half of the twentieth century, efforts to define the regional economic geography of metropolitan integration (Dickinson 1934), understand the regional subdivisions of nation-states (Ogilvie 1928; National Resources Committee 1935), or establish methods for regional survey (Hudson 1936) became key undertakings. Figure 2 shows a New Deal-era study of "natural community" boundaries that were meant to serve as the outlines for county planning districts in Oregon.

During the middle of the century, as geography turned toward statistical methods in its attempt to become a formal spatial science, and deepened its integration with governmental planning bureaucracies, work on the regionalization problem became ever more of an exercise in descriptive modeling—and geographers became gradually less interested in hazy terms like "community" that lent themselves poorly to statistical analysis. Functional geographical definitions such as the "metropolitan statistical area" have their origins in this line of research (Berry, Goheen, and Goldstein 1969; Berry 1964; Nystuen 1968; Bunge 1966). As innovative and sophisticated as these research programs were, they nevertheless began to drift away from the concept of a region as a socially or politically constitutive

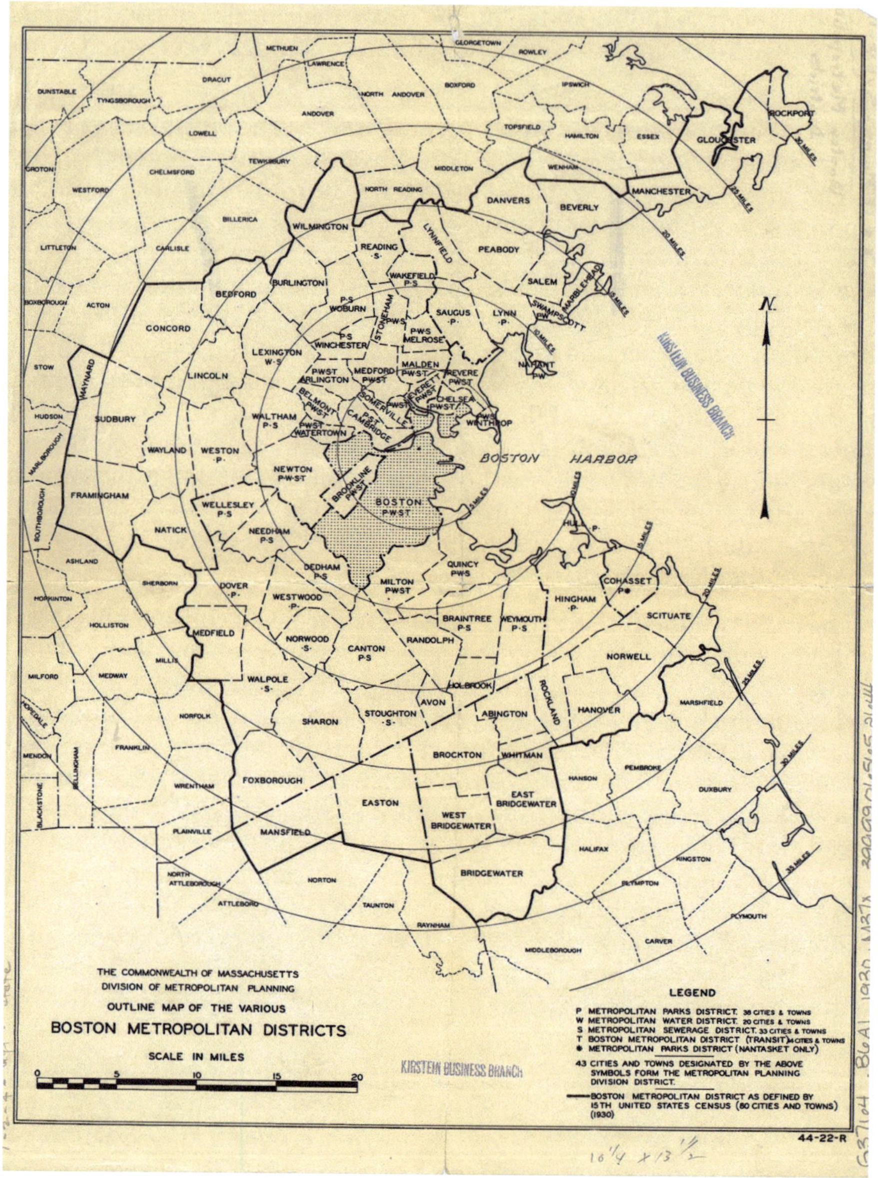

Figure 1: Multiple ways of districting metropolitan Boston, 1930s. From the Commonwealth of Massachusetts Division of Metropolitan Planning. Courtesy of the Leventhal Map & Education Center at the Boston Public Library.

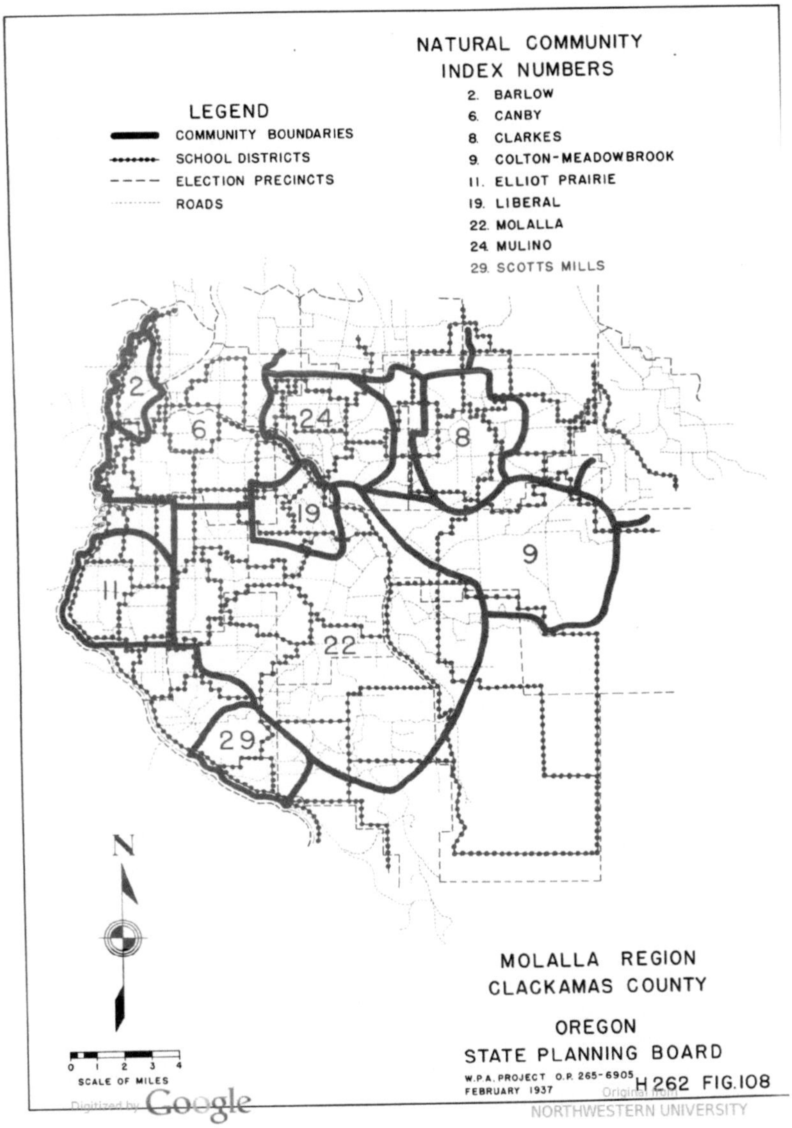

Figure 2: Districting in Oregon, from "A study of natural communities in three Oregon counties: an attempt to determine the importance of natural communities as a basis for community planning in country districts"

object: regions became merely objects of scientific categorization, rather than the building-blocks of group life that had been posited by earlier approaches steeped in cultural and historical studies.

4 GEOGRAPHIC COHERENCE: FROM EMPIRICAL EVIDENCE TO THE STRUCTURE OF A POLITY

A suggestive link between the regionalization problem and the larger theoretical question of whether and how spatiality structures group life can be found in Geddes's work, which included both an empirical study of emergent urban consolidation as well as a meditation on "the coming polity," in which he questioned what forms of spatial organization were most suitable for a democratic society transformed by the material conditions of modern life (Branford and Geddes 1917). The word *polity*, in fact, lies at the very heart of the conceptual intersection of territory, community, electoral representation. If common interests forged through the conditions of geographic proximity really do structure community life and form a mass of people into a meaningfully identifiable polity, then ensuring that these groups are represented within a representative or federal system becomes a matter of crucial importance. Should—and can—the boundaries of an electoral district be drawn so that they match the boundary lines of a polity?

This brings us back to the question of why and how we should organize our electoral constituencies according to geography. How can it be that, on the one hand, space and propinquity are losing their relevance as structuring factors for group life in a globalized world, and, on the other, that places and boundaries still retain their essential importance in defining political jurisdictions? Part of the reason is historical lag: our concepts of political authority and citizenship have not yet caught up to the scrambled spatial conditions of the present day. Yet part of the reason is that, in spite of airline travel and the Internet, places still are enormously important in defining interdependence and mutuality.

Moreover, the relationship between place and representation remains deeply embedded in the theoretical and legal framework of modern democracies. "The roots of Anglo-American political representation lie in the representation of communities, not individuals," writes Gardner (2002, 1243; see also Gardner 2006). "Originally, representation in Parliament was a metaphorical representation *of the land itself*" [emphasis added]. As Gardner argues, the trend in twentieth-century liberal philosophy has elevated the atomized individual over the constitutive community as the basic quantum of political representation—a theoretical shift very much in parallel with the observation of the social scientists like Wirth or Kimble who saw geographic community as a relic of premodernity. Yet, as Gardner argues, by way of recourse to John Dewey, Nancy Schwartz, and Hannah Arendt, it is the common political action of a community, situated in place, acting through its representatives, that "continually reconstitutes the polity" (2002, 1248).

This principle of hewing electoral districts according to geographic patterns of social structure is given voice in the numerous legal mandates to respect the so-called "community of interest" when drawing electoral maps (Brennan Center for Justice 2010). When explaining what exactly is meant by a "community of interest," lawmakers have usually invoked the same confusing slippage between spatial and nonspatial forms of group life that has been the discussion of this chapter. The Colorado constitution requires that "communities of interest, including ethnic, cultural, economic, trade area, geographic, and demographic factors, shall be preserved within a single district wherever possible." Vermont statute requires "recognition and maintenance of patterns of geography, social interaction, trade, political ties, and common interests." The California constitution requires that "the geographic integrity of any city, county, city and county, local neighborhood, or local community of interest shall be respected in a manner that minimizes their division to the extent possible," and goes on to note that "a community of interest is a contiguous population which shares common social and economic interests."

If respecting of communities of interest remains a legal desideratum for electoral maps, the actual definition of *where* communities of interest exist is elusive, and, due to this ambiguity, violations of the community of interest principle have been difficult to prove in court. As one legal scholar writes, "despite the widespread application of the concept, most states fail to define communities of interest thoroughly, rendering such statutes difficult to enforce" (Malone 1997, 467). Although the concept of a community of interest may make intuitive sense, when pressed to locate where exactly a community of interest begins and ends, its practical utility begins to fall apart. First of all, there is no standard for what dimensions of community ought to be prioritized. An economic area defined in terms of a labor-market area may be very different from a cultural region, which may in turn be different from a media-market circulation area. Second, even if the large set of potential community variables were narrowed down to just a single one, the spatial distribution of nearly every variable is fuzzy-edged, shading off at the periphery without any clearly demarcated border.

As the geographer Richard L. Morrill writes, the community of interest requirements express the belief that one basis for representation in a democracy is "territorial—not of arbitrary aggregations of geography for the purpose of conducting elections, but as meaningful entities which have legitimate collective interests that arise from citizens identifying themselves with real places and areas" (Morrill 1987, 253). Of course, the casual reference to so-called *real* places and areas is much easier to mention in passing than it is to rigorously define: is a commuter megaregion real? What about a neighborhood with no functional status but a strong sense of community identity? A statistical tabulation area? A historical culture-region?

It may be wise to sidestep the thorny ontological question of what constitutes "real places and areas" and to turn instead to a principle of geographic *coherence*, which can, first of all, be meaningfully tested against empirical measures of spatial structure, and, second of all, provide the theoretical outline for a flexible but nevertheless durable concept of geographic polity. If we accept that it is impossible to draw an absolute border around a perfectly self-enclosed social and political unit, it does not necessarily follow that *any* border we draw is utterly arbitrary. There are indeed meaningful—and detectible—patterns of group integration that we should seek to respect when drawing electoral boundaries.

How, exactly, are these "detectible"? On the one hand, statistical techniques in network science and clustering, many of them drawn from work undertaken on the regionalization problem, offer frameworks for delineating functional regions that do not require an *a priori* use of existing political borders. To take just one example, Alasdair Rae and I have shown how a massively complex network of commuting patterns resolves into sensible regional groupings when subjected to a community detection algorithm (Nelson and Rae 2016). Commuter flows are just one of many types of interactions that connect people to one another and bind them into place-bound communities, but measuring how well or how poorly a proposed districting system respects these functional threads of human co-dependence offers one possible test for a map's correspondence with at least one dimension of community interest. I have suggested a measure called the "ELBRIDGE score"—electoral boundary resemblance to identifiable geography—that tests how many threads of functional connection a districting scheme interrupts, with the assumption that a better map will sever fewer (Nelson 2018b). Figure 3 shows the Congressional districts of Indiana overlaid on top of commuter flow lines of commuters who both live and work inside the state's borders. Under the principle of geographic coherence, a good district map will try to preserve as many of these functionally integrated areas as possible, so that district lines also enclose interdependent webs of commuter-based community patterns.

But it is important to recognize that not all patterns of geographic community will lend themselves so easily to statistical interpretation. In addition to these mathematical measures of coherence, we must consider the intuitive sense of citizens' own geographical fields of belonging and mutuality. As Mac Donald and Cain (2013) write regarding the gathering of public testimony for California's redistricting commission, individuals reveal important preferences about their sense of community that cannot be derived from demographic measures alone. Taking into account these qualitative definitions of geographic coherence is necessary if an electoral map wishes to promote citizens feeling as though their district boundaries match with boundaries that are meaningful in their actual lives.

However, geographic coherence alone cannot provide the sole principle for a just electoral system. Place-bound communities undoubtedly have genuine stakes within a representative democracy; they define the limits of important sets of common problem and interests, and they form the outlines for meaningful classifications of self-identification. Yet, due to the myriad ways in which geography has served as an anti-equalitarian tool—most persistently and infamously in racial segregation, but also in numerous other forms of spatial concentration of privilege and exclusion—it is crucial not to exclusively valorize geographic community over all other desiderata in a representative system. Wherever possible, geographic coherence should be respected in a districting scheme in order to preserve and promote communities of interest. But if such a principle has the knock-on effect of reinscribing patterns of unequal community standing, then it should be carefully weighed against competing principles such as the right of minority representation.

The geographer Carl Sauer, in a little-noticed 1918 research article, sought to "show

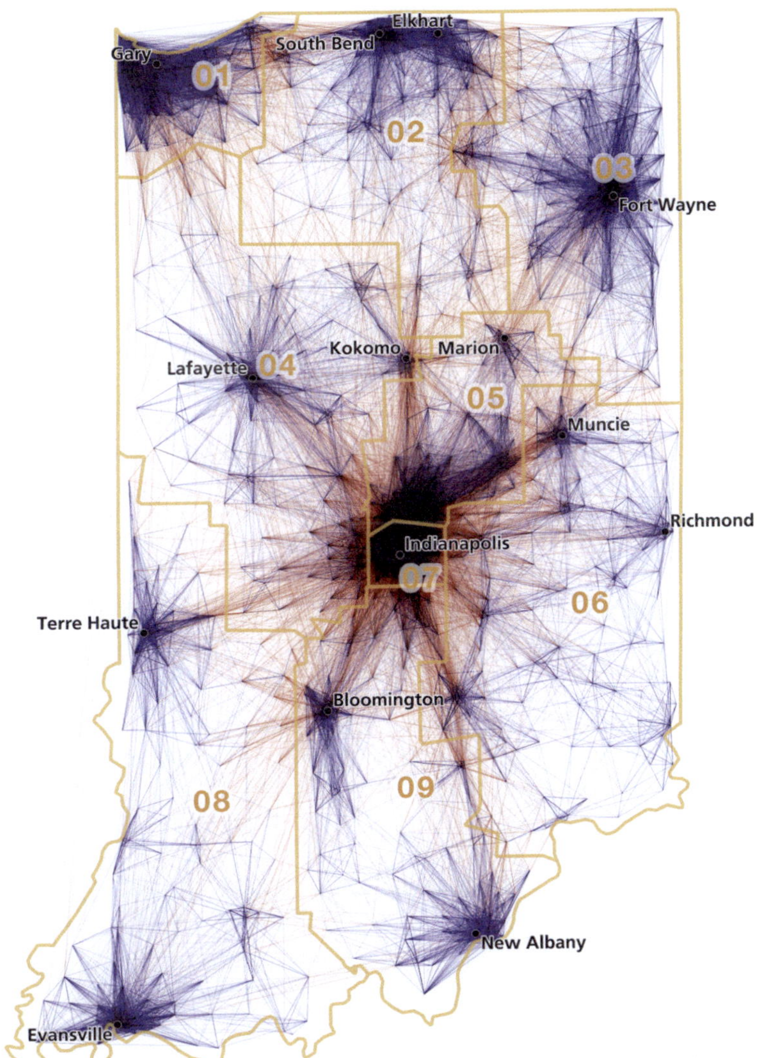

Figure 3: Commuter flows as a potential basis for regions. Commuter flows are overlaid on top of electoral district boundaries; purple flows begin and end in the same district (thus matching functional geography with electoral geography) whereas orange flows cross a district boundary (suggesting that a district "severs" a natural community). Indiana represents a state with a relatively high "ELBRIDGE score."

the gerrymander to be a violation of the geographic unity of regions and to indicate the possibilities of equable representation by reorganizing electoral districts on a geographic basis" (Sauer 1918, 404). Sauer was writing at a time when the organic principle of geographic "unity" still carried more currency than today. Yet we can retain Sauer's principle even if we reject the idea that social structure perfectly segments into a neat partition of spatial units: "reorganizing electoral districts on a geographic basis" would mean developing both a set of empirics and, perhaps more importantly, a set of political principles that identify the spatially structured polity—which is to say, a more substantive synonym for "community of interest"—as a key building block for political representation. The reason for doing so is not simply because appealing to geographically coherent communities of interest offers one tool for mitigating against partisan gerrymandering, but more broadly because such a method recognizes the important ways in which communities, representatives, places, and political action reciprocally structure one another.

REFERENCES

Baxter, Sylvester. 1891. Greater Boston: a federalized municipality to include the surrounding cities and towns. Boston: A. J. Philpott & Co.

Berry, BJL. 1964. "Approaches to regional analysis: a synthesis." Annals of the Association of American Geographers.

Berry, Brian J L, Peter G Goheen, and Harold Goldstein. 1969. Metropolitan area definition; a re-evaluation of concept and statistical practice. Washington, D.C.: U.S. Bureau of the Census.

Branford, Victor, and Patrick Geddes. 1917. The coming polity; a study in reconstruction. The Making of the Future. London: Williams & Norgate. http://hdl.handle.net/2027/mdp.39015028392283.

Brennan Center for Justice. 2010. "Communities of interest."

Bunge, William W. 1966. "Gerrymandering, Geography, and Grouping." Geographical Review 56 (2): 256–63. https://doi.org/10.2307/212882.

Chaskin, Robert J. 1997. "Perspectives on Neighborhood and Community: A Review of the Literature." Social Service Review 71 (4): 521–47. http://www.jstor.org/stable/30012640.

Coler, Bird S. 1899. "The Government of Greater New York." The North American Review 169 (512): 90–100. http://www.jstor.org/stable/25104849.

Coombes, M G, J S Dixon, J B Goddard, S Openshaw, and P J Taylor. 1982. "Functional regions for the population census of Great Britain." In Geography and the Urban Environment, edited by D T Herbert and R J Johnston, 63–112. London: Wiley.

Crary, Douglas. 1959. "A geographer looks at the landscape." Landscape 9 (1).

Dickinson, Robert E. 1934. "The Metropolitan Regions of the United States." Geographical Review 24 (2): 278–91. https://doi.org/10.2307/208793.

Gardner, James. 2002. "One person, one vote and the possibility of political community." North Carolina Law Review 80(4): 1237.

Gardner, James. 2006. "Foreword: Representation without Party: Lessons from State Constitutional Attempts to Control Gerrymandering." Rutgers Law Journal 37: 881.

Geddes, Patrick. 1915. Cities in evolution: an introduction to the town planning movement and to the study of civics. London: Williams & Norgate.

Glick, Joshua. 2003. "On the Road: The Supreme Court and the History of Circuit Riding Comment." Cardozo Law Review, no. 4: 1753–1844.

Granö, J G. 1997. Pure geography. Edited by Anssi Paasi and Olavi Granö. Baltimore: Johns Hopkins University Press.

Henschel, Albert Edward. 1895. Municipal consolidation: historical sketch of the greater New York. New York: Stettiner, Lambert, & Co.

Hudson, G Donald. 1936. "Methods Employed by Geographers in Regional Surveys." Economic Geography 12 (1): 98–104. https://doi.org/10.2307/140268

Jackson, John Brinckerhoff. 1964. "The meanings of 'landscape'." Kulturgeografi 16 (88): 1614–8.

Kimble, George H T. 1951. "The inadequacy of the regional concept." In London Essays in Geography , edited by L Dudley Stamp and S W Woolridge, 151–74. Harvard University Press.

Looker, Benjamin. 2015. A nation of neighborhoods : imagining cities, communities, and democracy in postwar America. Chicago; London: University of Chicago Press.

Mac Donald, Karin and Bruce E. Cain. 2013. "Community of interest methodology and public testimony." UC Irvine Law Review 3: 609.

Malone, Stephen J. 1997. "Recognizing Communities of Interest in a Legislative Apportionment Plan." Virginia Law Review 83 (2): 461. https://doi.org/10.2307/1073783.

Minar, David W, and Scott A Greer. 1969. The concept of community: readings with interpretations. Chicago: Aldine.

Morrill, Richard L. 1987. "Redistricting, region, and representation." Political Geography Quarterly 6 (3).

National Resources Committee. 1935. Regional factors in national planning and development. Washington: Government Printing Office. http://hdl.handle.net/2027/mdp.39015009271282.

Nelson, Garrett Dash. 2018a. "'The town was us': how the New England town became the mythical landscape of American democracy." Places Journal. https://doi.org/10.22269/180731.

Nelson, Garrett Dash. 2018b. "What would a good electoral map even look like?" CityLab

Nelson, Garrett Dash, and Alasdair Rae. 2016. "An Economic Geography of the United States: From Commutes to Megaregions." PLOS One 11 (11): e0166083. https://doi.org/10.1371/journal.pone.0166083.

Noronha, Valerian T, and Michael F Goodchild. 1992. "Modeling Interregional Interaction: Implications for Defining Functional Regions." Annals of the Association of American Geographers 82 (1): 86–102. https://doi.org/10.1111/j.1467-8306.1992.tb01899.x.

Nystuen, John D. 1968. "Identification of some fundamental spatial concepts." In Spatial Analysis , edited by Brian J L Berry and Duane Marble, 35–41. Englewood Cliffs, NJ: Prentice Hall.

Ogilvie, Alan G. 1928. Great Britain : essays in regional geography by twenty-six authors. Cambridge: Cambridge University Press.

Olwig, Kenneth. 1996. "Recovering the Substantive Nature of Landscape." Annals of the Association of American Geographers 86 (4): 630–53. http://www.jstor.org/stable/2564345.

Paasi, Anssi. 1991. "Deconstructing regions: notes on the scales of spatial life." Environment and Planning A 23: 239–56.

Saint, Andrew, ed. 1989. Politics and the people of London: the London County Council, 1889-1965. London: Hambledon.

Sauer, Carl O. 1918. "Geography and the Gerrymander." The American Political Science Review 12 (3): 403–26. https://doi.org/10.2307/1946093.

Stephanopolous, Nicholas O. 2012. "Redistricting and the territorial community." University of Pennsylvania Law Review 160: 1379.

Whyte, Nicola. 2007. "Landscape, Memory and Custom: Parish Identities c. 1550-1700." Social History 32 (2): 166–86. http://www.jstor.org/stable/4287424.

Wirth, Louis. 1938. "Urbanism as a Way of Life." American Journal of Sociology 44 (1): 1–24. http://www.jstor.org/stable/2768119

Wolff, Francis. 2014. "Polis." Dictionary of Untranslatables: A Philosophical Lexicon, ed. Barbara Cassin Princeton: Princeton UP: 801–802.

12 *Explainer: Communities of interest*

HEATHER ROSENFELD AND MOON DUCHIN

> "I have called my tiny community a world, and so its isolation made it; and yet there was among us but a half-awakened common consciousness, sprung from common joy and grief, at burial, birth, or wedding; from a common hardship in poverty, poor land, and low wages; and, above all, from the sight of the Veil that hung between us and Opportunity."
>
> –W.E.B. Du Bois, *The Souls of Black Folk.*

BACKGROUND: THE COMMUNITIES OF INTEREST CRITERION

One of the most enigmatic of the "traditional districting criteria" is that districts should try to preserve so-called *communities of interest*, or COIs. Though this principle is not national in scope (like population balance and minority electoral opportunity), the importance of drawing district boundaries to promote better representation for coherent communities is invoked in many states. As Chapter 11 recognizes and as Du Bois elaborates above, the question of what unites people in a place has many answers, and these answers are not always conscious or articulated. And indeed, COI rules are not always found in one place. Sometimes they are defined in state law or in non-binding guidelines published by line-drawing bodies, but other times they emerge in litigation, are carefully considered by mapmakers and community activists without any laws compelling them, or are celebrated (whether sincerely or not) in narrative descriptions formed entirely after the district creation process is complete.

Within these rules, norms, and practices, what is a COI, and what is its significance for redistricting? The answers to these questions are necessarily multiple, ambiguous, and sometimes contradictory. This explainer does not attempt to smooth this over—the fuzziness comes with the territory. What it does attempt, though, is to clarify some points of law, history, and current practice around community thinking in redistricting. Even when it stays somewhat fuzzy and imprecise, consideration of communities can be crucial to developing meaningful districts. And there are increasing prospects for handling COI consideration more concretely.

M. Duchin, O. Walch (eds.), *Political Geometry*, https://doi.org/10.1007/978-3-319-69161-9_12

COMMUNITIES OF INTEREST LANGUAGE IN THE RULES

Explicit language about preserving COIs is present in rules and redistricting guidelines in many states, as Figure 1 illustrates. Of these, sixteen states call for preserving COIs in Congressional redistricting, and twenty-five cite COIs for state legislative districts. Additionally, while Rhode Island has no explicit mention of the phrase "communities of interest," they have a right to "fair representation and equal access to the political process," which functions similarly in practice. Likewise, Oklahoma redistricting laws have no explicit reference to COIs, but their language about "economic and political interests" carries the same connotation [1, 10].

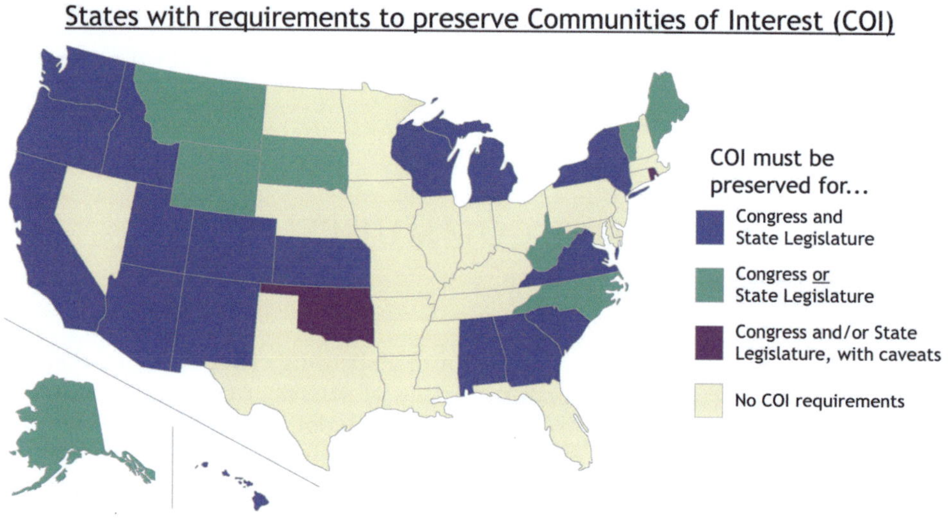

Figure 1: Community of Interest language in the rules for redistricting. Data from [1, 10].

A first point worth noting about COIs, in formal rules and in informal organizing, is that they are more stubbornly qualitative than other districting criteria. Most other such criteria can be more readily measured, usually in multiple ways with their own challenges.[1] COIs, by contrast, have primarily been based on narrative description—not entirely unlike that presented by Du Bois above.

Before contemplating the prospects for turning narrative into data, it is worth looking at the definitions of COIs in state laws and constitutions. References to communities of interest include variously detailed definitions along the lines of the following [1]:

- Alaska: "...a relatively integrated socio-economic area."
- California: "A community of interest is a contiguous population which shares common social and economic interests that should be included within a

[1] For instance, a close examination of the Census complicates the straightforward notion of population balance, and "compactness" has multiple inequivalent definitions (see Chapter 1).

single district for purposes of its effective and fair representation. Examples of such shared interests are those common to an urban area, a rural area, an industrial area, or an agricultural area, and those common to areas in which the people share similar living standards, use the same transportation facilities, have similar work opportunities, or have access to the same media of communication relevant to the election process. Communities of interest shall not include relationships with political parties, incumbents, or political candidates."

- Colorado: "communities of interest, including ethnic, cultural, economic, trade area, geographic, and demographic factors, shall be preserved within a single district wherever possible."
- Alabama: "community of interest is defined as an area with recognized similarities of interests, including but not limited to racial, ethnic, geographic, governmental, regional, social, cultural, partisan, or historic interests; county, municipal, or voting precinct boundaries; and commonality of communications."

We can observe from these examples that COIs are frequently defined in terms of economic commonality or shared industry. Echoing citations in various rule frameworks, COI references can be found repeatedly in defense or justification of a plan. For example, the Colorado redistricting commission based one district on the desire to connect several ski areas, and a court affirmed that this was a legitimate COI consideration. Another Colorado district was permitted to break county lines to unite a technology-oriented business district: the Denver Tech Center [3]. To this end, Du Bois' community might be considered a COI if its members were commonly employed by one or only a few specific industries. The references to cultural, racial, and ethnic similarities are also salient and would speak to Du Bois' community, although as we shall see, COIs cannot be characterized primarily in terms of race.

Just as often, however, the term is left as self-explanatory, as in the following states' statutes:

- Idaho: "To the maximum extent possible, districts shall preserve traditional neighborhoods and local communities of interest."
- Oregon: "Each district, as nearly as practicable, shall ... Not divide communities of common interest."
- South Dakota: "Protection of communities of interest by means of compact and contiguous districts."

The latter set of references offers much less information, lacking details that would be needed for practical guidance. Perhaps the most interesting case is that of South Dakota, where it seems to be suggested that just drawing nice shapes, while following other districting criteria, will tend to capture common interests all on its own.

PRINCIPLES FOR COMMUNITIES OF INTEREST

Having established that many states prioritize COIs, we turn to the question of how to find them and figure out their boundaries. Ideally, COIs should be co-extensive with districts—if the goal is to make districts respect the structure of these communities—but of course this is rarely possible on the nose. The size of districts is prescribed by law, and organic communities won't be exactly the right size, and only sometimes have clear boundaries. Where COIs are smaller than districts, one interpretation of the rules is that *a community should not be split by district lines.* We can justify this by a desire to give the community group a strong voice with the district's representative by making up a significant share of the constituents, or at least not splitting the community such that it has to use divided resources to reach multiple representatives.

Several authors have taken the logic farther and argued that where districts are large enough to contain multiple communities, mapmakers should attempt to unite those with similar features and avoid combining different ones. Law professor Nick Stephanopoulos writes that "district and community boundaries should coincide 'to the extent possible' because the one-person, one-vote rule makes perfect congruence impossible. When communities must be disrupted, however, the disruption should be minimized—for instance, by joining groups that are as similar in their interests and affiliations as is practicable." That is, good districting should avoid both the "unnecessary merger of disparate communities and [the] division of unified communities" [14].

As another example of like-with-like thinking, a court in a Virginia case wrote that "[t]he evidence shows a greater community of interest among the people of the Tenth District, which is a part of metropolitan Washington, than between the people of that district and those of the adjoining Eighth." The contrast here is between interests "primarily centered in Washington" as a whole and "largely agricultural" concerns in the Eighth district (Wilkins v. Davis 139 S.E.2d 849, 853 (Va. 1965)). Here, community identity is a sliding-scale assessment for a district as a whole, not a smaller geographic region preserved within a district.

OK, but how can mapmakers identify a COI? You can ask people, and many good districting practitioners do just that! (See Chapter 18.) Still others have argued that the identification of communities should follow guidelines that are clear enough that community detection can be automated, or just read off of an appropriate dataset. In this book, Garrett Nelson discusses the use of commuting patterns to infer relevant regionalizations, and other authors have argued for the use of cell phone data. In one influential example, Stephanopoulos proposed a "top-down" COI identification method that could be algorithmically driven, based on socio-economic and demographic data found in the American Community Survey and patterns of voting support for ballot initiatives [14]. Political scientists Karin Mac Donald and Bruce Cain have argued that this is unworkable, at least in California, because of rapid change and because of the poor correspondence of ACS data with the lived reality of communities [11]. They write, "[t]he 'interest' in a COI is not merely a clustering of some measurable social or economic characteristic. Residents in that area have to perceive and acknowledge that a social, cultural, or economic interest is politically relevant." Scholars and practitioners who reject top-down identification often advocate for public mapping initiatives as an alternative—in other words, there's no getting around asking people if you want to know how they think about where they live.

THE GAP BETWEEN NARRATIVE AND DATA

In terms of meaningful public input, how has it worked to ask people to define their communities for the purpose of drawing districts? At its most precise, public mapping might be able to show us communities as regions delineated on a map accompanied by a clear description of shared interests. It's worth thinking about what kinds of themes come up in public mapping, and how they fit in, and can fit in, to the practices of redistricting.

As one major example, the California Citizens Redistricting Commission (CRC), an independent redistricting commission, coordinated a substantial public mapping effort after the 2010 Decennial Census to get community input for drawing districts. Table 12.1 shows the most common themes that were identified in COI submissions.

COI Theme	Count
Environmental concerns	495
Common culture/cultural community	440
Recreation	251
Fire danger/services	220
Ethnic community	164
High-tech industry	104
Aerospace industry	97
Religious community	62
Air quality	38

Table 12.1: Top themes in 12,425 written COI submissions to California Citizens Redistricting Commission, by frequency [11].

The diversity of themes here cuts both ways: on one hand, it can be seen as a testament to the success of the CRC in terms of the breadth of public input. On the other hand, this list might suggest that the sheer number of possible COIs makes the concept less viable. The Maryland Court of Appeals came to the latter conclusion when considering the wrangling of testimony into data, arguing that the COI concept is "nebulous and unworkable" because "the number of such communities is virtually unlimited and no reasonable standard could possibly be devised to afford them recognition in the formulation of districts." (In re Legislative Districting of State, 475 A.2d 428, 445 (Md. 1984).)

12.1 CASE LAW

It's instructive to review what courts have found about the definitions and salience of communities of interest. This is drawn largely from Gardner [7].

The following descriptions have counted as COIs:

- Justice Sandra Day O'Connor, writing for the Supreme Court plurality, cites "for example, shared broadcast and print media, public transport infrastructure, and institutions such as schools and churches" as arguable "manifestations of community of interest," but warns that these are not easily disentangled from race (Bush v. Vera, 517 U.S. at 964 (1996))
- A "predominantly urban, low-income population" (Lawyer v. Dept of Justice, 521 U.S. 567, 581–82 (1997));
- Satisfactory evidence of shared socio-economic status (Chen v. City of Houston, 206 F.3d 502, 513 (5th Cir. 2000));
- "[L]ess-educated" citizens, or those "more often unemployed" have "common social and economic needs" suitable for a COI (Theriot v. Parish of Jefferson, 185 F.3d 477, 486 (5th Cir. 1999));
- "There are no doubt religious, class, and social communities of interest that cross county lines and whose protection might be a legitimate consideration in districting decisions." (Kelley v. Bennett, 96 F. Supp. 2d 1301 (M.D. Ala.))
- Lower courts have affirmed that Latino or Hispanic groups can constitute a COI (Barnett v. City of Chicago, 141 F.3d 699, 704 (7th Cir. 1998) and Meza v. Galvin, 322 F. Supp. 2d 52, 75 (D. Mass. 2004)), but there are ample indications that the Supreme Court would not agree (Miller, Session, etc.).
- Census tracts can serve as evidence: the California Supreme Court approved the plan of an outside expert (or "special master"), in part on the reasoning that building it out of whole census tracts contributed to preserving COIs (Wilson v. Eu, 823 P.2d 545, 556 (Cal. 1992)).

On the other hand, the following have been rejected as a basis for COIs:

- Keeping urban apart from rural areas (In re Legislative Districting of Gen. Assembly, 193 N.W.2d 784, 789 (Iowa 1972));
- A court itself may not "define what a community of interest is and where its boundaries are" (In re Legislative Districting of State, 805 A.2d 292, 297, 298 (Md. 2002)).

Finally, heterogeneity itself—just the fact that the district is not cohesive—is sometimes taken to show that race or ethnicity predominated over COIs as criteria for drawing districts, as in the following:

- Evidence of "fractured political, social, and economic interests" argues against specific COIs (Miller v. Johnson, 515 U.S. 900, 919 (1995));
- "Plaintiffs presented evidence of differences in socio-economic status, education, employment, health, and other characteristics between Hispanics who live near

> Texas's southern border and those who reside in Central Texas" in order to argue that ethnicity had predominated over COIs in forming the districts (Session v. Perry, 298 F. Supp. 2d 451, 512 (E.D. Tex. 2004));
>
> These are hints, but, as promised, they don't add up to a very clear picture! In fact, legal scholar James Gardner argues that only Alaska, Colorado, New Jersey (pre-1966), Vermont, and Virginia even have clear precedent in law for COI considerations at all [7].

As in the case of Maryland, the open-endedness of the task of collecting public input has most often led states in the past to decline even to try to define and locate communities formally. The defendants in the major Supreme Court case, Bush v. Vera, were chastised for this omission: the decision stated that "it is, however, *evidentially* significant that at the time of the redistricting, the State had compiled detailed racial data for use in redistricting, but made no apparent attempt to compile, and did not refer specifically to, equivalent data regarding communities of interest." (emph. original).

This discussion might well lead us to think that new technologies will help bridge from vague to tangible, by supplementing testimony with digital mapping. Building good COI mapping software is a complex task, and political geographer Benjamin Forest sounds a note of warning. His claim is that when using mapping software to identify communities of interest, the data available when you draw the lines—such as race and political party—can subtly dictate which COIs are mapped, and which maps are accepted [5]. Bush v. Vera tells us that race cannot predominate over other considerations, so it might follow that racial data cannot be the primary data shown in tools for redistricting, and specifically in tools for community input.

On the other hand, abandoning mapping altogether is not a solution either, of course. A court case in Hawaii makes the connection clear: if you can't map COIs, you can't tell a mapper to preserve them. As the court stated: "[t]he lack of defined boundaries precludes reapportionment based upon a strict recognition of community interests" (Kawamoto v. Okata, 868 P.2d 1183, 1187 (Haw. 1994)).

Public mapping tools have exploded in popularity in the 2020 census cycle, with multiple platforms helping to collect substantial public input around the nation. Their utility rose steeply in the face of the coronavirus pandemic. With in-person meetings for collecting community feedback being shut down, state and community groups collecting COIs have increasingly turned to web apps.

The MGGG Redistricting Lab developed mapping software called Districtr that is free to users, with a strong emphasis on COI identification. The Lab helped collect public input in many states—Michigan, New Mexico, Pennsylvania, Wisconsin, Ohio, Missouri, and more—on behalf of commissions, government offices, and community groups. Other software packages supporting COI collection include Representable and Dave's Redistricting App; even the commercial vendors like Maptitude, ESRI, and CityGate have gotten in on the action.

POST HOC COMMUNITY

Though not all states have a mandate to preserve COIs, sometimes districts are explained using the *language* of community. At best, this can identify or even create meaningful logic in district plans, but in some cases it can also engender concerns about gaming the system.

In one fairly persuasive example, after California's commission collected public testimony, they wove language about community intricately together with other districting criteria to describe their plan. For example, this passage combines COIs, population balance, political boundaries, and the VRA: "CD 16 includes all of Merced County and portions of Madera and Fresno counties. The city of Fresno is split in this district to achieve population equality and in light of the Section 5 benchmark for Merced County. The western valley portion of Madera County is included in this district, as well as many of the Highway 99 communities from Merced County into the city of Fresno, such as Livingston, Atwater, Chowchilla, and the city of Madera. Communities in this district share the common links of agriculture, water, and air issues, along with the serving as the main transportation routes connecting northern and southern California" [2].

Another recent example of crafting the language of community after the fact was the sprint to remake Congressional redistricting through the Pennsylvania Supreme Court in 2018. Democratic Governor Tom Wolf filed a benchmark map with the court laden with COI-heavy narrative. There had been no time to collect public testimony, so the descriptions drew on personal and professional knowledge from the Governor's office. For example, the court filing described the district comprising Pittsburgh and some of its suburbs as follows: "These communities share significant interests economically, including an evolving technology sector and strong educational and medical institutions." Likewise, the description for an Eastern Pennsylvania district began: "District 15 combines the region's four third-class cities of Allentown, Bethlehem, Easton, and Reading, with their shared heritage of manufacturing and common interests, into one district. This map recognizes the communities are similar in economies and histories and should be together" [15]. Pennsylvania has no requirement to preserve COI, so the narratives here served a softer, persuasive purpose: they were intended to confer legitimacy and consolidate support for the plan, either with the public or possibly with the court.[2]

WHOSE VOICES? WHOSE STAKES?

In the 1920s, a group of sociologists at the University of Chicago divided the city into 72 pieces, calling them "community areas." These community areas were used—and are still used—for studying neighborhood characteristics and change over time. While originally intended as a way to group Census data "to reflect real, not arbitrary, divisions within the city," the classification was contested from the

[2]Happy postscript: Grassroots organizations like Pennsylvania Voice stepped up to help in the 2020 cycle, using the longer timeframe to collect hundreds of maps through a grassroots initiative around the state, then using those maps to inform their redistricting advocacy.

start. The Encyclopedia of Chicago reflects some popular skepticism about the community areas, saying that "[f]rom the beginning they only unevenly reflected the actual experience of community within the spaces, and over time many of them have become even less indicative of the perceptions of their residents, whose characteristics have shifted considerably due to migration. As ossified zones, they capture neither individual community identity nor the territorial reality of social groups" [12]. So these areas were defined by an academic elite with knowledge of the city that was of course partial at best, but they have an enduring life in city statistics and a surprisingly resilient presence in everyday neighborhood talk. The layered story of the Chicago community areas encapsulates the themes of this explainer: how, when, and by whom community is defined and demarcated all matter. The stakes become very tangible in redistricting.

Elite edicts are problematic, but on the other hand soliciting public input can at worst be a completely empty gesture. In a study of community participation in development planning in Boston, sociologist Jeremy Levine observed that "residents' membership in 'the community' affords them legitimacy and recognition." But these invocations of "community" and especially "community participation" are promissory—and such promises do not always bear fruit. In practice, Levine finds that community participation did not lead to residents having concrete influence. In this way, community was constructed in a hollow and ultimately cynical way, enabling redevelopment authorities to claim community input through holding meetings, while they ultimately retained control over decisionmaking [4, 9].

To conclude, it is worth stepping back and situating communities of interest in broader debates about community as a component of representative democracy. We've seen that many people in redistricting want to interpret community in terms of industrial sectors, others want to define it after drawing districts to support or describe plans, and that despite attempts to define it, it remains nebulous. Martin Luther King, Jr. had a very different vision. Throughout his writings, King, Jr. voiced a commitment to what he called Beloved Community, a sense of community "in which all persons are respected and enjoy the economic benefits of the nation" [8]. The respect of all persons included affirming difference, especially racial difference, and addressing poverty and inequality.

The whole concept of community exists with a fundamental tension: bringing people together across difference and working to address inequality, as in King, Jr.'s Beloved Community, or just affirming sameness in a narrow sense. As geographer Lynn Staeheli writes, the concept of community risks "the construction of sameness, rather than a recognition of what is common; it [community] can create something totalizing, rather than something based on sharing" [13]. Because of this tension, some are fundamentally skeptical of the concept of COIs in redistricting. Gardner, for instance, writes: "[t]o speak of a community of interest is to presuppose a thin form of community based on nothing more than shared interests—we are all poor, perhaps, and thus share an interest in poverty programs; or we are all riparian homeowners and thus share an interest in flood protection. A shared interest is central to an advocacy organization, perhaps, but it hardly describes a three-dimensional political community" [6].

Others see the promise of fairer outcomes. For the 2020 redistricting cycle, more states have turned to official and unofficial citizens' and people's redistricting commissions. Using public mapping tools and working with community-based organizations, these commissions are centering the collection of COI maps and narratives in their efforts to democratize redistricting. To that end, we'll close with a much more hopeful note from Staeheli, who writes, "the powers to define community and to exclude on the basis of that definition are also the powers to reorder the public, and we can imagine the possibility of doing this to *enhance* democratic citizenship" (emph. added) [13]. So the stakes are high: community lets us reimagine coalitions and alignments, and therefore lets us reimagine representation. More than ever before, this cycle has had a starring role for grassroots mapping efforts. A challenge for the coming cycle will be the development of increasingly sophisticated summary and synthesis methods that negotiate the conflicting logics of community.

WHERE TO READ MORE

In addition to the references discussed above, several chapters in this book add insight to the questions raised here. Readers may be interested in the description of redistricting rules in Chapter 2, which situates COIs among other criteria. Chapter 11 speaks to the inherent tensions between people and place in what come to be called communities, and Chapter 10 gestures toward the significance of defining community for addressing racial inequality. Chapter 18 offers practitioner perspectives on districting, including the challenge of collecting community input for COIs. Finally, Chapter 21 discusses how communities of color were able to gain representation, not through COIs, but through coalition claims.

Finally, you can read about the MGGG Lab's extensive efforts to collect and synthesize public input at `https://mggg.org/COIs`.

REFERENCES

[1] Brennan Center, *50 State Guide to Redistricting.* https://www.brennancenter.org/sites/default/files/publications/2019%20Citations%2050%20State%20Guide%20Final.pdf

[2] State of California Citizens Redistricting Commission, *Final Report on 2011 Redistricting.* https://wedrawthelines.ca.gov/wp-content/uploads/sites/64/2011/08/crc_20110815_2final_report.pdf

[3] Colorado Reapportionment Commission, *Staff summary of meeting,* https://www.colorado.gov/pacific/sites/default/files/August%2016%20Summary%20-%20Littleton.pdf.pdf

[4] Bill Cook and Uma Kothari, *Participation: The New Tyranny?* Chicago: University of Chicago Press (2001).

[5] Benjamin Forest, *Information sovereignty and GIS: the evolution of "communities of interest" in political redistricting*, Political Geography 23:425–451 (2004).

[6] James A. Gardner, *One Person, One Vote and the Possibility of Political Community*, North Carolina Law Review 80:1237 (2002).

[7] James A. Gardner, *Foreword: Representation without Party: Lessons from State Constitutional Attempts to Control Gerrymandering*, Rutgers Law Journal 37:881–970 (2006).

[8] Joshua F.J. Inwood, *Searching for the Promised Land: Examining Dr Martin Luther King's Concept of the Beloved Community*, Antipode 41:487–508 (2009).

[9] Jeremy R. Levine, *The Paradox of Community Power: Cultural Processes and Elite Authority in Participatory Governance*, Social Forces 95:1155–1179 (2017).

[10] Justin Levitt, *All About Redistricting*. `http://redistricting.lls.edu/where.php`

[11] Karin Mac Donald and Bruce E. Cain, *Community of Interest Methodology and Public Testimony*, U.C. Irvine Law Review 3:609 (2013).

[12] Amanda Seligman, *Community Areas*, The Encyclopedia of Chicago. `http://www.encyclopedia.chicagohistory.org/pages/319.html`

[13] Lynn A. Staeheli, *Citizenship and the problem of community*, Political Geography 27:5–21 (2008).

[14] Nicholas Stephanopoulos, *Redistricting and the Territorial Community*, UPenn Law Review 160:1379 (2012).

[15] Thomas Wolf, *Statement of respondent Thomas W. Wolf in support of his proposed Remedial Congressional Map pursuant to court's orders of January 22 and January 26, 2018*. `https://www.pubintlaw.org/wp-content/uploads/2017/06/2018-02-15-Statement-in-Support-of-Proposed-Remedial-Congressional-Map.pdf`.

Chapter 13

Geography as data

LEE HACHADOORIAN AND RUTH BUCK

CHAPTER SUMMARY

The U.S. Census Bureau was established to enumerate all of the residents of the country every ten years. Its geographic units, and the counts of people attached to them, are the basic stuff that districts are built from. This chapter will tell us about Census and electoral data products and the spatial tools that let us manipulate them.

1 INTRODUCTION

The last two chapters discussed key concepts in geography and geographic thought as they relate to U.S. redistricting. This chapter offers an introduction to the data side of the story. We'll introduce common terminology, survey the data sources, and discuss the use of geographic information systems (GIS) software and other mapping tools to make it all come together.

From achieving population balance between districts, as required by the Supreme Court, to determining whether or not a districting plan complies with the Voting Rights Act, we find Census and electoral data, coupled with the powerful capabilities of GIS, at the heart of redistricting practices.

This discussion begins with a look at the Census. Census data act as the ground truth about what kinds of people live in what locations for most redistricting purposes. It must be brought together with a second and much messier data source: precinct and election results data from state and local authorities. After that, we spend a fair amount of time introducing the GIS tools used for contemporary geospatial data analysis and display. Finally, we review some puzzles and challenges that are specific to the geospatial data of redistricting.

M. Duchin, O. Walch (eds.), *Political Geometry*, https://doi.org/10.1007/978-3-319-69161-9_13

WHO'S WHO

It's important to remember that most of the materials and practices we describe in this chapter go far beyond the application to redistricting. Demographers, human geographers, spatial statisticians, public health scholars, and urban planners all make heavy use of census data and GIS.

In fact, GIS more broadly encompasses the entire system of tools and technologies that can be used to work with spatial data, including satellites, GPS devices, drones, web-mapping servers, spatial databases, and geoprocessing in a variety of programming languages. Our discussion here focuses on a subset of domain-relevant technologies and methods.

2 THE CENSUS AND ITS PRODUCTS

The Census divides up the nation into geographies from coarse (states) to fine (census blocks). In essentially every case, electoral districts are made from these geographic units. The relationship of the Census to redistricting goes back to the nation's founding: Article 1, Section 2 of the U.S. Constitution mandates an enumeration of the population every ten years for apportioning membership in the House of Representatives. Who is counted and how they are classified has changed over the decades, subject to both technical advances and broader social and political changes.

Margo Anderson, in *The American Census: A Social History*, gives us a look at the Census as a record of American social classification practices [1]. For example, the first Census, in 1790, asked for the total number of people in each household, according to the following categories: free White males under sixteen; free White males sixteen or older; free White females; "other free persons"; and slaves. Only the name of the head of the household was collected.

Since then, the Census has changed dramatically. Over the course of the 19th and 20th centuries, a number of economic and social questions were added. Concern over its length and infrequency led to the separation of the Decennial Census into a short form and a long form, with the short form covering only basic questions about age, race, and ethnicity for each member of the household.[1] The long form surveys culminated in the development of the American Community Survey (ACS), a detailed annual survey of about 1.5% of the population, which began in 2005 and has become the nation's leading source of socioeconomic data.[2] Unlike the ACS, which is based on sampling the population, the Decennial Census attempts to create a person-level database of the *entire* population.

From a nuts-and-bolts perspective, the Census Bureau begins with a Master Ad-

[1]The Decennial Census also includes information on housing and *group quarters* like dorms, prisons, and military bases, but for the purposes of this chapter we will focus on the information it collects regarding population.

[2]The ACS asks respondents many pages of detailed questions regarding income, education, access to cars and the internet, and so on. Because it is based on a survey, the ACS data are frequently used in rolling five-year averages.

dress File (MAF), a list of residential addresses where Census forms will be mailed. Households return the forms by mail, or, beginning with the 2020 Census, they can complete the form online. In the event of a non-response, Census canvassers will visit an address to determine who, if anyone, is living there. Everything the Census collects begins here, on the ground, and everything it assembles from these gathered data is considered a *product*, which is then *released* on a schedule. Current privacy law protects the full records from disclosure for 72 years, so the Census releases the data as counts that are aggregated into various geographic units.

2.1 CENSUS GEOGRAPHIES

HIERARCHY AND CONCEPTS

Census data products take a variety of forms, one of which is *geographies*: spatially described subsets of territory. Users may call them geographic areas, geographic units, or geographic entities.[3] This section will sketch out their structure and some of their many uses.

The Census Bureau provides demographic and socioeconomic data for a staggeringly large number of geographic areas. A hierarchical structure, depicted in Figure 1, helps to keep track of all the scales and interrelationships. The center line in the figure is called the *spine*, especially the six-level structure

NATION—STATE—COUNTY—TRACT—BLOCK GROUP—BLOCK.

The smallest units, census blocks, completely cover the territory of the United States. These nest inside block groups, which nest inside counties, and so on up to the nation. Geographies that fall outside of the spine may not nest neatly and may not entirely cover the larger unit to which they belong. For example, Congressional districts fall wholly within states, but do not necessarily honor any other geographic boundaries until the smallest unit, blocks. A particular redistricting plan may strive to keep counties whole, but this is not guaranteed, so counties do not appear in the hierarchy under Congressional Districts.[4]

Off-spine geographies include legal or administrative geographic units such as Federal American Indian Reservations and school districts. In addition to units like these that are decided externally, the Census Bureau and other federal agencies *create* some geographic units for the purpose of data dissemination. For example, the Office of Management and Budget defines *metropolitan / micropolitan statistical areas* (MSAs) based on commuting patterns and uses those to report statistics that are useful to researchers and planners. Another example is Zip Code Tabulation Areas (ZCTAs). The Post Office creates Zip Codes *only* for the purpose of delivery route planning, but because there is now a wealth of industry and marketing data available by Zip Code, the Census Bureau has built corresponding ZCTAs for which demographic data are reported. These are examples of geographies off the central

[3]Or you might see hybrid terms like *areal units* for units pertaining to areas.

[4]The Census Bureau releases data for many "part geographies," so it is possible to download demographic and socioeconomic data for "Counties split by Congressional Districts," for instance. This would show demographics for the part of a county that falls in a specific Congressional district.

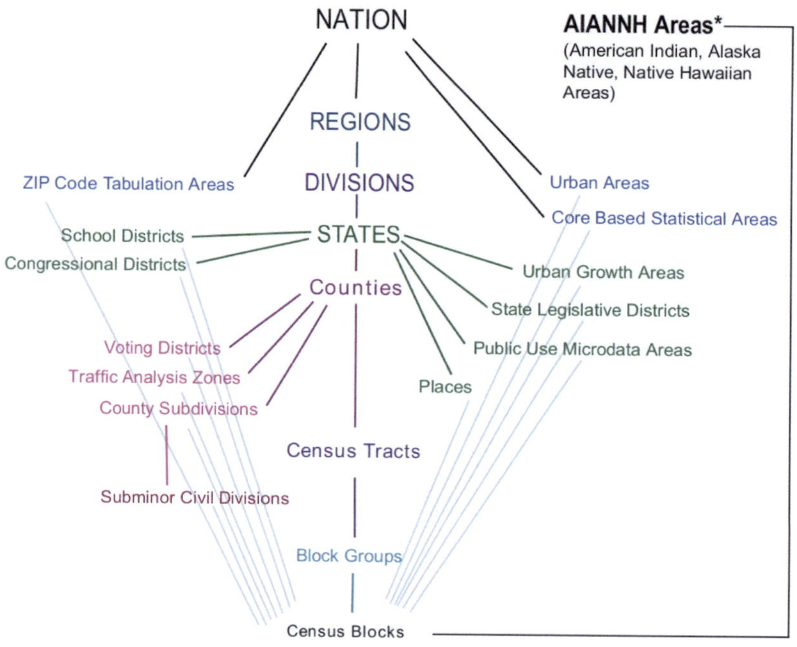

Figure 1: Hierarchy of Census geographies, from census.gov [24]

spine; they are made up of census blocks but do not necessarily nest in any of the other important units.

The hierarchical structure is reflected in the system of unique identifier codes attached to the geographic units, denoted GEOIDs. For example, states are identified by a two-digit FIPS (Federal Information Processing Standards) code (e.g., "01" = Alabama, "42" = Pennsylvania). Congressional Districts use an additional two digits; for instance, 4207 labels the 7th District in Pennsylvania.

GEOGRAPHIC PRODUCTS

Ultimately, all of the geographies described here are stored in a massive database called the TIGER system and are made available as a set of *TIGER/Line files* and related products.[5] If you want anything like a canonical set of mapping products for American geography, this is it. When a new geography is supplied by a state, it is standardized and processed into TIGER/Line format, correcting errors (e.g., gaps in coverage) and aligning units.

[5]TIGER stands for Topologically Integrated Geographic Encoding and Referencing. As the Bureau documentation says, "The TIGER/Line Shapefiles are the fully supported, core geographic product from the U.S. Census Bureau."

13.1 WHAT'S IN A BLOCK?

The *census block* is the basic building block of all census geographic units. Block boundaries are keyed to streets, roads, railroads, water bodies, legal boundaries, and "other visible physical and cultural features" [23]. In a city, a block might be coincident with a city block, while in a rural area a block might be a larger plot of farmland.

Given that the Census counts people by residence (i.e., where they sleep), and block construction is determined by the physical and social landscape, some census blocks may have odd shapes or zero population. In some cases, a census block picks out a traffic circle or a winding segment of a multi-lane highway. This helps to illustrate the difficult balancing act between the role of blocks in logically segmenting the territory of the country and their role in finely enumerating the population.

The use of census blocks to cover the entire country was initiated in the 1990 Census. By the time of the 2010 Census, the country was divided into over 11 million census blocks, of which over 40% (more than 4.8 million) have no reported population at all. Over half a million blocks (541,776 out of 11,078,297, to be precise) are wholly composed of water. Figure 2 shows census blocks in Philadelphia.

In populated areas, as blocks often divide from each other along streets, districts built out of blocks will also have the property that along their edges, people are separated from their across-the-street neighbors—a benign consequence of this choice of units, but somewhat counterintuitive. In California public meetings, some voters vocally objected to being in separate blocks from co-residents of a housing development that is separated by a street, while being joined across rear lot lines with another housing development [17]. Urban researchers have for some time questioned the usefulness of these census units for local analysis, noting that dividing lines *behind* homes are more natural for understanding how people think about neighborhoods [5].

Figure 2: Census blocks in Philadelphia. The Northeast Philadelphia Airport (yellow) is its own census block, population 2. (It's unclear who is considered to live there; perhaps homeless people.) The most populous block in Philadelphia is the Riverside Correctional Facility (pink), population 4535. But most census blocks are city blocks, like the five marked in the Germantown neighborhood (blue), which have between one hundred and three hundred people each.

Depending on the purpose of a map, factors such as scale, appearance, or even storage size, may require a reduction in the level of detail. The process of simplifying geographic information—removing detail, either manually or algorithmically—is called *generalization*. The objective of generalization is to reduce the precision of geographic data in order to suppress information that is irrelevant or inessential to a map's purpose without sacrificing clarity or accuracy of relevant spatial relationships [27].

Figure 3: Three different levels of generalization of Rhode Island.

In addition to the full-resolution TIGER/Line files, the Census Bureau also provides generalized cartographic boundary files at 1:500,000, 1:5,000,000, and 1:20,000,000 resolutions. Figure 3 shows the state of Rhode Island at these three levels of resolution. The purpose of these files is to be suitable for printed maps, which are not well served by accuracy down to feet or inches.

2.2 WHO GETS COUNTED, AND HOW?

The aspiration of the Census is to enumerate all the people resident in the United States on April 1 ("Census Day") of each year ending in 0. This is an increasingly difficult task as the population has grown to over 300 million, and as you can imagine, it was especially thorny in 2020, when the coronavirus pandemic shut down most of the country in mid-March.

Population counts will ultimately be reported in *tabular* form—i.e., in tables or spreadsheets. Besides the main tables, they also offer *cross-tabulations* ("cross-tabs") showing intersection counts for pairs of primary variables, and *special tabulations* ("special tabs") for alternative aggregations answering common queries that are not addressed in the main tables. In this section we introduce some of the complexity of how the numbers are produced.

13.2 PRODUCING RACE COUNTS

Among the demographic data that the Census collects is self-identified race and ethnicity, using categories that reveal changing social conceptions of race. For example, in the late 19th century, the Census included a "Chinese" racial category that subsumed all East Asians. South Asians were considered "White," but later were assigned to a new "Hindu" category, which included Muslim Pakistanis. Now, South Asians and East Asians are grouped in the racial category "Asian."

Figure 4 shows the question asking about race in the 2010 Decennial Census.

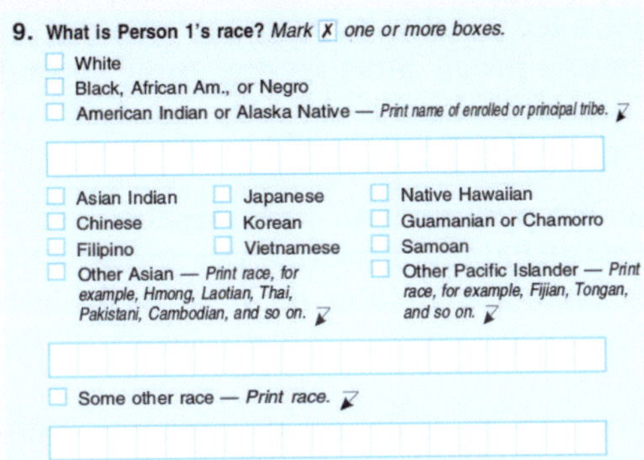
9. What is Person 1's race? *Mark* X *one or more boxes.*

- White
- Black, African Am., or Negro
- American Indian or Alaska Native — *Print name of enrolled or principal tribe.*
- Asian Indian
- Chinese
- Filipino
- Other Asian — *Print race, for example, Hmong, Laotian, Thai, Pakistani, Cambodian, and so on.*
- Japanese
- Korean
- Vietnamese
- Native Hawaiian
- Guamanian or Chamorro
- Samoan
- Other Pacific Islander — *Print race, for example, Fijian, Tongan, and so on.*
- Some other race — *Print race.*

Figure 4: "What is Person 1's race?" from the 2010 Census.

If a race is not reported, it may be inferred or *imputed* from data gathered about other household members, or from the previous census household record. The Census processes this information and then reports race counts by the following six categories.

- White
- Black or African American
- American Indian and Alaskan Native
- Asian
- Native Hawaiian and Other Pacific Islander
- Some Other Race

Since the 2000 Decennial Census, the questionnaire has allowed respondents to select multiple races. Since there are six racial categories, there are $2^6 - 1$ or 63 possible combinations (you can be in or not in each of the six, but you can't be enumerated in no category at all). In 2010, the great majority of respondents were enumerated in some single race category; only 2.9% of respondents were ascribed two or more races.

The form produces some difficulties in self-reporting because it does not map perfectly onto American racial discourse. For example, many Americans think of Hispanic/Latino as a race, but because the Census treats it in a separate ethnicity question, the form forces Hispanic-identified people to make a race choice from this list. Nationally, in 2010, about 65% of Hispanic people identified as White and 27% as Some Other Race (an option selected by only 0.3% of non-Hispanic people). This phenomenon had significant regional variation. For another example, the lack of a Middle Eastern/North African category leads to MENA self-identification spread over multiple categories (Asian, African American, White, and Some Other Race), making it ultimately harder to conduct social science research on this group, even though it has strong social and racial salience.

UNDERCOUNT AND OVERCOUNT

While the Decennial Census is intended to be a complete enumeration of the population, it obviously can't achieve perfect accuracy. The *undercount* is the number of persons resident in the United States on April 1 of a census year who are not enumerated. *Overcount* occurs when a reported number is too high, such as if a person is included twice.

The stakes for a correct count have increased considerably over the course of the last half-century. The growth of federal aid programs, which often distribute resources in proportion to population, has meant that state and local governments have a considerable financial stake in the accuracy of the Census. Civil rights legislation has included employment provisions that rely upon a count of racial minorities. And enforcement of the Voting Rights Act hinges on counts of population by race and sometimes by language group.

Since 1950, the Bureau has published an assessment of overcount and undercount for each Census, using both demographic analysis and a post-enumeration survey.[6] They estimated approximately 16 million omissions in the 2010 Census (5.3% of the population). While it may seem obvious that the Census Bureau will miss counting some individuals, the source of the overcount is, at first glance, puzzling. 85% of erroneous enumerations in the 2010 Census—persons counted who should not have been counted—were duplications, usually due to individuals whose residence is ambiguous, such as college students living away from home, incarcerated or military population, children with two custodial parents, and households with multiple properties. 15% were erroneously enumerated for other reasons, including persons who were born after or died before April 1 and "fictitious persons."[7]

The deeper issue is not merely obtaining more accurate overall numbers. The problem is also one of a *differential undercount*, deeply correlated with demography and geography. The Bureau's own estimates of undercount disaggregate as follows:

- Black: 2.07%
- Hispanic: 1.54%
- Native Hawaiian/Pacific Islander: 1.34%
- Asian 0.08%
- Native American: 4.88%/−1.95%
- White: −0.84% (a net overcount)

Notably, the Native American net undercount breaks down as an undercount of 4.88% on reservations, but −1.95% (a net overcount) off reservations.

The Bureau's report finds that many other characteristics also correlate with count inaccuracies. Residents of owner-occupied housing are consistently overcounted, while renters are undercounted. Very young children are undercounted, while teenagers are overcounted. Both males and females who are 50 years and older were overcounted in the past three Censuses. At younger ages, the pattern diverges: 30- to 49-year-old women are overcounted, while 18- to 49-year-old men

[6] Read about the Census Bureau's Coverage Measurement techniques: www.census.gov/programs-surveys/decennial-census/about/coverage-measurement.html.

[7] In 2010, there were 10 million erroneous enumerations and 6 million whole-person imputations—people from whom the Bureau did not collect sufficient information, but inferred characteristics to include in the count. The overcount thus almost exactly offset the undercount, yielding a net overcount of 0.01% [19].

are consistently undercounted.[8]

Because we have a great deal of external evidence that points to patterns in the undercount, statisticians at the Census Bureau could adjust the data to compensate for known disparities. But in 1999, the Supreme Court ruled (in *Department of Commerce v. United States House*) that statistical adjustment could not be used when deciding how many members of Congress are apportioned to each state. Intriguingly, they left the door open to the possibility of statistical adjustment used at the state level, such as for redistricting, and several states are now considering their options for count correction in an unprecedentedly challenging and contested census year.

2.3 SPECIFIC PRODUCTS FOR REDISTRICTING

The malapportionment court decisions of the 1960s put significant pressure on the Census to tabulate population counts for small-area geographies. To achieve the principle of "One Person, One Vote" in practice, and to judge compliance with the Voting Rights Act, public officials wanted more and more detailed census data, at the finest spatial resolutions possible, as quickly as possible.

In 1975, Congress passed a bill called Public Law 94-171 in order to meet these needs. In response, the Bureau took up the mission of "provid[ing] states the opportunity to specify the small geographic areas for which they wish to receive decennial population totals for the purpose of reapportionment and redistricting" [25].

THE PL 94-171

Since this legal shift in the 1970s, the Census Bureau has been required to provide redistricting data to the designated authority in each state by April 1 of the year following the Decennial Census. These tables are now eponymously referred to as the *PL 94-171 data*; they contain counts by race and ethnicity in every census block in the nation, with a second table reporting the same counts among voting age population.

States generally complete their redistricting process by the end of the year ending in 1, drawing new Congressional and legislative districts for use in the primary and general elections of the year ending in 2. In addition to these products, the Census Bureau issues a special data release in the year ending in 3 containing information about the newly enacted districts. In 2021, everything is expected to happen late because of pandemic-related data collection problems and an unprecedented level of interference from the White House.[9]

[8] Note that geographic imprecision when locating addresses can lead to under- or overcounts at the block level that disappear at larger levels of geography. An estimated 2,039,000 people, or 0.7% of the United States population, are enumerated in the correct county of residence, but the wrong "block cluster" (group of nearby census blocks). So the practice of fine-tuning population to ±1-person population deviation between districts may very well be shifting blocks whose measurement errors are larger than the population differences they are trying to correct.

[9] The most recent PL 94-171 was delayed until August 2021, causing many states to scramble in order to meet their timelines to create new maps.

13.3 THE CITIZENSHIP QUESTION

The Census Bureau often releases revised or specialized data separately, in so-called special tabulations. An important one for redistricting is the Citizen Voting Age Population (CVAP) by Race and Ethnicity. This special tabulation is based on the American Community Survey, in which respondents are asked about their citizenship status. Because only citizens are eligible to vote in state and federal elections, CVAP provides a much better estimate of the demographic balance in the electorate than population or voting age population alone. When Voting Rights Act challenges are brought on behalf of Asian or Latino plaintiffs, CVAP is always used in the supporting data.

However, citizenship status has been carefully kept out of the Decennial Census "short form." This is because, even though census form responses are supposed to be firewalled from other government agencies, people who are in the U.S. with legal concerns about their residency are likely to be intimidated by an official form asking for their citizenship status, making them potentially far less likely to be enumerated—and for apportionment purposes, this can't be corrected statistically. Without the question on the short form, citizenship numbers have not been included in the PL 94-171 data.

This too has become highly politicized. The conventional wisdom holds that excluding non-citizens from redistricting would benefit Republicans, and so several "red states" have made moves to do their redistricting on the basis of citizen-only population. In 2016, the Supreme Court heard a case called *Evenwel v. Abbott* in which Texas sought to use citizenship data in this way; its finding re-affirmed that total population is the traditional basis for redistricting but left the door open for future options.

The Census Bureau is technically part of the Department of Commerce, whose leader is a presidential appointee. As part of the Trump Administration's far-reaching efforts to control the levers of government, the Secretary of Commerce demanded that the Census Bureau add a citizenship question to the short form. When enactment was halted by the Supreme Court, the Bureau was instructed to use other means (like administrative records) to estimate citizenship numbers and include them in the PL94-171—a clear attempt to pave the way for citizen-only population balancing. Citizenship numbers were once more blocked from the redistricting data in this cycle, but future litigation is certain.

THE REDISTRICTING DATA PROGRAM

The Redistricting Data Program is a small division within the Census Bureau that supports the redistricting-specific needs of state and local officials. In the years leading up to each official Census Day, they conduct two main tasks: collect suggestions for changes to census block boundaries, and collect a snapshot of precinct boundaries from the states to create a data product called VTDs (voting tabulation districts, sometimes confusingly called "voting districts"). Both tasks rely on liaisons in participating states to coordinate with the Census Bureau. The opportunity for the states to specify the geographic areas for which they wish to receive redistricting data is granted by law [26].

Both block boundary suggestion and VTD collection should be thought of as a back-and-forth between the state liaisons and the census staff. On the census side, the submitted geography will be cleaned and aligned to make it conform to the TIGER protocol so that it can be used seamlessly with other data products. Because the VTD process is part of the Decennial Census effort, the Census Bureau publishes VTDs only once every 10 years. The Census Bureau neither keeps the geographies updated nor publishes ACS data for the VTDs during the intercensal period.

The Redistricting Data Program makes a valiant effort to standardize the wildly varying election administration units in the states, at least at one timestamp in the ten-year census cycle. We turn to that broader question now.

3 ELECTION DATA AND THE PRECINCT PROBLEM

Census geographies are an elaborate and crucial resource for understanding the human geography of the U.S. in spatial terms, but there is another fundamental piece of the redistricting puzzle: election results. These are not only needed for a wide range of analytic tasks for redistricting—from competitiveness to partisan skew—but are also an ineliminable part of Voting Rights Act enforcement, which relies on a showing of racially polarized voting, linking electoral history to demographics.

3.1 PRECINCT BOUNDARIES

The smallest unit at which election results are recorded and released is called a *precinct*. Usually, each one has a single polling place where people physically go to vote, but this is not always a one-to-one match. Here, we use the term "precinct" to refer to the electoral geography: the area whose residents are all handled together in voting administration terms. Perhaps surprisingly, and unlike census geographies, precincts are not drawn or maintained in a standardized way across or even within states!

Several states do have clear laws regarding the management and data transparency of precinct boundaries. In Minnesota, for example, Election Law requires municipal clerks to notify the secretary of state within 30 days of any precinct boundary change, who must then update the statewide precinct boundary database [20].

However, in many other states, the state does not track precinct boundary changes between redistricting cycles. Ohio, for example, requires that the list of registered voters be updated and all affected voters be notified of any precinct changes, but it does not explicitly require that those boundary changes be reported to any central election authority [7]. And in other states, counties or county subdivisions have complete control regarding the election precincts within their jurisdiction. Precincts are split, merged, or completely redrawn with such regularity that there is no guarantee that Precinct A in County X covers the same territory in a special election in 2017 as it did in the general election in 2016.

This reflects a broader difference between what state elections officials often refer to as a "top-down" versus a "bottom-up" approach to election administration. The secretary of state's office in a top-down state (e.g., North Carolina, Wisconsin, and Minnesota) exercises significant control in how elections are administered and in how data from elections are stored and disseminated. In these states, precinct boundary data and corresponding election results can often be downloaded from the state geospatial data portal or from the Secretary of State's website.

In a bottom-up state like Ohio or Missouri, however, county and municipal governments are given almost complete autonomy in conducting elections. As these local governments have differing ordinances and widely varying mapping technology capabilities, precinct boundary data also vary in both quality and format. For example, during work on a project intended to collect the precinct boundaries used in Ohio's 2016 general election, members of the Voting Rights Data Institute (co-led by one of us, Ruth Buck) contacted officials from each of the eighty-eight counties. Forty-six provided data in GIS format, twenty-seven had PDF maps, eighteen mailed paper maps (sometimes held together with scotch tape, with precincts drawn in magic marker), and seven counties either refused to or could not provide precinct boundary data in any format. In those instances where the county could or would not share its data, we had to estimate precinct boundaries using addresses in the state voter file.[10]

To summarize: Reconstructing precinct geographies can be a complicated and labor-intensive post hoc process.

3.2 MAPPING ELECTION RESULTS

Another complexity in this work is that precinct-level election returns are generally provided in a tabular format such as a spreadsheet, showing total votes by precinct for each candidate in each election. This means that the names appearing in these tables of results must be matched up with names of units in a mapping format if we are to have any hope of visualizing the elections. This is unfortunately not always straightforward, even in states where precinct boundary data and election returns are both publicly available. There are almost always disparities in the names and even the total number of precincts between the precinct boundary data and the tabular election returns.

Once these disparities are resolved to the best of one's ability, there is rarely a *natural key* (i.e., a code that is meaningful, such as the two-letter state postal code, as a opposed to an arbitrary serial identifier) for connecting the datasets. Also, while states do have an obligation to provide certified election results to the Federal Election Commission (FEC), the manner in which states report and publish their precinct-level results varies wildly. In Mississippi, 2016 precinct-level

[10]A voter file, or voter registration file, is a database that usually contains the names, addresses, party affiliation, and precinct assignments of registered voters. Information included in the voter file varies from state to state, as do the cost and the bureaucratic obstructions to obtaining the file. Political campaigns will often buy processed voter files from commercial vendors to aid in targeting likely voters.

election results were reported by the Secretary of State's office as PDF scans of paper printouts. For the redistricting analyst, time-intensive data cleaning would be necessary before these results could be used.

Additionally, not all reported votes are tied to a precinct: ballots cast through absentee, provisional, or early voting are often reported only at the level of the election authority, such as a county board of elections. Federal overseas absentee votes, including ballots cast by U.S. military, are also often reported state- or county-wide rather than at the precinct level.[11] At the present time, it's obvious that these challenges will only expand. Elections in pandemic conditions drove people to alternative voting modalities in record numbers this year, with many estimates indicating that as many as 50% of votes cast nationally were early, absentee, or by mail. There are a variety of methods in spatial statistics that can be used for assigning votes reported at a coarser geography (like a county or a house district) to a finer geography (like a precinct). Unfortunately, there is no solution that is reliable across contexts.[12]

The picture is made more complicated by the difficulty of prescribing election procedures to the states, as the Constitution makes it explicitly a state affair. This could be addressed, for instance, with federal regulation requiring the reporting of election data in a modern, spatial format. Until then, the work will continue to rely on difficult and time-intensive data preparation.[13]

4 GIS: SHAPES AND ATTRIBUTES TOGETHER

4.1 WHAT IS GIS?

Contemporary mapping is heavily reliant on spatial software. The *International Encyclopedia of Geography* defines it this way:

> A geographic information system (GIS) is a computer system (desktop or web mapping software) for capturing, storing, querying, analyzing, and displaying geospatial data. Geospatial data describe both the location and attributes of spatial features. [4]

[11] In Alaska, absentee, provisional, and early votes are only reported at the house district level. Those votes, however, were about *one-third* of all votes cast in 2018 in both the gubernatorial and U.S. congressional elections in that state.

[12] This problem of data disaggregation is discussed further below in §4.2.

[13] Voting rights litigators must conduct this work each time they seek to press a Voting Rights Act case. Some academic and civic groups have attempted to collect and curate election geodata and make it publicly available. Several groups are currently working to collect and make publicly available high-quality election geodata. During the 2010 redistricting cycle, political scientists from Harvard, MIT, and Stanford created the Harvard Election Data Archive, which contains election geodata from most states (projects.iq.harvard.edu/eda [14]). In this cycle, efforts include MGGG States from the MGGG Redistricting Lab (github.com/mggg-states) and OpenPrecincts from the Princeton Gerrymandering Project (openprecincts.org). For tabular election results alone (with no geography), OpenElections has results in various stages of cleaning up to mid-cycle (openelections.net) but the most comprehensive and up-to-date data can be found from the MIT Election Data + Science Lab (electionlab.mit.edu/data).

GIS developed and spread in the late twentieth century alongside the rise of personal computing. Before this, redistricting (like many other planning operations) was conducted largely on paper. Now, GIS technology suffuses the redistricting process, from the gathering and dissemination of census data, to the creation of the electoral districts, to the analysis of the impacts of specific redistricting plans.

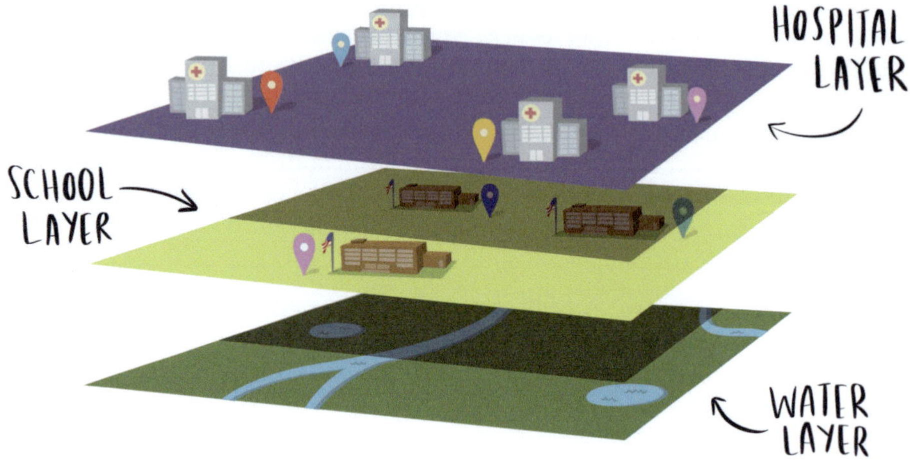

Figure 5: A schematic of data layers in GIS.

All map construction involves selection of what facts of the world to represent. Within GIS, maps are constructed by adding *layers* representing facts from different knowledge domains (concept illustrated in Figure 5). A map of a suburban street may be constructed of layers representing tax parcels, roadbeds, sidewalks, trees, building footprints, etc. A gas station layer might be important for a road map, but wouldn't be included on a map of religious affiliation.

MODELS, STRUCTURES, AND FORMATS

Our information will need to be organized. This is done on the one hand with a relational data model and on the other hand with a graphical interface.

GEOID	Name	Location	Biden votes	Trump votes
36059	Nassau County, NY	(list of vertices)	396,504	326,716
36081	Queens County, NY	(list of vertices)	569,038	212,665
36103	Suffolk County, NY	(list of vertices)	381,021	381,253

Table 13.1: A very small attribute table showing the relation between attributes of three entities. If we wanted to join further data, we would need it to be labeled in a common way, such as by the same GEOIDs, which can then be used as a key.

Here, the *entities* are counties, the *attributes* are the entries in the table, and the *relation* is the whole table's worth of information.[14] (See Harrington [13] for more

[14]Geographers tend to refer to nonspatial data as "attribute data," even though a geometry (such

on relational database design.)

Moving the discussion toward shapes, let's start with a distinction between *vector* and *raster* data (Figure 6). Vector formats store data as sets of coordinates that describe points, lines (sequences of endpoints with a line segment between each two successive ones), or polygons (a sequence of points that forms a closed loop). Raster formats store data as a grid of values. Geospatial phenomena can be represented in either model. For example, a forest could be represented as a polygon (vector) for the purposes of a road atlas, but could be represented by a specific value for grid cells in a land cover raster layer (where other values might represent grass, bare earth, etc.). The vector and raster models have different strengths; the vector model is more often used for hard-edged data, such as human-created political territories or tax parcels, while raster formats are better for the continuous or fuzzy-edged phenomena common in physical geography.

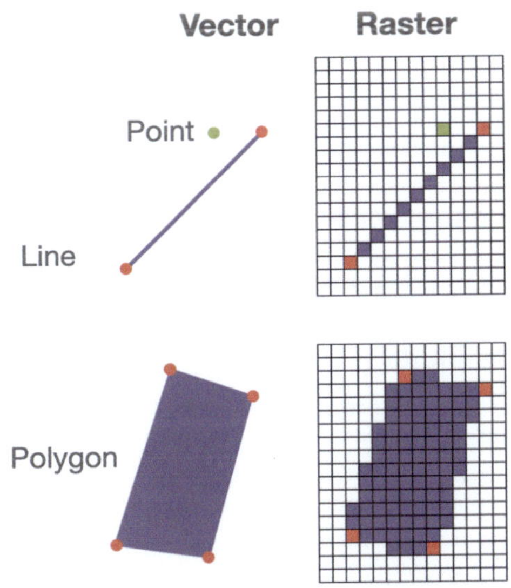

Figure 6: Vector vs. raster representations of points, lines, and polygons.

Census geographies are represented in GIS as vector polygons. Each element of a vector dataset is called a *feature*. For example, GIS might have a spatial layer for cities in which each individual city is a feature. Layers may be organized by type, scale, or theme. Roads are conceptually different from counties, and each would be stored as a separate layer file. Counties and states would also be stored as separate layers. Finally, the *attributes* ("facts") that one might want to know about an entity are virtually limitless, so to keep file sizes manageable, they may be separated thematically (e.g., into economic data and environmental data). As

as the shape of a Congressional District) is technically an attribute as well. Analysts from many fields might also refer to attributes as *variables*, and be interested in constructing models that demonstrate relationships among these variables.

the geometries (feature shapes) are usually the largest part of the file, geometries and attributes can be and frequently are stored separately.

LOCATION AND PROJECTION

The coordinates of a vector object are just numbers, and cannot be pinned to an earth location without specifying a *coordinate reference system (CRS)*. Latitude and longitude measurements are spherical coordinates that must be transformed for display on a flat computer screen or printed map using a *projection* method. Going from a sphere to a screen or map inevitably leads to distortion, but with knowledge of the specific projection method, it is possible to minimize and account for distortion in ways relevant to the project at hand.[15]

It is not uncommon for geographic data to be distributed without the CRS specified. This would be analogous to telling someone the temperature without indicating Celsius or Fahrenheit. The correct scale may be obvious from context, but using the wrong scale would lead to wildly incorrect conclusions. You cannot work with geographic data without interpreting the coordinates in *some* CRS, and assuming an incorrect CRS will lead to nonsensical or misleading results.[16]

All projections also have specific parameters; for instance, many projections are centered somewhere, so they have greatest accuracy close to the specified center point. Some CRS, such as the *state plane coordinate system*, are locally parameterized to have good accuracy on particular states. This makes them appropriate for use in mapping small areas, such as a county or group of counties within one state. Others, such as the USA Contiguous Albers Equal Area Conic are parameterized for mapping larger areas like the continental U.S.[17]

In going from the three-dimensional, uneven Earth to a flat map, all projections necessarily distort. Projection methods can be *conformal*, meaning that they preserve angles as measured at any point; *equivalent*, meaning that they preserve the relative areas of features; or *equidistant*, meaning that with respect to one or two special points, distances are faithfully represented.[18] Conformal projections

[15] To be more complete, here are some of the elements that are built into the process of projecting geographic data. Latitude and longitude coordinates must be interpreted with respect to a *geographic coordinate system* (GCS) and a *datum*. A GCS is a set of parameters that translate between the Earth's shape—an oblate spheroid—and a perfect sphere. Part of the GCS called the datum can be thought of as an anchor point for the GCS (so the precision of the projection will be greatest near that anchor point) [6]. GCS and datum considerations are unlikely to matter to most redistricting practitioners and researchers, but can be of life-or-death importance in engineering and military applications.

[16] For new GIS users, one of the most frequent mistakes is to *assign* a CRS to a spatial layer when one actually wants to *transform the coordinates* of the spatial layer to a new CRS. This is like changing "30 degrees Celsius" to "30 degrees Fahrenheit," rather than applying the arithmetic transformation to yield "86 degrees Fahrenheit." The telltale sign of this kind of mistake is having spatial layers that don't align with each other when viewed in GIS.

[17] Confusingly, detailed mapping is called "large scale" while zoomed-out mapping is called "small scale" in some geography language. To see why, look at Figure 3 and note that $1/500,000$ is a larger fraction than $1/20,000,000$, even though that figure is more detailed. Other CRSs like the Albers Equal Area Conic can work across the globe but parameters must be set based on the area and scale to be mapped.

[18] Interestingly, to achieve this, one further cheat is needed: a special point on the sphere will typically have to be represented by an arc on the flat representation.

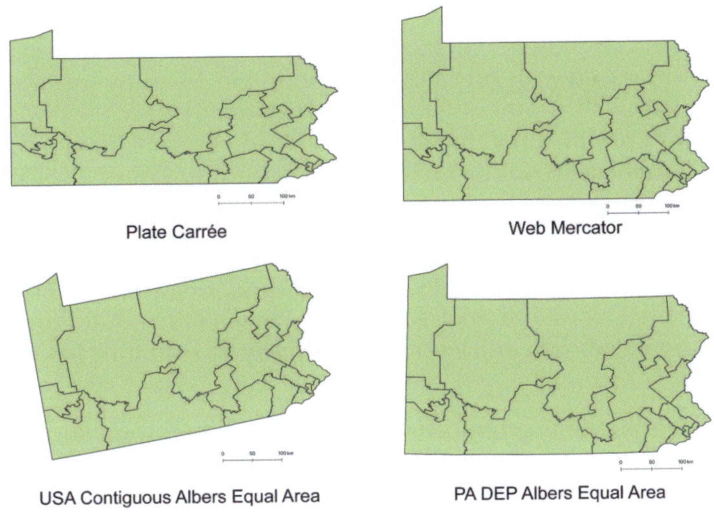

Figure 7: Pennsylvania in four map projections.

may be the best choice for navigation purposes because the right angle of a street intersection will appear as a right angle on the map; equivalent projections may be most appropriate for demographic mapping. There are compromise projections that try to balance different kinds of distortion, especially in zoomed-out mapping (e.g., maps of the world), but no projection can be conformal, equivalent, and equidistant at the same time. Of course, the curvature of the Earth is imperceptible over small distances, so for locally parameterized CRS, such as state plane zones, there is very little loss in this tradeoff. The state plane zones are all conformal, and they have a maximum scale distortion of 1 part in 10,000, for a maximum measurement error of 52.8 feet in 100 miles, which is generally considered to be acceptable for most applications.

FILES AND SOFTWARE

Broadly speaking, the industry giant in GIS is the multi-billion dollar ESRI corporation, which makes a package called ArcGIS that is so ubiquitous that for many casual users it is referenced interchangeably with the whole idea of GIS. The most common vector file format in redistricting analysis is the shapefile, a format created by ESRI in the 1990s and eventually published as a standard for data interoperability [10].[19] The format is widely supported by government agencies, including the Census Bureau.[20] A widely used free and open source alternative is QGIS.

[19] There is some ambiguity as to whether the shapefile format can be considered an "open standard." It is not included among the standards published by Open Geospatial Consortium, the major international geospatial standards body [21]. The published technical documentation is incomplete, and in 2012, geospatial programmers reverse-engineered the unpublished spatial indexing component of the format [15]. There are many other geospatial data formats, but the shapefile is so ubiquitous that there is a temptation (which should be avoided!) to think refer to *any* geospatial data set as "a shapefile."

[20] When gathering information for spatial entities such as political and administrative geographies, the possible attributes are limitless. However, many software applications are limited in the number

ESRI also makes packages specific to redistricting, and some states have built state-specific ArcGIS plugins, such as the RedAppl application in Texas, which is notoriously associated with ruthless redistricting in Texas by some authors [11]. Within redistricting, the dominant commercial software package is Maptitude for Redistricting (often shortened to Maptitude), made by a smaller company called Caliper.[21] And there are a number of others, notably autoBound (by CityGate GIS), that are selected by various localities for their line-drawing purposes.

It's important to realize that redistricting in these software programs is essentially all based on human selection (using keyboards and mouses), rather than relying on algorithms that draw the lines for you.[22] They display building blocks with data visible on a dashboard and let the user assign each building block to a district interactively.

Today there is growing momentum toward web-based, free, and open-source alternatives. Three of the most popular available tools are Dave's Redistricting App (davesredistricting.org), made by a volunteer team of Microsoft-affiliated engineers; DistrictBuilder (districtbuilder.org), made by a company called Azavea; Districtr (districtr.org, Figure 8), made by the MGGG Redistricting Lab. The impetus behind this open software push is to demystify and democratize redistricting for the public.

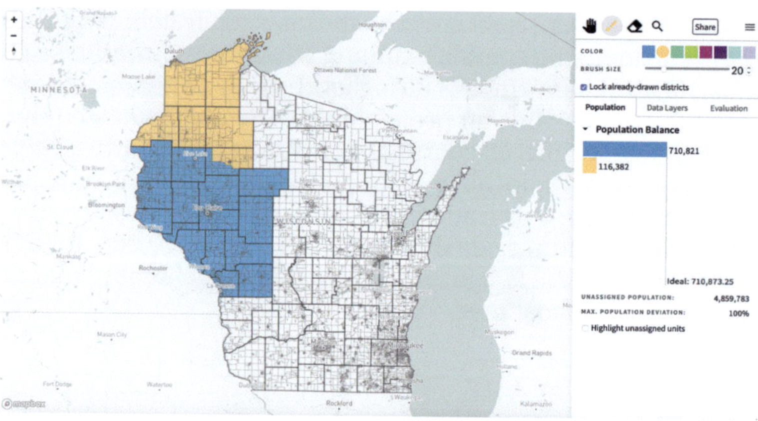

Figure 8: A screenshot from Districtr.

of columns in a single table. For the American Community Survey, for example, although the Census Bureau does indeed manage the entire (very large) dataset, the attribute data are split among many tables merely to keep file sizes manageable.

[21] As a further indication of ESRI's dominance of the GIS market, Caliper publishes Maptitude for Redistricting as both standalone software and as an extension for ArcGIS.

[22] In their 2020 release, Maptitude included some algorithmic districting options for the first time, but they are quite clearly not the primary mechanism for line-drawing. See caliper.com/mtredist.htm.

4.2 JOINING DATA TO GEOGRAPHIES

ONE-TO-ONE JOINS

We have already encountered a case where two kinds of data must be joined: the spatial data of precincts and the attribute data of vote totals (see Table 13.1). You can perform joins within desktop GIS, but for complicated operations many users prefer to work with spatial database systems or geospatial data packages within programming languages like Python and R.[23] All that is needed is a matching key, such as GEOIDs, that specifies the correspondence between the entities in the two datasets. This is another extraordinarily useful aspect of working with census data: they are painstakingly constructed to facilitate joining.

By contrast, electoral data are frequently a mess. Different agencies may use different and poorly documented conventions and abbreviations (even within the same state). Joins between these data sources may therefore be difficult without manual inspection, or by applying methods that are not usually available in desktop GIS such as natural language processing or fuzzy matching. Matching on names rather than unique identifiers also creates the potential hazard of multiple matches.

ONE-TO-MANY AND SPATIAL JOINS

A more complex but very useful operation is a one-to-many join with aggregation. This is when one row in the spatial data or tabular attribute data is joined to many rows in the other table, and then an additional operation is performed to combine some of the rows. For example, if the analyst has polling place wait times, and wants to create a map showing the average wait time by county, this requires both aggregation (grouping the polling places by county and averaging the wait times) and joining the resulting average to the county layer. These steps can be accomplished in desktop GIS, but can often be executed more easily and reliably in a programming language where you can write code to perform the operation systematically, saving the script for later inspection.

Finally, an analyst may have two sources of data that need to be joined spatially by common locations. For example, consider the polling place data mentioned above. Assigning each polling place to a Congressional District requires a *spatial join* that queries the districts to figure out where each polling place belongs. Spatial joins are more "expensive" (computationally intensive) operations than attribute joins; with large datasets such as census blocks, spatial joins can take minutes or hours to run on desktop-grade processors. Spatial joins can be sped up with proper spatial indexing (a technique that organizes spatial data) and by writing results to disk so that certain expensive operations don't have to be repeated.[24]

[23] PostGIS is a major spatial database program. The Python package called geopandas and the R package called sf are also popular choices.

[24] For instance, we were able to join Pennsylvania's 400,000+ census blocks to a set of hypothetical Congressional Districts in PostGIS in about 20 minutes without spatial indexes, or 2 minutes with properly defined spatial indexes.

INTERPOLATION

If data are provided for smaller units that nest in larger ones, then it can be joined to the larger units by aggregation, as we've seen. But if data exist on one set of geographic units (like CVAP on block groups) but needs to be joined to a different, unrelated set of geographic units (like precincts), then a more sophisticated operation is needed. The problem of estimating attribute data for a new set of *target* units based on data provided for a set of *source* units is known in geography as the areal interpolation problem.

Areal interpolation methods can be classified as *intelligent*, meaning they make use of ancillary information such as land use or census data, or as simple, which incorporate no additional data. Simple areal weighting, which allocates data to a target unit proportionate to the area of overlap between the target and source units, is widely used due to its simplicity [8]. Areal weighting requires no external data and can be performed using most desktop GIS software without special code or plugins. For many voting-related or demographic variables, however, areal weighting is a poor choice. When trying to interpolate CVAP from block groups to census blocks, for example, areal weighting will award the largest census block the most population despite the fact that the largest blocks by area often have the smallest populations (as in Sidebar 13.1).

Many official state election data products, such as in North Carolina, Wisconsin, and Texas, use an intelligent areal interpolation method that Schroeder terms "target count weighting" [22].[25] This method relies on a third, smaller *control* unit that nests in both the source and target units and acts like a common denominator to do the operation. The Wisconsin State Legislature's Legislative Technology Services Bureau, for example, releases a decade's worth of election data interpolated onto a single election year's precincts by essentially disaggregating votes from an older set of precincts (the source units) to census blocks (the control units) based on total population, and then aggregating the votes from the blocks up to the new set of precincts (the target units).

5 SOME SPECIFIC CHALLENGES

DATA VINTAGES

As geographies and associated data change over time and show no exact agreement between data sources, it is preferable to use geographies and demographic data from the same source (or at least from compatible sources) and also from the same time stamp. For example, American cities and municipalities often change borders due to annexation or secession. Official populations may grow or decline suddenly, and this growth or decline may be due to a gain or loss of legal territory.

[25] For examples in state data, see `https://www.ncleg.gov/Files/GIS/Base_Data/2016/Numeric/Data_Processing_Notes_2016.pdf` (NC), `https://redistricting.capitol.texas.gov/docs/pubs/Data_2011_Redistricting.pdf` (TX), and `https://data-ltsb.opendata.arcgis.com/datasets/2012-2020-election-data-with-2020-wards/explore` (WI).

Census geographies are released in *vintages* named for the release year, like wines! For the Decennial Census and ACS 1-year Estimates, the matching year is obvious. For ACS 5-year Estimates, there is a convention that the matching year is the final year of the period: for instance, the 2013–2017 ACS 5-year Estimates should be used with the 2017 vintage TIGER/Line files.

The Census Bureau continually updates the TIGER/Line files, and they publish information on changed geographies every year. For example, in 2016, Florida, Minnesota, North Carolina, and Virginia reported Congressional Districts that changed due to a mid-decade redistricting. These might be considered "real" changes to the geography of Congressional Districts. But on the other hand, if you compare the Congressional District geometries from the 2015 and 2016 TIGER/Line files, you would find that 386 of 444 features show changes between these years.[26] *These are not reflecting changes to the legal boundaries of the districts!* The updated geographies are entirely due to tiny adjustments to the census blocks that are regarded by the Bureau as accuracy improvements, which might add vertices to the polygons or move vertices by a matter of feet or inches. Thus, you cannot simply compare geographies through shapefiles directly to find out if they have changed from year to year. And you also have to be careful if your layers come from different vintages, because operations like intersection, subset, and shared boundary can go awry if the entities are recorded in a subtly mismatched way.

MEASURING COMPACTNESS

Earlier, in Chapter 1, you read about measures of *compactness*, which are aimed at describing the shapes of districts as either eccentric and convoluted or simple and regular. Certain compactness metrics, particularly those based on area versus perimeter, have been shown to be highly sensitive to changes in resolution and projection [2, 9, 16]. This is a practical impact of the choice of coordinate systems beyond "squashed" or distorted appearances (see Figure 7).

Each state might have a different best choice for its coordinate system and projection for redistricting. The Pennsylvania Department of Environmental Protection uses an Albers Equal Area projection that is locally parameterized to be suitable for displaying the entire state, making it reasonable to use for compactness measures (although the user must set some parameters to do so). The lack of a canonical choice across or even within states is a great reminder that even objective-sounding metrics actually require quite a substantial amount of user discretion.

WATER AND ADJACENCY

Speaking of choices that are under-specified, consider the question of whether to include bodies of water in electoral districts, and whether to consider areas separated by water as being adjacent. There is no universal standard about the

[26]This number includes 435 districts with voting members, 6 non-voting delegate districts, and 3 water areas that are legally parts of the territory of some state but not assigned to any Congressional district.

inclusion of water. To see the strange effects this can produce, look at the map of Michigan's congressional districts shown in Figure 9.

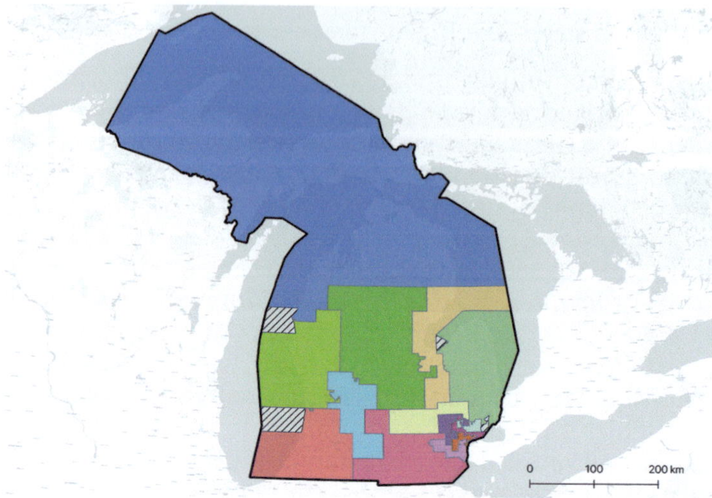

Figure 9: Michigan Congressional Districts drawn after the 2010 Census. Note that the districts zigzag in and out of the lake—the diagonal hatched regions are lake areas not assigned to any district in the TIGER/Line files.

The sawtooth pattern turns out to be explainable as a GIS artifact. Michigan's 6th District (bottom left in the figure) includes the entirety of Van Buren County, but only part of the area of Allegan County. The district geography is therefore constructed from all of Van Buren County, which legally includes the water area in Lake Michigan, and Allegan county subdivisions, tracts, or blocks, for which the water area is omitted.

Water is also important when it comes to thinking about what land area is next to what, which is important since districts are typically expected to be contiguous (i.e., made up of one connected piece). In some cases, redistricting seems to have been done without regard for physical geography, like when land areas separated by water are combined in a single district. But this can be necessary, such as for islands, and it can also be done to further other legitimate redistricting goals (such as complying with the Voting Rights Act or respecting communities of interest).[27]

For example, New York City covers part of the continental mainland and three islands (Staten Island, Manhattan Island, and Long Island), divided into five major pieces called boroughs. When drawing congressional districts after the 2010 Census, Staten Island was kept whole but had to be connected to *some* territory across water in order to get up to the needed population. On the other hand, Manhattan had more than two districts' worth of population. Rather than placing any

[27] When researching state House districts in Alaska, Caldera et al. found that the level of restrictiveness applied to water adjacency had profound implications for minority representation. When districts were randomly drawn on the more permissive adjacency network, there were significantly more House districts with a majority of Alaska Native population than when working with the restrictive network with fewer edges [3].

district wholly within Manhattan, the island was split among four congressional districts, all of which cross water to include significant territory and population from Brooklyn, Queens, or the Bronx.

So even when districts are required to be contiguous, the rules for crossing water are unclear for legislatures or for courts, and therefore for researchers trying to understand the space of possibilities. Islands off the coast may make up their own census units, but may also be part of a larger geography that includes the mainland and the water in between. When geographies are not physically adjacent, should allowable connections be based on distance measurements, or the transportation network, or demographic similarity? This is another case of interpretive ambiguity that becomes very visible at the level of geospatial data representation, but can't be resolved without deliberation on the larger goals of good redistricting.

6 CONCLUSION: TECHNOLOGY CUTS BOTH WAYS

There are widespread misconceptions about both the availability of reliable geospatial election data and the ease of manipulating it. It doesn't help that popular media sources publish many maps that look like precinct-level election results—for instance, on election night. Although these maps may look authoritative and reliable, the data sources for them are unclear at best: in many cases, no public sources *could have* provided the data that are displayed. This creates an impression of availability that combines with other factors—including confusion about the control of precincts and commercial incentives against widespread public access—to impede reform and modernization of data maintenance and transparency practices.

In this chapter, we have tried to show some of the ways that data you may have taken for granted are actually built and handled. Some are meticulously curated by the Census Bureau and some are scraped and digitized and re-shaped and matched by many hands—and until there is regulatory reform, election data will stay scattered and messy. Within best practices, there are still a substantial number of user choices to make.

We are often asked for our opinions on whether, on balance, the trend toward more powerful geospatial technology has made gerrymandering worse.[28] The blatant partisan gerrymanders following the 2000 and 2010 Censuses do seem to coincide with the increasingly widespread adoption of GIS. However, geospatial technology is just another tool being used to redistrict, and the sophistication attributed to gerrymanderers seems to us to be overblown. While many of the professional redistricting consultants use proprietary software and commercialized data, researchers and engaged members of the public can benefit from the current boom in the open data and civic tech movements, and from rising attention to redistricting from diverse academic researchers. We have considerable hope for the future.

[28] Many authors have considered this question, such as McGann et al., who point to the availability of computer software for redistricting prior to 2000, among other factors, in dismissing the idea [18].

REFERENCES

[1] Anderson, Margo J. 2015. *The American Census: A Social History.* 2nd ed. New Haven: Yale University Press.

[2] Barnes, Richard, and Justin Solomon. 2018 "Gerrymandering and Compactness: Implementation Flexibility and Abuse." *arXiv:1803.02857[Cs],* March. http://arxiv.org/abs/1803.02857

[3] S. Caldera, D. DeFord, M. Duchin, S. Gutenkust, and C. Nix. "Mathematics of Nested Districts: The Case of Alaska." Statistics and Public Policy (2020).

[4] Chang, Kang-Tsung. 2017. "Geographic Information System." Edited by D. Richardson. *International Encyclopedia of Geography: People, the Earth, Environment and Technology,* March. Wiley. https://doi.org/10.1002/9781118786352.wbieg0152.

[5] Clapp, John M., and Yazhen Wang. 2006. "Defining Neighborhood Boundaries: Are Census Tracts Obsolete?" *Journal of Urban Economics* 59 (2): 259–84.

[6] Clarke, Keith C. "Geodesy." In International Encyclopedia of Geography: People, the Earth, Environment and Technology, edited by Douglas Richardson, Noel Castree, Michael F. Goodchild, Audrey Kobayashi, Weidong Liu, and Richard A. Marston, 1–10. Oxford, UK: John Wiley & Sons, Ltd, 2017. https://doi.org/10.1002/9781118786352.wbieg0456.

[7] Code, Ohio Revised. 1977. "3503.17 Precinct Boundary Changes."

[8] Comber, A., W. Zeng, Spatial interpolation using areal features: A review of methods and opportunities using new forms of data with coded illustrations. Geography Compass. 2019; 13:e12465. https://doi.org/10.1111/gec3.12465

[9] Duchin, Moon and Bridget Tenner. *Discrete geometry for electoral geography,* preprint. Available at https://arxiv.org/abs/1808.05860.

[10] ESRI. 1998. "ESRI Shapefile Technical Description." White Paper J-7855. Redlands, CA: Environmental Systems Research Institute, Inc.

[11] Forest, Benjamin. 2004. "Information Sovereignty and GIS: The Evolution of 'Communities of Interest' in Political Redistricting." *Political Geography* 23 (4): 425–51. https://doi.org/10.1016/j.polgeo.2003.12.010.

[12] General Assembly of Pennsylvania. 2011. "Congressional Redistricting Act of 2011."

[13] Harrington, Jan L. 2016. *Relational Database Design and Implementation.* 4th ed. Amsterdam; Boston: Morgan Kaufmann/Elsevier.

[14] *Harvard Election Data Archive.* https://projects.iq.harvard.edu/eda.

[15] Lawhead, Joel. 2011. "Your Chance to Make GIS History." *GeospatialPython.com.*

[16] Li, Wenwen, Michael F. Goodchild, and Richard Church. 2013. "An Efficient Measure of Compactness for Two-Dimensional Shapes and Its Application in

Regionalization Problems." *International Journal of Geographical Information Science* 27 (6): 1227–50. https://doi.org/10.1080/13658816.2012.752093.

[17] MacDonald, Karin. 2019. "California's Statewide Database." Presentation at the Voting Rights Data Institute, Tufts University, July 10, 2019.

[18] McGann, Anthony J., Charles Anthony Smith, Michael Latner, and Alex Keena. 2016. *Gerrymandering in America: The House of Representatives, the Supreme Court, and the Future of Popular Sovereignty.* New York, NY: Cambridge University Press.

[19] Mule, Thomas. 2012. "Census Coverage Measurement Estimation Report: Summary of Estimates of Coverage for Persons in the United States." 2010-G01. Washington, DC: U.S. Census Bureau.

[20] Office of the Minnesota Secretary of State—Elections Division. 2019. "2019 Minnesota Election Laws."

[21] Open Geospatial Consortium. n.d. "Open Geospatial Consortium Standards." http://www.opengeospatial.org/docs/is.

[22] Schroeder, J.P. (2007), Target-Density Weighting Interpolation and Uncertainty Evaluation for Temporal Analysis of Census Data. Geographical Analysis, 39: 311–335. https://doi.org/10.1111/j.1538-4632.2007.00706.x

[23] U.S. Bureau of the Census. 1994. *Geographic Areas Reference Manual.* Washington, D.C.: U.S. Dept. of Commerce, Economics, and Statistics Administration, Bureau of the Census.

[24] United States Census Bureau. "Standard Hierarchy of Census Geographic Entities," October 27, 2010. https://www2.census.gov/geo/pdfs/reference/geodiagram.pdf?#.

[25] U.S. Bureau of the Census. 2018. "Redistricting Data Program Management." https://www.census.gov/programs-surveys/decennial-census/about/rdo/program-management.html

[26] U.S. Bureau of the Census. 2014. "Establishment of the 2020 Census Redistricting Data Program." 79 FR 41258.

[27] Weibel, Robert. 1997. "Generalization of Spatial Data: Principles and Selected Algorithms." In *Algorithmic Foundations of Geographic Information Systems*, edited by G. Goos, J. Hartmanis, J. Leeuwen, Marc Kreveld, Jürg Nievergelt, Thomas Roos, and Peter Widmayer, 1340:99–152. Berlin, Heidelberg: Springer Berlin Heidelberg. https://doi.org/10.1007/3-540-63818-0_5.

IV

Math and computer science

Chapter 14

Three applications of entropy to gerrymandering

LARRY GUTH, ARI NIEH, AND THOMAS WEIGHILL

CHAPTER SUMMARY

This chapter is an exploration in how a single mathematical idea—entropy—can be applied to redistricting in a number of ways. It's meant to be read not so much as calling for an entropy revolution in redistricting, but as an illustration of the power and limitations of fashioning mathematical "interventions."

1 INTRODUCTION

Redistricting primarily concerns data, typically data about people — which political jurisdictions they reside in, what demographic groups they belong to and how they vote, to name just a few. More than raw data, though, studying redistricting is about investigating the relationships between different kinds of data: for example, how is the demographic and vote data distributed amongst the districts in a state? This is not unique to redistricting — many scientific fields now find that their biggest questions revolve around studying relationships between large data sets. Mathematical tools originally designed for completely different purposes can help us to address these questions. In this chapter, we examine a tool of this kind: information theory, and more specifically, the notion of entropy.

Mathematician Claude Shannon originally developed information theory during the Second World War to find an upper bound on how much information could be transmitted through a communication channel. One can easily see why such a bound would be important for someone working on the secret communication sys-

M. Duchin, O. Walch (eds.), *Political Geometry*, https://doi.org/10.1007/978-3-319-69161-9_14

tem used by Churchill and Roosevelt. The basic idea is that information is related to uncertainty: the fewer patterns exist in the data, the harder it is to efficiently communicate that data — that is, there is more information we have to get across. The story goes that Shannon was persuaded by Von Neumann to call this quantity neither "information" nor "uncertainty" but rather *entropy*, inspired by the thermodynamic quantity of the same name.

Consider the following very simple example. Suppose that you want to know the party affiliation of every person in a town hall meeting. (For the sake of simplicity, pretend that only Republicans and Democrats are in attendance.) If you know beforehand that the vast majority of the attendees will be Republicans, then you can ask the Democrats to raise their hands and assume that those that didn't were Republican. Here, the effort to communicate the data is measured in how many people had to raise their hands. In this case, not much effort was required on average by the audience — the entropy in the party-affiliation data is low. If, on the other hand, the room was a fifty-fifty split, then more hands would have to be raised, resulting in a greater effort per person — in this case, the party-affiliation data have high entropy.

Figure 1: It takes a lot of bits to give a precise description of a chaotic configuration.

In order to make his notion of entropy precise, Shannon grounded his theory on the idea that information is conveyed by binary bits. For those interested in the precise mathematical definition in the appendix to this chapter, this is where all the $\log_2$ symbols come from. For our applications, it will be enough to think of data being conveyed by a series of two-way distinctions: are you Democrat or Republican, Hispanic or non-Hispanic? For facts about people that have more than one value (like their district), we should think of encoding that data into a binary string of 0s and 1s that a computer might use, using the most efficient possible such encoding.

We are interested in the relationship *between* datasets, not just datasets themselves. One way to quantify the relationship between two types of data associated with people is to ask how much information one tells you about the other. Returning to our town hall example, consider a situation where the room is still a fifty-fifty split, but you know that Democrats tend to sit on the left side of the hall whereas Republicans tend to sit on the right. You can save a lot of effort by asking all the Republicans on the left to raise their hands and asking the Democrats on the right to raise theirs, while still correctly determining the party of every participant. In this case, *where* an individual was told you a lot about their party affiliation, so less effort was required to transmit the information once we knew where everyone was sitting.

The amount of information one piece of data gives you about another can be measured using the information theory concept of *conditional entropy*, denoted

$$Ent(X|Y),$$

which is the answer to the question, "If I know Y, how much additional information, on average, do I need to also know X?" In our applications, conditional entropy will always be calculated between two partitions of a population (e.g., into racial categories or into districts) — in this case the data X and Y are just which piece of each partition an individual belongs to (e.g., their race category or their district). In this chapter, we present three examples of how conditional entropy can be applied to topics in redistricting: to measure how segregated a population is, how much counties are split up by a districting plan, or how similar two districting plans are to each other.

2 APPLICATION: MEASURING SEGREGATION

A key issue in redistricting is the question of minority representation. Segregation plays a role in minority representation in a number of ways. For example, cases involving the Voting Rights Act often hinge on showing that a certain population group is concentrated enough to secure representation if districts were drawn in a certain way (this is the first criterion in the so-called "Gingles test" established by the courts in *Thornburg v. Gingles*).

We can use conditional entropy to quantify geographical segregation by any demographic quality. In this section, we will focus on racial segregation. The segregation score we will end up with was introduced by Theil and Finizza in 1971 to study racial integration in school districts [4].[1] Let T be a partition of a geographic area into smaller units, and let R be the partition of voters by race. The quantity $\mathrm{Ent}(R|T)$ will tell us to what extent knowing where a voter resides predicts their race. However, for this measure to be comparable across different regions, we must account for the relative size of minority populations – if the overall population is largely homogenous, it's easy to guess a voter's race even if there is significant geographic segregation.

We therefore define the *absolute entropy* $\mathrm{Ent}(X)$ of a partition X to be the average amount of information we need to know X, measured in bits. For example, if X is a partition with 2^k parts of equal size, then $\mathrm{Ent}(X) = k$, as shown in Figure 2

[1] In fact, as a quirk of the way we derive the score, the score we come up with is technically equal to one minus Theil and Finizza's score.

Figure 2: Partitions with absolute entropy 1, 2, and 3, respectively.

(this is a consequence of measuring information in binary bits, as discussed in the Introduction). Now we can define an overall measure of segregation:

$$Seg(R,T) = \frac{Ent(R|T)}{Ent(R)}.$$

Note that $\text{Ent}(R|T)$ ranges from 0 to $\text{Ent}(R)$, which explains our choice of this normalization factor. (The real-world interpretation of this upper bound is that knowing a voter's location cannot make it any harder to guess their race.) Therefore, $\text{Seg}(R,T)$ must always be between 0 and 1. If $\text{Seg}(R,T) = 0$, then $\text{Ent}(R|T) = 0$, which means that knowing a voter's location is completely sufficient to determine their race — in other words, complete racial segregation. If $\text{Seg}(R,T) = 1$, then $\text{Ent}(R|T) = \text{Ent}(R)$, so the knowledge of a voter's residence has no effect on knowledge of their race. In this case, every geographic unit has the same proportion of voters of each race, and there is no geographical segregation.

To try and get an idea of how our score behaves, we consider four fictional states – Aregon, Barkansas, Cattachusetts, and Ducklahoma, shown in Figure 3. Aregon and Barkansas each have a statewide minority population share of 1/16th, whereas Cattachusetts and Ducklahoma each has a statewide minority share of 1/8th. Aregon is visibly more segregated than both Barkansas and Ducklahoma, and the scores indicate this. On the other hand, Aregon could be considered less segregated than Cattachusetts because, despite an increase in the minority population, the region to which the minority population is confined has not increased; the scores agree with this intuition. We also want to investigate how two states with the same segregation score but with vastly different statewide minority populations can look. We thus construct two more states, New Cattachusetts and New Ducklahoma, each designed to have the same score as their predecessors but to have equal pink and white populations. Although the difference in segregation between Cattachusetts and Ducklahoma is obvious to the eye, the difference between New Cattachusetts and New Ducklahoma is more subtle, despite having the same score difference. This serves as a cautionary tale about interpreting the absolute difference between segregation scores, which will be important to keep in mind when we move to real-world data.

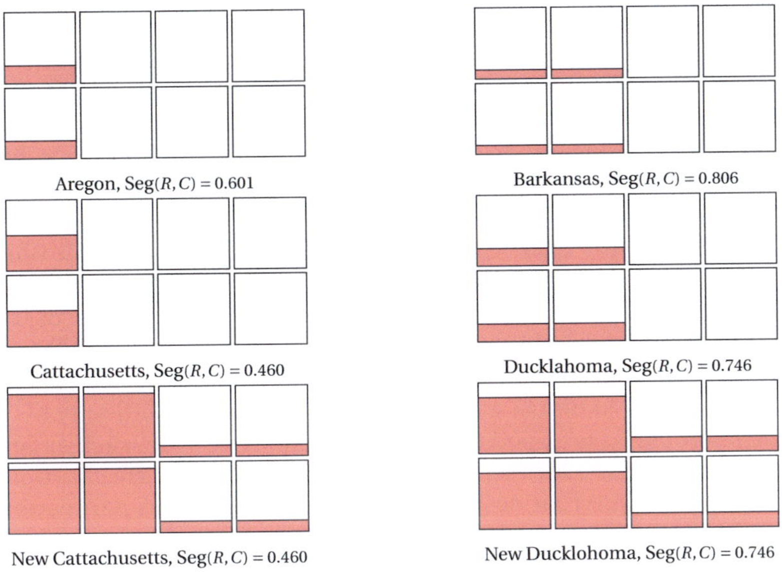

Figure 3: Six imaginary states and their segregation scores

Year	Black	White	Hispanic
1990	0.262	0.514	0.596
2000	0.315	0.589	0.591
2010	0.377	0.625	0.610

Table 14.1: Segregation score by race category and year for Chicago

To illustrate this measure on a real-world example, we calculate racial segregation scores in the city of Chicago, using Census demographic data from 1990, 2000, and 2010. [2] For each category, we report the segregation $\mathrm{Seg}(R, T)$ where T is the partition of the residents of Chicago into census tracts and R is the bipartition given by whether or not residents are in the specified category. Table 14.1 gives the results. The maps in Figure 4 show the racial demographics of the Chicago tracts for comparison (such maps are called choropleths) and Table 14.2 shows demographic breakdowns by year.

[2] In this survey, Hispanic is considered an ethnicity and can be selected independently of race variables like Black and White. For our calculations, we therefore categorize as Hispanic all persons who selected a Hispanic ethnicity, and denote as Black and White only people who selected non-Hispanic ethnicities and a single race category. Also, since some census tracts may cross the city boundary, we take as our domain of study all tracts that intersect the city.

Year	Black	White	Hispanic
1990	35.7%	41.7%	18.9%
2000	34.0%	34.3%	25.5%
2010	30.2%	34.1%	28.7%

Table 14.2: City-wide racial demographics of Chicago census tracts

Recall that a higher score indicates less segregation. For all the years considered, Black/non-Black segregation was higher than Hispanic/non-Hispanic – in information theory terms, knowing where someone resides improves one's chances of guessing whether they are Black or not more than it improves one's chances of guessing whether they are Hispanic or not. Table 14.1 seems to indicate that the Black/non-Black and White/non-White segregation has decreased over time, whereas the segregation of the Hispanic population has remained fairly steady, decreasing only slightly from 1990 to 2010. The decrease in White/non-White segregation is observable from the choropleths in Figure 4, whereas the decrease in Black/non-Black segregation is less apparent. The number of census tracts with less than 1% Black population went from 252 in 1990, to 165 in 2000, and to 91 in 2010 (each out of about one thousand total census tracts), which gives some evidence for decreased Black/non-Black segregation. The difficulty of detecting this on the choropleths is a good motivation for a precise score such as $\text{Seg}(R, T)$ to measure segregation. As for the Hispanic population, both the choropleths and entropy scores indicate that despite a significant increase in population overall (from 19% in 1990 to 29% in 2010), the Hispanic/non-Hispanic segregation has persisted to a large degree.

An important feature of the entropy segregation score is that it is highly dependent on the geographic units chosen — in this case, census tracts. Indeed, for the finest units imaginable — individual persons — knowing which geographic unit someone belongs to completely determines that person's race. On the coarsest scale, knowing that someone is in Chicago does not give you any information about their race beyond that given by the city-wide demographics, i.e., the data captured by $\text{Ent}(R)$. Using this score thus requires a careful choice of particular *scale* on which one is trying to detect segregation. This is a downside if one is interested in a scale-free measure, but it can also be an advantage. For example, in congressional redistricting one may be more concerned about a coarser scale of segregation than when dealing with public policy at the level of a single city.

Measures of segregation have an extensive history, which we will not be able to cover in this section — see for example Massey and Denton [10] for a classic survey of some 20 scores, including the entropy score studied in this section. Some of these scores, such as the Dissimilarity score, the Frey index, and the segregation Gini index, are calculated using a specific choice of geographic unit, but beyond that do not take geography (e.g., adjacency of units) into account, just like the entropy score in this section. The most widely used measure that does take geographic adjacency into account is Moran's I, introduced in 1950 by P. A. P. Moran. However, Moran's I still depends heavily on the geographic units chosen. For a modern method based on network science that is less sensitive to choice of geographic unit, see Duchin et al.[1].

3 APPLICATION: SPLITTING COUNTIES

Many state constitutions prohibit voting districts from dividing up counties. In cases when counties are too populous to completely avoid splitting, there is often a legal requirement to minimize the amount of division. Sometimes this is measured by the number of counties split, but in other cases there is no specified metric.

Conditional entropy gives us a precise and meaningful way to quantify the extent to which a districting plan D divides up the county partition C. The quantity $\text{Ent}(D|C)$ tells us how much we can deduce about a voter's district given their county of residence. If this quantity is low, then counties are not split very much; if it is high, they are split drastically.

In this simple example, our hypothetical state consists of four counties of equal size, denoted with colors. We will calculate the conditional entropy for several different districting plans.

Consider one extreme case, shown on the left in Figure 5, where no counties are split. In this case, knowing a voter's county fully determines their voting district, and therefore no additional information is needed. In this case, $\text{Ent}(D|C) = 0$.

Suppose, however, that our plan is not able to do so well. Take the case where each county is divided evenly into two districts, as on the right side of Figure 5. Then, if we know which county a voter resides in, we can narrow down their district into two options. We would need exactly one more binary bit of information (i.e., one more two-way distinction) to determine their district. In this case, $\text{Ent}(D|C) = 1$.

To take the opposite extreme, suppose that all of the counties are evenly divided into four pieces by the districting plan, as shown on the left side of Figure 6. In this case, knowing a voter's county tells us nothing at all about their district. We would need to be told outright which of the four districts they lived in, which requires 2 bits of information. In this case, $\text{Ent}(D|C) = 2$.

Last, consider the plan on the right side of Figure 6, which splits two counties but leaves two intact. For a voter in the blue or green county, one additional bit of information is needed to specify their district. However, for a resident of the yellow or red county, no information is needed. Thus, averaging over the entire state, we see that $\text{Ent}(D|C) = 0.5$.

To see how this plays out in practice, we compute $\text{Ent}(D|C)$ for eight different congressional districting plans for Pennsylvania drawn by various groups and individuals, as summarized below and shown in Figure 7.

- *2011*: The congressional districting plan enacted by the Republican-controlled state legislature and signed into law by Governor Tom Corbett (R) in December 2011. The map was challenged in court and in January 2018 the Pennsylvania Supreme Court struck down the map for being an illegal partisan gerrymander, giving parties to the suit until 15 February 2018 to propose new maps to the court.
- *TS*: The plan submitted directly to Governor Tom Wolf (D) by House Speaker Mike Turzai (R) and Senate President Pro Tem Joe Scarnati (R). Governor Wolf declined to submit the map to the PA Supreme Court, saying that it was a Republican gerrymander.
- *GOV*: The plan submitted by Governor Tom Wolf (D) to the PA Supreme Court.
- *REM*: The remedial map drawn by Stanford Professor Nate Persily at the

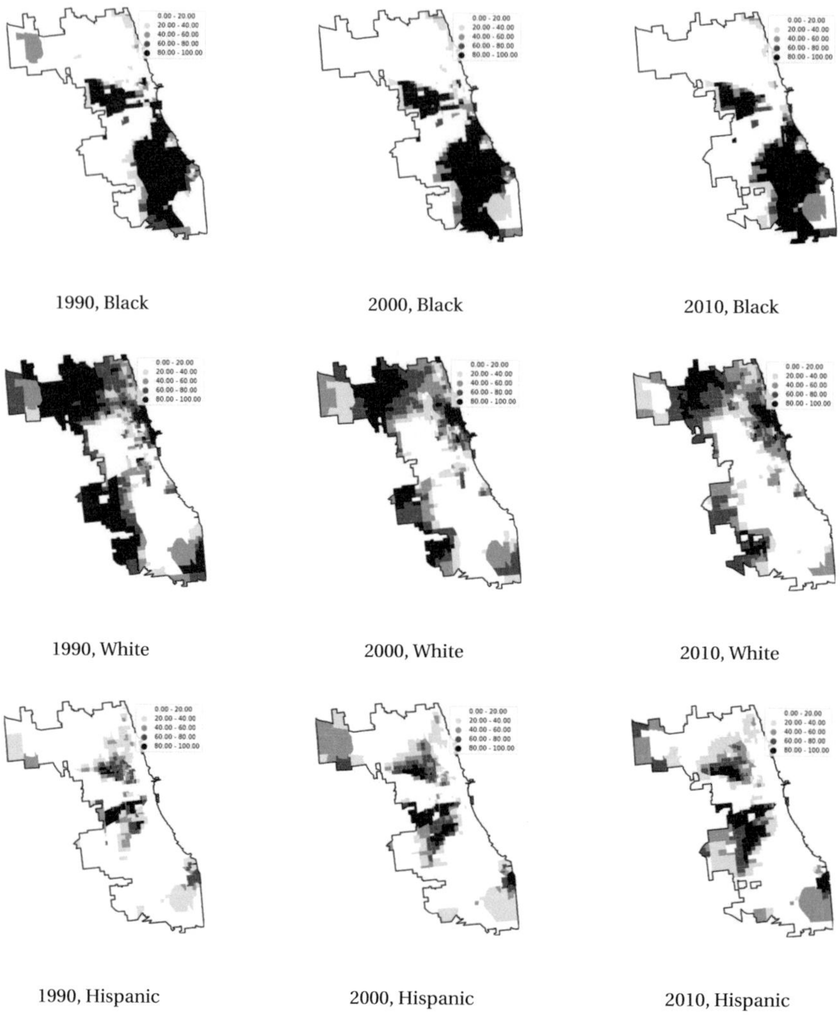

Figure 4: Race category choropleths for Chicago 1990–2010

direction of the PA Supreme Court. This map was adopted by the court on February 19, 2018 and was still in place at the time this chapter was written.

- *8th*: Districting plan drawn by a class of eighth-graders (who were not involved in the lawsuit around the 2011 plan).
- Three plans from FiveThirtyEight's "The Atlas of Redistricting," a collection of congressional maps for every state optimizing various criteria [5].
  - *538 CPCT*: plan optimized for compactness.
  - *538 DEM*: plan optimized for Democrats.
  - *538 GOP*: plan optimized for Republicans.

Figure 5: The left-hand plan splits no counties and has a conditional entropy of $\text{Ent}(D|C) = 0$. The right-hand plan bisects each county and has conditional entropy $\text{Ent}(D|C) = 1$.

Figure 6: For a districting plan that cuts each county into four equal pieces, we get $\text{Ent}(D|C) = 2$. The plan on the right is mixed, but averages out to a conditional entropy of $\text{Ent}(D|C) = 0.5$.

In addition to our entropy score, we also compute two more natural measures of county splitting for each plan:

- *splits*, the total number of counties that intersect more than one district, and
- *pieces*, which we compute by overlaying the district boundaries and county boundaries and counting the number of connected pieces in the resulting partition of the state

Table 14.3 shows all three county splitting scores for the plans.

Although *splits* care only about the number of counties split and not how they are split (e.g., 50/50 or 99/1), entropy takes the nature of the splits into account. This can be considered an advantage or a disadvantage. On the one hand, $\mathrm{Ent}(D|C)$ is a finer measure of county splitting that is grounded in a real-world question (if I know my county, how well do I know my district?). On the other hand, the entropy score barely penalizes "nibbles," where a district includes a tiny fraction of a county. Intuitively, this is because most of the county can still easily guess their district correctly. Whether such nibbles should be considered negligible, a significant downside, or even the worst possible way to split a county depends on the perspectives of stakeholders.

In Pennsylvania, as in many states, county populations vary wildly — from about 1.5 million in Philadelphia County to a mere 5,000 or so in Cameron County. The entropy splitting score naturally up-weights more populous counties: splitting Philadelphia County in half adds 300 times more to the entropy score than splitting Cameron county. On the other hand, *splits* and *pieces* treat all counties the same regardless of population. This is again a question of priorities — but observe that Iowa's redistricting code, for example, requires that "[w]hen there is a choice between dividing local political subdivisions, the more populous subdivisions shall be divided before the less populous" [6].

Other ways to measure county splitting proposed in the literature include entropy-like scores such as the one proposed in footnote 6 of DeFord and Duchin [2], which adapts the entropy score to be more sensitive to the above-mentioned "nibbles." In their Markov Chain Monte Carlo sampling method, Mattingly et al. impose a penalty for counties that are split between two districts and a much larger penalty for counties that are split between three or more districts, with weight factors depending on the size of the intersections between a county and any extra districts [9]. Beyond the realm of computational methods (which are often benefited by finer, real-valued measures such as entropy), most expert reports dealing with county splitting use discrete measures such as *splits*. See for example Kennedy's expert report in *League of Women Voters of PA v. Pennsylvania*, which places a great deal of emphasis on the number of counties split (i.e., the *splits* score) by the with its predecessors [7], or the court's emphasis on the low number of counties split in their decision to adopt the REM plan [8].

Plan	Ent(D|C)	*splits*	*pieces*
REM	0.474	14	85
538 CPCT	0.482	12	83
GOV	0.579	18	88
TS	0.601	15	87
538 GOP	0.816	21	102
8TH	0.833	42	138
2011	0.868	28	111
538 DEM	0.920	31	117

Table 14.3: County-splitting scores for Pennsylvania plans.
(*splits* = # counties split by districts, *pieces* = # connected pieces when county and district boundaries are overlaid)

4 APPLICATION: DISTANCE BETWEEN PLANS

In the redistricting process, it is often useful to ask the question "how different are two districting plans?" For example, a redistricting process may legally require one to satisfy certain constraints while also remaining as "close" as possible to the existing districts. Or one may want to know when a particular plan is an outlier compared with a collection of other plans, either human-drawn or generated by some algorithm.

Conditional entropy provides us with a mathematically precise answer to this question. If we have two redistricting plans D_1 and D_2, we can look at their conditional entropy $\text{Ent}(D_1|D_2)$ to see how much they coincide.[3] To get some notion of what this number means, we may look at two extreme cases. If $D_1 = D_2$, then knowing a voter's district under D_2 is completely sufficient to know that voter's district under D_1, and therefore $\text{Ent}(D_1|D_2) = 0$. At the other end, the maximum possible value of $\text{Ent}(D_1|D_2)$ is simply $\text{Ent}(D_1)$ (which itself is bounded above by $\log_2(k)$ where k is the number of districts), which corresponds to the case that the two plans are as different as they could possibly be: knowing a voter's district under D_2 tells you nothing about their district under D_1, and vice versa.

To illustrate the usefulness of conditional entropy in comparing plans, we consider three different redistrictings for the state of Maryland.[4] Figure 8 shows the four relevant districting plans, as well as the conditional entropy between the old and new plans, for each time the districts were redrawn.

The entropy calculations seem to indicate that the 2011 redistricting brought the greatest amount of change to the congressional districts, a fact that is also visible

[3]Although it is not generally true that $\text{Ent}(X|Y) = \text{Ent}(Y|X)$ (see Section 5) it is possible to prove that $\text{Ent}(D_1|D_2) = \text{Ent}(D_2|D_1)$ so long as each plan has the same number of districts, and the districts are perfectly population-balanced. In practice we use the average of $\text{Ent}(D_1|D_2)$ and $\text{Ent}(D_2|D_1)$ when computing the distance

[4]Census data obtained from NHGIS [13]. Congressional districts obtained from Lewis et al. [12] and matched to census blocks using the maup package (`github.com/mggg/maup`).

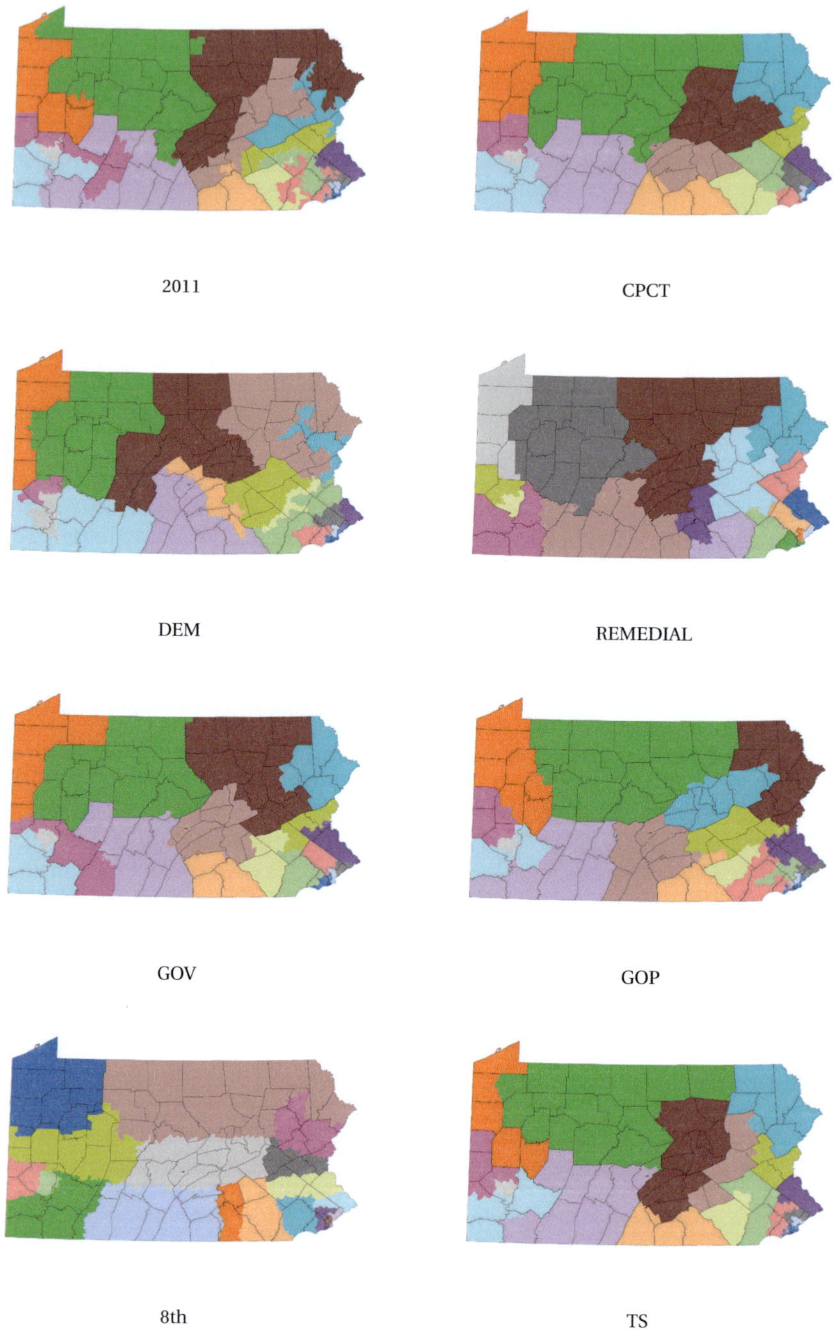

Figure 7: Districts (shaded regions) over county boundaries (in black)

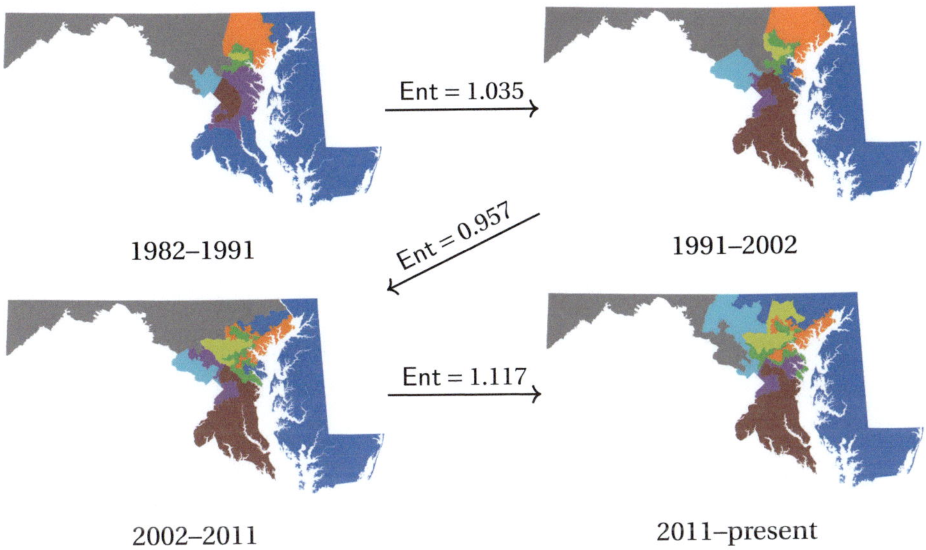

Figure 8: Maryland congressional districts over three cycles with entropy distance

on the maps to some extent. (Note also that this redistricting was the subject of a Supreme Court case.) The relationship between the 1991 and 2002 redistrictings is less clear just by looking at the maps. Remember that the entropy score weights regions strictly by population, not land area, so that smaller geographic details around large cities have a large influence on the entropy score, and these are hard to see with the naked eye.

Returning to the eight Pennsylvania plans from the previous section, we can also compute the entropy distance of each plan to the 2011 plan. Table 14.4 shows the results. The result for TS, GOV, and REM are particularly interesting, since these three plans were all designed to replace the 2011 plan. Among these three, the TS plan has the lowest distance — that is, it introduces the least change in districts (as measured in this particular way). In fact, the map-drawers of the TS plan made special mention of the fact that "[the TS map] retains 68.8% of the populations of existing districts in the same districts, which will help to reduce overall voter confusion" [15].

The difficulty of visually determining the extent of the difference between maps motivates the use of a precise mathematical score like conditional entropy. It is important to remember, however, that conditional entropy is insensitive to strictly geographic dissimilarity between plans because it is based only on population overlaps. It would fall to stakeholders in a given scenario to decide whether geographic or population-based similarity is more important. Even guidelines such as Nebraska's requirement that new districts "preserve the cores of prior districts" [11] do not make the choice clear.

Another possible application of entropy as a distance between plans is to try to detect outliers. To demonstrate this, we consider the analysis of congressional redistricting in North Carolina performed by Mattingly et al. in [9]. Using a large

Plan P	Ent(P|2011)
2011	0
TS	1.144
GOV	1.340
REM	1.320
8TH	1.541
538 CPCT	1.247
538 GOP	1.314
538 DEM	1.668

Table 14.4: Comparing Pennsylvania plans with the 2011 enacted plan

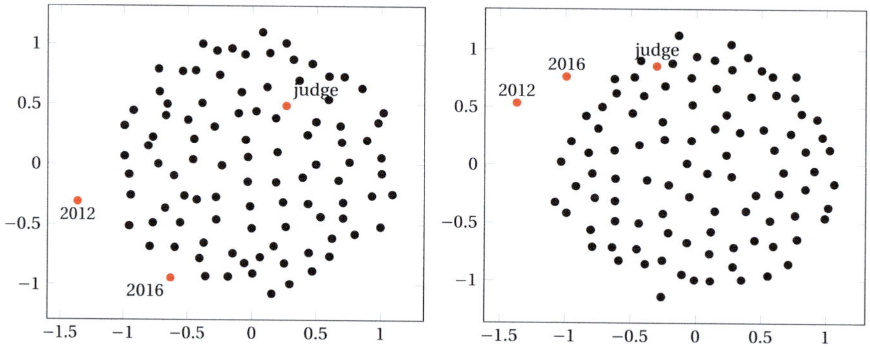

(a) Ensemble from Herschlag et al. [9]

(b) ReCom ensemble

Figure 9: Human (red) and computer-generated districting plans for North Carolina

ensemble of computer generated plans, the authors demonstrate that the districting plans enacted in 2012 and 2016 were highly atypical outliers in terms of their partisan statistics (e.g., seats won for either party), whereas a third plan proposed by a panel of judges represents the ensemble of plans much more closely. Figure 9 shows the three human-made plans (2012, 2016, and the judges' plan) as well as one hundred plans drawn from the authors' ensemble using conditional entropy as the distance. The embedding is done using multi-dimensional scaling. We also repeat the analysis with an ensemble generated using a recombination Markov chain [3]. [5] For more details on Markov chains and ensemble methods, see Chapter 17 in this book.

Figure 9 seems to indicate that in addition to being outliers in *partisan statistics*, the 2012 and 2016 plans are atypical (compared with the ensembles) with respect to how they divide up the population, whereas the judges' plan is more representative in this sense. Note that no vote data are involved in this kind of analysis since it depends only on the populations of the geographic units used, so that the outlier status of the 2012 and 2016 plans is independent of how the votes fall in any given election. This may or may not be an advantage of this kind of analysis, since a gerrymander that is geographically similar to the ensemble but manages to produce extreme partisan outcomes would not be flagged as an outlier.

[5] `github.com/mggg/gerrychain`

Other distances between plans that can be found in the literature include a distance based on ideas from optimal transport [14]. This distance is sensitive not just to the size of the overlap between pairs of districts (which completely determines the entropy score), but also to the geographic distance between pairs of districts, making it a far more sensitive measure of dissimilarity. The cost is that it takes longer to compute since an optimization problem must be solved to calculate the distance between two plans. A restriction to this distance (and others that require a perfect matching between the two plans) is that it is only defined for two plans with the same number of districts. Conditional entropy, on the other hand, does not require this condition and thus can be used in situations where a state gains or loses a congressional seat.

5 CONCLUSION: MATH FOR DEMOCRACY

When Shannon developed his theory of communication at Bell Laboratories, he probably wasn't thinking about redistricting. Then again, neither was Markov, Bayes or any of the other great mathematicians and scientists whose work forms the basis of the modern study of redistricting, some of which is outlined in this book. Although no mathematical concept is likely to be the perfect analytical framework for every possible situation — this includes entropy, as we have remarked at various points in this chapter — they can often provide valuable tools for analyzing the complex issues that arise every day in a democracy.

APPENDIX: FORMAL DEFINITIONS

Our previous discussion has relied on the idea of an average number of bits of information, but we can also give a definition that is more straightforward to calculate.

First, we define entropy for a single partition X. Suppose that X divides the region into pieces of sizes p_i, normalized such that $\sum_i p_i = 1$. That is, p_i is the fraction of voters in part i. Then, the entropy of X is

$$Ent(X) = \sum_i p_i \log_2 \left(\frac{1}{p_i}\right).$$

Intuitively, a higher value of $Ent(X)$ means that the partition divides the region into smaller pieces. As mentioned above, we can also think of $Ent(X)$ as the average number of bits needed to communicate which part of X a voter is in.

What about conditional entropy? For two partitions X and Y, the conditional entropy $\text{Ent}(X|Y)$ is a weighted sum of entropies calculated over each part of Y. This definition is a bit technical, but the underlying concept is simple: just take the entropy of each piece, and add them up with weights proportional to the population.

For each part Y_j of Y, the partition X also induces a partition of Y_j. The parts of this induced partition are $X_i \cap Y_j$ for each part X_i of X. (For example, if X is a districting plan, and Y is the county partition, the induced partition describes how the districting plan partitions county j.)

We define $\text{Ent}(X|Y_j)$ to be the entropy of this induced partition. That is,

$$Ent(X|Y_j) = \sum_i p_{ij} \log_2 \left(\frac{1}{p_{ij}}\right)$$

where p_{ij} is the fraction of the population of Y_j that lies in X_i. You can think of $\text{Ent}(X|Y_j)$ as sort of a "local entropy" of the plan X when restricted to only Y_j.

Now, to calculate the conditional entropy, we simply add up the local entropies weighted by population:

$$Ent(X|Y) = \sum_j q_j \, \text{Ent}(X|Y_j)$$

where the coefficients q_j are the sizes of the parts Y_j.

For a concrete example, consider the districts and counties application we had earlier. To calculate $\text{Ent}(D|C)$ for districts D and counties C, we can use the formula

$$Ent(D|C) = \sum_{c \in \text{counties}} \frac{\text{pop}(c)}{T} \sum_{d \in \text{districts}} \frac{\text{pop}(c \cap d)}{\text{pop}(c)} \log_2 \left(\frac{\text{pop}(c)}{\text{pop}(c \cap d)}\right)$$

where T denotes the total population of the state.

Finally, a note on $\text{Ent}(X|Y)$ versus $\text{Ent}(Y|X)$. These two quantities differ in general: the former asks how well Y predicts X whereas the latter asks how well X predicts Y. In the case of counties and districts, for example, $\text{Ent}(D|C)$ will be zero if every county is contained in only one district, but $\text{Ent}(C|D)$ will typically be nonzero since each district may have more than one county contained in it. Nonetheless, there is the following relationship:

$$(\textit{Bayes' Rule}) \quad \text{Ent}(X|Y) + \text{Ent}(Y) = \text{Ent}(Y|X) + \text{Ent}(X)$$

Intuitively, this holds because both sides of the equation are the number of bits of information required to encode both X *and* Y.

ACKNOWLEDGMENTS

The authors would like to thank Colby Brown and Brandon Kolstoe, two students at the Voting Rights Data Institute 2019, for their help with the experiments in this chapter. We would also like to thank Ruth Buck, former MGGG geographer, for help with the geographic data required for the experiments. We would also like to thank Olivia Walch for the illustrations. Thomas Weighill acknowledges the support of the NSF Convergence Accelerator Grant No. OIA-1937095.

REFERENCES

[1] Alvarez, Emilia, Moon Duchin, Everett Meike & Marshall Mueller. 2018. "Clustering propensity: A mathematical framework for measuring segregation." https://mggg.org/Capy.pdf

[2] DeFord, Daryl & Moon Duchin. 2019. "Redistricting Reform in Virginia: Districting Criteria in Context." *Virginia Policy Review* 12(2):120–146.

[3] DeFord, Daryl, Moon Duchin & Justin Solomon. 2021. "Recombination: A family of Markov chains for redistricting." *Harvard Data Science Review* 3.1.

[4] Theil, Henri & Anthony J. Finizza. 1971. "A note on the measurement of racial integration of schools by means of informational concepts." *Journal of Mathematical Sociology* 1, 187–194.

[5] FiveThirtyEight, *The Atlas of Redistricting*, https://projects.fivethirtyeight.com/redistricting-maps/ [accessed 3 October 2019].

[6] Iowa Code, Chapter 42 — Redistricting General Assembly and Congressional Districts.

[7] J. J. Kennedy, *Report on Pennsylvania's Congressional Districts*, available at http://redistricting.lls.edu/.

[8] League of Women Voters of Pennsylvania vs Pennsylvania, No. 159 MM 2017, Decided: 7 February 2018.

[9] Herschlag, Gregory, Han Sung Kang, Justin Luo, Christy Vaughn Graves, Sachet Bangia, Robert Ravier, & Jonathan C. Mattingly. 2020."Quantifying gerrymandering in North Carolina." *Statistics and Public Policy* 7, no. 1:30–38.

[10] Douglas S. Massey & Nancy A. Denton. 1988. "The Dimensions of Residential Segregation", *Social Forces*. 67(2), 281–315.

[11] Legislative Resolution No. 102, adopted by the Nebraska Legislature, 8 April 2011

[12] Lewis, Jeffrey B., Brandon DeVine, Lincoln Pitcher & Kenneth C. Martis. 2013. Digital Boundary Definitions of United States Congressional Districts, 1789–2012, Retrieved from http://cdmaps.polisci.ucla.edu on 4 September, 2019.

[13] Manson, Steven, Jonathan Schroeder, David Van Riper, and Steven Ruggles. 2019. IPUMS National Historical Geographic Information System: Version 14.0. Minneapolis, MN: IPUMS. http://doi.org/10.18128/D050.V14.0.

[14] Abrishami, Tara, Nestor Guillen, Parker Rule, Zachary Schutzman, Justin Solomon, Thomas Weighill & Si Wu. 2020. "Geometry of graph partitions via optimal transport." *SIAM Journal on Scientific Computing* 42, no. 5: A3340-A3366.

[15] Office of Senator J. Scarnati, *Joint Submission Congressional District Map*, 9 Feb 2018, http://www.senatorscarnati.com/2018/02/09/

joint-submission-congressional-district-map/ [Available via Wayback Machine at archive.org].

15 Explainer: Measuring clustering and segregation

MOON DUCHIN AND JAMES M. MURPHY

This explainer will focus on a statistic—"Moran's I"—that is so ubiquitously used in geography to measure spatial structure in a dataset that it has become almost interchangeable with the *concept* of spatial structure in that field. This explainer assumes math-major background in some places, but we hope it is still accessible at a high level for all readers. We begin with a little bit of history.

BACKSTORY

Though P.A.P. Moran developed it slightly earlier, the statistic called Moran's I was brought into geography during the rise of *spatial analysis*, a subdiscipline that emerged during the late 1940s. Before that turn, geography as a university discipline had been framed as a study of places and regions, with an emphasis on description and characterization. After World War II, universities became increasingly entangled with what Eisenhower had famously dubbed the "military-industrial complex," which led to increased research emphasis in areas connected to defense, planning, and decision science. This brought a so-called "quantitative revolution" to geography, among many other domains.

By the 1950s, with a boost from Red Scare politics, a new muscularly mathematized toolset had pushed cultural and Marxist geography to the margins, sometimes seeing geography departments entirely eliminated in the course of postwar modernization. In the 1970s, the pendulum began to swing back, and critiques of the spatial analysis framework—as reductive, politically, and culturally disconnected, and too far from the more descriptive geography of the early twentieth century—became more audible. But by then, the rise of computing meant that the calculational spirit of spatial analysis was fairly entrenched. Metrics like I, which were developed to measure the degree of spatial patterning in data, could now become instantly accessible in spatial software. Now one could load a dataset with population demographics for Chicago and, at the push of a button, learn that I = .884 for Black population and I = .828 for Latino population, both very high numbers in a city where random population distributions would yield scores closer to zero. With this ease of use, the straight-up comparison of one score to another, across different localities and time periods and contexts, became unavoidably tempting.

What is a score like this trying to measure? Arthur Getis, one of the standard-bearers

M. Duchin, O. Walch (eds.), *Political Geometry*, https://doi.org/10.1007/978-3-319-69161-9_15

of the spatial analysis school, cites the following definition:

> Given a set S containing n geographical units, *spatial autocorrelation* refers to the relationship between some variable observed in each of the n localities and a measure of geographical proximity defined for all $n(n-1)$ pairs chosen from n. [1]

This is essentially just a quantification of the common-sense maxim called Tobler's First Law of Geography: *Everything is related to everything else, but near things are more related than distant things.* As we will see, the `I` metric attempts to take this literally by measuring how much the values in a unit are like the values at the neighbors. When the score was built to focus on literally adjacent units, as in the early work of Moran and his contemporaries, it was sometimes called a *contiguity ratio.* In its more general form, it was given the lasting name of spatial autocorrelation in an influential conference paper by Cliff and Ord in 1968, followed by a monograph in the early 1970s [2, 3].

So the story of `I` is a story of a chalkboard-math intervention in spatial statistics that caught on because of the academic politics of its era. Just as the need to prove geography's mathematical bona fides was starting to fade, the rise of computers made it easy to crunch numbers on larger and larger datasets, keeping `I` alive for the next generation. And once a formula becomes a button to press, the meaning of what it measures can fade away from the forefront of debate.

Below, we'll turn back the clock and look at measurements of clustering (a.k.a. segregation), poking them with a mathematical stick to see what we find.

WHAT IS SEGREGATION?

A classical problem in quantitative social science is to define a measurement of segregation that matches up with the ways that people talk about their communities. More precisely, given a geography with location data about a demographic subgroup, the problem is to quantify how much the group is separated rather than undifferentiated from the rest of the population in terms of residential patterns.

Figure 1: How interspersed or separated are the black and white colors—i.e., how segregated are these cities? Segregation scores attempt to answer this quantitatively.

In Chapter 10, Chris Fowler talked about this question, and noted that geographers think of this as a multiscale measurement problem—as Figure 1 makes clear, the answer might depend on how the pattern falls against a chosen set of units. We'll develop that idea mathematically here.

(DIS)SIMILARITY ACROSS GEOGRAPHICAL UNITS

One approach is to demarcate the geography into smaller units and analyze the demographics on these units. For instance, we can divide up a city into its census tracts and look at demographic population proportion by tract. Suppose the units are numbered 1 through n, and we want to study the population of a group B relative to the total population. We can denote the number of B-type people and the total population of unit i as b_i and p_i, respectively, and write the totals for the large geography as $B = \sum_{i=1}^{n} b_i$ and $P = \sum_{i=1}^{n} p_i$. Let's write $\rho_i = b_i / p_i$ for the share of B population in unit i and $\rho = B/P$ for the global ratio. Then, if the local population shares ρ_i are the same for all i, and therefore equal to the global ratio ρ, we would declare the city completely unsegregated at the scale of the units we have chosen. But if there is one part of the city where the local ratio is far higher and another part of the city where it's far lower, that sounds segregated.

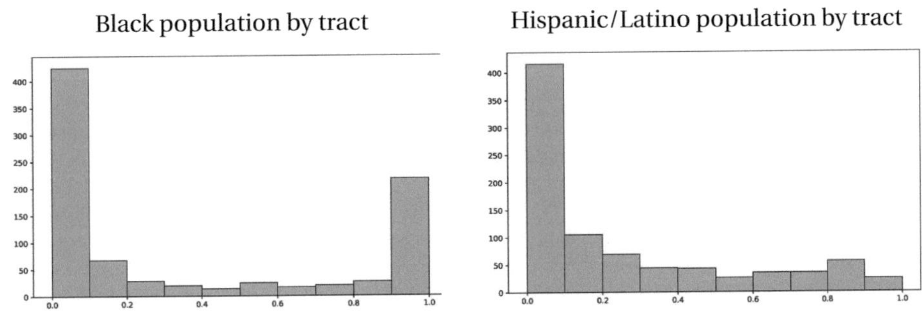

Figure 2: These histograms show how the Black population (left) and Hispanic/Latino population (right) are distributed across the 853 census tracts in Chicago. There are over 200 tracts that are more than 90% Black, but no comparably large number of heavily Latino tracts.

Reasoning this way, we can define the *dissimilarity index* by comparing b_i/p_i to the global ratio B/P in each unit, which turns out to be equivalent to comparing b_i/B to p_i/P. We define:

$$D = \frac{1}{2}\sum_{i=1}^{n}\left|\frac{b_i}{B} - \frac{p_i}{P}\right| = \frac{1}{2\rho}\sum_{i=1}^{n}\frac{p_i}{P}\cdot\left|\rho_i - \rho\right|.$$

That means a geographical unit makes no contribution to dissimilarity if the population share in that unit, ρ_i, is the same as the share in the whole geography, ρ. But if one unit has very different population proportions than the region overall, it contributes significantly to the overall dissimilarity.[1,2]

We can write the population shares in an ordered list, or vector; we can then subtract off the average to record how much each unit deviates from the overall average.

$$(\rho_1, \rho_2, \ldots, \rho_n) \longrightarrow (\rho_1 - \rho,\ \rho_2 - \rho, \ldots,\ \rho_n - \rho).$$

If we call this *deviation vector* $\mathsf{x} = (x_1, \ldots, x_n)$, so that $x_i = \rho_i - \rho$, then we see that dissimilarity D can be interpreted as the average magnitude of deviation—it begins with the average of the $|x_i|$, weighted by the population of each unit. (Then there's a normalization by a factor out front.)

For example, let's consider Black population in Chicago, according to the 2010 decennial census. There are 853 (populated) census tracts in the city, and we can make a vector of length 853 recording the Black population share by tract. The citywide Black population was 32.2%, so we can subtract off .322 from each coordinate to get our deviation vector. We compute $D = .54$ for Black population. The Hispanic population share citywide is .288 and $D = .45$. If you rescale these to get dissimilarity on a zero-one scale, you'd see that the Black population is scored by D as having roughly 80% of the maximum possible dissimilarity for a population of that size, while the Hispanic/Latino dissimilarity is 63% of its maximum.

You might have noticed a major limitation of dissimilarity for understanding segregation: each unit is treated separately, with no spatiality taken into account, so the score makes no distinction between a left/right split and a checkerboard (see Figure 1). That's not a great fit for how we talk about segregation, where the former is clearly more segregated than the latter.

SPATIALIZING SEGREGATION

Next, we can treat the geography as a *spatial network*, recording the spatial relationships by placing edges between the nodes when the units they represent are adjacent. Possibly the simplest network model is an undirected graph $G = (V, E)$ where the vertices correspond to geographical units (like the census tracts of Chicago) and the edge set encodes spatial adjacency of units. We show Chicago as a *dual graph* (a graph dual to the tracts) in the right column of Figure 3.[3]

[1] There is a large body of literature on the dissimilarity index. This expression for D matches the one used in Frey and Myers [4]; an expression with a different normalization coefficient is cited in Massey and Denton [5], where it is noted that the formulations have varied in the literature.

[2] More generally, the dissimilarity index allows us to *compare* any two populations B and C with an exactly similar formula, with C in place of P. What is presented here is the special case that C is the total population. For this case, let's examine the scale or range of values for the above formalism. Consider a group with population share $\rho = B/P$. It's clear that the lowest possible dissimilarity would be $D = 0$ when every unit has ρ share. The highest possible would occur if some units (roughly ρn of them) have share approaching 1 and the rest have share 0. With the normalization shown here, this yields $D = 1 - \rho$. That is, small populations can register as very segregated, but large populations can't get a very high D score. This is not crazy, as a reflection of how we talk about segregation! But if you want to rescale to get D ranging from zero to one, you would divide by $1 - \rho$.

[3] Just to fix terminology: we use "graph" and "network" interchangeably, and we use "node" and "vertex" interchangeably, as is common in the literature.

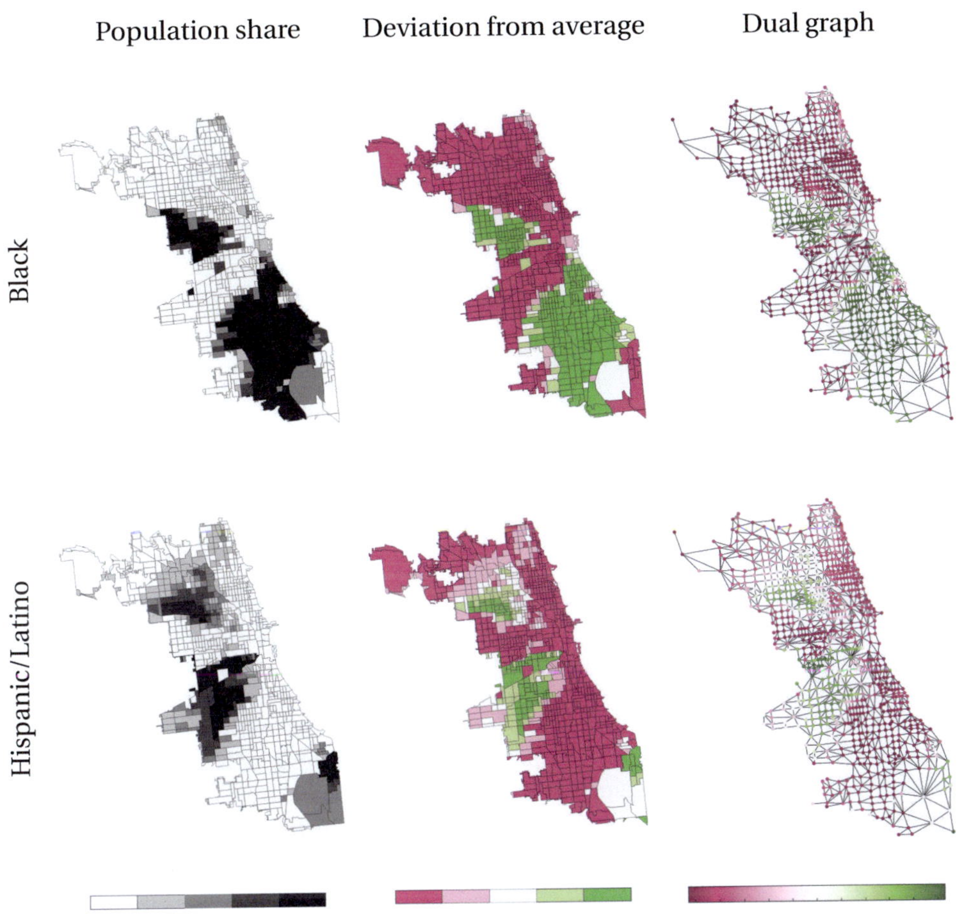

Figure 3: Demographics in spatial context: population share of minority group, deviation from average, and dual graph. Green is above average, purple below average.

Now quantifying how much a subgroup is segregated is a question not only of statistics but also of the graph structure (sometimes called the *network topology*). The connectivity of the underlying graph determines to a large extent what kinds of patterns will count as segregated. In this sense, meaningful measures of segregation must not only account for fluctuations of ρ_i around its mean, but must also relate to the structure of the underlying network.

Finally we can define *Moran's* I [6, 7]. If we start with any numerical values associated with the nodes, such as the population shares ρ_i, Moran's I returns a real number, usually between −1 and 1. The standard interpretation is that values near 1 indicate extreme segregation, values near zero indicate no pattern, and negative values flag a kind of "anti-segregation," where the units alternate between one population subgroup and the other (as in a checkerboard). Just as before, we're going to start with a vector of population shares by unit and subtract off the average to get a deviation vector. We'll do one thing differently this time: we'll assume that all we know is the *share* at each node, and not the total population. When we take an average, we'll have to do so weighting all nodes equally. This time, if $(\rho_1, \ldots, \rho_n)$ is the share by unit, and $\beta = \frac{1}{n}\sum_{i=1}^{n} \rho_i$ is the average of these values, we get the deviation vector $x_i = \rho_i - \beta$. For census tracts in Chicago, for instance, this way of averaging makes four percentage points of difference: the average Black population in a census tract is $\beta = .362$ rather than the citywide $\rho = .322$ Black population share, indicating that tracts with high Black percentages are relatively underpopulated.

Now suppose the graph/network G has m pairs of adjacent units (i.e., the graph has m edges) in all. Then we can define

$$\mathtt{I}(x_1, \ldots, x_n) = \frac{n}{m} \frac{\sum_{i \sim j} x_i x_j}{\sum_i x_i{}^2}$$

where $i \sim j$ if spatial units i and j are adjacent. This is asking how the average product of neighboring values compares to the average product of a value with itself. Computing for the Black and Hispanic population in Chicago, we get $\mathtt{I} = .881$ and $\mathtt{I} = .825$, respectively.

We can interpret $\mathtt{I}$ as looking for patterns in the locations where the x values are positive or negative. To see why $\mathtt{I}$ detects patterns, think again about the left/right configuration versus the checkerboard. If the units are chosen so that they have solid, alternating colors (the checkerboard situation), each term in the numerator will be negative (because x_i and x_j have different signs), making I negative overall. In the left/right division, most terms will contribute positively to the numerator because they will have negative next to negative or positive next to positive, making the expression positive overall. And if there is no pattern, we will tend to see a lot of cancellation. So $\mathtt{I}$ is telling us whether there is a coarse pattern (positive), a fine pattern (negative), or no pattern at all.

Compare that to dissimilarity values for Black and Latino population that were, respectively, .8 and .63 of their max. So in a sense, these populations look *more* segregated when you consider spatial patterns (via $\mathtt{I}$) and not just the existence of units with large deviation (via D).

I AS A SLOPE

There are two intuitive interpretations of $\mathtt{I}$ that bear mentioning. First, $\mathtt{I}$ is the slope of the best-fit line relating the value at a node to the value at the neighbors (which is called the *lagged* value).[4]

So if tracts are just like their neighbors except in a small transitional area, $\mathtt{I} \approx 1$. If everything is the opposite of its neighbor, then there will be clusters in the northwest and southeast corners of the scatterplot, and $\mathtt{I} \approx -1$. And if there are no patterns at all, then the average of your neighbors will tend to be the same as the overall average, no matter what your value is, which makes the fit line flat, or $\mathtt{I} \approx 0$.

Perhaps a few limitations of this approach are now visible. By just reporting the *slope* of the fit line, it loses a great deal of information from the scatterplot (see

[4] This would be perfectly accurate for regular graphs, see [8].

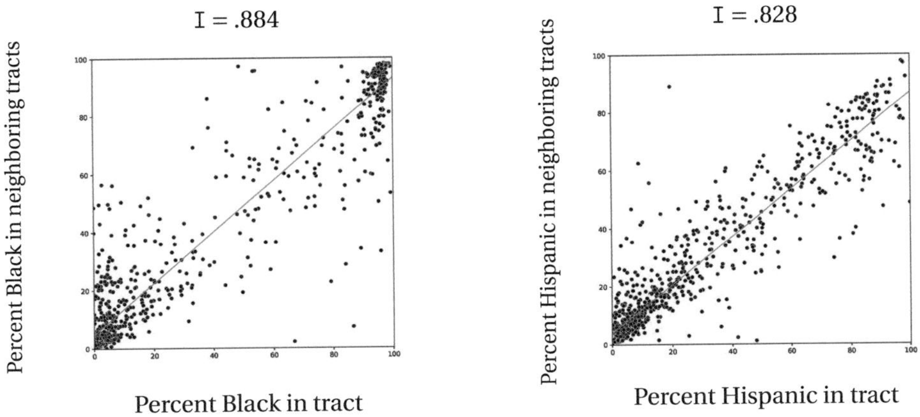

Figure 4: Moran scatterplots for the Black and Hispanic population of Chicago—note that the projection of these plots to the x-axis would give back the histograms in Figure 2. Segregation in Chicago is captured by the fit line being nearly diagonal: tracts tend to have neighboring tracts with similar demographics.

Figure 4 for scatterplots and lines of best fit for the Black and Hispanic population of Chicago). It fails to adequately distinguish the bimodal distribution of Black population in Chicago, which is much more characteristic of the way we think about segregation, from the less concentrated Latino population. Another drawback is made clear by this way of visualizing the score: it's going to give meaningless answers for a very uniformly distributed population, because all the data points will be in one small area of the scatterplot. When you fit a line through a small ball of points, its slope does not have much meaning! So in the case of a very unsegregated population, the score is more noise than signal, and in the limit when the population is exactly even over the units, the score is undefined.

I VIA LINEAR ALGEBRA

There's another ready interpretation of the I formula that sheds a lot of light on what it's doing, from a mathematician's perspective. Let A be the adjacency matrix of the graph: an $n \times n$ matrix that has a zero in position i, j if the i and j units are not adjacent, and a one if they are. Let x be the vector $(x_1, \ldots, x_n)$ of deviations from the mean, as above. Then there's a neat way to write the calculation in matrix notation: $\mathtt{I}(\mathsf{x}) = \frac{n}{2m}\left(\frac{\mathsf{x}A\mathsf{x}^\mathsf{T}}{\mathsf{x}\mathsf{x}^\mathsf{T}}\right)$. Those who have some linear algebra background will recognize this expression in parentheses as a *Rayleigh quotient*: it is exactly what you maximize or minimize to get the largest and smallest eigenvalues of A, and the values of x where these occur are the corresponding eigenvectors.[5] The study of eigenvalues for matrices coming from graphs, like our adjacency matrix A, belongs to the kind of math called *spectral graph theory*.

[5]If the effect of A is to stretch a vector v by a factor λ, then λ is called an eigenvalue and v is called an eigenvector—the list of eigenvalues is called the *spectrum* of A. Eigenvalues come up absolutely all over pure and applied mathematics.

A major theme in spectral graph theory is to relate the connection structure of the graph G to the eigenvalue spectrum of associated matrices. In particular, when the graph is *regular* (same number of edges incident to every node), the eigenvectors of A associated with the largest eigenvalues are known to capture latent cluster structure in the data [9].[6]

This suggests that a population deviation vector x will give a large, positive Moran's I score precisely when it puts its positive and negative values into highly interconnected graph clusters, which are often colorfully called "communities."[7] Back to plain English: this segregation score maxes out when you can find two parts of the graph that are relatively well connected internally but relatively separate from each other, and your population group mainly lives in one of these clusters. (See Figure 5. Ever since Figure 1, we have wanted a score that can tell a checkerboard from a left-right pattern, and now we've done it!)

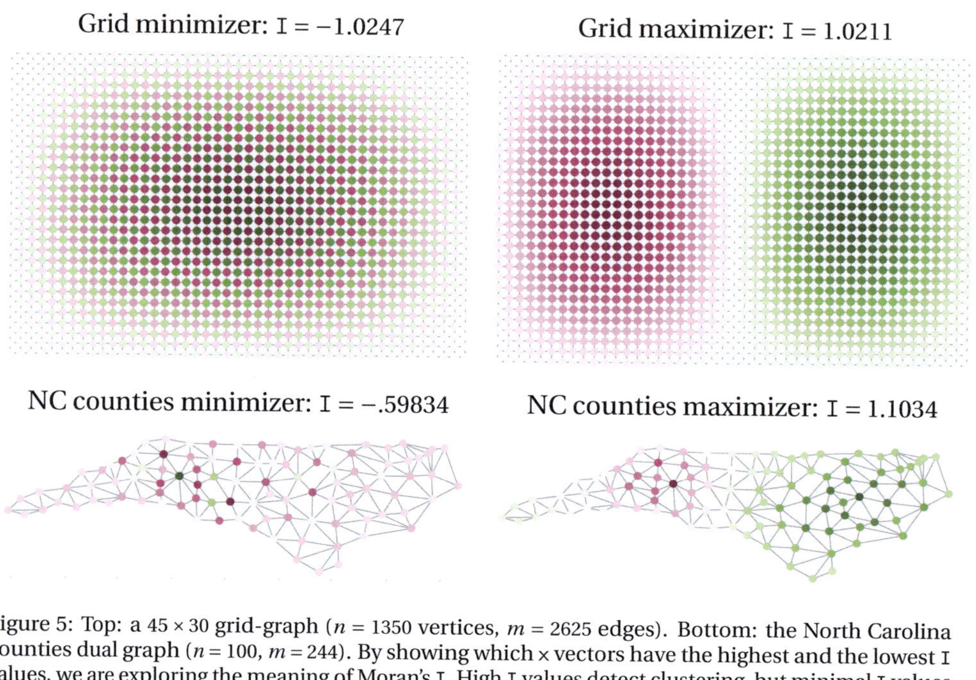

Figure 5: Top: a 45×30 grid-graph ($n = 1350$ vertices, $m = 2625$ edges). Bottom: the North Carolina counties dual graph ($n = 100$, $m = 244$). By showing which x vectors have the highest and the lowest I values, we are exploring the meaning of Moran's I. High I values detect clustering, but minimal I values are less interpretable once we depart from the world of grids.

As we've seen, I maximizers tend to concentrate the group's population in a cluster, and this remains true on a grid or a real-world graph like the counties of North Carolina (Figure 5). On grid-graphs, for example, rectangular grids, the pattern

[6] In spectral graph theory, it's more common to study eigenvalues of a related matrix called the graph *Laplacian* rather than the adjacency matrix; when the graph is regular of degree d, the Laplacian is $L = dI - A$, so the spectrum of L is related to the spectrum of A by translation and reflection.

[7] Translating this into the related language of Fourier theory, the patterns that correspond to large eigenvalues have low frequency, so they may just have one negative area and one positive area, while the ones for the low eigenvalues have a high frequency, which may correspond to fast oscillation from positive to negative. Think of this as being like a sinusoid function that takes a long time to complete a period (low frequency) versus another that oscillates rapidly (high frequency).

minimizing I forms a checkerboard pattern. However, most dual graphs are not bipartite (i.e., admitting an alternating pattern). The eigenvectors of A with smallest eigenvalues may be hard to interpret, both on abstract graphs [10] and on the irregular graph structures that come up in practice.

For our part, we conclude that it's not advisable to read too much into negative I values—which in any case are very rare in practice for real-world demographic data. And, more problematically for overall usability, we have no good plain-English interpretation for intermediate values of I across graphs. That is, what does an $I = .6$ residential pattern in one city or state have in common with an $I = .6$ pattern in another? It's not so clear.

SO... WHAT IS SEGREGATION?

This explainer has explored perhaps the two most prominent metrics in the social science literature for measuring clustering/segregation: dissimilarity and Moran's I. Both are very widely used, with D appearing much more commonly in cross-city comparisons in the popular press[8] and I in technical work in GIS and in fields as diverse as epidemiology, urban planning, and environmental studies [1].

As for the latter, we should look at how people actually use Moran's I in the social science literature. In the examples we have found, authors usually apply a kind of significance testing for I to see if the answer is larger than you should expect [11, 12, 13, 14]. That is, does the observed demographic data get a score indicating that it is more segregated than would be expected under a "null model," where values are distributed at random according to a normal or some other distribution? By comparing to randomized values on the same fixed graph, this kind of inference controls for the role of graph connectivity. We explore these themes further in a paper we recently wrote with Thomas Weighill [8].

Are there other ways to measure segregation? Of course! Many mathy readers are probably itching to play around with or replace the definition entirely, such as by using an idea like the probability that your neighbor belongs to your own group—this family of ideas is called *assortativity* in network science, and it plays out a bit differently than the two we saw here [15]. But we hope that this brief intro models a few good practices: First, when it comes to metrics, an open-minded mathematical inspection can give you insight into how best to use the scores (and what to avoid!) and whether their meaning is stable across contexts. Also, and crucially, to do good interdisciplinary work you must engage the literature in other fields than your home discipline, rather than thinking you're painting on a blank canvas.

As for our motivating question, "what is segregation?", we think that looking hard at the notions picked out by different metrics shows us that our shared intuitions don't fully specify an answer: both D and I leave something to be desired. So there is still both conceptual and measurement work to be done!

---

[8] For just a few recent examples, check out two posts from the data blog 538 on diversity vs. segregation (`fivethirtyeight.com/features/the-most-diverse-cities-are-often-the-most-segregated`) and partisan dissimilarity (`projects.fivethirtyeight.com/republicans-democrats-cities`).

ACKNOWLEDGMENTS

Thanks to Heather Rosenfeld and Thomas Weighill for enormously helpful research collaborations on this material, and to Gabe Schoenbach for creating notebooks—available in our GitHub (`github.com/political-geometry`)—to explore the scores and data.

REFERENCES

[1] A. Getis. A history of the concept of spatial autocorrelation: A geographer's perspective. *Geographical Analysis*, 40(3):297–309, 2008.

[2] A.D. Cliff and J.K. Ord. The problem of spatial autocorrelation. *Studies in Regional Science London Papers in Regional Science*, page 25–55, 1969.

[3] A. D. Cliff and Ord. J.K. *Spatial Autocorrelation.* 1973.

[4] William H. Frey and Dowell Myers. Racial segregation in U.S. metropolitan areas and cities, 1990–2000: Patterns, trends, and explanations. *University of Michigan Population Studies Center Research Report*, 2005.

[5] Douglas S. Massey and Nancy A. Denton. The dimensions of residential segregation. *Social Forces*, 67(2):281–315, 1988.

[6] P.A.P. Moran. Notes on continuous stochastic phenomena. *Biometrika*, 37(1/2):17–23, 1950.

[7] M. Tiefelsdorf and B. Boots. The exact distribution of Moran's I. *Environment and Planning A*, 27(6):985–999, 1995.

[8] M. Duchin, J.M. Murphy, and T. Weighill. Measures of segregation and analysis on graphs. *Preprint*, 2021.

[9] U. Von Luxburg. A tutorial on spectral clustering. *Statistics and Computing*, 17(4):395–416, 2007.

[10] M. Desai and V. Rao. A characterization of the smallest eigenvalue of a graph. *Journal of Graph Theory*, 18(2):181–194, 1994.

[11] P. De Jong, C. Sprenger, and F. Van Veen. On extreme values of Moran's I and Geary's c. *Geographical Analysis*, 16(1):17–24, 1984.

[12] L. Anselin. The Moran scatterplot as an ESDA tool to assess local instability in spatial association. volume 4, page 111. CRC Press, 1996.

[13] T. J. Barnes. Spatial analysis. *The Sage Handbook of Geographical Knowledge edited by John Agnew and David Livingstone*, pages 380–391, 2011.

[14] B. J. L. Berry and D. F. Marble. *Spatial analysis: a reader in statistical geography.* Prentice-Hall, 1968.

[15] E. Alvarez, M. Duchin, E. Meike, and M. Mueller. Clustering propensity: A mathematical framework for measuring segregation. *Preprint*, 2018.

Chapter 16

Redistricting algorithms

AMARIAH BECKER AND JUSTIN SOLOMON

CHAPTER SUMMARY

Why not have a computer just draw the best map? For many people, this is the first and only reasonable approach to the problem of gerrymandering. But there are more than a few reasons to be skeptical of this idea. In this chapter, two computer scientists survey what's been done in algorithmic redistricting, with an eye to what computers can and can't do.

1 INTRODUCTION

Given frustrations with human behavior when drawing and assessing district boundaries, technologically minded citizens often call for *algorithms* to serve as unbiased arbiters in the redistricting process. Projecting this optimism about the objectivity of computers, popular science articles regularly trumpet a programmer who has "solved gerrymandering in his spare time" [27] or claim that a "tech revolution... could fix America's broken voting districts" [46]; one blogger even opined that Google could "quickly create a neutral, non-gerrymandered election map... in a few weeks" [23].

Enthusiasm for algorithmic redistricting dates back at least to the 1960s. Even in the early days of computer technology, multiple authors recognized its potential value for redistricting. In 1961, Vickrey wrote in favor of what he called "procedural fairness" [50]:

> "If there is thus no available criterion of substantive fairness, it is necessary, if there is to be any attempt at all to purify the electoral machinery in this respect, to resort to some kind of procedural fairness. This means,

M. Duchin, O. Walch (eds.), *Political Geometry*, https://doi.org/10.1007/978-3-319-69161-9_16

> in view of the subtle possibilities for favoritism, that the human element must be removed as completely as possible from the redistricting process. In part, this means that the process should be completely mechanical, so that once set up, there is no room at all for human choice. More than this, it means that the selection of the process, which must itself be at least initiated by human action, should be as far removed from the results of the process as possible, in the sense that it should not be possible to predict in any detail the outcome of the process."

Writing at nearly the same time, Edward Forrest [18] advocated for using computers for unbiased redistricting in a behavioral science journal in 1964:

> "Since the computer doesn't know how to gerrymander—because two plus two always equals four—the electronically generated map can't be anything but unbiased. Its validity is immediately acceptable to responsible political leaders and the courts."

Since that time, many algorithms have been designed to support redistricting. Software has become a ubiquitous partner in the design and analysis of districting plans, including sophisticated tools that leverage census data, electoral returns, and highly detailed maps.

The introduction of computer technology brings a new set of ethical and philosophical—as well as mathematical—challenges to redistricting. Modern algorithms make it possible to engineer districts with remarkable precision and control, providing opportunities to gerrymander in subtle ways. Questions of allowable data and procedures are complicating long-standing conversations about "traditional districting criteria." On the technical side, fundamental limits involving the computational complexity of certain redistricting problems reveal that Vickrey's and Forrest's dream of perfect redistricting through algorithms may be practically unrealizable.

On the other hand, recent progress has made algorithms into very promising partners in redistricting reform. This chapter explores the ways in which computing has been used in redistricting and presents a survey of redistricting algorithms. We categorize algorithms into those that are used to *generate* new plans and those that are used to *assess* proposed plans. We conclude with a general discussion about best practices for algorithmic tools in the redistricting process.

1.1 WHAT IS AN ALGORITHM?

An *algorithm* is a procedure or set of instructions that a computer uses to solve a problem. Generally, algorithms take *input* data and produce an *output* solution. For example, an algorithm that generates a districting plan might take as input the populations and geographies of census units (e.g., precincts, census blocks, census block groups) as well as the desired number of districts and produce as output a plan listing which census units are assigned to each district.

Although a computer may ultimately be carrying out an algorithm, *humans* write the instructions and make the algorithmic design decisions. For example, for computers to help identify *good* districting plans, a human must first define what it means for one plan to be better than another. A computer has no built-in method for assessing plans or anything else; it simply follows the user's instructions. In this sense, an algorithm or piece of software easily can inherit the biases and assumptions of its human designer.

Our chapter focuses on techniques that generate district boundaries, and we leave the discussion of how to use those for other parts of the book. *Enumeration* algorithms list every possible way to district a given piece of geography. These algorithms have the advantage that no stone is left unturned, but even small municipalities can have unfathomably huge numbers of possible plans, putting this hope out of practical reach. So this section is brief. We divide the subsequent discussion into two variants of this problem:

- *Sampling* algorithms also generate collections of districting plans, but the intention is not to be exhaustive. When carried out well, these methods can provide an overview of the properties of possible plans.
- *Optimization* algorithms attempt to identify a single plan that extremizes some quality score. These automated redistricting tools are effective in some scenarios but require everyone to agree on a single scoring function—a difficult task since so many metrics are used to evaluate proposed districting plans.

We examine various algorithms aimed at sampling or optimization, including scenarios where they are likely to perform well or poorly.

When it comes to quality, some of the algorithms described in this chapter are specifically built around particular measures (e.g., Plan A is better than Plan B if and only if it has a smaller population deviation), while others allow the user to specify a score function (e.g., $\alpha \cdot$ county splits $+ \beta \cdot$ population deviation where α and β are left as user parameters). The decision of how to weight various measures when comparing plans is expressly human, and the consequences of adjusting the weighting even a little can be drastic. For the rest of this chapter, unless otherwise noted, comparative terms like *better plan* and *best plan* are assumed to be with respect to whichever plan score is being used, and should not be interpreted as endorsements of fairness.

1.2 COMPARING ALGORITHMS

Algorithms are primarily analyzed by two considerations: the *quality* of the solutions they identify and their *efficiency*. The former addresses how good or usable an output is, and the latter addresses how long it takes to generate output. Usually there is a trade-off between quality and runtime: it takes more time to find better solutions. Typically, however, one algorithm will outperform another for some problem instances but not for others. Moreover, many of the algorithms we present are designed to accomplish slightly different objectives from each other and may not be suitable for direct comparison. Ultimately, which algorithm is *right* or *best* depends on the priorities of the user.

In redistricting, major properties of interest for sampling algorithms include whether the algorithm is efficient enough to be practical on (large) real-world instances, whether it is actually used as a sampler in practice, whether it generates or can generate every valid plan, whether it tends to generate more compact plans (with nicely shaped districts), and whether it targets a known probability distribution (i.e., if it can be tuned to weight a certain plan more than another by a factor that we control). Table 16.1 summarizes several sampling algorithms presented in this chapter along these axes. For some algorithms the given property is true with a caveat (indicated with yellow squares), explained in the caption.

	Generates *every* valid plan	*Can* generate any valid plan	Can sample real-world instances	Used as a sampler in practice	Promotes compact plans	Targets a known distribution
Enumeration	✓	✓	–	–	–	✓
Random-Unit Assignment	–	✓	–	–	–	✓
Flood Fill	–	✓	–	–	✓	–
Iterative Merging	–	–	✓	✓	✓	–
Flip Step Walk	–	✓	✓	✓	✓	✓
Recombination Walk	–	✓	✓	✓	✓	✓
Power Diagrams	–	–	✓	–	✓	–

Table 16.1: General properties of redistricting sampling algorithms; note that each of these methods admits many variations that may disagree with this table. Caveats are indicated in yellow: power diagrams can handle large instances but generate geometric partitions rather than plans built from census units; the ability of flip step walks and recombination walks to generate *any* valid plan depends on the particular redistricting constraints and underlying geography; some flood fill variants promote compactness; and flip step walks can target particular distributions (including ones that favor compactness) but often lack evidence of convergence.

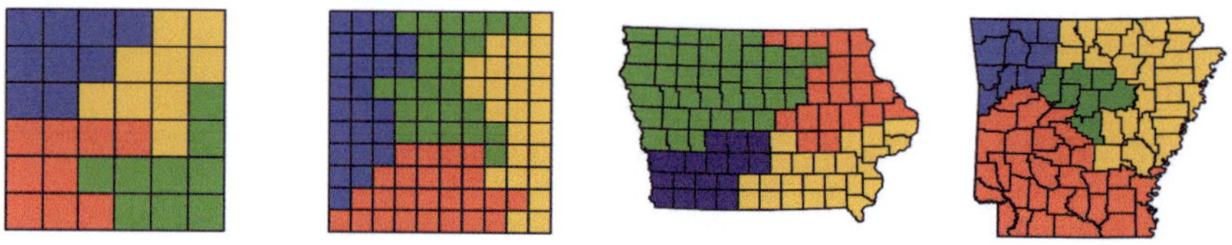

Figure 1: Example districting plans for our four test cases.

For optimization algorithms, in addition to whether it is efficient enough to be practical for real-world instances, we are further interested in whether the algorithm identifies global optima or only local optima and whether the algorithm can handle customized scoring (objective) functions. Table 16.2 summarizes these considerations for several optimization algorithms presented in this chapter. A key takeaway highlighted in this table is that there are no algorithms that are efficient enough to use in practice that can identify *best possible* plans (global optima) for nontrivial scoring functions. This makes it difficult to assess solution quality: If we

	Practical at scale	Finds global optima	Finds local optima	Customizable
Enumeration	–	✓	✓	✓
Power Diagrams	✓	–	✓	–
Metaheuristics/Random Walks	✓	–	✓	✓
Integer Programming	–	✓	✓	✓

Table 16.2: General properties of redistricting optimization algorithms; note that each of these methods admits many variations that may disagree with this table. Caveats: power diagrams are geometric and do not directly generate plans built from census units; and metaheuristics are not guaranteed to be practical, but are often efficient in practice.

do not know the best possible score, we have no yardstick with which to measure other solutions.

BENCHMARKS

The remainder of this chapter explores these algorithmic properties in detail. Throughout the chapter, we illustrate algorithms using figures and experiments. In some cases, we rely on simplified ("toy") examples, like partitioning the cells of a small grid into contiguous pieces. Our four most frequent test cases are the 6×6 grid, the 10×10 grid, the 99 counties of Iowa, and the 75 counties of Arkansas (see Figure 1).

Iowa is a particularly useful choice because it is grid-like but has a variable population in the units and a manageable number of pieces.[1]

As an objective function, we often use a compactness metric called *cut edges* to compare algorithmic techniques (Figure 2). By definition, the cut edge score counts how many pairs of neighboring units are assigned to different districts in a given plan.

2 GENERATING ALL PLANS: ENUMERATION

A natural algorithmic strategy in redistricting is simply to enumerate *all* valid plans. That is, given a list of rules determining which plans are valid, the computer is tasked with generating a list of every possible compliant plan. In this section, we explain why enumeration is impossible in practical terms.

If we could enumerate all plans, we would have a straightforward optimization algorithm: score all possible plans to identify the best one (this is called a *brute force* algorithm). This approach to optimization is *exact* because it considers every

[1] Unlike most states, which build plans out of much smaller census units (like census blocks), Iowa, by law, builds congressional districts out of whole counties.

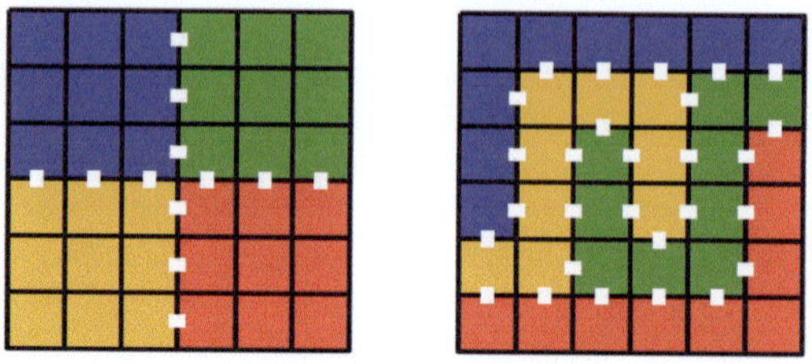

Figure 2: A 6×6 grid has a total of 60 pairs of neighboring units, and we could imagine drawing an edge between neighbors. A plan dividing the grid into quadrants would cut just 12 of these edges. On the other hand, a plan with winding borders could cut up to 28 edges out of 60. The white markers indicate the cut edges for the two plans shown.

n	# plans
1	1
2	2
3	10
4	117
5	4,006
6	451,206
7	158,753,814
8	187,497,290,034
9	706,152,947,468,301

Table 16.3: The number of ways to divide an $n \times n$ grid into n contiguous parts with n units each [14, 24].

possible alternative, so it is guaranteed to find the best one. Random plan generation is also straightforward with enumeration: from the set of all valid plans, select one uniformly at random, or one weighted by some desired probability distribution. Because of its conceptual advantages, enumeration has been proposed as a strategy to identify and evaluate plans for decades [22, 29, 43, 44].

If enumeration is so powerful, why is it not used broadly in redistricting? There are two key issues. First is the sheer number of ways we can draw district lines, making the list of valid plans unfathomably large in practice. For this reason, in practice it can only be used on very small instances. Put differently, redistricting famously suffers from *combinatorial explosion*: as the problem gets larger, the number of valid solutions increases exponentially, quickly exceeding the practical limits of computing power and data storage.

To get a sense of how quickly the number of valid solutions increases, consider the simple problem of partitioning an $n \times n$ grid into n equal-sized districts [14, 24]. The number of partitions as a function of n is given in Table 16.3, for small values of n, and shown in Figure 3 for the $n = 3$ case. Even for this relatively simple redistricting problem, these numbers quickly become too large for enumeration to be practical. This combinatorial explosion is not unique to grids: Enumerating plans for actual states is out of the question in nearly any context. For example, the number of ways to build four congressional districts out of the 99 counties of Iowa is estimated to be about 10^{24} (or a trillion trillions) [16], but the exact number is not known. This

Figure 3: Complete enumeration of ways to divide a 3×3 grid into three equal-sized, rook-contiguous districts. Note that each of the ten plans has the same number of cut edges (6 out of the 12 neighboring units are cut).

estimate is tiny compared to the number of plans that can be built from the finer units like precincts or census blocks that are typically used. At the precinct level, the still-small problem of dividing 250 contiguous Florida precincts into two districts has approximately 5×10^{39} different valid solutions [16], which in turn is minuscule compared to the full problem of dividing Florida's roughly 6,000 precincts into its 27 congressional districts.

Even when the full list of plans for a given geographic area is small enough to store on a computer, we have to consider the amount of time it takes to *generate* such a list. That is, not only is the list of plans extremely long, each plan on the list can take a long time to find.

Another problem arises from combinatorial explosion. For the 6×6 grid with four districts, recall that plans can have anywhere from 12 to 28 cut edges. Complete enumeration shows that over 93% of all these plans have 21–28 cut edges. Like other districting issues, this imbalance only accelerates as the size of the problem grows. In a full-sized problem, more than 99.999% of balanced, contiguous plans are so wildly shaped that they would never be considered in practice.[2] So if you are trying to use the enumeration to get an overview of possibilities, you may not get a very good picture if you simply weight them all equally.

Given that enumeration is neither computationally tractable nor sufficient for understanding real-world redistricting problems, we need other strategies for generating and assessing plans.

[2] This is another matter of counting: there are more winding lines than straight lines, so there are far more noncompact than compact plans. See DeFord et al. [12] for more discussion.

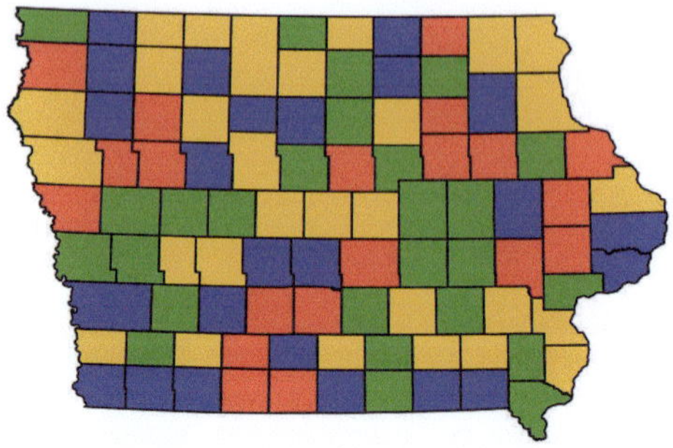

Figure 4: Random assignment of Iowa counties; each color represents a different district. Unsurprisingly, the resulting plan has disconnected districts and balances neither the number of counties nor the district populations.

3 GENERATING MANY PLANS: SAMPLING

Enumeration is an example of an *exhaustive search* technique, in which we visit every corner of the space of districting plans to get a complete understanding. But as we have discussed, computational impediments make enumeration impractical when processing real-world data. For this reason, most algorithms related to redistricting generate and analyze a relatively small set of districting plans, essentially targeting their search. With this motivation in mind, in this section we introduce *sampling* algorithms, whose job is to produce a short but useful list of options.

Enumeration algorithms are *deterministic* in nature, meaning that every time the same piece of code is run we receive the same result. In contrast, sampling algorithms tend to be *randomized*, meaning that they have the ability to make different decisions every time they are run. Randomized—also known as *stochastic* or *nondeterministic*—algorithms can be extremely powerful in their simplicity and efficiency.

We focus on *random plan generators*, randomized algorithms that generate samples of redistricting plans given a fixed piece of geography. These methods produce output that can either be analyzed on its own or used as a subroutine for other algorithms in redistricting. For example, optimization algorithms designed to extract high-quality plans frequently use random plans as starting points and then employ a number of strategies to improve the quality of the starting plan to shape it into the optimized output.

There are several crucial questions to ask when evaluating random plan generators:

- **What is the *distribution* of the generated plans?**
  By nature, sampling methods only generate a subset of possible plans. For this reason, we must understand both the distributional design and unintentional biases of these methods. Your first thought might be to sample from a *uniform* probability distribution, in which all valid districting plans are equally likely to be included in the sample.[3] But some sampling meth-

[3]An efficient algorithm probably does not exist that can draw uniform samples efficiently [41].

ods might instead be weighted toward more compact plans, or tilted toward a particular partisan balance. Whether intentional or unintentional, this weighting can have substantial consequences if sampling is used to summarize the population of alternative plans.

- **Can the sampler generate any possible plan?**
Even though sampling might not be uniform, we might want to know whether there is *some* nonzero probability of generating every possible plan. Some sampling methods restrict their consideration to plans with certain shapes or other properties, which makes it easier to traverse the space of plans but may unintentionally exclude plans relevant to a given redistricting task.
- **Do the samples accurately capture priorities and constraints?**
Redistricting rules can be complex, placing many restrictions on the properties of acceptable plans. It can be difficult to customize a new sampling algorithm to each set of rules and regulations. *Winnowing*, in which samples are generated using a first method and then noncompliant plans are discarded, can repair a sampler after the fact, but few if any plans may be left in an ensemble after this cleanup step, and it can limit control over the probability distribution.

3.1 GENERATING PLANS FROM SCRATCH

Many random plan generators start from a blank slate, taking as input the parameters of a redistricting task: the desired number of districts, the population of each census unit, and the adjacency of these units (i.e., which units share a border). The algorithm then outputs a random plan that assigns these units to districts or describes where to draw the lines between districts.

RANDOM ASSIGNMENT AND REJECTION SAMPLING

Perhaps the simplest approach to generating a plan is the random assignment algorithm, which is one of several approaches implemented in the BARD redistricting software package [2]. This algorithm divides a region into k districts by randomly and independently assigning each unit a district label from 1 through k. Figure 4 shows a typical random assignment; unsurprisingly, it does not satisfy any of the familiar constraints, such as contiguity or population balance. One way to rectify this problem is to repeatedly generate candidate plans, discarding invalid plans until a valid one is produced. This is our first encounter with the tactic called *rejection sampling*, shown in Algorithm 1.

This is not very efficient. In fact, it is so unlikely that random assignment of census units results in a valid plan that we would expect to discard an astronomically large number of proposed candidate plans before finding a single valid one.

Algorithm 1 Random-Unit Assignment

```
for each census unit i do
    District assignment(unit i) ←RANDOM(1,2,...,k)
```

Algorithm 2 Random-Unit Assignment with Rejection

```
while plan is invalid do
    Plan ←RANDOM-UNIT ASSIGNMENT
```

Random assignment is the easiest algorithm to analyze theoretically and has the favorable property that *every* possible plan can be generated with some nonzero (but vanishingly small) probability, but most of the analysis simply reveals that it is ineffective. Instead, in practice it is desirable for samplers to produce valid plans with reasonably high probability. In the remainder of this section, we stay attentive to the question of rejection rate.

FLOOD FILL AND AGGLOMERATION

A widely discussed family of practical plan-generating algorithms employs a *flood fill* strategy. These algorithms *grow* districts from seed units by gluing together adjacent units until the districts reach the desired population. Many flood fill variants have been proposed, including [8, 37, 43, 44, 48, 50]; we highlight a few examples below.

Algorithm 3 District-by-District Flood Fill

```
for each district i do
    seed←RANDOM(unassigned census unit)
    District assignment(seed) ← i
    while Population(district i) ≤ target_population do
        spread←RANDOM(unassigned NEIGHBOR(district i))
        if Pop(district i) + Pop(spread) ≤ target_population then
            District assignment(spread) ← i
```

Many flood fill algorithms build one district at a time, as outlined in Algorithm 3; Figure 5 shows an example. In this case, a single unit is selected as a "seed" to start growing a district (the red county in the example in Figure 5). Then, the algorithm iteratively glues units onto its side until the district reaches a desired size. At each step there are multiple options for which unit to add next (pink counties in the example in Figure 5); the simplest way to make this decision might be to choose randomly, which is the version in the pseudocode for Algorithm 3.

Rather than making this decision completely randomly, we can make strategic choices that encourage the partially built district to have particular properties. For example, one variant preferentially chooses units that lie within the *bounding box* of the currently growing district to improve the compactness of the plans [8]; Figure 6a illustrates this heuristic on our running example. Another variant also suggested in Cirincione et al. [8] preferentially chooses units that belong to census tracts or counties that are already included in the growing district (see Figure 6b).

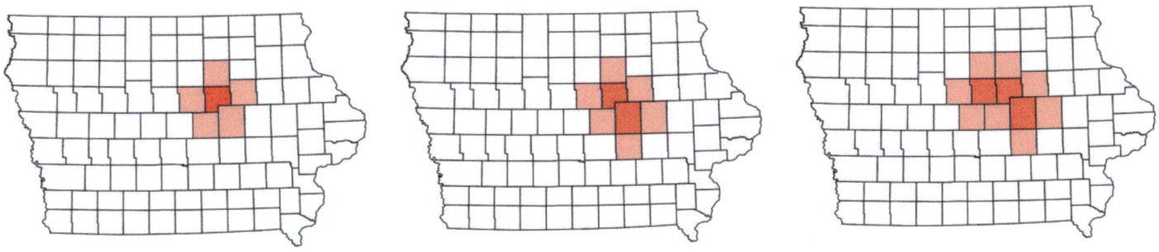

Figure 5: An example of flood fill on Iowa counties. Counties are colored red as they are added to the growing district. The pink counties indicate candidate neighboring counties to annex at each step.

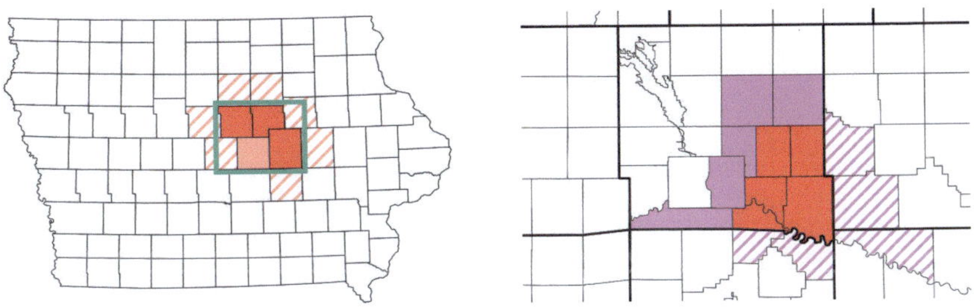

(a) Bounding-box variant

(b) County-preserving variant

Figure 6: (a) Bounding-box flood fill variant [8]; the solid pink county lies entirely within the bounding box of the growing red district and is preferentially chosen over the striped pink neighboring counties. (b) County-respecting flood fill variant [8]; the solid pink county subunits lie within the county of the growing red district and are preferentially chosen over the striped pink neighboring subunits in different counties.

A challenge with flood fill algorithms is that they can get stuck. Some districts grow in such a way that does not leave enough space for the remaining districts (see Figure 7). Typically at this point, the plan is rejected and the algorithm starts over, repeating this process until a valid plan is generated, although a few algorithms can *adjust* initially invalid plans until they are valid (see Section 4.4). Though this refinement strategy has a lower rejection rate, these adjustments are often quite computationally intensive and time-consuming.

A second approach to flood fill builds all the districts simultaneously, as depicted in Algorithm 4. Rather than building one complete district, fixing it, and moving on to the next district, this algorithm grows all the districts in the state in parallel. In each iteration, the algorithm now has to make two decisions: which district

Algorithm 4 Whole Plan Flood Fill

```
for each district i do
    seed i ←RANDOM(unassigned building-block unit)
    District assignment(seed i) ← i
while some district is still underpopulated do
    district j ←RANDOM(underpopulated district)
    spread ←RANDOM(unassigned NEIGHBOR(district j))
    if Population(district j) + Population(spread) ≤ target_population then
        District assignment(spread) ← j
```

to grow, and which adjacent unit to add onto that district. A primary advantage of this approach is that all the districts are treated symmetrically. In contrast, in the one-at-a-time strategy from Algorithm 3, the shape of the first district drawn might be quite different than the shape of the last district because as the algorithm proceeds there are fewer options for how to grow a district outward.

One variant of the whole-plan flood fill method suggested in Liu et al.[32] starts from k seed units along the boundary of the state. Another variant chooses one seed from each of k predefined *zones* across the state, promoting a more uniform distribution of the growing districts [32, 44]. Figure 8 compares these options. Such variations may be designed to reduce the rejection rate or to tailor the sampling distribution (e.g., to generate samples with more compact plans).

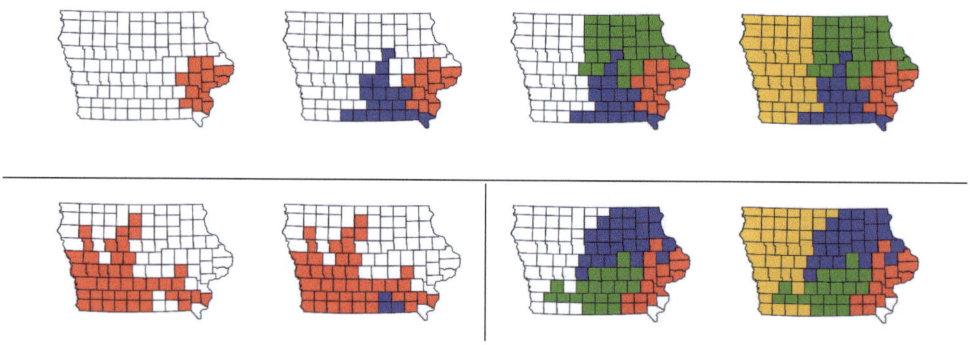

Figure 7: The flood fill algorithm grows the districts one at a time. The top example arrives at a complete plan. The bottom two examples lead to rejection because there is no way to complete the plan with contiguous districts and population balance.

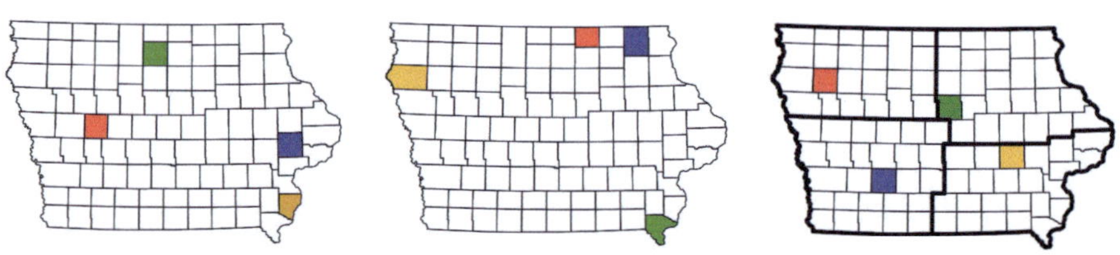

Figure 8: Left: district seeds were chosen uniformly at random from all counties. Center: district seeds were chosen uniformly at random from the boundary counties. Right: one district seed was chosen uniformly at random from each predefined zone.

Another method with a similar flavor to flood fill is an iterative merging approach [9, 10]: at each step, a geographic unit is chosen at random and merged with an adjacent unit to form a new aggregate unit. This process is repeated until the number of aggregate units is equal to the desired number of districts. These resulting pieces (composed of many of the original census units merged together) may have unbalanced populations, so the algorithm might then make small refinements until the populations are balanced (see Section 4.4). The merging process is efficient, and by choosing the closest unit to merge at each step (measured by distance between the units' centroids), this method promotes compactness. The population-rebalancing process, however, can be inefficient and may degrade compactness. This flood fill variant has been used many times in recent redistricting litigation, including, for example, the 2018 *League of Women Voters v. Commonwealth of Pennsylvania* case.

For many flood fill variants, every valid plan has a *chance* of being generated in theory. It is not clear, however, how *likely* some plans are to be generated in practice, and there has been little focus in the literature on the distribution of plans from which flood fill algorithms sample.

We can empirically demonstrate the non-uniformity of the flood fill method. Figure 9 depicts the distribution of the *number* of cut edges across sampled plans generated by different flood fill variants[4] versus complete enumeration, which is possible because we can check all 442,791 ways to partition a 6×6 grid into four equal-sized districts. Figure 10 shows the corresponding distributions of *where* these cut edges occur most frequently across the samples.

Recall that low numbers of cut edges indicate that the districts have short boundaries. If a flood fill method sampled uniformly among all valid plans, a large sample of generated plans would be expected to have a nearly identical distribution of cut edges as the full enumeration. These experiments demonstrate that—as we would expect—flood fill does not uniformly sample from the set of valid plans and that different flood fill variants give different distributions of generated plans.

This is our first example of a non-uniform distribution over plans (i.e., some districting plans are more likely to appear than others). We have already seen that non-uniformity can be desirable or even necessary for a useful sampling method, but it could easily be the case that innocuous modeling decisions significantly affect the behavior of the resulting samples. This *distributional design* question comes into play if we want to perform statistical calculations, e.g., comparing a human-drawn plan with the "average" computer-drawn plan generated using a sampling method. Without an understanding of the sampling distribution, a districting plan being "typical" or being an "outlier" holds little meaning.

3.2 GENERATING PLANS FROM PLANS

A different sampling strategy generates a random plan by editing an existing plan, rather than starting from a blank slate. This strategy could be *perturbative*, for example, generating a new plan by wiggling the boundaries of an old plan, or it could take larger steps, for example, merging several districts in an existing plan and then redistricting that region using a method from Section 3.1. In either case, we are likely to "see" part of the previous plan in the edited plan, but if we repeat this process enough times in a *random walk*, this *autocorrelation* decreases: after many steps, we should see a plan that has little in common with the initial one.

There are many reasons why random walks might be preferable to generating a completely new plan for each sample. As we have discussed, the space of potential plans is huge and includes many undesirable examples. If we find a good plan,

[4] For this demonstration, each flood fill variant was run 300,000 times. In our implementation, the success rate of flood fill on a 6×6 grid ranged between 5 and 10% for the standard and whole-plan variants and was close to 40% for the bounding-box variant.

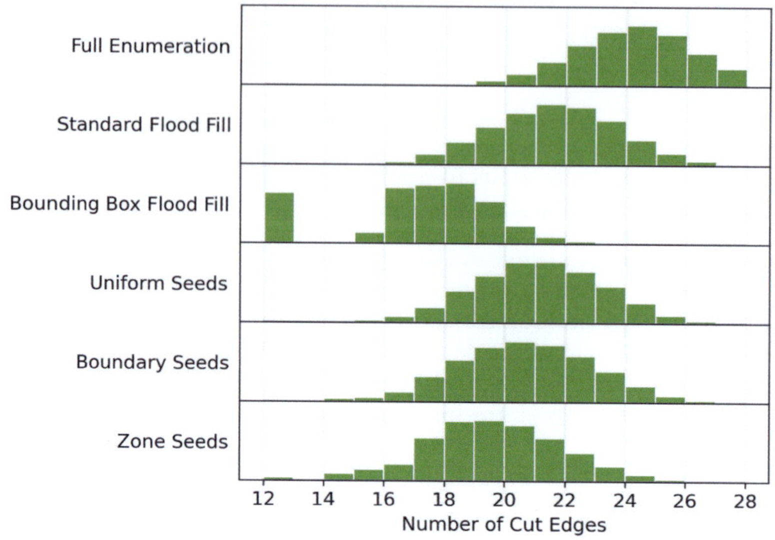

Figure 9: Cut-edge comparison for flood fill methods that divide a 6×6 grid into four equal districts. The bounding box method succeeds at favoring more compact plans.

we might want to see if we can edit it to find others (or to make it even better) rather than obliterating it. Furthermore, editing steps are often more efficient to implement than generating a plan from scratch. On the other hand, suppose that each time we generate a new plan, we do so by making a tiny perturbation in the boundary of a plan we have already generated (cf. the "flip" method described below). Then, we will need many, many steps of this random walk before the plans we generate look significantly different from one another. This "explore–exploit" trade-off—between the potential for large *exploring* steps to find a completely new, effective plan and the potential for small perturbative changes, *exploiting* a good plan to make it better—is a typical one in randomized search.

To introduce some terminology, the new plan generated by editing an existing plan corresponds to a *step* of the algorithm. The set of plans that can be generated in one step from some Plan A define the *neighbors* of Plan A in the state space. A *random walk* is a process for taking a sequence of these steps, where at each step a neighboring plan is generated and chosen to be the next plan from which to continue. Random walk algorithms are motivated and discussed in more detail in Chapter 17. Here, we briefly describe them mostly as points of comparison with other district generation methods.

FLIP AND RECOMBINATION

Perhaps the most minimalistic way to generate a new plan from an existing plan is to change the district assignment of a single unit at a time. For instance, if a census block in District A lies on the boundary between District A and District B, changing the assignment of this block so that it is in District B results in a slightly different

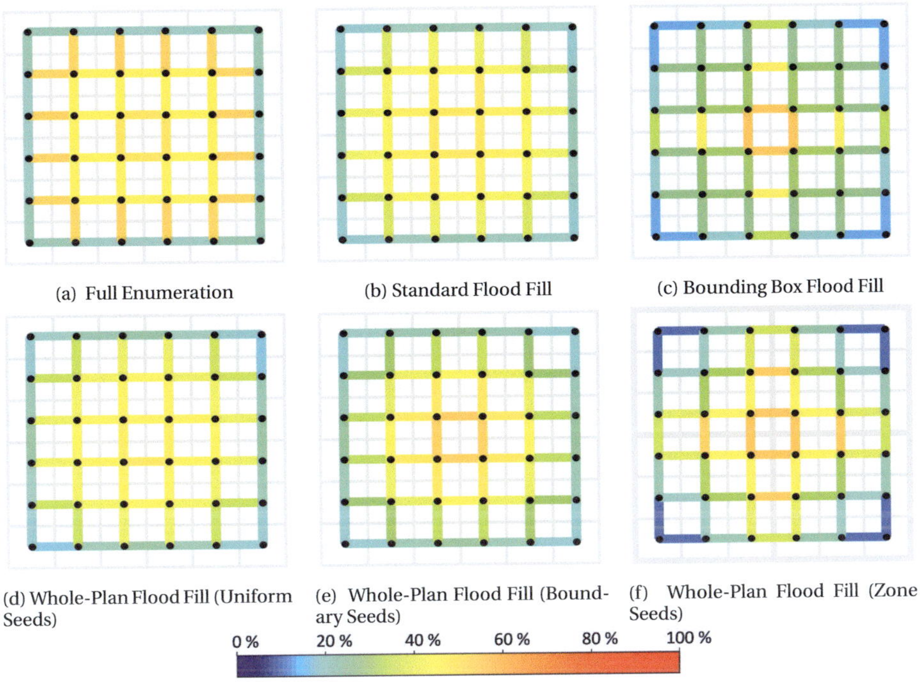

(a) Full Enumeration

(b) Standard Flood Fill

(c) Bounding Box Flood Fill

(d) Whole-Plan Flood Fill (Uniform Seeds)

(e) Whole-Plan Flood Fill (Boundary Seeds)

(f) Whole-Plan Flood Fill (Zone Seeds)

Figure 10: Cut-edge heatmaps corresponding to samples in Figure 9. The color indicates the percentage of plans in the sample in which that edge is a cut edge. Figure 9 shows that all of the flood fill variants tend to have fewer cut edges than would be expected from a uniform sample (compare the top distribution to the bottom five). These heatmaps show that the relative distribution of *where* the cut edges occur is similar for several of the variants (compare (b) with (d) and (e)). The plans made by the bounding box method (c) tend to have *substantially* fewer cut edges than would be expected from a uniform sample (see Figure 9), and we see here that edge cut frequency corresponds to proximity to the center of the grid. All three whole-plan variants tend to produce samples with a smaller number of cut edges than the full enumeration, where edge cut frequency increases closer to the center of the grid, and in the zone-seeded variant the edges close to the zone boundaries are cut substantially more frequently than the edges far from the zone boundaries.

districting plan. We proceed if this simple edit known as *flipping* preserves the key property that Districts A and B are contiguous.

Flip-step algorithms change the assignments of randomly chosen units along district boundaries; they are used widely in the redistricting literature [6, 15, 38, 43, 45]. Algorithm 5 details a single flip step, illustrated in Figure 11. Specifically, a candidate flip unit is chosen randomly from geographic units on the boundaries of two or more districts. The assignment of this candidate unit is then *flipped* from its current district to that of a neighboring district, unless doing so results in an invalid plan (e.g., districts becoming discontiguous or resulting population deviation larger than allowed). A single flip step creates a new plan that is nearly—but not completely—identical to the previous plan, so typically this step is iterated many times (up to millions or billions) to generate the next plan in a random walk.

Modifications to the flip step can be made to anticipate and fix issues that appear with the naive version. For instance, to promote plans with a balanced population,

Algorithm 5 Basic Flip Step

flip unit ← RANDOM(boundary unit)
district i ← District assignment(flip unit)
district j ← District assignment(neighboring unit not in district i)
if reassigning flip unit from i to j results in a contiguous and population-balanced plan **then**
    District assignment(flip unit) ← district j

Nagel [38] proposes swapping the assignments of two units on either side of a shared boundary, rather than just flipping a unit from one side to another. We can call this a *swap* step. To take larger coherent steps than a single-unit flip, some techniques [15] flip clusters of contiguous units along the same boundary. Introducing a probabilistic weighting can promote compactness or any other desired priority.

Another method of stepping from one valid plan to a neighboring valid plan is to merge two or more neighboring districts and re-partition them into new districts, keeping the rest of the plan the same; see Figure 12 for an example. This strategy has been named *recombination* in recent work (see DeFord et al. [12] for a survey).

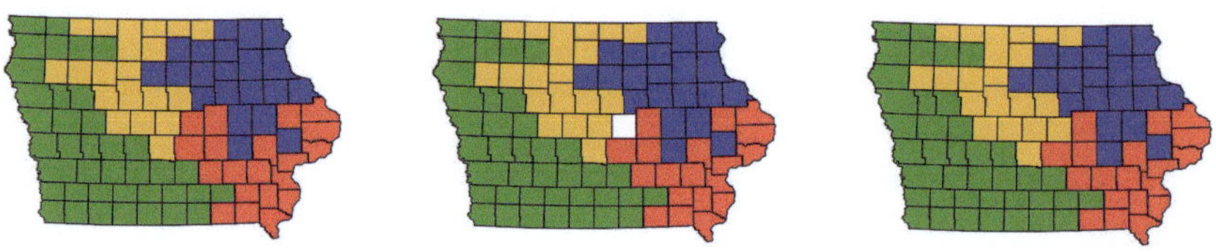

Figure 11: In this illustration of a **flip step**, the white node in the middle figure is the randomly chosen district boundary unit. The left figure shows the original plan and the right figure shows the plan after the white unit 'flips' from red to yellow.

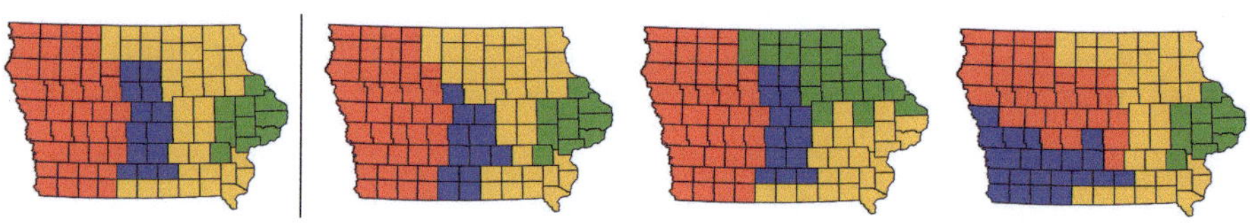

Figure 12: The figures on the right illustrate three different potential outcomes from taking a single **recombination** step from the starting plan shown in the left figure.

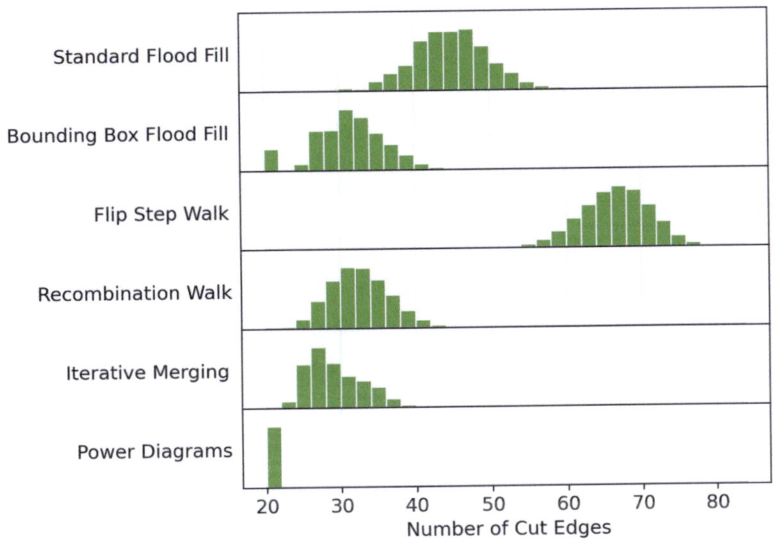

Figure 13: Cut-edge distribution comparison for generating four-district plans with up to 5% population deviation for a 10×10 grid.

3.3 COMPARING SAMPLERS

We conclude this section by comparing several different sampling methods for generating four-district plans from the 10×10 grid and the Iowa counties. Again, as we did for the flood fill variants, we compare the distributions of both the number of cut edges (Figures 13, 14) and the location of where the cut edges occur most frequently (Figures 15, 16). This time, since the problem is too big for complete enumeration, we use our samplers: standard flood fill, bounding box flood fill, a random walk using flip steps, a random walk using recombination steps, iterative merging, and a power diagram method (described below in Section 4.1).[5]

For both the grid and Iowa, the samples generated by flip-step walks have substantially more cut edges than those generated by the other variants. The samplers that tend to generate more compact plans also tend to have cut edges more highly concentrated closer to the center and very few cut edges near the corners and boundary. Interestingly, the edges near the higher populated Iowa counties are also substantially more frequently cut than the less populated counties (see Figure 17). Again we see that for both the 10×10 grid and Iowa, the bounding-box flood fill variant tends to generate plans that have fewer cut edges than the standard flood fill variant.

Though samples derived from recombination walks, iterative merging, and power diagrams tend to have more compact plans (fewer cut edges) than the other vari-

[5] As we will see below, power diagrams are *geometric* partitions in which district boundaries may split census units. We assign each of these split units to the district with the most populous share of the unit, rejecting any plan that fails to maintain contiguity and population balance. This last method helps to illustrate the blurry line between sampling and optimization.

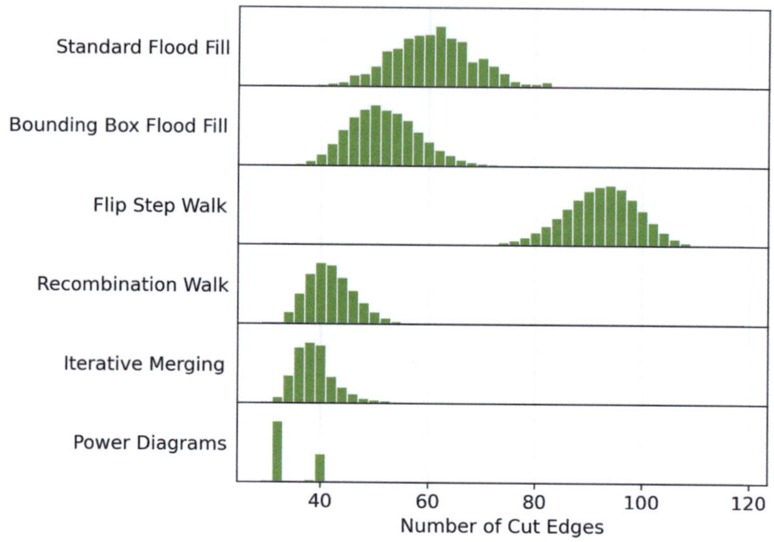

Figure 14: Cut-edge distribution for generating four-district plans with up to 5% population deviation using Iowa counties with various methods. The enacted plan (shown in Figure 1) has 47 cut edges and 0.005 percent population deviation.

ants, the power diagram samples in particular contain only a small number of *distinct* solutions (for the grid, there was only *one* unique solution in a sample of 100,000). This illustrates that even algorithms with some randomness might not generate a particularly diverse, let alone *representative*, sample.

4 SEEKING "BEST" PLANS: OPTIMIZATION

Perhaps the most obvious application of computation to redistricting involves *optimization* of districting plans. Optimization algorithms *extremize* (maximize or minimize) an *objective function* while satisfying some set of *constraints*. In the context of redistricting, an optimization algorithm might be designed to maximize measures such as the Polsby–Popper or Reock compactness scores (see Chapter 1) or the number of competitive districts, or to minimize measures such as population deviation, the number of county splits, or the number of cut edges. The objective function might be one of these single measures, or it can be a composite that combines several. Accompanying the objective function, the constraints likely include legal requirements, such as contiguity and population balance. Other requirements such as VRA compliance can be difficult to express as formal mathematical constraints.

Unfortunately, redistricting optimization problems are not easily solved in practice. Nearly any way of phrasing optimization for redistricting suggests that it belongs to a class of problems called *NP-hard*. For this reason, we should not expect optimization to extract the *best possible* solution to a redistricting problem.

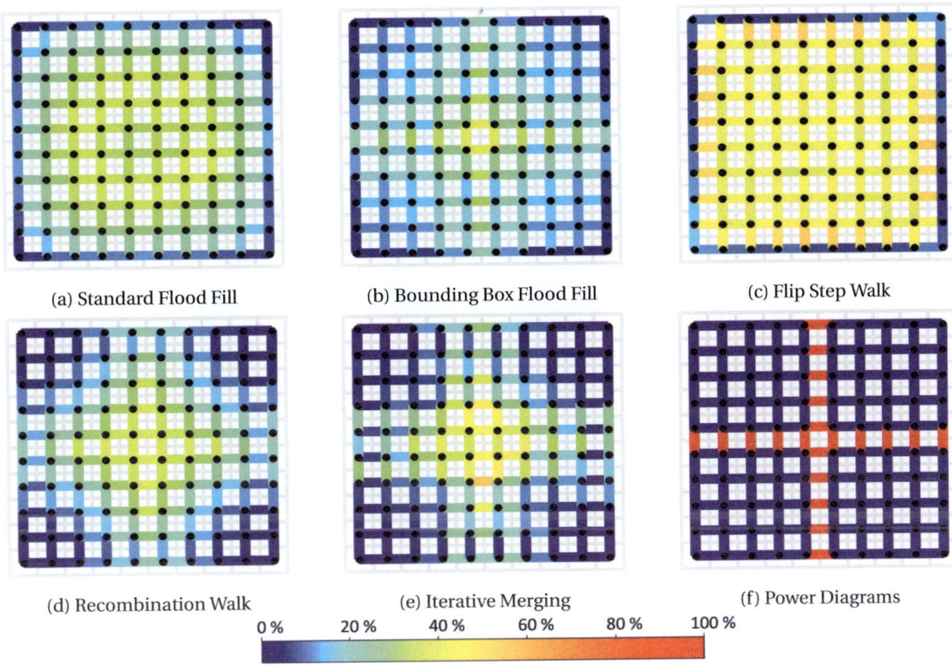

Figure 15: Cut-edge heatmap corresponding to samples in Figure 13. The color indicates the percentage of plans in the sample in which that edge is a cut edge.

Complexity limitations have been known throughout the history of algorithmic redistricting: In Nagel 1965 [38], acknowledged that their proposed algorithm "will not guarantee that the criterion is as low as mathematically possible, though it should be low enough to satisfy the political and judicial powers that be."

A reasonable expectation of optimization algorithms is that they can identify *good enough* plans and make improvements to proposed plans. With this looser goal in mind, in this section we describe several algorithmic optimization techniques that have been proposed for redistricting.

4.1 CLUSTERING AND VORONOI APPROACHES

Most methods in this chapter construct plans by assigning labels to a fixed set of census or other geographical units. A different class of methods for sampling and designing plans ignores these unit boundaries in favor of drawing lines directly on a map of the underlying geography. These *geometric* methods typically lead to plans with compact boundaries, although sometimes this compactness comes at the cost of other redistricting criteria that are harder to express geometrically.

Note that although there is a huge-but-finite number of plans when we construct them out of census units, there is an *infinite* number of ways we can draw geometric dividing lines on a map. For this reason, we cannot expect these algorithms to have a positive probability of generating every possible plan. Rather, they often make local decisions intended to promote generation of attractive plans (e.g., that

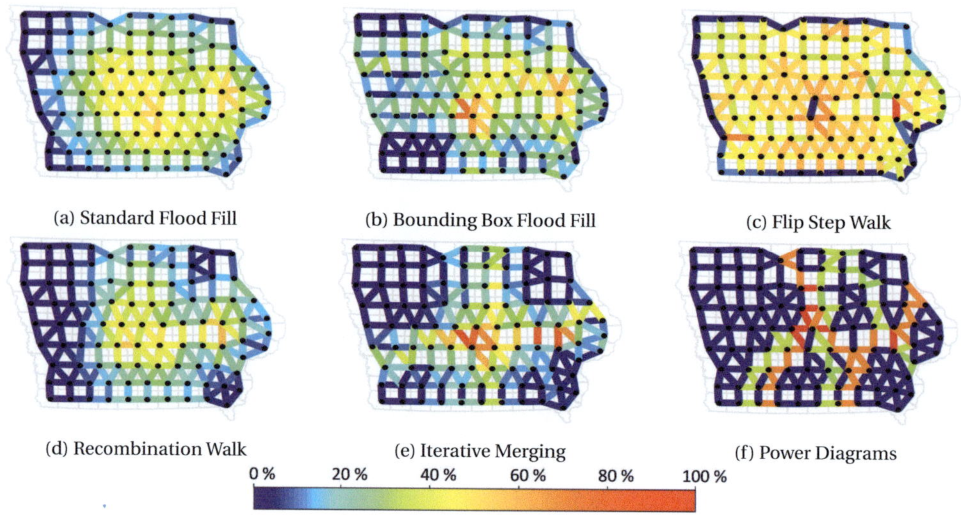

(a) Standard Flood Fill (b) Bounding Box Flood Fill (c) Flip Step Walk

(d) Recombination Walk (e) Iterative Merging (f) Power Diagrams

Figure 16: Cut-edge frequency comparison corresponding to samples in Figure 14. The color indicates the percentage of plans in the sample in which that edge is a cut edge.

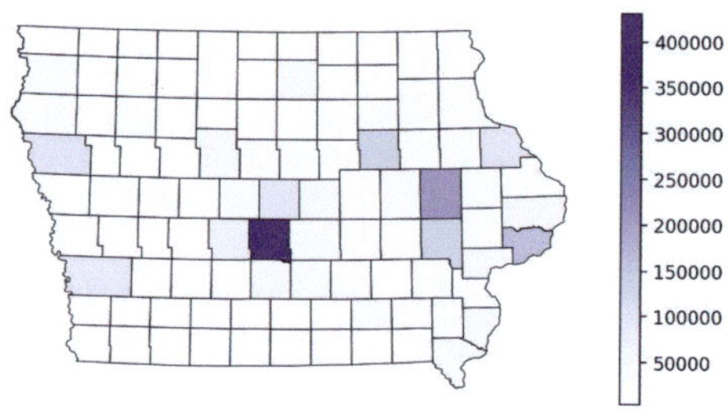

Figure 17: Iowa population map.

the boundaries between districts must be straight lines) but might assemble these local decisions in a stochastic fashion.

A central example of a geometric approach to redistricting draws inspiration from another part of computer science. A common task in statistics and machine learning involves *clustering* data based on proximity. Clustering methods are often geometric in nature: their job is to find the dividing lines between different groups of data points. Along these lines we can cluster units on a map into districts based on geographic and other considerations.

For example, a "splitline" algorithm [20] repeatedly divides regions into two subregions, starting with the entire state and ending with districts, at each step identifying a line that evenly splits the population (see Figure 18). The result is a fairly compact districting plan. A sampling variant might randomly draw from the set of valid splitlines in each bipartitioning step, whereas an optimization variant might choose the shortest splitline each time.

One of the simplest and most common approaches to clustering uses *Voronoi*

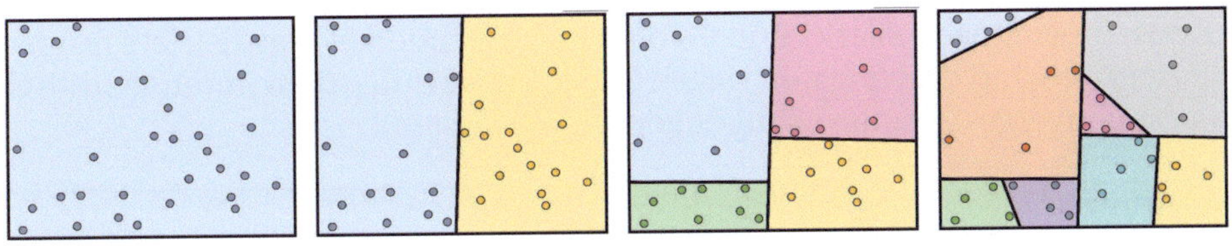

Figure 18: In this illustration of the shortest-splitline algorithm, the left figure shows the population distribution of a fictional region. The subsequent figures show the regions being bisected using the shortest line that evenly divides the region's population.

diagrams, such as the ones illustrated in Figures 19 and 20. In a Voronoi diagram, we choose a set of k points on the map to be district *hubs*. Then, the map is divided into regions based on proximity: The *Voronoi cell* associated with a particular district hub is the set of points on the map closer to that hub than to any other hub. That means that the cells are built for efficiency in distance terms, which obviously promotes compactness. This, in addition to straight-line boundaries and convex shapes, has made the Voronoi approach to redistricting attractive to several teams of researchers [13, 33, 35, 40, 45, 47].

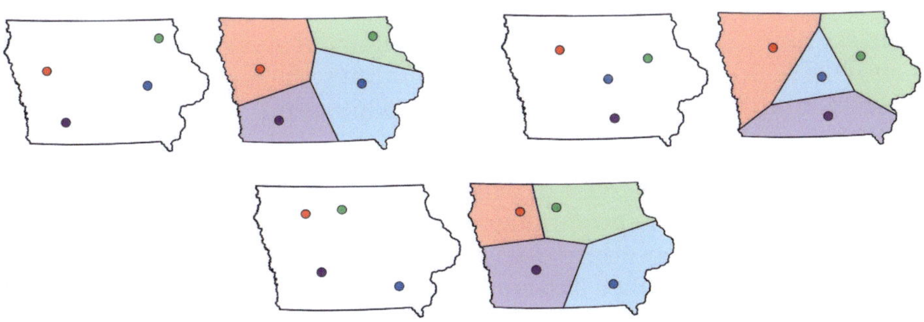

Figure 19: Three different ways to draw four district hubs in Iowa and the corresponding Voronoi diagrams.

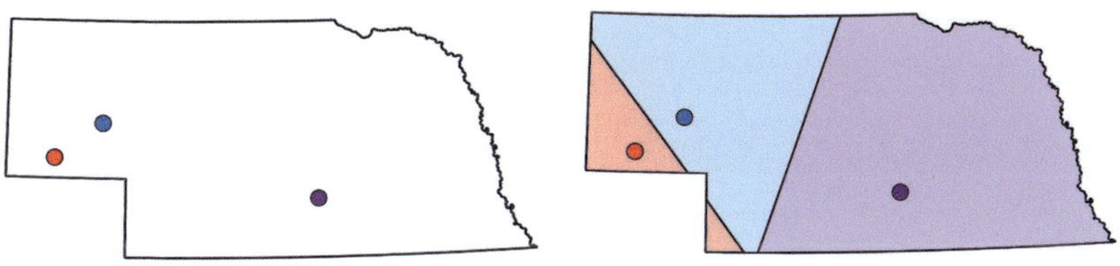

Figure 20: The three centroids on the left correspond to the Voronoi diagram on the right. Because Nebraska's geography is nonconvex, the red district is disconnected.

Note that the Voronoi process is only deterministic after the location of the hubs has been fixed (see Figure 19); a reasonable question to ask when using Voronoi diagrams for redistricting is where to place the k hubs to optimize the quality of the diagram. The most popular formulation is called a *k-means problem*, seeking to place district hubs to minimize the average squared distance between a resident and their district's hub [2, 7, 21]. A simple and often effective algorithm for k-means is *Lloyd's algorithm* [19, 34], detailed in Algorithm 6, which alternates between moving each district hub to the average location of that district's residents (the *centroid*) and drawing new districts with those hubs, then iterating until this process converges, i.e., the changes get arbitrarily small.

Algorithm 6 Lloyd's k-Means Algorithm

```
Identify k initial district hubs: hub 1, hub 2, ..., hub k (also called means).
while process has not yet converged do
    for each geometric point i do
        District assignment(point i) ← District assignment(hub closest to point i)
    for each district j do
        hub j ← Centroid(district j)
```

Optimal placement for the hubs is famously NP-hard, or likely to be computationally intractable, and Lloyd's algorithm often converges on a *local optimum* (the best in its neighborhood) rather than a *global optimum*.

There are several fundamental challenges for Voronoi-type algorithms in this setting: first, we must decide on a notion of distance. Should we measure distance to the nearest hub based on geographic distance, travel time, or something else? Moreover, these Voronoi diagrams have lines that cut across building blocks like census blocks and precincts. Finally, the algorithm so far is completely targeted to distance minimization and does not balance population.

A few of the issues above can be addressed. For example, *power diagrams* are generalizations of Voronoi diagrams in which each hub also has an associated weight (Figure 21) [7, 17].[6] In one power diagram implementation based on a modified Lloyd's algorithm, Cohen-Addad, Klein, and Young [7] construct districting plans that are population-balanced and compact.

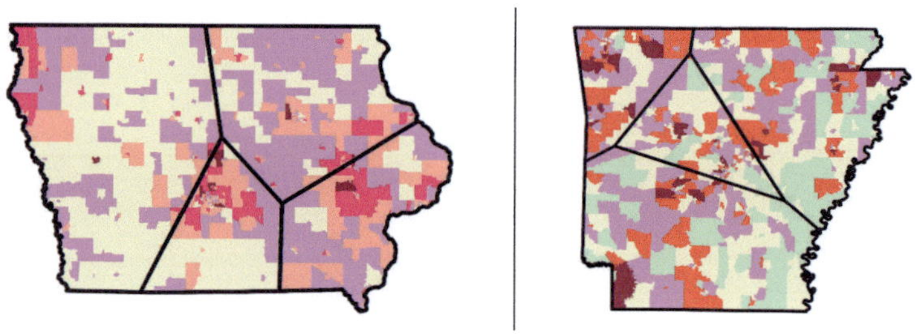

Figure 21: Power diagrams are shown for Iowa and Arkansas. The shading of each map represents population density. Figures provided by Richard Barnes.

Moreover, all of the cells are convex polygons, and they have at most six sides on average.

Although the districts induced by power diagram cells are balanced and compact, they still face the issue of units. To achieve population balance, the polygons often split units through their centroids, where their entire population is assumed to be

[6]The power diagram cell for a hub h with weight w_h and distance function d is the set of points x such that $d(x,h)^2 - w_h \le d(x,h')^2 - w_{h'}$ for every other hub $h' \neq h$.

located. This means that assigning census units based on these idealized polygon districts is not straightforward. Modifying these plans to respect unit boundaries may ultimately require sacrificing compactness, population balance, and even contiguity. We discuss refinement issues more generally in Section 4.4.

4.2 METAHEURISTICS AND RANDOM WALK VARIANTS

When perfect optimization is elusive, computer scientists often turn to *heuristics*, which accept approximate or local solutions instead of exact or global solutions. In practice, a good heuristic can often identify strong solutions quickly. Although some heuristics are specialized to redistricting, *metaheuristics* are general strategies that can be applied "out of the box" to optimization problems drawn from many different domains. Various computational redistricting methods have adapted well-known metaheuristic algorithms to map-drawing.

Many common metaheuristics employ random walks of the kind discussed in Section 3.2, which explore the *state space* (the set of all valid plans) by starting with some plan and making an edit. Since we begin with a plan and compare it with neighbors, we can also call this strategy a *local search*. A basic local search algorithm known as *hill climbing* is illustrated in Figure 22. In each step, the algorithm considers replacing the current plan with a proposed neighbor, and proceeds with the replacement if the new plan has a better score. For example, the neighborhood of a Plan A may be composed of all plans that can be created from Plan A by a single flip step, swap step, or recombination step (see Section 3.2). In addition, we now need an objective function and a rule for determining acceptance of each neighbor based on its score.

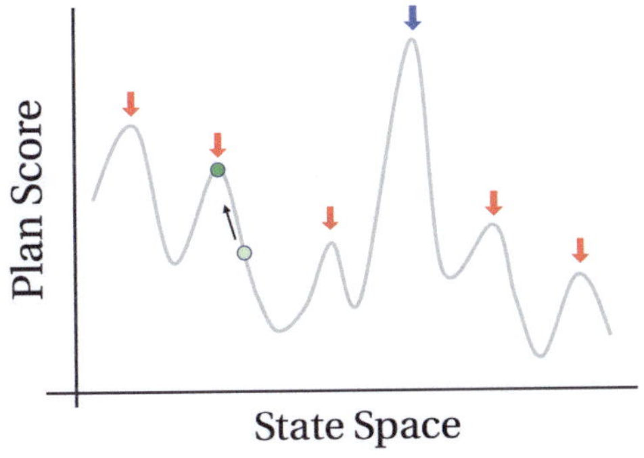

Figure 22: In this visualization of the state space, the global maximum is indicated with a blue arrow and local maxima are indicated with red arrows. If hill-climbing optimization begins at the light green circle the algorithm will identify the local maxima at the dark green circle, but will not find the global maximum.

Careful engineering is required to design an effective local search algorithm. If the objective functions take a lot of time to evaluate or if the algorithm has to evaluate, many neighboring plans at each step, it can take a lot of time to carry out a single step of local search. On the other hand, if plans do not have many neighbors, it could take many steps before a poorly performing plan is improved to an acceptable level.

Several local-search variants have been used to design districting plans:

- **Hill climbing**, as already mentioned, only accepts improvements until it reaches a plan whose neighbors are all worse, which is necessarily a local optimum. (This is used in [2, 31, 45].) Hill climbing often gets stuck in local optima, as illustrated in Figure 22. To improve the likelihood of success, hill-climbing algorithms often call a sampler to draw many different random starting plans and restart the process several times, keeping the best-performing local optimum among the different runs.
- **Simulated annealing** (used in [2, 6, 15, 30, 45]) attempts to avoid local optima by sometimes allowing moves to worse scores. Inspired by certain physical processes, this stochastic algorithm maintains an additional *temperature* parameter that starts "hot" (high parameter) and "cools" (decreasing the parameter) slowly over the course of successive steps. The probability of transitioning to a worse neighboring plan is controlled by the temperature: while the temperature is hot, worse plans are accepted and the algorithm can explore the state space; and as the temperature cools, hill climbing kicks in and the algorithm can improve to a high-quality plan.
- **Tabu search** (used in [2, 3, 28, 45]) keeps a memory of the plans that it has recently visited and avoids returning to these already-visited plans. This strategy encourages broader exploration of the state space by preferring unvisited neighbors.
- **Evolutionary algorithms** (used in [2, 32]) draw inspiration from biology to quickly create a diverse collection of plans. In this technique, a *population* of plans evolves over the course of the algorithm by a combination *mutating*, or taking a basic random walk step, and *crossover*, which combines two plans in a more drastic move to generate one or more *child* plans with traits of both *parents*.

The crossover method that we implemented in our evolutionary algorithm, depicted in Figure 23, is based on the approach in Liu et al. [32]. Two parent plans are drawn from the population and their *common refinement* (the regions resulting from overlaying the two parent plans) is calculated, as in Figure 23c. The common refinement likely has many more regions than the desired number of districts, so the next step is to merge these smaller regions together (using a similar method to Chen et al. [9, 10]) until there is a correct number of districts (see Figure 23d). At this point the districts may have unbalanced populations, so the final step is to make small adjustments to balance out the population (see Figure 23e).

COMPARING METAHEURISTICS

To offer a coarse comparison of these optimization methods, we apply several of them to seek the minimum number of cut edges in an Iowa congressional districting plan. We apply hill climbing, simulated annealing, and an evolutionary algorithm, together with a flip step random walk. In each case we allow a population deviation

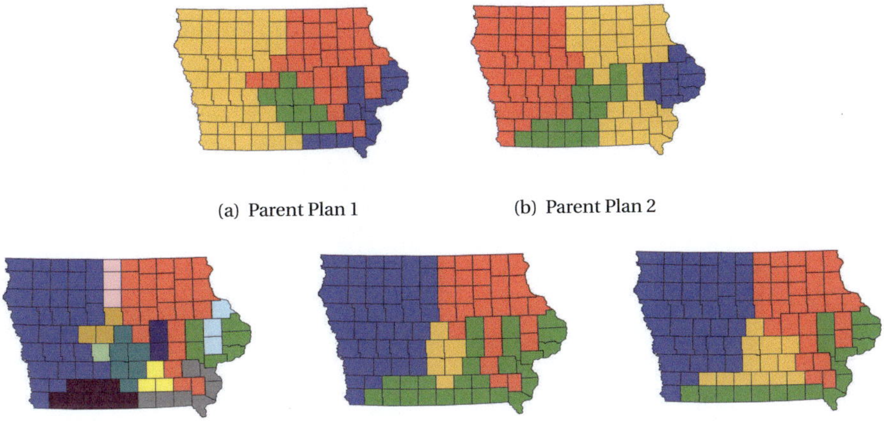

(a) Parent Plan 1 (b) Parent Plan 2

(c) Common Refinement (d) Merged Districts (e) Child Plan

Figure 23: (a-e) Example of a crossover step similar to the one used in Liu et al. [32]. Two *parent* plans are chosen from the population. Their common refinement is computed and the resulting regions are merged until there are four districts. The merged districts are adjusted to achieve population balance, and the resulting child plan is added to the population.

of up to 5% from the ideal district size. In Figure 24 we show the outcomes of running the same suite of metaheuristic algorithms with the same starting plan two different times.

Some details of these comparisons are found in the figure captions, but here are a few themes. Hill climbing quickly (within a few hundred steps) finds a local minimum in each run, whereas simulated annealing fluctuates but eventually outperforms the strictly greedy method. One risk with simulated annealing is that it may pass up a promising solution early in hopes of finding something better, only to end at a poor-quality solution.

To illustrate the evolutionary algorithm, we show the maximum score (light green) and minimum score (dark green) over the population of ten plans at each step. (The same starting plan from hill climbing and simulated annealing is included in the starting population for the evolutionary runs.)

The relative performance of these metaheuristic approaches, however, depends heavily on user choices. They could always have been run for longer, or cleverly implemented and tuned. Nonetheless, we hope to have illustrated some of the issues and tradeoffs with basic implementations. In these short and untuned runs, each method identified plans with a small number of cut edges, but none of them found the global minimum. Figure 25 shows an example of a plan with only 29 cut edges and less than 5% population deviation, which we show below to be the true minimum.

Metaheuristics are largely agnostic to the particular objective function in a redistricting problem. This is both a positive and a negative aspect of these algorithms: On the one hand, they are easily adapted to the particularities of a given state or set of district criteria, but on the other hand, they are unable to leverage structure in a specific objective function that might make optimization faster.

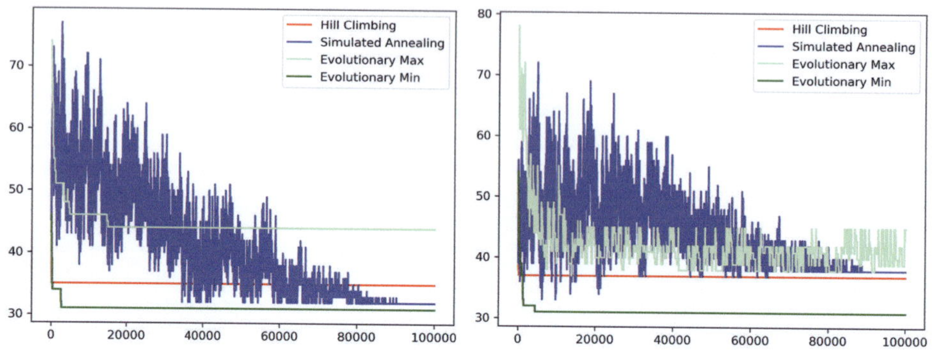

Figure 24: These plots compare two different runs of three metaheuristic strategies: hill climbing (red), simulated annealing (blue), and an evolutionary heuristic (dark green is the population minimum and light green is the population maximum number of cut edges)

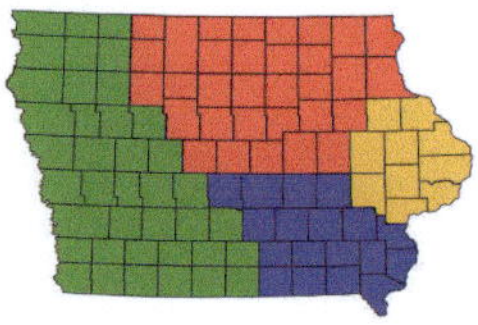

Figure 25: This plan has the fewest number of cut edges (29) for partitioning Iowa into four districts with at most 5% population deviation from ideal (this plan has a population deviation of less than 3.5%).

4.3 INTEGER PROGRAMMING

Somewhere between applied mathematics and computer science is the discipline of **operations research**, which is built around approaches to difficult optimization problems, from maximizing profit while satisfying demand to scheduling tasks on a computer to maximize throughput. Starting from these problems as central examples, these fields have developed a taxonomy to classify optimization problems based on their objective functions and constraints. Once we recognize a problem within that taxonomy, we can leverage appropriate general-purpose strategies that solve similar problems efficiently.

Within this taxonomy, many redistricting problems can be understood as *integer programs*, which can be written in the following form:

$$\begin{aligned} \text{minimize}_x \quad & f(x) \\ \text{subject to} \quad & g(x) \geq 0 \\ & x \in \mathbb{Z}^n. \end{aligned}$$

Here, x denotes the set of *decision variables* used to encode districting plans. The function $f(x)$ gives the value of the objective function evaluated at x. The function $g(x)$ encapsulates constraints on x (e.g., that the population of each district must not deviate substantially from the ideal); we can think of $g(x)$ as *vector-valued*, meaning that we can enforce more than one constraint at a time. Finally, the constraint $x \in \mathbb{Z}^n$ is mathematical notation denoting that x is a tuple of n *integers*; that is, the unknown in an integer problem is a list of n numbers without fractional parts. The notation above is extremely generic: Nearly any computational problem whose output consists of integers can be written in this form.[7]

Not unlike the metaheuristics in Section 4.2, algorithms for solving integer programs attempt to cut down the state space of possible solutions to a manageable size by using the constraints and bounding the objective in various ways. For example, a common strategy is *relaxation*, whereby constraints are removed from an integer program to make it easier; if one of the relaxed constraints is violated in the resulting solution, it is added back to the integer program and solving is restarted [25]. Similarly, *branch-and-bound* algorithms may drop the integer constraint, resulting in a much easier (and typically convex) problem to solve algorithmically, as well as a bound on the best possible objective value; then, various variables are pinned to nearby integers until the solution satisfies all the constraints [36, 49].

Integer programming algorithms and local search metaheuristics are similar in the sense that both navigate the space of feasible solutions while looking to improve an objective function. The main difference is that integer programming algorithms are typically (but not always) aimed at extracting a *global* optimum via carefully designed bounding and search strategies. That is, the heuristics used in integer programming are generally conservative, ordering potential x values to try based on their likelihood of solving the integer program but never throwing one away until it can safely be ruled out.For this reason, we can be confident in the output of integer programming tools, but they can take an extremely long time to terminate.

Integer programming formulations have a long history in redistricting, going back again to the 1960s and the work of Hess, Weaver, and collaborators [4, 5, 7, 11, 25, 26, 30, 36, 39, 49]. In the section below, we give an example of how one could phrase redistricting as an integer program.

DISTRICTING AS AN INTEGER PROGRAM

Suppose we wish to design d districts in a state with n census units; our task is to assign each unit to one of the d districts. We can introduce a binary variable x_{ij} for each census unit i (ranging from 1 to n) and district j (ranging from 1 to d). We interpret x_{ij} as follows:

$$x_{ij} = \begin{cases} 1 & \text{if unit } i \text{ is in district } j \\ 0 & \text{if unit } i \text{ is } \textit{not} \text{ in district } j. \end{cases}$$

[7] In a standard trick, an inequality constraint like $2x \leq 5$ can be written as $-2x+5 \geq 0$ and an equality constraint like $x = 10$ can be written as $x \geq 10, -x \geq -10$. We can put all three together in vector form as $(-2x+5, x, -x) \geq (0, 10, -10)$. This lets us use $g(x) \geq 0$ as a canonical form for systems of equalities and inequalities.

Our goal is to assign each x_{ij} a value of 0 or 1 in such a way that satisfies certain constraints and corresponds to the best possible plan for a certain objective function.

We have to add several constraints to make sure that x is reasonable. Each variable x_{ij} always takes on one of two values, 0 or 1. These are integers, but to avoid nonsensical outputs like $x_{ij} = 5$ we additionally constrain $0 \le x_{ij} \le 1$ for all units i and districts j. Similarly, we want to make sure to assign each unit to exactly one district. Consider a single unit i. If we sum the x_{ij} values for all districts j, we have computed the number of districts to which unit i was assigned. For example, if there are four districts, then $x_{i1}+x_{i2}+x_{i3}+x_{i4}$ equals the number of districts to which unit i is assigned. Hence, we need to enforce the constraint

$$\sum_{j=1}^{d} x_{ij} = 1, \text{ for all units } i.$$

Next, suppose we want to ensure that every district has a population between a lower bound ℓ and an upper bound u. We write p_i for the population of unit i; for example, $p_8 = 100{,}000$ means that there are 100,000 people in unit 8. Given the variables x_{ij} above, this means that $\sum_{i=1}^{n} p_i x_{ij}$ records the population of district j. (Since x_{ij} has a value of zero if unit i is not in district j, the sum excludes the populations of units i that are not assigned to district j, leaving behind just the relevant populations that are used to construct the district.) This gives

$$\ell \le \sum_{i=1}^{n} p_i x_{ij} \le u, \text{ for all districts } j.$$

Finally, we need to design an objective function. There are many possible objective functions that are relevant to redistricting, such as minimizing the sum of distances between voters and their district's center [26, 36, 49], minimizing the number of counties that are split into different districts [4], or optimizing a measure of compactness [30]. In our example, we minimize the number of cut edges by introducing another variable x', with components x'_{ab} for each pair of adjacent units a and b. These x' variables encode whether or not a given edge is a cut edge in the assignment:

$$x'_{ab} = \begin{cases} 1 & \text{if units } a \text{ and } b \text{ are in different districts} \\ 0 & \text{if units } a \text{ and } b \text{ are in the same district.} \end{cases}$$

The number of cut edges in a plan can then be written as the sum $\sum_{ab} x'_{ab}$, and the objective is to minimize this sum.

These new x' variables require additional constraints. They too must be constrained to take on integer values between 0 and 1. We also need to ensure that the x'_{ab} can only take a value of 0 if a and b are *actually* assigned to the same district; otherwise the algorithm would assign 0 to all of the x' to achieve a minimum objective value of zero. That is, there must be some district j such that $x_{aj} = 1$ *and*

$x_{bj} = 1$ for x'_{ab} to take a value of 0, and otherwise the constraints must require x_{aj} to take a value of (at least) 1. We achieve this by constraining

$$\left.\begin{array}{l} x'_{ab} \geq x_{aj} - x_{bj} \\ x'_{ab} \geq x_{bj} - x_{aj} \end{array}\right\} \text{ for all adjacent units } a \text{ and } b \text{ and all districts } j$$

These constraints ensure that x'_{ab} can only equal 0 (reflecting that the edge from a to b is not cut, which happens when a and b are in the same district) if $x_{aj} = x_{bj}$ for all districts j. (For instance if $x_{a1} = x_{b1} = x_{a3} = x_{b3} = x_{a4} = x_{b4} = 0$ whereas $x_{a2} = x_{b2} = 1$, this records that a and b are both in district 2.) If $x_{aj} \neq x_{bj}$ for some j, then the inequalities force x'_{ab} to be at least one (indicating a cut edge).

Letting m denote the number of pairs of adjacent units, we can put all these pieces together, arriving at an integer program:

$$\begin{array}{rll} \text{minimize}_{x'} & \sum_{ab} x'_{ab} & \\ \text{subject to} & 0 \leq x_{ij} \leq 1 & \text{for all units } i \text{ and districts } j, \\ & 0 \leq x'_{ab} \leq 1 & \text{for all adjacent units } a \text{ and } b, \\ & \sum_j x_{ij} = 1 & \text{for all units } i, \\ & \ell \leq \sum_i p_i x_{ij} \leq u & \text{for all districts } j, \\ & x'_{ab} \geq x_{aj} - x_{bj} & \text{for all adjacent units } a \text{ and } b \text{ and districts } j, \\ & x'_{ab} \geq x_{bj} - x_{aj} & \text{for all adjacent units } a \text{ and } b \text{ and districts } j, \\ & x \in \mathbb{Z}^{n \times d} & \\ & x' \in \mathbb{Z}^m. & \end{array}$$

This problem is nothing more than careful, unambiguous mathematical notation for our map-partitioning problem. Once our problem is written in this form, it can be handed over to powerful *solvers*, pieces of software designed to tackle problems in a specific form efficiently.

There are many properties to notice about the problem above. In simple notation, we are able to capture many of the demands of a redistricting problem in a fashion that is easy to communicate to a computer. More importantly, the objective and constraints are *linear*, a valuable property that can help integer programming algorithms to succeed. Once expressed as an integer program, our problem can be run through an integer programming solver, software specifically designed to optimize these instances. Many such solvers are available commercially and open source. These solvers tell you when they definitively identify global optima, but they may take an extremely long time to do so.

Our example neglects important criteria for districting plans, some of which are difficult or cumbersome to express in the formalism above. High up on that list is district contiguity. But something is working in our favor here: although contiguity is not explicitly enforced in the constraints of our integer program, because we minimize cut edges, the optimal plans identified by the program tend to be contiguous. That is, districts that are split into several parts usually have more cut edges than contiguous districts and so are unlikely to be identified as optimal. In Figure 26, we see four example outputs of the above integer program using counties as building block units for two different values of population deviation allowance and two different states. Three of the identified plans are contiguous and the fourth has only one discontiguous district. The idea that we dropped the difficult contiguity constraint and got contiguous districts anyway is a successful *relaxation*, which we discuss in general terms below.

When dealing with multiple scores in an optimization framework, one solution x is said to *dominate* another solution x' if x is at least as good as x' in each score. In our setting, we can say that plan P_1 dominates P_2 if it has no more cut edges *and* no greater population deviation. The plans identified by our integer program lie on the *Pareto frontier*, the set of solutions that are not dominated by any other solutions. *All* other plans must lie above the curve formed by this frontier.

Notably, we found that the global minimum in Iowa at $\leq 5\%$ deviation is 29 cut edges (Figure 27), which beats the local minima identified by metaheuristic runs in Section 4.2. But beware of two major caveats: first, Iowa is much smaller than the typical redistricting problem in combinatorial terms, and this program would not have run to completion on any state's precincts or census blocks. Second, the choice of cut edges as an objective function gave us contiguity more or less "for free." Explicit contiguity constraints have been included in some approaches [42, 49] so that they can optimize other objective functions, at the cost of increased size and complexity for the integer program.

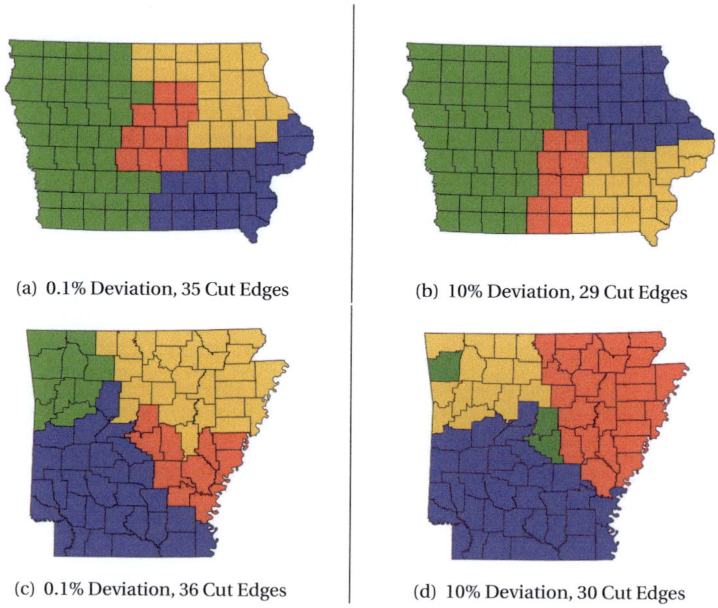

(a) 0.1% Deviation, 35 Cut Edges

(b) 10% Deviation, 29 Cut Edges

(c) 0.1% Deviation, 36 Cut Edges

(d) 10% Deviation, 30 Cut Edges

Figure 26: Plans built out of Iowa counties with minimum cut edges for allowed population deviations of 0.1% and 10% (top), and analogous plans built out of Arkansas counties (bottom).

4.4 RELAXATION AND REFINEMENT

We have seen several examples where it is strategic to "relax" constraints, i.e., temporarily loosen them or drop them altogether. Relaxations can make a difficult problem more tractable, and solutions to these relaxed problems can then be

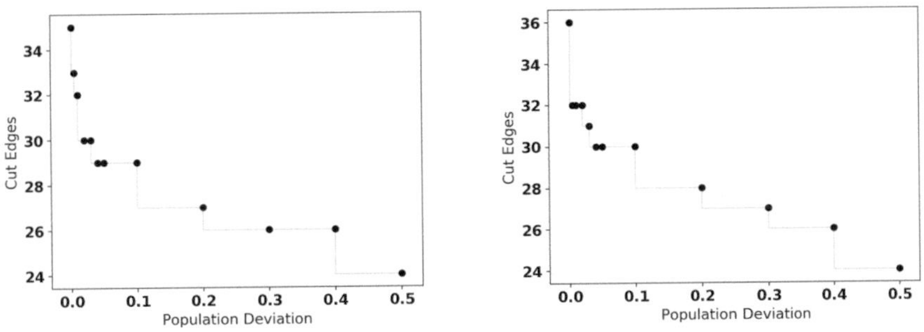

(a) Iowa (b) Arkansas

Figure 27: The relationship between allowed population deviation and minimum cut edges in districting plans built out of Iowa counties (left) and Arkansas counties (right). The black dots represent resulting runs of our example integer program for various values of population deviation. The gray line shows a lower bound on the minimum number of cut edges. For example, the minimum value is not known for deviations between 0.1 and 0.2 (i.e., 10% and 20%), but the value must be between 27 and 29 in Iowa and between 28 and 30 in Arkansas.

refined, if needed, to make them valid. Alternatively, these refinement steps can occur at intermediary stages throughout the algorithm.

Given a solution that has been produced by an optimization procedure, we can refine it in several ways: passing to discrete units, passing from coarser to finer units, or exchanging units to better meet some goals [9, 10, 32, 37]. The most common refinement strategy employs local search methods such as flip and swap steps (Section 3.2); see Levin and Friedler [33] for an example. Indeed, many metaheuristic local search methods can be thought of as iterative refinement.

For a coarse-to-fine example, integer programming may be too slow to run at the census block level, so we can first optimize a plan at the census tract level (with bigger pieces) and then try to break down a small number of tracts to tune the solution to better population balance at the block level. For a discretization example, the geometric methods in Section 4.1 generate districts as polygonal shapes. To transform these plans into partitions of the census units, Cohen-Addad et al. [7] assigns the divided blocks to one of the neighboring districts with attention to population deviation (see Figure 28).

Refinements can occur as mid-course adjustments. Consider that agglomerative methods (Section 3.1) often fail because they build partial plans that have no connected, population-balanced completions. Instead of restarting the process each time this problem is encountered, a refined procedure can backtrack from near a dead end to look for a sequence of choices that can be completed. Another example is found in Jacobs and Walch [30], where a population-balancing auction is run at every iteration of their energy-minimizing algorithm.

Refinements help to ameliorate high rejection rates. There is no guarantee, however, that small refinements can repair invalid plans or correct course effectively in mid-run, and the repair time may end up being longer than the time it takes to run the initial algorithm repeatedly.

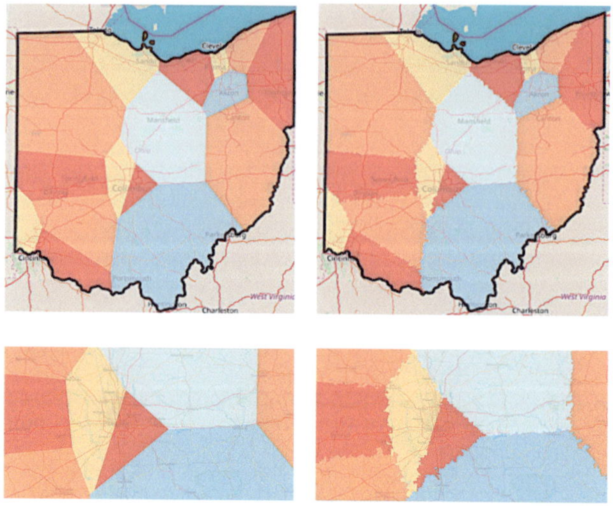

Figure 28: The figures on the left are power diagrams for Ohio (generated as described in Cohen-Addad et al. [7]) and the figures on the right show how these geometric districts can be adjusted to respect the boundaries of census blocks while still maintaining population balance and contiguity. Figures courtesy of Philip Klein.

5 CONCLUSION: THE FUTURE OF REDISTRICTING ALGORITHMS

Algorithmic tools are already extremely useful in redistricting. They can identify promising plans and refine proposed plans. They can generate samples of many thousands or millions of plans to help contextualize a candidate plan.

The active field of redistricting algorithm design continues to produce improved techniques, and mapmakers and analysts can make swift use of these advances. But as with all powerful tools, it is important to understand the limitations of redistricting algorithms in order to use them effectively and responsibly. In this section we discuss several of these limitations.

In the early days of optimistic outlooks on computational redistricting, it looked like rapid advances in computing would soon overcome the difficulties in the problem. In 1963, Weaver and Hess [51] wrote:

> "No available programs or computer techniques are known which will give a single, best answer to the districting problem, though such a solution seems possible if enough funds and efforts are put to the problem, especially considering the rapid advances in size and sophistication of available computers."

Real-world redistricting problems, however, are likely to be *forever* too complex for computers to solve optimally. Even if there were consensus as to what makes one plan better than another, not only does computational intractability limit our ability to identify the best solution in current practice, but the underlying reasons might keep that prospect out of reach, despite advances in computing. Instead, we must settle for plans that may be far from optimal.

There are roles both for humans and for computers in redistricting. Algorithms can efficiently produce potential plans, evaluate their properties, and suggest new ways to divide up a region, but they are limited by the accuracy of the input data, tractability issues, and fidelity of the computational model to the realities of redistricting. Humans also are subject to the same tractability issues but have a better understanding of the populations affected by potential districting plans and the assorted criteria at play when designing voting districts. Hence, a key piece of the puzzle is how to mediate the relationship between human and machine. Subtle issues are at play when designing redistricting systems that citizens and legislators trust—but not trust *too* much.

5.1 ABUSE AND GAMING

Computers and algorithms do not remove humans from redistricting and therefore do not remove human bias and error from the process. On the contrary, algorithms sometimes *amplify* human bias, whether intentionally (e.g., an optimization algorithm can be used to maximize the number of seats for a political party) or unintentionally. One risk of putting unfounded trust in computer-identified plans is that actors can hide their bias behind the justification of a seemingly neutral process. If the algorithm generates a random plan, a user can repeatedly re-run it until it yields a favorable result. Similarly, if the output depends on the starting point, then a user can re-run the algorithm from different starting points, looking for a favorable answer. The more user choices available in an algorithm, the more opportunities to turn the knobs to try to control the answer. The user can disingenuously defend the cherry-picked outcome as being neutrally generated by a computer.

Some argue that the opportunities to game the rules would be avoided by the use of optimization: if the *best* plan is mandated, then a user does not have an opportunity to advance an agenda. Beyond the usual problem of finding some common notion of *best*, there remains the possibility that many dissimilar plans may earn indistinguishably good scores. An agenda-driven user then can just choose their favorite plan among those tied for best.

Before placing value on algorithmically generated districting plans, it is crucial to understand the design decisions and underlying assumptions and simplifications of the algorithm and the effects that these factors have on the resulting outcomes. It is important to be cautious with techniques that are not accompanied by an explanation of these decisions and, when possible, to replicate and perform sensitivity analyses on techniques before advocating for or building off of them.

5.2 BEST PRACTICES AND A CALL TO ACTION

Far more lawyers, legislators, and everyday citizens are capable of using redistricting software than writing it, which means that they must place a degree of trust in developers. A few concrete steps can begin to counteract this asymmetry in understanding of what is "under the hood."

- Expert consultants and other users of redistricting software should provide with their reports code and detailed, unambiguous descriptions of the procedures used to arrive at their conclusions. This will help others to assess potential bias in their analysis caused by distributional design, under-sampling, instability in computation, or the choice of heuristic.
- Maps and other datasets used for analysis should be publicly available to make the analysis reproducible.
- Academic and commercial tools for redistricting should be released under open source license to reinforce trust in redistricting procedures.

Redistricting problems are not only *too big*, but also *too human*, to be completely addressed by computers. Algorithms require constraints and objectives to be precisely defined, but in real-world instances this is not straightforward. Even seemingly simple constraints like contiguity and population balance are not always easy to define: How is geographic adjacency handled for islands and bays? How much population deviation is too much? How do we define compactness? Abstract goals like preserving communities of interest and legal constraints like VRA compliance are nearly impossible to quantify precisely enough for a computer to operationalize. Many of the legal, social, and political aspects of what makes a valid plan—let alone a good plan—are dependent on *context* and subtleties that are better understood by humans than machines. Given that there are real, human consequences of redistricting decisions, these complexities should not be entrusted to an algorithm alone.

Computer technology, mathematical theory, and the political landscape continue to co-evolve. As technology improves, it holds potential to make the tradeoffs involved in choosing districting plans more transparent, using a wide range of districting possibilities as a tool in assessment.

Even though algorithmic techniques for redistricting have been used for decades, many of the mathematical and computational aspects of redistricting are not yet fully understood, and existing techniques have considerable room for improvement. As these areas continue to grow, there are many opportunities for involvement, whether it's analyzing existing algorithmic techniques, replicating published findings, contributing to open-source projects, drawing up strong-performing benchmark plans, or designing new redistricting algorithms and analysis tools.

ACKNOWLEDGMENTS

There are several people who supported the production of this chapter. Daryl DeFord provided coding support for our local search hill climbing and simulated annealing experiments in Section 4.2. Richard Barnes provided Figure 21 using a modification of the algorithms from Cohen-Addad et al. [7]. Phillip Klein provided Figure 28. We also acknowledge the generous support of the Prof. Amar G. Bose Research Grant and the Jonathan M. Tisch College of Civic Life. Amariah Becker acknowledges the support of the NSF Convergence Accelerator Grant No. OIA-1937095.

REFERENCES

[1] Micah Altman and Michael McDonald. The promise and perils of computers in redistricting. *Duke J. Const. L. & Pub. Pol'y*, 5:69, 2010.

[2] Micah Altman, Michael P McDonald, et al. BARD: Better automated redistricting. *Journal of Statistical Software. Forthcoming, URL http://www. jstatsoft. org*, 42(4):1–28, 2011.

[3] Burcin Bozkaya, Erhan Erkut, and Gilbert Laporte. A tabu search heuristic and adaptive memory procedure for political districting. *European Journal of Operational Research*, 144(1):12–26, 2003.

[4] John R Birge. Redistricting to maximize the preservation of political boundaries. *Social Science Research*, 12(3):205–214, 1983.

[5] Allan Borodin, Omer Lev, Nisarg Shah, and Tyrone Strangway. Big city vs. the great outdoors: Voter distribution and how it affects gerrymandering. In *IJCAI*, pages 98–104, 2018.

[6] Michelle H Browdy. Simulated annealing: an improved computer model for political redistricting. *Yale Law & Policy Review*, 8(1):163–179, 1990.

[7] Vincent Cohen-Addad, Philip N Klein, and Neal E Young. Balanced centroidal power diagrams for redistricting. In *Proceedings of the 26th ACM SIGSPATIAL International Conference on Advances in Geographic Information Systems*, pages 389–396. ACM, 2018.

[8] Carmen Cirincione, Thomas A Darling, and Timothy G O'Rourke. Assessing South Carolina's 1990s congressional districting. *Political Geography*, 19(2):189–211, 2000.

[9] Jowei Chen, Jonathan Rodden, et al. Unintentional gerrymandering: Political geography and electoral bias in legislatures. *Quarterly Journal of Political Science*, 8(3):239–269, 2013.

[10] Jowei Chen and Jonathan Rodden. The loser's bonus: Political geography and minority party representation, 2016.

[11] Felipe Caro, Takeshi Shirabe, Monique Guignard, and Andrés Weintraub. School redistricting: Embedding GIS tools with integer programming. *Journal of the Operational Research Society*, 55(8):836–849, 2004.

[12] Daryl DeFord, Moon Duchin, and Justin Solomon. Recombination: A family of Markov chains for redistricting. *arXiv*, 1911(05725):1–28, 2019.

[13] Adam Dobrin. A review of properties and variations of Voronoi diagrams. *Whitman College*, pages 1949–3053, 2005.

[14] Johan de Ruiter. On jigsaw sudoku puzzles and related topics, bachelor thesis. 2010.

[15] Benjamin Fifield, Michael Higgins, Kosuke Imai, and Alexander Tarr. A new automated redistricting simulator using Markov chain Monte Carlo. *Work. Pap., Princeton Univ., Princeton, NJ*, 2015.

[16] Benjamin Fifield, Kosuke Imai, Jun Kawahara, and Christopher T Kenny. The essential role of empirical validation in legislative redistricting simulation. 2019.

[17] Roland G Fryer Jr and Richard Holden. Measuring the compactness of political districting plans. *The Journal of Law and Economics*, 54(3):493–535, 2011.

[18] Edward Forrest. Apportionment by computer. *The American Behavioral Scientist (pre-1986)*, 8(4):23, 1964.

[19] Edward W Forgy. Cluster analysis of multivariate data: efficiency versus interpretability of classifications. *Biometrics*, 21:768–769, 1965.

[20] Center for Range Voting. Gerrymandering and a cure for it—the shortest splitline algorithm (executive summary), 2019.

[21] Olivia Guest, Frank J Kanayet, and Bradley C Love. Gerrymandering and computational redistricting. *Journal of Computational Social Science*, 2(2):119–131, 2019.

[22] Robert S Garfinkel and George L Nemhauser. Optimal political districting by implicit enumeration techniques. *Management Science*, 16(8):B–495, 1970.

[23] David Grace. How to quickly create a neutral, non-gerrymandered election map. *Decentralize Today*, 2017.

[24] Bob Harris. Counting nonomino tilings and other things of that ilk. 2010. For G4G9.

[25] Mehran Hojati. Optimal political districting. *Computers & Operations Research*, 23(12):1147–1161, 1996.

[26] Sidney Wayne Hess, JB Weaver, HJ Siegfeldt, JN Whelan, and PA Zitlau. Nonpartisan political redistricting by computer. *Operations Research*, 13(6):998–1006, 1965.

[27] Christopher Ingraham. This computer programmer solved gerrymandering in his spare time. *Washington Post Wonkblog*, 2014.

[28] Hai Jin. *Spatial Optimization Methods and System for Redistricting Problems*. PhD thesis, University of South Carolina, 2017.

[29] Michael A Jenkins and John W Shepherd. Decentralizing high school administration in Detroit: an evaluation of alternative strategies of political control. *Economic Geography*, 48(1):95–106, 1972.

[30] Matt Jacobs and Olivia Walch. A partial differential equations approach to defeating partisan gerrymandering. *arXiv preprint arXiv:1806.07725*, 2018.

[31] Myung Jin Kim. *Optimization approaches to political redistricting problems.* PhD thesis, The Ohio State University, 2011.

[32] Yan Y Liu, Wendy K Tam Cho, and Shaowen Wang. PEAR: a massively parallel evolutionary computation approach for political redistricting optimization and analysis. *Swarm and Evolutionary Computation*, 30:78–92, 2016.

[33] Harry A Levin and Sorelle A Friedler. Automated congressional redistricting. *Journal of Experimental Algorithmics (JEA)*, 24(1):1–10, 2019.

[34] Stuart Lloyd. Least squares quantization in PCM. *IEEE Transactions on Information Theory*, 28(2):129–137, 1982.

[35] Stacy Miller. The problem of redistricting: the use of centroidal Voronoi diagrams to build unbiased congressional districts. *Senior project, Whitman College*, 2007.

[36] Anuj Mehrotra, Ellis L Johnson, and George L Nemhauser. An optimization based heuristic for political districting. *Management Science*, 44(8):1100–1114, 1998.

[37] Daniel B Magleby and Daniel B Mosesson. A new approach for developing neutral redistricting plans. *Political Analysis*, 26(2):147–167, 2018.

[38] Stuart S Nagel. Simplified bipartisan computer redistricting. *Stanford Law Review*, pages 863–899, 1965.

[39] Susan K Norman and Jeffrey D Camm. The Kentucky redistricting problem: Mixed-integer programming model: Working paper series–03-04. 2003.

[40] Antonio GN Novaes, JE Souza de Cursi, Arinei CL da Silva, and João C Souza. Solving continuous location–districting problems with Voronoi diagrams. *Computers & Operations Research*, 36(1):40–59, 2009.

[41] Elle Najt, Daryl DeFord, and Justin Solomon. Complexity and geometry of sampling connected graph partitions. *arXiv*, 1908(08881):1–58, 2019.

[42] Johannes Oehrlein and Jan-Henrik Haunert. A cutting-plane method for contiguity-constrained spatial aggregation. *Journal of Spatial Information Science*, 2017(15):89–120, 2017.

[43] John O'Loughlin and Anne-Marie Taylor. Choices in redistricting and electoral outcomes: The case of Mobile, Alabama. *Political Geography Quarterly*, 1(4):317–339, 1982.

[44] David J Rossiter and Ronald J Johnston. Program GROUP: the identification of all possible solutions to a constituency-delimitation problem. *Environment and Planning A*, 13(2):231–238, 1981.

[45] Federica Ricca and Bruno Simeone. Local search algorithms for political districting. *European Journal of Operational Research*, 189(3):1409–1426, 2008.

[46] Aaron Sankin. The tech revolution that could fix America's broken voting districts. *The Daily Dot*, 2016.

[47] Lukas Svec, Sam Burden, and Aaron Dilley. Applying Voronoi diagrams to the redistricting problem. *The UMAP Journal*, 28(3):313–329, 2007.

[48] James D Thoreson and John M Liittschwager. Computers in behavioral science: Legislative districting by computer simulation. *Behavioral Science*, 12(3):237–247, 1967.

[49] Hamidreza Validi, Austin Buchanan, and Eugene Lykhovyd. Imposing contiguity constraints in political districting models. Preprint, 2019.

[50] William Vickrey. On the prevention of gerrymandering. *Political Science Quarterly*, 76(1):105–110, 1961.

[51] James B Weaver and Sidney W Hess. A procedure for nonpartisan districting: Development of computer techiques. *Yale LJ*, 73:288, 1963.

Chapter 17

Random walks and the universe of districting plans

DARYL DEFORD AND MOON DUCHIN

CHAPTER SUMMARY

Random sampling is a key idea across this book, and a leading way to do that is to let a "random walker" loose in your universe to collect samples as they explore. The mathematical framework for this is called Markov chains. This chapter is the place where we dig into Markov chains and MCMC: the motivation, the theory, and the application to redistricting.

1 OVERVIEW: NOT A SOLVED PROBLEM

This book has already described many ways in which the modern computing era has revolutionized redistricting: on one hand, an explosion in the sheer amount and diversity of data that map drawers are able to integrate into their methodology; on the other hand, serious algorithmic innovations and expanded computing power for actually constructing plans. It is now easily possible to generate millions of distinct, reasonable plans on a standard laptop in an afternoon, something that would have been unthinkable a few years ago. As access to these data and software has become more widespread, new theoretical developments and applications have changed the way we think about redistricting.

In this chapter we will explore uses of *Markov chains* or random walk methods for generating large collections of districting plans and applications of the resulting ensembles. These techniques have been successfully applied in court cases and legislative reform efforts and are playing an increasingly large role in the design

M. Duchin, O. Walch (eds.), *Political Geometry*, https://doi.org/10.1007/978-3-319-69161-9_17

of both plans and legislation. The underlying mathematical concepts are widely used in many other scientific fields and transferring these techniques to this new setting has led to some great successes in studying redistricting plans.

At a high level, Markov chain Monte Carlo (MCMC) attempts to generate districting plans from a distribution that is "tuned" to satisfy some version of each state's legal criteria, without incorporating explicit partisan biases. The new plans are generated by making iterative changes to a given initial plan while continuing to satisfy the legislative rules. We outline the robust mathematical theory that guarantees that good samples can be constructed, given sufficient time. This gives us an approach to what you might think of as the holy grail for understanding districting plans in context: *baseline ranges* for all kinds of plan metrics that incorporate state rules and voter geography and help us understand the properties of "typical" "reasonable" plans. Back in the 1940s–1960s, when the U.S. courts were trying to figure out how and whether to engage with redistricting, this baseline challenge was laid out by Justice Felix Frankfurter as a prerequisite for thinking clearly about gerrymandering.

Since we're already talking about the holy grail, it's time to introduce our research team's religious mantra:

This is intended both as a reminder and as an exhortation. We *do* have some wisdom about what does and does not work after several years of focused work on this problem, but we do *not* have all the answers about the best ways to apply Markov chain methodologies to our redistricting problems. There are many challenging math problems yet to be tackled—both some relatively low-hanging fruit and some devilishly hard questions—and we hope to provide pointers to researchers looking to enter this area.

Finally, a major goal of this chapter will be to debunk the notion that all computer techniques, or all random map generation techniques, are created equal. In math we like to call a choice "canonical" (echoing religion again) if it is dictated in a standardized and unique way. We'll see that there's very little canon in redistricting and an ineliminable array of modeling choices. We will highlight opportunities for greater community consensus on methods and practices that will promote more consistent, reliable, and repeatable results. Redistricting is a relatively new application domain for these methods and there are many challenging questions at the boundary of current research in this area.

The remainder of this chapter is organized as follows: First, we provide high-level motivation for the focus on sampling rather than optimization. Next, we describe the basic underlying idea of Markov chains, Monte Carlo, and MCMC. Finally, we connect this methodology to its current state-of-the-art applications to court cases and reform efforts, also highlighting some of the exciting new avenues for future work.

A GLOBAL VIEW OF THE LANDSCAPE OF PLANS

The previous chapter gave a great guided tour from the history to the present day in computational redistricting, from punch-card methods in the 1960s through more modern integer programming or power diagrams. Many of the algorithms developed for redistricting operate by attempting to optimize a particular score or metric, but do not aspire to generate representative samples from the enormous space of possible districting plans. Even the methods that include some *stochasticity* (or random steps) mostly do not provide guarantees about the diversity or distribution of plans that are generated.

The last decade's litigation around partisan gerrymanders has spawned a particular type of counterfactual argument based on the neutrality of random maps. Suppose that a randomized algorithm which is not provided with partisan information, constrained only by (some instantiation of) the traditional districting principles, is shown to never, or at least rarely, generate a map whose partisan measurements are as extreme as those in the challenged plan. We are invited to conclude that the challenged plan is an impermissible partisan gerrymander. In order to justify this argument, we must be persuaded that the sampling methodology is generating *representative* districting plans. To see the perils of mistaking random for representative, imagine that I have a favorite districting plan. I can instruct a computer to select one census unit in the plan that is on the border between districts 1 and 2, and to randomly assign that unit either to district one or district two by a coin flip. I then run this algorithm 100 times and, behold! it gives me back 47 plans with the unit assigned to district 1 and 53 with the unit assigned to district 2. It would be obviously unreasonable to conclude anything at all from this highly specialized collection of 100 plans, even though they have indeed been randomly generated by a computer.

This is where Markov chains enter the scene. The thing that differentiates MCMC methods from other algorithms is the explicit focus on a particular *distribution* over all permissible districting plans. Additionally, the ergodic theorem (Sidebar 17.4) states that if we can generate sufficiently many samples, the distributions of statistics that we are interested in will converge to a stable distribution over the full universe of possibilities. This is what makes it reasonable to describe a given plan as a statistical outlier.

Although the application that motivated much of this research developed in the adversarial court setting, recent analyses have used the same technique to evaluate and assist reform efforts. Here the question is not, "Was a specific map drawn with improper purpose?" but rather "How would changing the rules change the underlying distribution?" shifting the evaluation from comparing a single map to a distribution to comparing how distributions result from the design of the rules. This evolution has introduced many new research questions that subtly depend on details of the implementations and methodology.

The underlying premise of both of these research directions—outlier analysis and rule design—is that MCMC can be used to discover neutral baselines of arbitrary metrics across the space of districting plans. Even simply comparing these baselines to each other, across elections or states, is already offering new insights into the geospatial structure of American elections and redistricting. It has also guided understanding of the fundamental properties of the metrics that have been proposed in the past as proxies for good redistricting quality.

To begin, we need to address the following questions:

- What is a districting plan?
- How do we know that a districting plan is permissible?
- How do we know that a districting plan is desirable, or even plausible?
- How do we define a distribution that prioritizes plausible or desirable districting plans?
- How can we sample from such a distribution?

We shouldn't expect punchy, universal answers to these questions. Each state has different rules and laws that govern the redistricting process, as well as different political geography that shapes the landscape of possibilities. This chapter will explore how MCMC sets us up for a promising suite of approaches.

2 INTRODUCTION TO MCMC

Let's dive in with a friendly introduction to the ideas and background of Markov chain sampling on discrete state spaces. Applications to political districting have created a renewed interest in these methods among mathematicians, political scientists, geographers, computer scientists, and legal scholars (among others) and this introduction is aimed at presenting the underlying mathematical material in an intuitive fashion for all of those audiences. The goal is to present the key ideas without the need for a significant amount of mathematical background or formalism. Math-ier information will mostly be in sidebars. For additional tools exploring these ideas see the GitHub repository associated with this book (`https://github.com/political-geometry/`).

17.1 EXPECTATION AND SAMPLING

A *probability distribution* is a function that assigns a probability or likelihood to various events. A *random variable* is a variable whose value is determined as the result of a draw from the distribution. We'll focus on the case that there are finitely many possible outcomes, so that the sum of their probabilities is one. We'll call each outcome a *state* and the universe of possible outcomes the *state space*.

An example is rolling a fair die: the state space is $\{1,2,3,4,5,6\}$ and each value has a 1/6 chance of being on top when the die stops moving. This is an example of a *uniform distribution*, where each outcome has exactly the same probability of occurring. An example of a non-uniform distribution is picking a random letter out of the previous sentence, which gives E and I the highest weight and gives J, Q, Y, Z no weight at all.

The *expected value* of a random variable is a weighted average of the values of the state space: we multiply each value by its probability and add them up. (Notice that the uniform distribution just gives back the usual average.) So for the fair die roll, we get the expectation

$$\mathbb{E}(X) = \frac{1}{6}\cdot 1 + \frac{1}{6}\cdot 2 + \frac{1}{6}\cdot 3 + \frac{1}{6}\cdot 4 + \frac{1}{6}\cdot 5 + \frac{1}{6}\cdot 6 = \frac{21}{6} = 3.5$$

Notice that although we will never actually roll a 3.5 on a six-sided die, it does represent a type of average value: if we rolled the die many times and recorded a large *sample* of random outcomes, the sample average would converge to 3.5.

The same calculation approach applies when the probabilities are not equal, which changes the weights on the values. For example, if we have a loaded die that is weighted so that 3 comes up 3/10 of the time, 4 and 5 come up 1/10 of the time each, and 6 comes up half of the time, then the long-term expectation would be $.3(3) + .1(4) + .1(5) + .5(6) = 4.8$.

One of the fundamental results of all of mathematics is the **Central Limit Theorem**. Suppose a random variable is drawn from a probability distribution with true expectation μ and variance var. It does not matter what the shape of that distribution is! If the variable is sampled independently from that distribution, and we let μ_n be the average value of n observations, then the distribution of μ_n converges to a normal with mean μ and variance var$/n$. This means that if you can only study a random variable by sampling in a "black box" fashion, the experimental evidence helps you to estimate the true expectation. The larger your experiment, the less variance in your estimate.

2.1 MONTE CARLO METHODS

Monte Carlo methods study the aggregate properties of random samples. The origin story for formal Monte Carlo analysis is a deterministic solitaire game played by mathematician Stanislav Ulam while he was sick in bed [1]. ("Deterministic" games are those with no choices to make, such as the game of War—the winner is just determined by the shuffle.) Ulam wanted to figure out how often a randomly shuffled deck would lead to a win. The exact calculation was out of reach, since there are $52! = 80,658,175,170,943,878,571,660,636,856,403,766,975,289,505,$-

440, 883, 277, 824, 000, 000, 000, 000 possible shuffles. But once you've analyzed what makes a winning shuffle, you can have a computer repeatedly carry out a sequence of games and see how often you win. This is exactly the type of task that computers are excellent for, since they execute instructions exactly and do not complain of boredom (or of repetitive strain injury).

The same general outline that we applied here is common to most examples of Monte Carlo methods. In sketch:

1. Draw an (independent) sample from the set of all possibilities;
2. Extract some data for each sample;
3. Repeat many times;
4. Average/aggregate the derived data.

Following this procedure offers a way to generate *approximate solutions to difficult problems* by aggregating a large number of *random solutions to easier problems.*

Like many other elite mathematicians in the 1940s, Ulam was working for the war effort, in his case the Manhattan Project in Los Alamos. Ulam's idea was quickly adopted by others in the project, notably John von Neumann, for modeling the behavior of particles released by subatomic processes. Enormous strides in computing power after the war allowed researchers to run a much larger number of trials than would have been possible by hand and provided access to efficient pseudo-random number generation. In the intervening decades, these methods have been applied to problems in physics, chemistry, and computer science as well as in purely mathematical settings.

2.2 DEFINING A MARKOV CHAIN

A Markov chain is a process that moves from state to state in a randomized way in a state space. Its defining property is that the probability of moving to each state at a certain time is determined by your current position. One example is the children's game Snakes and Ladders, where the probability of landing on a particular square on your turn is completely determined by your current square. This kind of process is also the secret sauce in Google's original PageRank algorithm, which works by estimating the *importance* of a website as the probability that a web-surfer would land there after following totally random links for a long time.

Let's build three simple examples to start to understand this. All of them will be random walks on a space with 27 states consisting of the letters of the alphabet plus a space. (For graph representations of these chains, see Figure 2.)

1. **Alphabet Path**: Only allowed moves are from a letter to the ones before or after it in the alphabet, with SPACE after Z. So from A, your next move is definitely B, but from G, you could move to either F or H with equal chances.
2. **Alphabet Cycle**: Same, but now SPACE is also connected to A. Now every state has two "neighbors" and picks one at random.

3. **Keyboard Walk**: starting with any letter, you are equally likely to move to any of its physical neighbors on a standard (QWERTY-style) keyboard. So from H you can transition to any of Y,G,B,N,J, or U with a probability of 1/6 each, while from Q you are equally likely to transition to A or W.

17.2 MONTE CARLO GEOMETRY

Here is a geometric question that can be tackled with Monte Carlo analysis: What is the expected distance between two points randomly drawn in a unit cube? Although this problem has a mathematical formulation

$$\int_0^1\int_0^1\int_0^1\int_0^1\int_0^1\int_0^1 \sqrt{((x_1-y_1)^2+(x_2-y_2)^2+(x_3-y_3)^2}dx_1dx_2dx_3dy_1dy_2dy_3$$

and a mysterious looking exact solution

$$\frac{4+17\sqrt{2}-6\sqrt{3}+21\log(1+\sqrt{2})+42\log(2+\sqrt{3})-7\pi}{105},$$

this is a perfect problem for trying out the Monte Carlo method. If we sample pairs of points with coordinates uniformly random in $[0,1]$, we can report the average. The first run of 1000 trials gave us about .67122, the second run gave .66921 and the third gave .65919. A run of 1,000,000 trials gave .66157. These are not so far off from the theoretical value of .662959... and would continue to improve with longer runs.

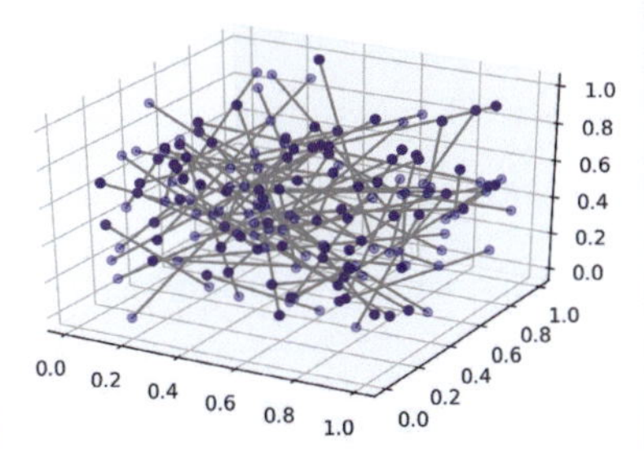

Similarly, it may be hard to visualize a ball of radius 1 in five-dimensional space, but it's easy to estimate its volume! I'll just sample n points $(x_1, x_2, x_3, x_4, x_5)$ by randomizing their coordinates in $[0,1]$ and see what proportion of them satisfy $x_1^2+x_2^2+x_3^2+x_4^2+x_5^2 \le 1$. This is the part of the ball with positive coordinates. Since the coordinates can have any combination of signs, there are $2^5 = 32$ similar sections of the ball, so I can multiply the ratio of hits by 32, and voilà! Turns out this volume is about 5.264. It's integration without integrals.[a]

[a]Fun fact! Dimension 5 is the peak volume for the unit ball. The volume decays to zero faster than exponentially in the dimension n, even though the unit n-ball fits snugly in an n-cube whose volume grows exponentially. This seems totally unreasonable until you think about how unlikely it is that $x_1^2+\cdots+x_n^2 \le 1$ for large n.

17.3 TEXT GENERATION

An early application of Markov chains was in the analysis of text passages, trying to predict the next letter that would appear in a book written by a given author.[a] Symbols in text are not distributed uniformly: for instance, `q` is almost always followed by `u`, periods are followed by spaces, and the letter `e` is most commonly found at the end of a word. Given a long passage of text, we can compute how often each symbol follows each other symbol and use these proportions to generate new text probabilistically.

This is indeed a Markov chain: the probability for choosing the next letter only depends on the current letter. Let's call this the 1DS chain (one-digit sequences). We can similarly define a 2DS chain that takes into account that `Al` is likely to be followed by `ad`, a 3DS chain that sees that `sec` is frequently followed by `ret` or `ond`, and so on. The longer the strings you consider, the more the output looks like language, at least until you try to figure out what it means. The tradeoff is that the size of the transition matrix grows quickly: if there are n characters in the alphabet, then there are n^2 two-digit sequences, n^3 three-digit sequences, and so on.

Below are some examples generated from letter patterns in the story "Aladdin and the Magic Lamp" from the Arabian Nights.[b] Each of these is a single sample path of the Markov chain induced by the letter sequences. The first line is 50 characters chosen uniformly, for comparison, and 0DS generates letters in proportion to their frequency in the text.

(uniform) ,Kni;;.RgkY:f;;.?ACKKDFtjaBD-vjalAezAFO-hOzOe?NAm

(0DS) idaleuiupefeibseauitisavisrogeme,aob,aWtosde

(1DS) y mpo fewathe he m, main, wime touliance handddd

(2DS) If ho rembeautil wind was nearsell ith sins. He don the whimsels hed his the my mign for atim, but

(3DS) but powerful not half-circle he great the say woman, and carriage, she sup window." He said feast father; "I am riding that him the laden, while

(4DS) as he cried the palace displeasant stone came to him that would not said: "Where which was very day carry him a rocs egg, and horseback." Then the might fetched a napkin, which were hunting in the

There turns out to be a significant amount of interesting structure in this type of analysis. The transition matrices alone are often enough to distinguish authors from each other, or poetry from prose (see Figure 1).

[a] Similar methods are used for auto–complete functions on smartphones!

[b] All three texts cited here are available from Project Gutenberg (`https://www.gutenberg.org/`).

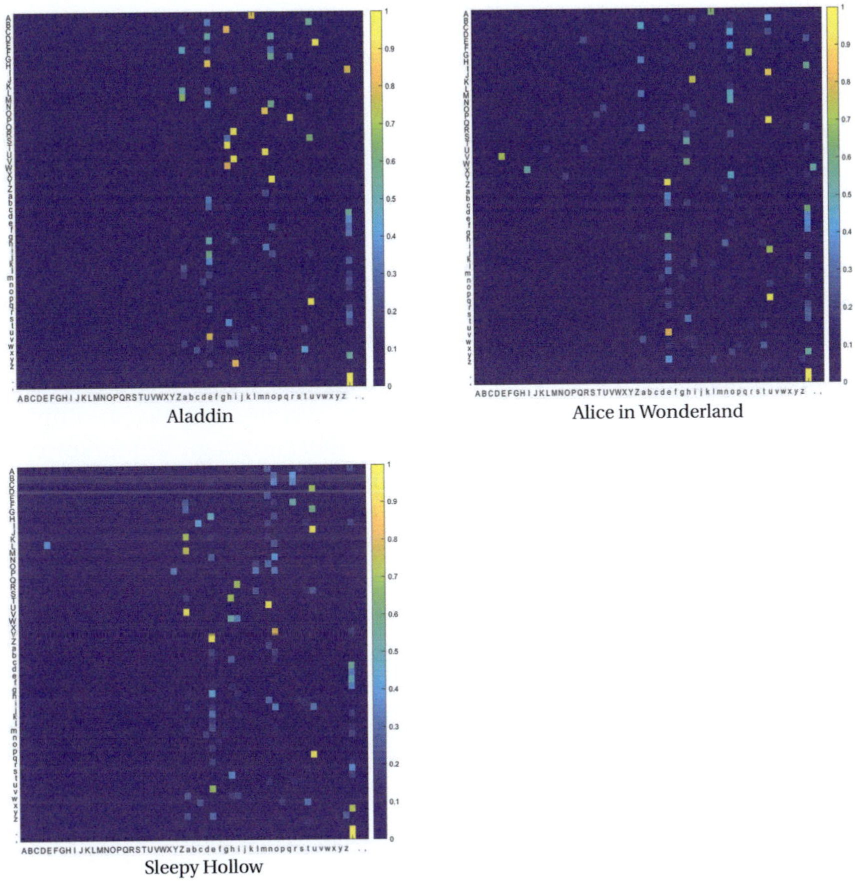

Figure 1: Transition matrices for *Aladdin, Alice in Wonderland,* and *Sleepy Hollow.* Each row and column corresponds to a pair of letters, and the brightness of the matrix at that point captures the likelihood of a transition from the first letter to the next in the text.

2.3 APPROACHING A STEADY STATE

We can visualize the random walk on the state space by imagining a person, or an ant, who is crawling from state to state. If the state space is finite, we can use nodes to represent the states and draw edges to represent the possible transitions; this gives us a graph representation of the random walk, as in Figure 2. If the instructions of the random walk amount to, from any node, choosing from among the incident edges with equal probability, then we call it a *simple random walk*—our alphabet path, alphabet cycle, and keyboard walk are all simple in this sense.

The mathematical formalism gets nicer and more unified if we instead consider the evolution of a *probabilistic* position vector. For instance, if the random walk on the keyboard begins at letter Q, then we can record that one step later, its probability vector has 1/2 weight at W and 1/2 weight at A.

	initial prob.	step 1	step 2
Q	1	0	1/4
W	0	1/2	1/8
A	0	1/2	1/8
Z	0	0	1/8
S	0	0	1/4
E	0	0	1/8

In this way, the probabilities keep diffusing through the state space. (At the next step, nonzero probabilities will expand to X, D, and R.)

Thinking about Markov chains in this probabilistic way allows us to study questions about long-term behavior. In general, dynamical systems can have multiple states that are attractors and others that are repellers. But the magic of Markov chains is that there exists a unique attractor—everything is drawn to it. That is, *for a (suitably designed) Markov chain, any initial position will converge to a unique steady state.* This steady state is also called a *stationary distribution.*

If there are finitely many states, say n, then we can formalize this with an $n \times n$ transition matrix M whose (i, j) entry records the probability of transitioning from state i to state j in one step. Let us call its transpose $P = M^{\mathsf{T}}$ the *iteration matrix* of the system. We call it this because it has a nice property: for a position vector v, the matrix product Pv records the probability of being at each position one step later in the process. So if v is your initial position, then $P^N v$ is a complete description of your position at time N.

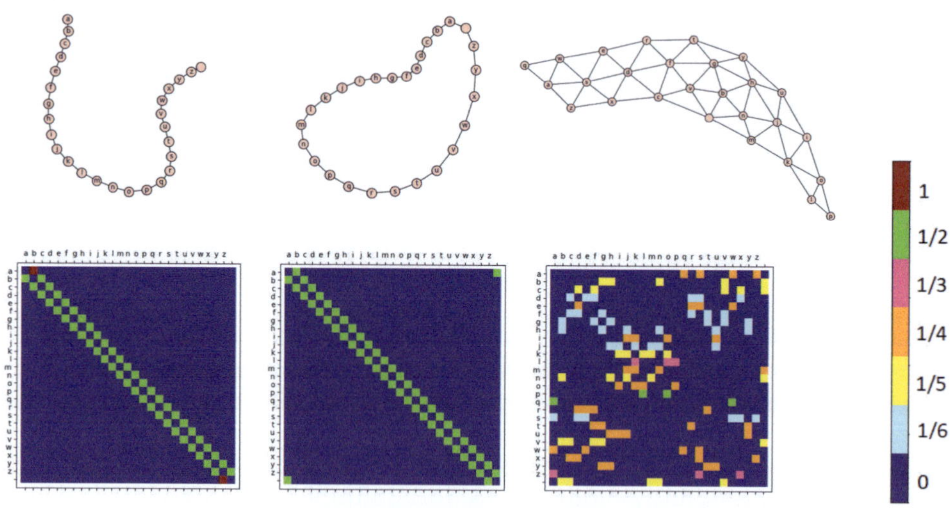

Figure 2: The top row shows the state space as 27 nodes in a graph, with edges for allowed transitions. The bottom row shows the transition matrices in visual format, allowing you to scan the likelihood of going from a letter to any other latter in one step for each of the chains.

17.4 THE FUNDAMENTAL THEOREM OF MARKOV CHAINS

We need to define a few properties of Markov chains to state the fundamental theorem and some surrounding facts. The first adjective we will consider is *periodic*. The period of a Markov chain is the greatest common divisor of all cycle lengths (paths that start and end at the same state). A chain is said to be *aperiodic* if the period is one. Looking at our example chains, we can see that the keyboard is aperiodic (Q–W–Q has length two and Q–W–A–Q has length three, and these have no nontrivial common divisors), and the alphabet cycle is aperiodic (A–B–A has length two and the full tour around the alphabet has length 27), but the alphabet path is periodic because any path starting and ending at the same letter has even length. A common trick to make a walk aperiodic is to add a small probability of remaining in place. This is picturesquely called a "lazy" random walk.

Next, a Markov chain is called *irreducible* if each state can be reached from any other state in a finite number of steps. All of the examples that we have encountered so far have this property. A link-following random walk on the internet does not have this property because some sites have no outgoing links.[a] Markov chains that are both aperiodic and irreducible are called *ergodic*.

A Markov chain is *reversible* if it satisfies a symmetry condition known as "detailed balance." This condition states that in the steady state, the probability of being at state i and transitioning to state j is equal to the probability of being at state j and transitioning to state i. In mathematical notation, if w represents the steady state vector and P the iteration matrix, this condition reads

$$w_i P_{ij} = w_j P_{ji} \quad \forall i, j.$$

The Aladdin text chain in Sidebar 17.3 is an example of a nonreversible chain, since the probability of transitioning from I to A is zero, while the string Aladdin itself shows that the reverse probability is nonzero. Reversible chains have many nice properties and this symmetry condition means that the steady-state distributions are particularly easy to analyze.

Finally, a quick note about measuring success. To say how close one probability distribution is to another, a natural notion is the *total variation* distance between the two measures. Given two distribution vectors u and w, the total variation distance between them is $d_{TV}(u, w) = \frac{1}{2}\sum_i |u_i - w_i|$. This is just adding up the differences in weight over each state in the state space, normalized so that the distance between any two measures is always between zero and one.

Fundamental Theorem of Markov Chains:

1. Any ergodic Markov chain has a unique stationary distribution. That is, if the iteration matrix is P, then there exists a unique probability vector w (entries summing to 1) such that $Pw = w$.
2. For any probability vector v, its iterates converge to w. That is, $d_{TV}(P^N v, w) \to 0$ as $N \to \infty$.
3. Every Markov chain can be represented by a random walk on a graph—possibly

a directed graph with weights on the edges. For the case of simple random walk on a finite graph, the stationary probability of being at position i is proportional to the *degree* of vertex i. That is, where d_i is the number of edges leading to vertex i and $D = \sum_i d_i$, the steady state vector has coordinates $w_i = d_i/D$.

This explains why the steady-state probability is uniform for the alphabet cycle, while the endpoint vertices have half the long-term weight of the others in the alphabet path walk (Figure 2). Note also that it easily follows from the Fundamental Theorem that all simple random walks on undirected graphs (where from each vertex you choose an incident edge with equal probability) are reversible.

Building on this theory, the Markov chain Central Limit Theorem and its various refinements guarantee that given any real-valued function F on our state space, we can estimate its statistics over the state space as a whole by simply collecting samples from a random walk and averaging the values of F on the states in the sample.

This is the sense in which the Markov chain theory is so well suited to redistricting. For years, it has been a burning question to find the normal range of metrics in nongerrymandered plans. If we can find a Markov chain with a suitable steady state, we can use samples to estimate these baselines.

The amount of time that it takes to be guaranteed that $P^N v$ is within a prescribed (total variation) distance of the steady-state w is called the *mixing time* of the Markov chain. There is very beautiful theory when it comes to mixing times, but it is almost never possible to bound mixing times in scientific applications.

To read more, see Levin et al., Aldous and Fill, and Geyer [2, 3, 4].

[a] To make a walk on a finite state space irreducible, one hack is to add a small probability of teleporting anywhere at each step. PageRank works this way.

We can use the three simple chains we introduced in the previous section (the path, cycle, and keyboard walks). Instead of considering a particular sequence of visits to individual states, we instead use the equation above to compute the exact probabilities of arriving at each of the other states. We find that even though the three chains are defined on the same state space, their steady states are different! In the keyboard walk, some states are weighted three times as high as others in the long term, while in the alphabet cycle all states are equally weighted.

2.4 BASELINES WITH MARKOV CHAINS

We can test out the main theorem by seeing how well the Markov chain approximates a numerical "summary score" of the state space. We'll look at two functions from the state space to the real numbers (also called *functionals*).

- **Ascending:** Score is based on position in alphabet.

$$A \mapsto 1,\ B \mapsto 2, \ldots,\ Z \mapsto 26,\ \text{SPACE} \mapsto 27$$

- **Vowel-weighted:** Assign 1 to each consonant, 100 to each vowel, and 50 to Y.

Table 17.1 below compares the theoretical expected values to the estimates obtained from each Markov chain with increasing sample length. All of these runs do a decent job,[1] but we can observe that some seem to converge faster than others, and the rate can depend on the score we choose!

Walk	Score	Experimental				Exact
		2k steps	10k steps	50k steps	100k steps	
Path	Ascending	15.75 (12.5%)	13.99 (0.07%)	14.04 (0.3%)	14.07 (0.5%)	14
	Vowel-Weighted	18.69 (6.6%)	19.70 (1.6%)	19.29 (3.6%)	20.03 (0.07%)	20.02
Cycle	Ascending	14.5 (3.5%)	14.32 (2.3%)	14.1 (0.7%)	13.88 (0.8%)	14
	Vowel-Weighted	21.36 (1.0%)	21.76 (2.9%)	20.97 (0.8%)	21.22 (0.4%)	21.15
Keyboard	Ascending	13.12 (1.2%)	13.34 (0.4%)	13.32 (0.2%)	13.30 (0.05%)	13.292
	Vowel-Weighted	21.02 (9.9%)	19.36 (1.2%)	19.53 (2.1%)	18.91 (1.2%)	19.13

Table 17.1: Experimental comparison for estimating scores. These are independent single runs beginning at the letter A and collecting every score visited by the random walker. Percent error in parentheses.

It is very rare to have rigorous control of the run design needed to get an estimator with provable guarantees. Instead, scientific applications typically use heuristic methods to determine whether or not an estimation has converged.[2]

This example also motivates the use of a common strategy called burning and subsampling. Frequently, practitioners will set a parameter b called "burn-in time" and a second value s called a "sub-sampling parameter." Then, instead of collecting every observed state, a sampling ensemble will be created by collecting states visited at time $b, b+s, b+2s$, and so on. If s is roughly the mixing time of the chain, then these samples will be approximately independent draws from the stationary distribution.[3] For chains where neighbors tend to have similar scores, like the ascending chain, this helps to counter the *auto-correlation*, or degree of similarity from one step to the next, which makes a full sample change its average value more slowly.

We also see a hugely important fact illustrated here that is worth emphasizing: the ground truth itself—what is the average value of the score over the whole universe of possibilities?—depends not only on the state space but also on the probability distribution, because it is a *weighted average*. So, if we're working with the stationary distribution for the keyboard walk, D is weighted three times as much as Q. This will be important below when we turn to redistricting.

[1]If you are following the details, the path walk is periodic. Probabilities proportional to degree is one stationary distribution. Exercise: find them all!

[2]A primary example, sometimes called the *multi-start heuristic*, is to start the chain from different initial states and check that the sample distributions agree. Of course, this can't ensure that you're getting the right answer, but if your multi-start experiment fails, then you can be sure that you're not running long enough.

[3]The use of burn-in in particular is somewhat controversial; see for instance Geyer [4].

2.5 TARGETING A DISTRIBUTION

So far, our examples begin with a process governed by some transition probabilities, then run until they approach stationarity. But in most MCMC applications, we start with a specific distribution we are trying to sample from, then create an irreducible, aperiodic Markov chain designed to target it. The same property that made problems tractable for Monte Carlo analysis—the relative ease of evaluating the properties of a sample rather than the whole—also turns out to be useful for drawing from a target distribution. We can design an appropriate Markov chain knowing only local comparisons for the target.

This was the key idea that was exploited by Metropolis and coauthors in 1953 [5]. As with early Monte Carlo techniques, the original application was to statistical mechanics of atomic particles. This idea was further developed by Hastings [6] and others and has come to be one of the most fundamental computational tools in all of computer science and statistics. In 2000, the IEEE described Metropolis-style MCMC sampling as one of the top 10 most important algorithms of the twentieth century [7].

One situation that calls for this kind of maneuver is when we have a score that makes us regard some states as "better" than others, and we want to prescribe a distribution that prioritizes or up-weights the higher-scoring states. This turns out to be a common situation in physics and Bayesian statistics. We essentially use the score to start with one Markov chain, then design a cleverly weighted coin and use a coin flip to accept or reject each proposed move. When the weighting is just so, it pulls the first Markov chain away from its own steady state and toward the desired distribution.

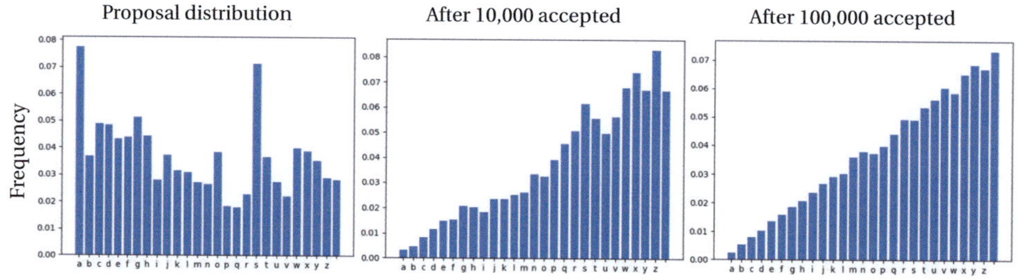

Figure 3: In this run, the proposals are generated according to the keyboard walk, but re-weighted in the Metropolis style to target the ascending distribution. The Metropolis rule is successfully pulling the distribution away from its stationary tendency (left) and toward the ascending shape (right).

This is well illustrated by Figure 3. On its own, the keyboard walk would approach the distribution on the left, but instead it is converging toward the ascending shape. In order to achieve this, many proposed transitions away from the high-scoring letters are rejected, while lower-scoring letters are more readily left behind. Our fidelity to the target distribution improves with longer runs.

17.5 METROPOLIS–HASTINGS

Begin with a score function s on our state space Ω, so that $s:\Omega\to\mathbb{R}$. For instance, the ascending score function has $s(A)=1$, $s(B)=2$, etc. We want to sample from the distribution where the states are weighted in proportion to their scores. For example, if we target the ascending scores, then the letter J (score 10) should be twice as likely as E (score 5). For any state $y\in\Omega$, we should therefore assign it probability $\mathbb{P}(y)=\frac{s(y)}{\sum_{x\in\Omega}s(x)}$. When Ω is too large to construct entirely, we won't be able to compute this denominator. However, notice that we can compute *ratios* of probabilities, since the denominators cancel:

$$\frac{\mathbb{P}(z)}{\mathbb{P}(y)}=\frac{s(z)}{s(y)}.$$

And that's good enough for the re-weighting we need.

We perform the Metropolis–Hastings procedure by beginning with a Markov chain to propose steps, and then using the score function to decide whether to accept them. We use the ratio of the new score to the old score to decide. It is this possibility of remaining in place that transforms the stationary distribution to our desired values.

More formally, we follow this sequence of steps:

1. From an initial state y, generate a proposed state z according to the Markov chain with transition matrix M;
2. Accept z with probability $\alpha=\min\left(1,\frac{s(z)}{s(y)}\frac{M_{zy}}{M_{yz}}\right)$;
3. The next state is z if it was accepted and remains y if not. Repeat.

As you can see, a proposed move to z is likely to be accepted if $s(z)>s(y)$, and unlikely if $s(z)$ is significantly lower. This new Markov chain—which has all the possible transitions of the M chain but re-weighted—is ergodic and reversible with steady-state distribution proportional to s.

2.6 TEMPERATURE VARIATION: EXPLORE AND EXPLOIT

In physics, we often have systems that can be modeled with simple (but very large) state spaces, where we want to explore high-energy and low-energy configurations. For instance we can try to understand magnetic systems, or the chemical structure of glass. Randomized models like the one we describe here have been so successful that they've birthed a whole field, called *statistical physics.*

We'll use the classic *Ising model* to illustrate, with math details set aside to the sidebar. We'll describe system configurations as states σ (appearing as red/blue patterns in our pictures), then define a score $s(\sigma)$ called an "energy" which distinguishes between chaotic and clustered states.

Let's imagine that we want a random sample of clustered states—those where red cells are likely to have other red cells as their neighbors and blue cells are likely to have other blues. And suppose we've set up the energy-based score $s(\sigma)$ to reward this with a higher score for clustered states. How do we sample? For starters, we can use a weighted Metropolis run as described in the last section to try to up-weight states based on s. But this won't work well off the shelf. The Markov chain can get stuck in meta-stable configurations (local optima) that are well-separated from other configurations with similar energy scores, or may even reach global optima that are difficult to escape, which makes it hard to see the diversity of clustered configurations.

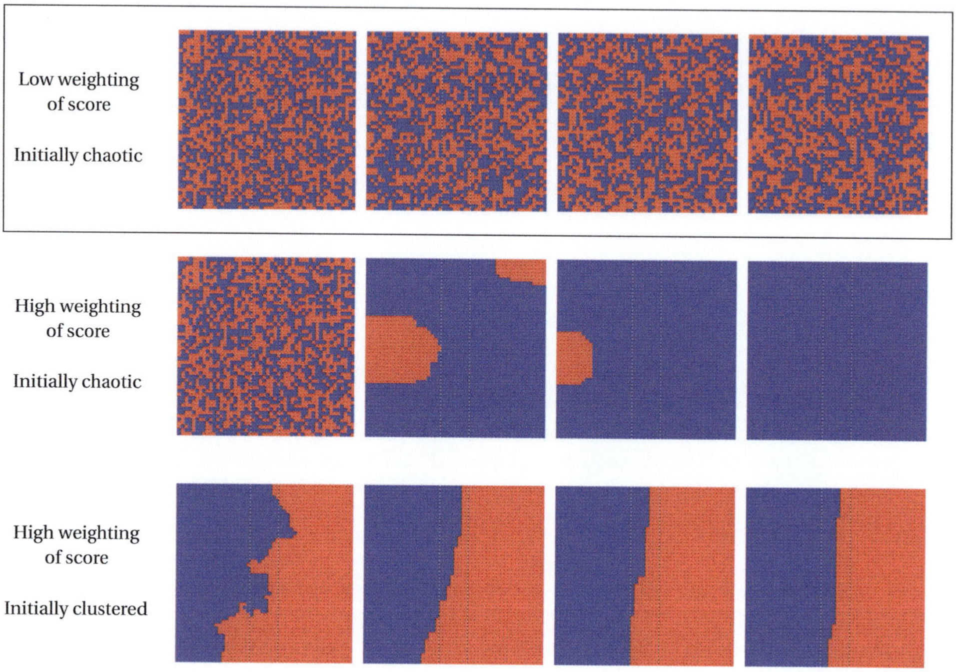

Figure 4: Exploring clustered configurations in a grid with three runs, each at fixed temperature (which controls how the clustering score is weighted during the run). Snapshots are 250,000 steps apart, reading from left to right. These are not efficient strategies for producing a diversity of clustered states.

We see some of the problems in Figure 4. If we run without weighting by the score (first run), clustered configurations are so unlikely that we would not expect to find them at random. But if we run with a high preference for clustering (second and third runs), we will have trouble finding diversity. Notice in particular that the all-red and the all-blue state are both globally optimal for clustering (all neighbors have matching colors), but it would take a truly enormous number of unlikely steps to travel from one to the other while penalizing neighbor differences at every proposed shift. So the second run is stuck at a global optimum, while the third one is stuck in a meta-stable local optimum.

17.6 ISING MODEL: SPECS

The Ising model of ferromagnetism is a mathematical abstraction of a physical system: a network of 'sites,' each of which can be in one of two 'spin states,' usually represented with labels in $\{\pm 1\}$, corresponding to the red and blue colors in the figures here. This is one of the most commonly studied models in statistical physics and also one of the big success stories in the field of MCMC sampling.

Most commonly, the sites are arranged in an $n \times n$ grid; we will denote the assignment of a sign to each node by σ so that σ_i is the spin of node i. Then the 2^{n^2} possible configurations σ make up the states in the state space we will study. Viewing the spin states as magnetic poles that interact with their neighboring sites, we can define an expression called a *Hamiltonian* that represents the total energy of the configuration of spins. In a simple setup this might be written

$$H(\sigma) = -J \sum_{i \sim j} \sigma_i \sigma_j,$$

where $i \sim j$ means that the nodes are adjacent in the grid and J is a term for the interaction strength, here assumed to be constant over the grid. This is designed to distinguish between various kinds of spatial arrangement: if the σ_i are random, the sum will have many positive and many negative terms and will often cancel down to near zero. If the $+1$ and the -1 nodes are highly clustered, most terms in the sum will be $+1$; if they are in a checkerboard pattern the terms will be -1.

The goal is then to sample from a probability distribution over the states given by setting $\mathbb{P}_\beta(\sigma)$ proportional to $e^{-\beta H(\sigma)}$, where $\beta \geq 0$ is a parameter called the inverse temperature. For high values of β (low temperature), this will put a lot of weight on the configurations σ with a large negative $H(\sigma)$, which corresponds to clustering when $J > 0$; on the other hand, for β near zero (high temperature), the probability will be near-uniform.The sum $Z(\beta) = \sum_\sigma e^{-\beta H(\sigma)}$ is sometimes called the *partition function*, which then allows us to write $\mathbb{P}_\beta(\sigma) = \frac{e^{-\beta H(\sigma)}}{Z(\beta)}$.

In his Ph.D. thesis, Ernst Ising solved the one-dimensional model (i.e., on a path graph) exactly, showing that correlations between spins decay exponentially with the distance between the sites. In higher dimensions, we get a more interesting model, finding a sharp *phase transition* as the inverse temperature value β varies. That is, at a critical value of β, we observe a sudden shift between complex, disorganized states for small β and structured, clustered states for large β. Exact solutions are not known for these cases but this is a perfect setting for exploring with MCMC.

Directly sampling from $\mathbb{P}_\beta$ is challenging for high β, despite the fact that it is easy at $\beta = 0$ by assigning the spin at each site uniformly. Instead, we will start with an arbitrary assignment and use MCMC to construct samples from the desired distribution. To move between states, we define transitions where at each step we propose to flip the assignment of a single, randomly chosen node to the opposite sign—this is called *Glauber dynamics*. Following the Metropolis–Hastings procedure with fixed β, we would accept a proposed transition from σ to τ with probability equal to $e^{-\beta(H(\tau)-H(\sigma))}$. The examples in this section illustrate that we get superior results with temperature variation than by running either hot (here, at inverse temperature $\beta = .1$) or cool ($\beta = 3$) alone.

A technique called *simulated annealing* was developed to deal with this sort of phenomenon—its name is motivated by the physical process of heating and re-cooling metal to change its structure. Following this analogy, we will introduce a *temperature* parameter such that steps are more wild and random at high temperature, then settle into aggressive score optimization at low temperature. In the *cooling* regime, we pay more attention to the relative differences between plans, demanding that almost every accepted state must have a higher score than the previous one. Sometimes this distinction in behavior is referred to as "explore/exploit" since running hot lets us *explore* the state space more freely, while cooling then forces us to stay in a smaller neighborhood of the best thing we've found recently, thereby exploiting the high score locally.

The temperature variation over time, also called the *annealing schedule*, is set before the run, prescribing cycles of heating and cooling. Let's see an example of annealing in action, again with the goal of viewing a diversity of clustered states.

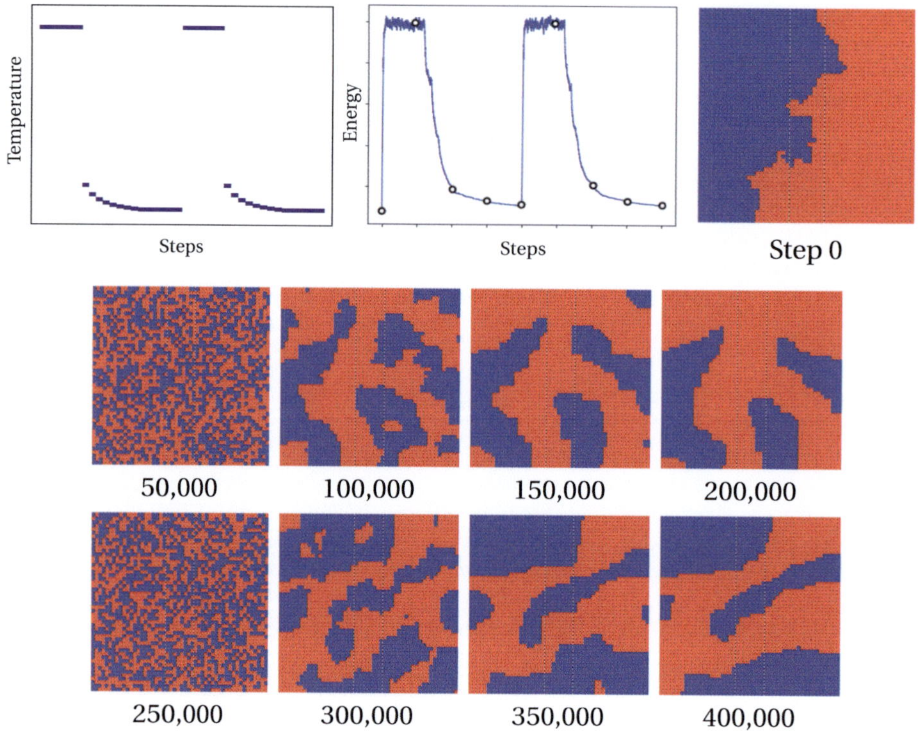

Figure 5: The annealing schedule is shown at the top, with timestamps marked every 50,000 steps and corresponding snapshots below. Each cycle of heating and cooling lets us reset with uniform sampling, then settle into a different clustered configuration by increasingly weighting the score as the temperature drops.

Figure 5 shows annealing performing exactly as advertised. Note that we were able to move between qualitatively distinct configurations in a relatively small number of steps compared to the fixed-temperature runs above.

However, as with many of the methods discussed in this chapter, setting an annealing schedule requires choices and does not come with a canonical strategy. We will return to this idea in the redistricting application below.

3 MCMC FOR REDISTRICTING

We now turn our attention to the application of MCMC methods for political redistricting. First, we describe a formalization of the district-drawing problem in the language of *graph theory*, and then discuss possible formalizations of rules and laws around redistricting. We will introduce several styles of MCMC sampling and survey the state of the art.

3.1 GRAPH PARTITIONS

Although many people think of gerrymandering in terms of lines or curvy boundaries drawn on a map, political redistricting is naturally modeled as a discrete problem where a collection of separate units, like census blocks or voting precincts, are partitioned into districts. This fits the real-world problem, as the Census Bureau reports population and demographic data at the level of census blocks and most states report election results aggregated at the level of precincts. In virtually all cases, you can regard a plan as being built out of census blocks, in the sense that it does not split them.[4] And in many states (like Massachusetts, Louisiana, or Minnesota), state or local redistricting plans are built out of whole precincts. This viewpoint allows us to study redistricting as a discrete problem, using the MCMC tools introduced above.

The object to partition will be a *dual graph* of the chosen units covering the state, which represents each individual unit with a node and places an edge between two nodes if they are adjacent.[5] For instance, Figure 6 shows the counties of Arkansas and the corresponding dual graph. There are 75 counties in Arkansas; on the other hand, there are 186,211 census blocks. In reality, Arkansas's four Congressional districts are built of these smaller pieces. This gives a sense of the gigantic scale of the computational problem and why sampling-based procedures have become so important in this area.

As we've seen, the first step toward defining a Markov chain is to identify the state space. Once we've fixed a dual graph, we'll let the states in our state space be districting plans, i.e., partitions of the dual graph. By a partition, we mean a division of the vertices into groups, which in this case are the districts of the plan. So the partition of Arkansas counties in the last picture of Figure 6 is one state in a space consisting of many trillions of possible plans. Our random walks will wander from one plan to the next. In the next section, we will discuss how to decide whether a partition constitutes a *valid* plan.

---

[4]In fact, the official description of the districts available from the Census Bureau is given by a *block assignment file*, a table mapping the individual census blocks to their assigned districts.

[5]You have to make a decision about whether to include corner adjacencies. And when you're doing this on real data, you also have to make decisions about what counts as being adjacent across water—for instance, what are islands next to?

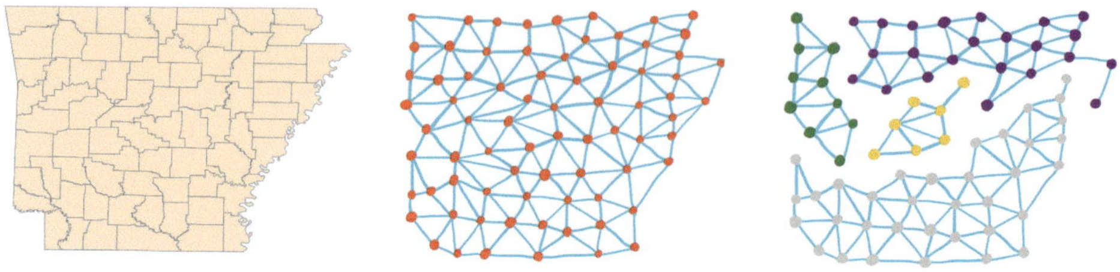

Figure 6: The 75 counties of Arkansas (left), the corresponding dual graph (center), and a districting plan (right).

3.2 DEFINING VALID PLANS

Clearly, many graph partitions do not correspond to reasonable districting plans. Unfortunately, there are far more poorly behaved partitions than reasonable ones. In order to address this issue, we need to enforce some constraints that cut down our state space. The purpose of this section is not to describe the perfect state space for all redistricting problems—no such thing exists!—but rather to highlight the decisions that must be made so that the Markov chain samples are producing reasonable plans for comparisons.

We begin by discussing some commonly enforced *traditional districting principles.* (See Sidebar 0.2.) We'll touch on contiguity, population balance, compactness, county splitting, communities of interest (COI), and the Voting Rights Act (VRA).

To get an algorithm to take an idea into account, you must *operationalize* it, or render it in a formulation that can be handled by a computer. This is one of the steps that is easy to take for granted when making a mathematical model, but the devil is often in these details. Let's just give some examples of operationalizing the rules, which we can illustrate on our Arkansas plan from above.

- Contiguity: the pieces of the partition (the districts) are connected subgraphs.
- Population balance: for some threshold ϵ, each district has a total population between $(1-\epsilon)$ and $(1+\epsilon)$ times the ideal size. (For instance if we set $\epsilon = .05$, then the top to bottom deviation is no more than 10% of ideal.)
- Compactness: the number of edges that were cut to break up the graph into pieces is no more than a threshold.
- County splitting: no more than a threshold number of counties is split between multiple districts.

For most of these, our Arkansas plan from Figure 6 sails past the validity check: the districts are connected, the number of cut edges is just 44 (which is pretty good for this particular dual graph), and the number of split counties is zero (since the building blocks are counties).[6] But the population balance is not great: since it's

[6]Unfortunately, this is not always compatible with prioritizing the preservation of city boundaries, because plenty of cities, including in Arkansas, belong to more than one county.

made out of relatively large pieces (counties), each district is within 10% of ideal but not close to usual Congressional standards of balance.

The COI and VRA criteria are quite a bit harder to handle. One main obstacle to operationalizing COI is that almost no states have a concrete process for official recognition of qualifying communities. If you had those, with shapefiles that tell you their boundaries, then you could handle them with splitting rules like for counties or cities. But another fundamental obstacle is that it's not even clear if most places would prefer to handle COI quantitatively or qualitatively in the first place. (See Chapter 12.)

As for the VRA, the law around its invocation is so complex that it's fairly daunting to incorporate into a mathematical model. In particular, it is a widespread misunderstanding that the VRA requires a certain number of majority-minority districts; instead, it calls for the creation of districts in which minority communities have an opportunity to elect candidates of choice. This can often be roughly gauged by the share of population that belongs to a minority group, but this is not enough to ensure compliance. Nonetheless, several approaches are possible. One thing to keep in mind is that Markov chains will collect plans whose principal intended use is for *comparison*, not for enactment. If you have a quantitative approach to estimating whether a district will be effective, then a reasonable strategy would be to use the number of effective districts as a VRA validity proxy. We'll discuss an approach to gauging effective districts below in Section 3.4.

3.3 FLIP CHAINS FOR REDISTRICTING

In our introduction to Markov chains using the letters of the alphabet, we specified the chain by defining the transition probabilities between each pair of letters. Unfortunately, there are far too many partitions of a state-sized dual graph for us to attempt to compute or store all of the necessary probabilities. Instead, we can specify a Markov chain by describing a set of *elementary moves* that we can apply to a given state in order to generate proposed neighbors.

One theme we will encounter is that even when it is easy to describe a move, it is costly to computations that depend on all the neighbors and the neighbors' neighbors. For instance, let's do a very natural move: start with a plan (say the four-district Arkansas plan in Figure 6), pick a random node, and try to reassign that node a random color. For each of the 75 nodes, there are three new colors to try, so that's $75 \cdot 3$ possibilities. Most of the proposed changes will break contiguity. In an 18-district Congressional plan for Pennsylvania, built out of precincts, there are $9000 \cdot 17$ naive neighbors. What we'll see is that it's easier to try a move and then check validity rather than pre-computing the valid neighbors from each position and choosing among them. Trying and sometimes failing is called *rejection sampling*, and it can be quite efficient as long as the check is quick and the rejection rate is not too high.

DEFINING A FLIP

The very most natural thing to do, especially considering the great successes in the Ising model, is to flip a single node at a time. We'll call this a *flip* walk. This type of proposal has some computational advantages, in that it is easy to iteratively update the computations of score functions and to keep track of the set of nodes that can potentially be changed at each step while preserving contiguity. Figure 7 shows an example of this proposal, where one of the nodes on the boundary between two districts changes its assignment.[7]

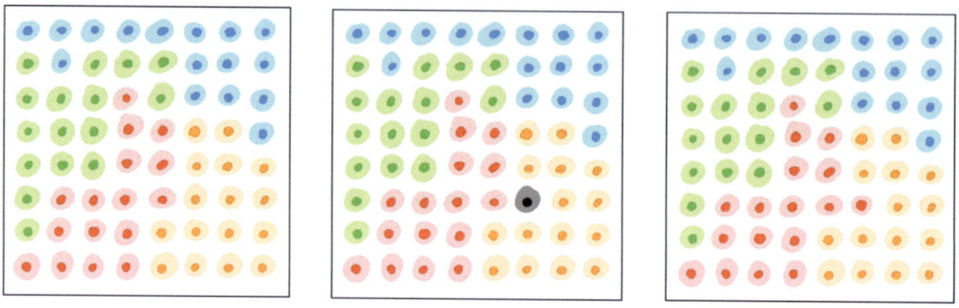

Figure 7: The basic flip step: one unit flips from one district to another.

Even this simple-sounding flip proposal can be implemented in subtly different ways. One option would be to select an edge whose endpoints are in different districts and randomly change the assignment of one endpoint (also at random) to match the other. Alternatively, a boundary node could be chosen and its assignment changed to match one of its neighbors at random. Another version might instead keep track of all of the (node, district) pairs that could be changed to remain contiguous and sample uniformly from that set. Exercise for the reader: confirm that these proposals do not have the same steady-state distribution!

Depending on the formulation of the state space, implementations of the flip proposal can suffer from pathological behavior without careful tuning [8] and in particular have a preference for non-compact plans. Additionally, as only a single node changes assignment at each step, Markov chains using this proposal can require an enormous number of steps to construct approximately uncorrelated samples. In Najt et al. [9], explicit families of graphs were constructed where this proposal exhibits slow mixing. This does not mean that the proposal is wrong for all applications, simply that care must be taken in choosing the sampling methodology and the parameters of the walk to generate useful samples.

TARGETING AND ACCELERATING FLIP CHAINS

Because simply running a flip chain on its own would take astronomically long to converge, and would draw from an undesirable distribution, it seems very natural

[7]One subtlety: if we re-assign a boundary node, we can be sure that the new district that node joins is connected, but it may disconnect its old district by its removal. This makes it slightly trickier to count the neighboring partitions.

to use some of the techniques from the last section to *target* a different distribution of your choice, and to *accelerate* the progress.

At first, it may seem reasonable to target the uniform distribution, equally weighting all plans that pass validity checks. But there are major reasons not to do this. First, this distribution is even more undesirable than one that flip began with, in terms of being overpowered by the least compact plans. Second, the complexity obstructions that suggest very slow convergence for flip chains also apply to the uniform distribution.[8]

Instead, inspired once again by the physics examples, we can target a distribution that is proportional to some score of quality.

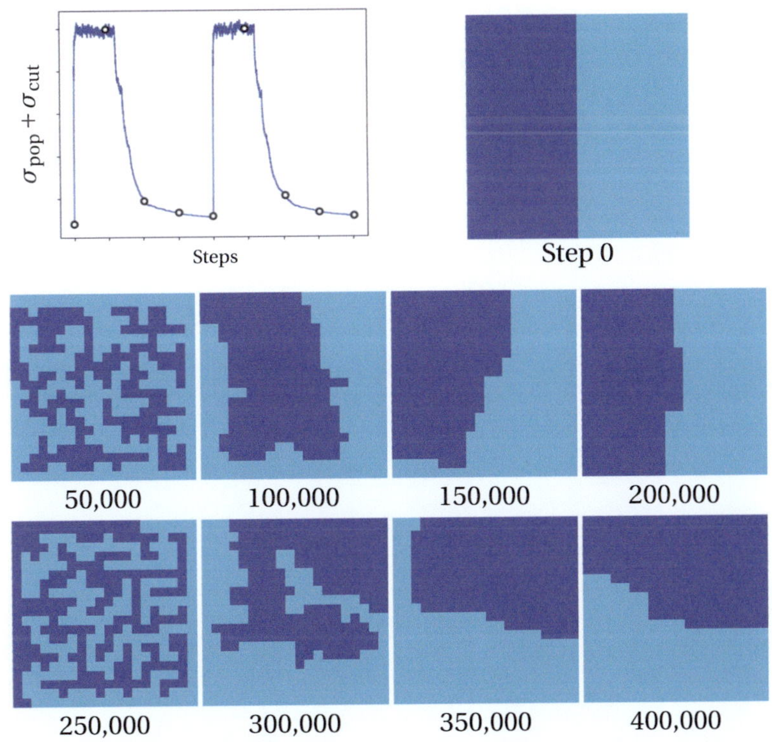

Figure 8: A success for simulated annealing: we sample contiguous two-district plans for a 20×20 grid using a flip walk, with a score $s(P)$ that combines population balance and compactness. The annealing schedule is reflected in the energy trace shown at the top, with timestamps marked every 50,000 steps and corresponding snapshots below. The high-energy states look "fractal," but this level of cooling is successful at recovering short boundaries. Even better, we are sometimes able to traverse from one side of the state space to the other, moving from the vertical split to the horizontal split.

Choosing an appropriate score function for plans, like defining the state space and selecting a proposal distribution, is not a problem with a definitively correct answer. We've already seen that operationalizing the criteria is slippery and subtle.

[8]Some authors, like Fifield et al. [10], first collect a sample with one method and then try to re-weight it after the fact to approximate the uniform distribution. This does not circumvent the complexity obstructions and is likely instead to give poor summary statistics.

To make matters worse, we will now need to combine all of those elements into a single numerical value to serve as the "energy function," summarizing all of the relevant properties of a given plan. This is usually done by a linear combination of several metrics. But choosing the coefficients requires not only deciding on relative importance, but also contending with different units of measurement. How much additional leeway should we permit a plan in population balance in order to make it more compact? A responsible modeler will not only justify these choices, but will also offer a robustness analysis showing that the answers produced by the model are not very sensitive to these decisions.

Let's test out the physics approach in a simple districting application that partitions graphs into $k = 2$ parts. As an example of an energy function we will consider both population balance and compactness. Given a partition $P = (A, B)$ into districts A and B, we set $\sigma_{\text{pop}}(P) = \big||A| - |B|\big|$ to be the difference in the sizes of the districts. (In a grid, the size of a district is just the number of nodes; in a dual graph, it is the population.) Sampling proportional to $e^{-\sigma_{\text{pop}}(P)}$ means that a plan with exactly balanced populations should be drawn with approximately $e^{10} \approx 22,026$ times the chance as a plan where one district has 10 more nodes than the other. Next, we'll use the number of cut edges as a compactness proxy. Given a partition $P = (A, B)$ we set $\sigma_{\text{cut}}(P)$ to be the number of cut edges and then can combine the scores into the score function

$$s(P) = e^{-\left(\sigma_{\text{pop}}(P) + \sigma_{\text{cut}}(P)\right)}.$$

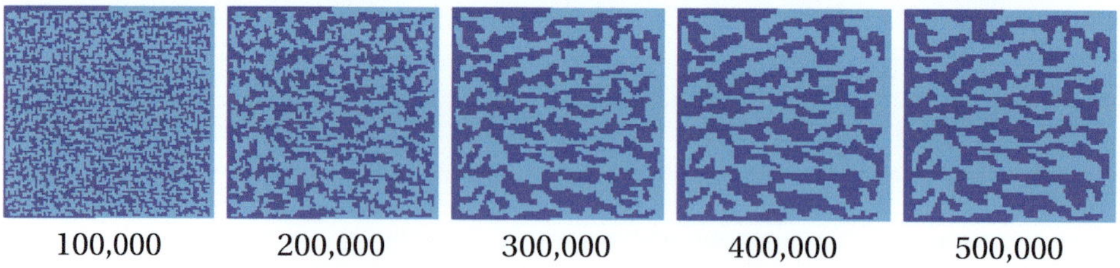

Figure 9: On the 100×100 grid, it is much less obvious how to design a successful annealing cycle. Even after heating and cooling, the boundary assignments have barely changed, and further cooling may not succeed in a reasonable time. A different energy function might be needed.

This example highlights some of the difficulties in making principled decisions about this type of sampling. In a real-world redistricting scenario, designing useful score functions is not a simple task.

Next, having selected a distribution that we would *like* to converge to, we are left with the problem of how to make sure that we actually get there, overcoming any bottlenecks in the state space. So it is natural to try temperature variations to

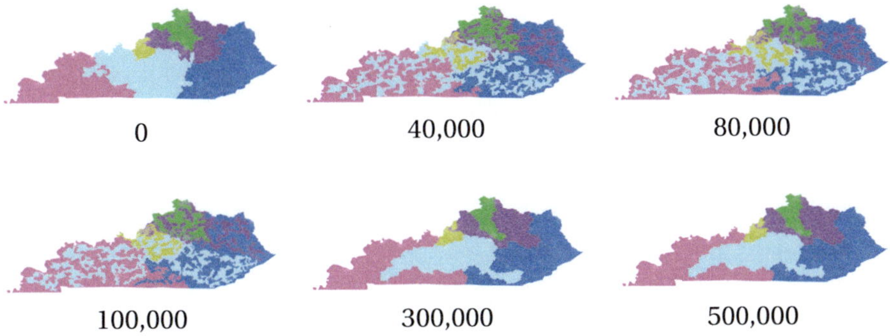

Figure 10: Kentucky's 3,285 block groups tell a similar story to the large grid in Figure 9. This time the cooling has nearly succeeded in returning to a plan as compact as the original, but we can see that annealing has failed to effect a major change. The boundary assignments are still stubbornly persistent.

achieve diversity and accelerate convergence.

Temperature strategies for redistricting flip chains turn out to encounter major problems that were not present in the motivating examples from statistical physics. Like before, the space of plans is huge. But this time, unlike the physics examples, we will need to get lucky enough to select a huge number of changes in a particular order to avoid breaking contiguity—and this is true at every temperature. Second, since the flip procedure itself inclines toward a distribution that is highly noncompact, any compactness preference we implement will be fighting directly against the tendencies of the proposal distribution. Finally, adding cutoff constraints to limit the noncompactness or the population deviation can disconnect the state space entirely: if the limits are too tight, it is easy to construct examples of partitions that cannot be reached from each other using this procedure.

Figures 8, 9, and 10 show annealing runs on a small grid, a larger grid, and the block groups of Kentucky. (For more extensive examples, see DeFord et al. and Najt et al. [8, 9].) Naive annealing was quite successful on the 20×20, but the problems described above began to have real bite once the number of units got to the thousands. One lesson to draw is that it is dangerous to validate techniques on small examples only, since much of the difficulty only kicks in at scale.

CASE STUDY: FLIP WALKS IN NORTH CAROLINA

To see how all these techniques can be combined, let's look at the work of Duke mathematician Jonathan Mattingly and his team, the Quantifying Gerrymandering Group. They participated in federal and state litigation in North Carolina as well as providing model analysis on other states such as Wisconsin.

Both federal and state courts have found their methodology to be persuasive and it was part of the basis of the invalidation by the state Supreme Court of the NC legislative plans. It formed a fundamental part of the evidence before the U.S. Supreme Court in Rucho v. Common Cause (2019). Here we focus on the methodology as presented in Herschlag et al. [11], which analyzed Congressional districts and parallels Mattingly's expert report in that case.

The dual graph was constructed from 2692 Census Voting Districts (VTDs), which approximate the precincts in the state. The score function has terms that relate to the North-Carolina-specific districting criteria. In particular, NC has a very strong rule requiring the preservation of counties, and it also has a significant Black population, triggering VRA scrutiny for Congressional districts.

The score terms are:

- $\sigma_{\text{pop}} = \sqrt{\sum_i (p_i - I)^2}/I$, where I is the ideal population of a district and p_i is the population of district i. This is zero if every district is exactly the ideal size.
- $\sigma_{\text{compact}} = \sum_i P_i^2/A_i$, where A_i and P_i are the area and perimeter of district i respectively. This is minimized when the districts are nearly round, which would give an ideal value of $4\pi \approx 12.6$ for any particular district.
- $\sigma_{\text{county}} = f(C_2) + M \cdot f(C_3)$, where C_2 is the set of counties belonging to two districts and C_3 is the set of counties belonging to three or more districts, and f is a function rigged to report 0 if and only if the set is empty. The authors say that M is a large constant but do not report its value.[9]
- $\sigma_{\text{VRA}} = \sqrt{\min(0, 44.48 - B_1)} + \sqrt{\min(0, 36.2 - B_2)}$, where B_1 and B_2 are the highest and second-highest percentages of Black population in any district. This score is zero if and only if $B_1 \geq 44.48$ and $B_2 \geq 36.2$, which are values obtained from an existing map that was approved by a court.[10]

Putting it all together, they attempt to sample proportional to the score function

$$s = e^{-\beta\cdot\left(3000\,\sigma_{\text{pop}} + 2.5\,\sigma_{\text{compact}} + .4\,\sigma_{\text{county}} + 800\,\sigma_{\text{VRA}}\right)}.$$

Here we start to see the dizzying array of choices that go into an analysis like this. Why is the population score weighted 1200 times as heavily as the compactness score and 7500 times as heavily as the county score? Let's continue to describe the setup and hold that question for a discussion of the *robustness* of the findings.

Proposal generation. Select a cut edge uniformly; change the assignment of one of its endpoints to match the other with probability $\frac{1}{2}$.

[9] Suppose county i is in two districts and it has s_i share of its geographical units in the district with the largest share. If it is in three or more districts, let s_i be its share of units in the two districts with the largest share. Then, Mattingly's function is $f(C) = |C| \cdot \sum_C \sqrt{1 - s_i}$. Note that $f(C_2) \leq |C_2|^2 \cdot \sqrt{1/2}$ and $f(C_3) \leq |C_3|^2 \cdot \sqrt{1/3}$, with slowly decaying penalties for cutting off smaller pieces. This is related to an entropy score, but with a number of ad hoc customizations. See Chapter 14 for a discussion of how to use entropy to measure county splitting.

[10] To be clear, this is a massive shortcut to the VRA. There is no basis in law for requiring maps to retain the demographic percentages of an existing map. It may nonetheless be passable for a court if the ensemble is used for comparisons only, and regarded as containing plans that took the VRA into account rather than plans that are certified compliant.

Acceptance probability. Automatically reject discontiguous proposals. Accept contiguous proposals with the Metropolis probability associated with the score s.

Annealing schedule. Initialize temperature parameter β at 0 for the first 40,000 steps, then gradually increase to $\beta = 1$ over the next 60,000 steps. Take an additional 20,000 steps with $\beta = 1$ and then add the final map to the ensemble. This means that a total of 120,000 flips have been proposed between maps in the ensemble.

Winnowing. Remove all plans with population deviation greater than 1%, compactness score of any district worse than 60, any county split four or more ways, or African-American population share falling below $B_1 = 40$ or $B_2 = 33.5$. In their experiment, about one-sixth of the generated plans survived the winnowing step.

Using all of these settings, they generated an ensemble of 24,518 North Carolina districting plans made out of VTDs. Then they compared the distribution of partisan statistics over the ensemble to the statistics observed in the enacted plans from 2012 and 2016, using various recent elections for the voting baseline.

They found that the enacted plans display extreme behavior favoring Republicans, whether measured with partisan bias, efficiency gap, the number of seats won by each party, or a variety of new metrics they devise. (See Chapter 2 for an overview of partisan metrics.) By contrast, there is a plan proposed by a bipartisan panel of retired judges, built to model the work of an independent commission. The judges' plan performs well in line with their ensemble of neutrally generated alternatives. (See Figure 12 for some of the Duke output, together with a replication study.)

To account for the dozens of detailed choices that went into this approach, the authors offer several convergence heuristics and sensitivity analyses to argue that the analysis is robust to the arbitrary choices in its setup. For instance, they tried exchanging their Polsby-Popper compactness score for an alternative dispersion-based compactness score or changing the coefficients in the score function and found that their bottom-line results were qualitatively similar. Any approach with so many choices to make must contend with worries about gameability, so a suite of strong robustness checks of this kind is needed to raise our confidence in the reliability and replicability of this kind of analysis.[11]

3.4 RECOMBINATION

We've seen that flip-based walks can be quite powerful in the redistricting application, but that they are subtle to manage in terms of the centrality of user-chosen specifications. Our research group, the Metric Geometry and Gerrymandering Group (MGGG), based at Tufts and MIT, has spent several years refining a markedly different approach, surveyed in DeFord et al. [8]. It revolves around a fundamental graph theory concept called a *spanning tree*.

[11]The Duke team did not publicly share the code used in this study and report, but a second-generation package is available if you'd like to try your hand at sensitivity analysis. You can find materials in Greg Herschlag's git repo at `https://git.math.duke.edu/gitlab/gjh`.

17.7 "CAREFUL CRAFTING": A LOCAL TEST

A completely different—and very elegant—use of Markov chains has also been developed for redistricting applications, proposed by Chikina–Frieze–Pegden (CFP) [12]. This approach was applied by Pegden (and later by Duchin) in the Pennsylvania Supreme Court litigation over the Congressional map. The theory is developed further in Chikina et al. [13], and earlier work revolving around the same ideas can be found in Besag and Clifford [14].

Suppose your state space is Ω and you fix any functional $f:\Omega \to \mathbb{R}$, any reversible Markov chain, and any value $0<\epsilon<1$. Now suppose that a sequence of consecutive states visited by the chain is

$$P_0, P_1, P_2, \ldots, P_N.$$

Then we can consider the set of scores $\{f(P_i)\}$ observed over that sample. The theorem states that the probability that $f(P_0)$ is in the most extreme ϵ fraction of the $\{f(P_i)\}$ is at most $\sqrt{2\epsilon}$. Notice that this result does not require any statement about the convergence of the chain, only that the proposal is reversible. It applies to very short runs as well as long runs, but provides a weaker conclusion. And it also does not require ergodicity—the state space need not be connected, and the theorem applies just as much when the sample is collected from a small connected component, but again with a possibly weaker conclusion.

This suggests a rigorous gerrymandering test, which is very powerful because it does not require a demonstration of ergodicity or convergence. Start a Markov chain at a given plan and let it run for a large number of steps. If the initial state scores worse than the vast bulk of observed variants, then you can be extremely confident that it was not chosen from the stationary distribution!

For the Pennsylvania example described in the CFP paper, the dual graph is made out of the approximately 9000 precincts in Pennsylvania and the Markov chain is a "lazy" flip run designed to have a uniform steady state. In Pegden's report, he didn't use scores for weighting but instead thresholded the various measurable criteria (county splitting, population deviation, compactness) to be in a reasonable range, which amounts to a uniform walk on a restricted state space. Unlike the score-based approach above, the chain is not up-weighting plans that are better aligned with the districting principles. It means all allowable plans are equally likely in the steady state.

This is exactly what is needed to apply the CFP theorem and test. In the paper, the null hypothesis that the enacted plan was drawn from the uniform distribution is rejected with p-values between .0001 and .0000001 depending on the chosen partisan metric, using chains of approximately a trillion steps. And Pegden's expert report found even more eye-popping p-values in litigation. That means you can be very confident that the enacted plan wasn't chosen uniformly at random because its partisan behavior is a great deal more Republican-favoring than the other plans that were found by the chain.

But so what? After all, the Republican legislators never claimed that they were choosing a plan blindly from all the possibilities, and humans are terrible at imitating uniform distributions even when they try. In order for this to have strong persuasive power, you'd want to be sure that this test doesn't merely tell you that a plan was made by people rather than a computer! In Duchin's use of the CFP test while consulting for the Pennsylvania governor, she included evidence that a plan constructed by the governor's map-making team (without her involvement) *passed* the test—it had partisan properties typical of the observations over long chains. On the other hand, a new compact plan created by members of the legislature failed just as badly as the original it was vying to replace.

Doing this kind of double-check—making sure you have not inadvertently set up a test that only computers can pass—is essential for rolling out this test on a wider scale, as is more study of its gameability.[a] Nonetheless, this way of arguing that a plan has been "carefully crafted" to be much more favorable to the party that controlled the process than a great bulk of similar plans has an unmistakable appeal, and has been found persuasive by the court in Pennsylvania, and later in North Carolina.

[a]This is particularly true since later improvements [13] have massively strengthened the sensitivity of the test by upgrading from a p-value on the order of $\sqrt{\epsilon}$ to one on the order of ϵ, heightening worries about false positives.

A recombination step will typically change the assignment of many nodes at once, which allows for far quicker traversal of the state space. At each step, two districts are merged, forming a graph of twice the desired size. Next, a spanning tree is chosen for that graph. Then, we seek an edge in the tree that we'll call a *balance edge*: when we cut it, the two new pieces that are formed should have equal size. Replacing the two merged districts with these two formed by cutting the tree gives us a new districting plan. This process is cartooned in Figure 11.

As with the flip walk, recombination admits many variants. For example, there are multiple ways to construct random spanning trees of the subgraph and multiple ways to seek and select a balance edge to cut. We could also merge more than two districts at a time (though at the possible cost of more complexity in the partition step). In the bigger picture, there is no need to use spanning trees at all, as any method for partitioning a merged subgraph could be used to generate the next plan. Nonetheless, we will stick with two-district-at-a-time spanning-tree-based methods for the rest of this exposition, and we will call that Markov chain ReCom.

ReCom has three major features that differentiate it from flip chains. One is that there is far less autocorrelation from one plan to the next, which promotes faster convergence and avoids some of the rigidity that made physics-motivated techniques less effective for flip steps. Second, it scales well with the size of the graph being partitioned, because its spanning tree step has polynomial complexity and the number of steps needed to touch all districts is proportional to the number of districts, not the number of units. Finally, it does not need careful weighting to obtain reasonably compact plans.

17.8 SPANNING TREES

As we've seen, a graph is a collection of vertices, together with edges that join some of them pairwise. A *tree* is just a graph with no cycles: there is no edge path that starts and ends at the same vertex without backtracking. And so for any graph, you can create a *spanning tree*—a tree that covers all of the vertices—just by removing edges that appear in cycles until there are none left.

To illustrate this, consider the 3×2 grid graph. It has seven edges, and you can make a spanning tree by removing any two of them while being careful not to disconnect the graph. There are exactly fifteen ways to do this:

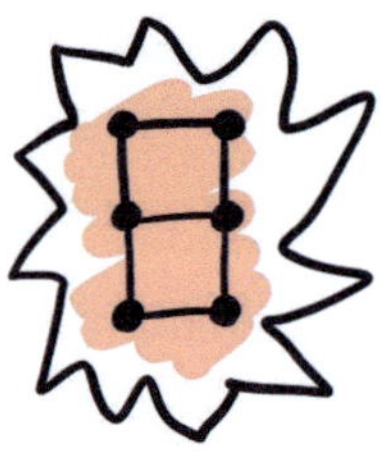

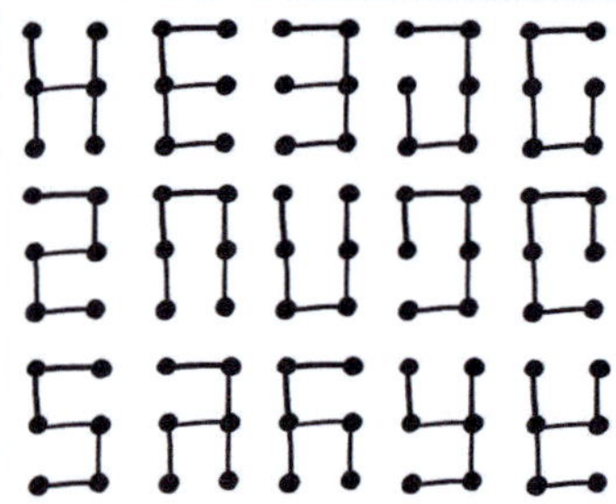

For a 3×3 grid graph, there are 192 possible spanning trees and for a 4×4 there are 100,352—the count grows fast! Let's define $\mathrm{sp}(G)$ to be the number of spanning trees of a graph G. There is a beautiful little formula that counts them for you. It is attributable to the physicist Gustav Kirchhoff, as part of his study of electrical circuits in the nineteenth century. First, form the $n \times n$ *graph Laplacian* L by putting the n vertex degrees on the diagonal and subtracting off the adjacency matrix of the graph. This matrix L encodes all sorts of fundamental information about our graph. If G is connected, then $\mathrm{sp}(G)$ is just $1/n$ times the product of the nonzero eigenvalues of L. You can compute this straight from L by eliminating any row and corresponding column to form an $(n-1) \times (n-1)$ minor, then taking its determinant! For our 3×2 example, we get

$$L = \begin{pmatrix} 2 & -1 & -1 & 0 & 0 & 0 \\ -1 & 2 & 0 & -1 & 0 & 0 \\ -1 & 0 & 3 & -1 & -1 & 0 \\ 0 & 0 & -1 & 0 & 2 & -1 \\ 0 & 0 & -1 & 0 & 2 & -1 \\ 0 & 0 & 0 & -1 & -1 & 2 \end{pmatrix}; \qquad \mathrm{sp}(G) = \det \begin{pmatrix} 2 & 0 & -1 & 0 & 0 \\ 0 & 3 & -1 & -1 & 0 \\ 0 & -1 & 0 & 2 & -1 \\ 0 & -1 & 0 & 2 & -1 \\ 0 & 0 & -1 & -1 & 2 \end{pmatrix} = 15.$$

For us, spanning trees will be a crucial device for partitioning because of a key feature: *if you cut any single edge of a tree, you have divided the graph into exactly two parts.*

One last thing to know about spanning trees before we move on: there are remarkably efficient algorithms for generating them randomly! In particular, Wilson's algorithm (based on loop-erased random walk) can be used to get near-uniform sampling of all the spanning trees of G in polynomial time. So when we need to find a spanning tree as a step in a recombination algorithm, we can usually do it fairly fast, even for large graphs.

Now let's consider what the spanning tree count sp(G) can be said to measure about a graph G. One thing to note is that sp(G) = 1 if and only if G is a tree. This means that a path has only one spanning tree, no matter how long it is. But on the other hand, the number of spanning trees of a grid-graph grows explosively. It's not hard to convince yourself that the spanning tree count is greater when a graph is "plumper," and that it's reduced dramatically by "tentacles" or "necks." From this point of view, it's reasonable to treat sp(G) as a compactness score for the graph!

To read more about the basics of spanning trees, most introductory texts in computer science or combinatorics cover properties and elementary algorithms, such as Cameron [15]. To read more about what spanning trees might have to do with compactness, check out Duchin and Tenner [16].

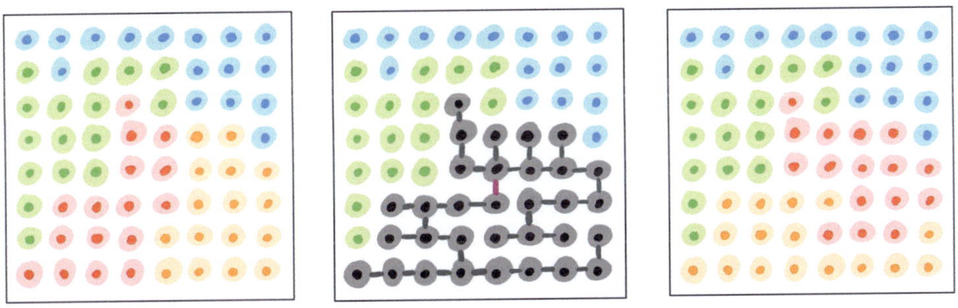

Figure 11: The basic recombination step: two districts are merged, a spanning tree is chosen, a balance edge is selected and cut, leaving two new districts.

Recently, Cannon et al. introduced a reversible variant of ReCom with a prescribed stationary distribution [17]—that is, we know exactly how much some plans are weighted relative to the others. It's even very easy to write down that stationary distribution in closed form. Recall from Sidebar 17.8 that sp(G) is the number of spanning trees of a graph G, which can be regarded as a kind of compactness score for the graph. Suppose a districting plan P is composed of districts $P_1, \ldots, P_k$. Cannon et al. [17] show that the stationary distribution of reversible ReCom puts a weight on P that is precisely proportional to $\prod_{i=1}^{k} \mathrm{sp}(P_i)$, the product of the spanning tree counts of its districts. That means that plans are naturally weighted by compactness!

Even though "regular" ReCom does not have exactly this stationary distribution, it draws a similar distribution of plans so that it creates compact ensembles without any tuning, and case studies have found extremely fast convergence in summary statistics.

RECOM CASE STUDIES

We will briefly describe two case studies with ReCom chains: a run on Congressional districts in North Carolina and another on state House districts in Virginia. The North Carolina study was focused on the partisan gerrymandering case Rucho v. Common Cause, intended to demonstrate whether different Markov chain methods might hope to give similar answers. And the Virginia study was directed at analyzing a racial gerrymandering case, Bethune-Hill vs. Virginia State Board of Elections (2019), which revolved around the manipulation of Black population in the state's legislative districts.

Let's start with North Carolina. There, we convened a team of mathematicians and law scholars to write a "friend of the court" brief aimed at helping the Supreme Court to understand the recent Markov chain breakthroughs.[12]

We aimed to see how a partisan-neutral ensemble of compact, contiguous, population-balanced plans would compare to the carefully tuned Duke output. To do this, we used precincts as the building blocks and ran a ReCom chain for 100,000 steps, allowing maximal population deviation of 2% from ideal.[13] That's it! In a few hours on a standard laptop, we get a large and diverse collection of plans.

Figure 12 shows that this very simple run gave outputs that are remarkably consonant with the Duke ensemble. In both, at least 50% of plans have Democrats winning districts indexed 9 through 13, which means 5 seats out of 13, and roughly 25% of plans have Democrats winning 6 seats. By contrast, the plans enacted in 2012 and 2016 had only 3 seats for Democrats in this vote pattern, which is in line with the notoriously brazen assertion by David Lewis, that he had commissioned map locking in a 10–3 Republican advantage only because he couldn't find a way to get an 11th seat.

We highlight this comparison because it is encouraging to see that two very different Markov chain methods give harmonious answers. To see more about North Carolina ensembles and the effects of layering in various districting criteria, visit the GitHub repo for this chapter [18].

Next we turn to Virginia's 100-seat House of Delegates, the subject of the long-running lawsuit Bethune-Hill v. Virginia State Board of Elections. Blocks, rather than precincts, were the natural choice of geographic units to analyze Virginia, because their House plans do not in fact keep precincts whole (and are only allowed 1% population deviation)—just as importantly, vote totals at the precinct level are less salient in a racial gerrymandering case. Thus, the dual graph of Virginia has 285,762 nodes, which is far out of range to expect good performance from a flip chain.

In this study, we used a ReCom ensemble to shed some light on VRA litigation. As we've heard, VRA law is very tricky, because it centers on the creation of *effective* districts for the minority group, in this case Black voters, to elect candidates of choice. To do that requires an estimate of voters' preferences by racial group, which

[12]Fully, it's the *Amicus brief of mathematicians, law professors, and students in support of appellees and affirmance.* As a historical note, we believe it to be the first Mathematicians' Brief (so named) for the Court. There was a Statisticians' Brief in Gonzalez v. Planned Parenthood (2007) arguing that the government had misrepresented p-values in its arguments about late-term abortion. And there was a Computer Scientists' Brief in Lotus v. Borland (1996) weighing in on whether certain elements of computer interfaces were more like languages or functions, for copyright purposes.

[13]For partisan cases, we consider it highly valuable to use precise cast vote totals, which means that precincts are the smallest usable unit. Since those are bigger than blocks, we will typically allow 1–2% population deviation in order to have the Markov chain move efficiently. A professional mapmaker can easily refine such a plan to zero balance.

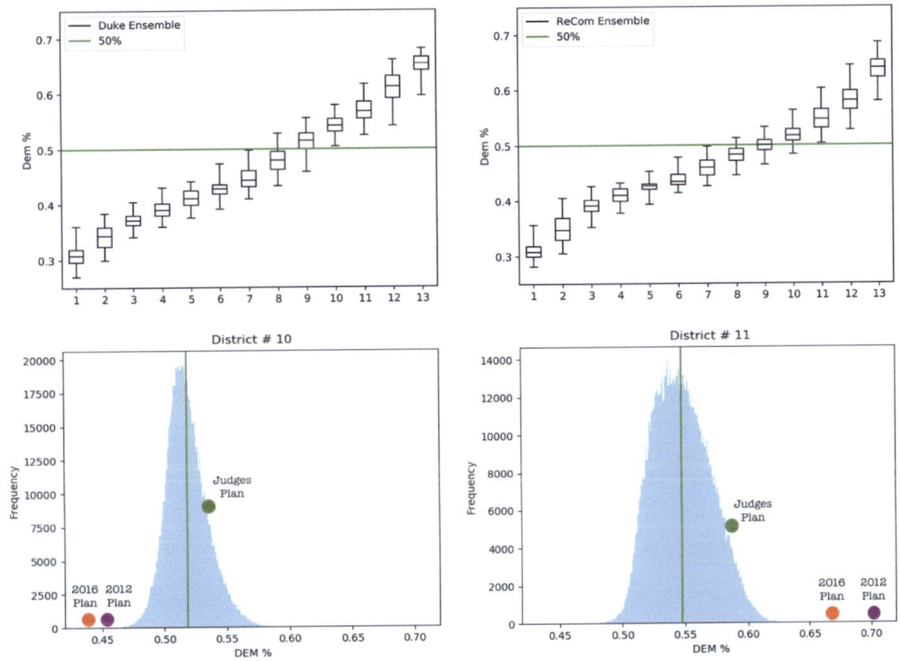

Figure 12: Images from the Mathematicians' Brief. Fixing a vote pattern (Senate 2016), we order the districts in each plan from the smallest (1) to the highest (13) Democratic share of the two-party vote. The boxes show the 25th–75th percentile vote share observed in the ensemble, and the whiskers show 1st–99th percentile. Top row shows the Duke ensemble compared to an untuned ReCom ensemble: not identical, but substantially similar. Bottom row highlights the districts indexed 10 and 11, and shows blatant packing and cracking in the legislatures' plans compared to the ReCom ensemble, or the bipartisan judges' plan. (The ensemble mean is marked with a line in the bottom row.)

is not immediate in a system with a secret ballot. The state of the art for racially polarized voting analysis is a method called "ecological inference" or EI; Bethune-Hill plaintiffs' expert Max Palmer performed EI in all of the house districts of the state, finding that every challenged district was expected to favor a candidate of choice for the Black community in a general election as long as its electorate was at least 45% Black by voting age population (BVAP) [19]. In fact, by Palmer's methods, just 37-38% BVAP would suffice in all but one district. None of the districts requires 55% BVAP, so we treated the range of 38–55% BVAP as a critical one for plausibly effective districts.

How many districts can be simultaneously created that are over 50% BVAP? How about over 37% BVAP? We created an ensemble of alternative plans that are compact, contiguous, and population-balanced to within the 1% deviation limit prescribed in Virginia law. We found that just taking 20,000 accepted ReCom steps and recording every single plan that was encountered gave statistics that were consistent across different starting points or further run lengthening.

We found that hundreds of plans in our race-neutral ensemble had 15 districts in

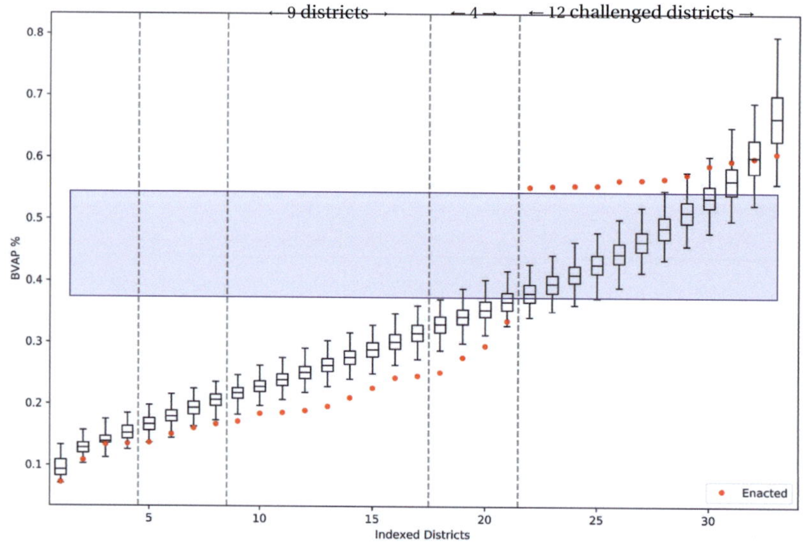

Figure 13: Black voting age percentage (BVAP) by district in an ensemble covering the region of Virginia affected by the Bethune-Hill lawsuit. Red shows the levels in the challenged plan. Boxes show the 25th–75th percentile of the ensemble and whiskers show 1st–99th percentile, as before. We see that packing in the 12 challenged districts has led to cracking in the next 4, and even the following 9.

our plausible range of effectiveness, as opposed to 12 districts in the Republican legislators' plan.... and 13 in the Democratic counter-proposal. So *both* sides are leaving opportunity districts on the table. More than that, the analysis reveals the costs of packing in the plan that was challenged by the lawsuit: the elevation of Black population in the first 12 districts is balanced out by depressed Black population that is not evenly distributed over the other districts, but instead concentrated in the ones with the highest prospects for Black voting power, alone or in coalition with other groups (see Figure 13).

These examples illustrate that ReCom lets you generate a large collection of plans with far less user choice (parameter-tuning, temperature manipulation). It puts Felix Frankfurter's haunting challenge—finding the neutral baseline—within reach of laptop computing.

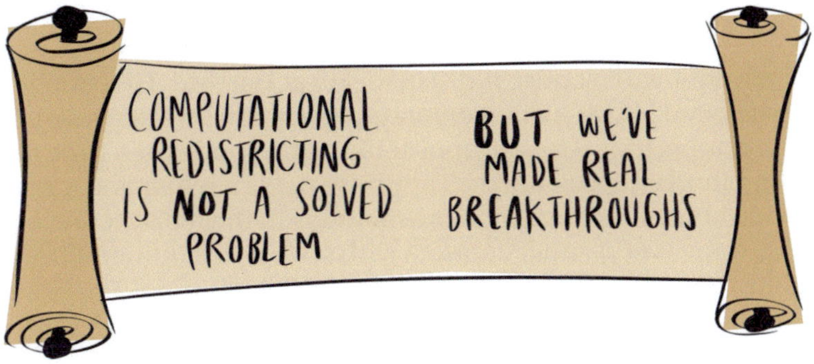

3.5 SURVEY OF OTHER SAMPLING APPROACHES

There are numerous other district-generation methods out there with various pros and cons. In particular, three political scientists have developed notable sampling methods, which we'll briefly describe.

Jowei Chen (University of Michigan) uses an agglomerative algorithm that is based on iterative merging (see Chapter 16), and has made numerous court appearances based on generated ensembles, typically containing 100–200 maps.[14] His method is intuitive to describe: growing and merging regions amoeba-style until they cover the space with the correct number of districts. These techniques are likely to be useful in the future for finding plans that can be used as starting points in various ways: for initializing a random walk, or as a jumping-off point for the deliberative work of a commission. On the other hand, the method comes with no control or description of the sample distribution, and so provides no grounding for statistical claims. Due to the high rejection rate, agglomerative techniques have difficulty generating diverse ensembles and sometimes have difficulty generating large ensembles at all. Given these various limitations, agglomerative methods are likely to have worryingly high false-positive rates if used for outlier analysis.

Wendy Cho leads a team at the University of Illinois, with supercomputing expert Yan Liu as a major collaborator. Theirs is an evolutionary algorithm that uses a flip step most of the time, then occasionally applies a crossover step based on the common refinement of two partitions. The main upside is that, as discussed in Chapter 16, evolutionary algorithms can do an excellent job of heuristic optimization—that is, they can find "good" plans—and managing multiple populations lets you successfully take advantage of many computing cores in parallel, so the algorithm runs very fast. On the other hand, you lose touch with the theory guaranteeing convergence to a steady state, so it's not clear how the quickly proliferating plans are distributed. Furthermore, the parallelization is carefully engineered for the Blue Waters computing environment (a research supercomputer at University of

[14] Even if drawing from a well-justified probability distribution, this very small sample sizes make unlikely events look impossible. For instance, if something occurs 1% of the time, there is a greater than one in three chance that it will be entirely absent from a collection of 100 maps. ($.99^{100} \approx .366$)

Illinois Urbana-Champaign) and the code is not public, making it quite hard to draw comparisons between this and other methods.

Kosuke Imai is a political scientist and statistician at Harvard. His team, like Duke's, does physics-inspired MCMC with temperature variation (in their case, a technique called parallel tempering). Their proposal flips a few nodes at a time rather than one, providing a modest acceleration. They have made a major and commendable effort to provide benchmarking for all the methods in the redistricting community by developing some (very) small datasets with complete enumeration.[15] Equally commendable: they make their code publicly available!

4 EXPLORING WITH ENSEMBLES

4.1 NOT JUST FOR LITIGATION!

The method of ensembles has many applications for redistricting analysis and reform beyond the adversarial setting of challenging plans in court.

Criteria tradeoffs. The Virginia study described above mainly relied on "vanilla" ensembles, made only with compactness, contiguity, and population balance. In DeFord and Duchin [21], we layer in many other criteria—tighter population balance, preference for keeping cities and counties intact, attention to voting rights share—to see how they interact. We found no basis for some "folk knowledge" that was circulating as Virginia considered a constitutional amendment for redistricting reform, such as the idea that requiring higher compactness would hurt minority representation or that keeping cities intact would favor Republicans. In fact, the preservation of cities and counties had the effect of narrowing observed partisan outcomes, reducing the frequency of maps that had the greatest advantage for either party.

Partisan metrics. Sometimes metrics are designed to measure one thing, but end up being sensitive to factors other than the ones that are advertised. For instance, Chapter 2 shows that the efficiency gap, advertised as a measure of packing and cracking, actually only depends on how many seats are won by each side. You can similarly study other metrics like partisan symmetry scores to see how they behave when tested on real data and many thousands of plausible districting plans [22]. We show that the partisan symmetry standard has many bugs in practice, including systematically reporting advantage for the wrong party in some realistic cases. (We dub this the "Utah Paradox"!)

Nesting. Alaska law requires the 40 House districts in their state legislature to nest 2-to-1 within the 20 Senate districts. (Nine other states had similar nesting rules in the last redistricting cycle.) Suppose you were handed the current House map and required to come up with a pairing of adjacent House districts. We found

[15] Of particular mathematical interest is their approach to approximate enumeration using a data structure called Zero-suppressed Binary Decision Diagram [20].

that there are 108,765 possible matchings in Alaska, which is an imperceptible sliver compared with the usual size of a redistricting problem. Nevertheless, we found that the ability to choose the matching gives almost as much control of the partisan outcome as if one were drawing the map from scratch [23].

Competitiveness. Here, ensemble methods are used to study a range of possible rules for promoting competitiveness in districting. Using two methods—winnowing to the most competitive plans in a neutral ensemble and hill-climbing with flip steps to preferentially create more competitive plans—we show that quantitatively prescriptive language that has been appearing in recent reform measures may be ineffective or generate unintended effects in the future [24].[16]

Least change. What if you have a reason to want a map that makes the least change from a previous one, such as under a rule favoring the preservation of district cores? Mattingly et al. [11] consider this with an experiment using local sampling. Short runs are carried out, rejecting proposals that have more than 40 nodes assigned to different districts than in the initial plan. In Chapter 14 an approach to this, by introducing an entropy-based metric on the space of plans, is also given.

Coalitions and alternative voting systems. Finally, what if you are not sure your community is well served by districts at all? We use single-member districts by law at the Congressional level, but counties and cities have much more latitude to design a system of election. MGGG has carried out tailored studies that look at Asian communities in Santa Clara, CA [26], Latino and Asian communities in Lowell, MA [27], and Black and Latino communities in Chicago, IL [28]. In all three cases, we recommended serious consideration for ranked choice voting in multi-member districts (see Chapter 20 and Chapter 21), finding that it performs at least as well as the districting options found by algorithmic means, but with enhanced opportunities for coalitional representation.

4.2 AN INVITATION

Beyond the Markov chain methods already developed, there is still significant room for creativity in designing elementary moves on graph partitions. New methods should strive for easy implementation, low rejection rate, adaptability to varied districting criteria, and of course theoretical properties like provable ergodicity or reversibility. Although probably extremely difficult, it would also be very valuable to prove mixing time bounds for any of the graph partition methods that are currently in use, where many problems are even open for grid graphs. Lower bounds are useful because they warn of likely failure of certain approaches at large scale; upper bounds would give better statistical guarantees.

With respect to the basic spanning tree ReCom proposal, there are many natural questions about the combinatorics. (What share of trees have a cut that partitions

[16] The districts found by very short and simple hill-climbing runs were similar to extremely competitive plans hand-made for the 538 Atlas of Redistricting project [25].

the nodes equally or near-equally? What kinds of planar graphs have the most spanning trees? and so on.)

Often, algorithms that have poor worst-case performance can still be efficient on nice classes of graphs. Moving beyond grid graphs and other lattices, it would be fundamentally interesting to understand the properties of the graphs realized as dual to Census and precinct geography.[17] This would be useful not only for complexity analysis of algorithms, but also to answer basic intriguing questions like whether different states and places have any artifacts of planning policy visible in the graphs themselves—or even within a state, whether cities are detectable from the abstract dual graph alone.

There would be immediate applications for efficient multi-resolution partitioning methods that start with larger units before refining with smaller sub-units. Prioritizing preservation of counties, cities, or COI could benefit from multi-resolution mapmaking, and a block-level tuning step could provide population balance at the end of a mapmaking process rather than foregrounding it at the beginning.

Last but not least, stability and robustness—demonstrating that consonant results are obtained within and across techniques as user choices vary—are paramount concerns as the toolkit continues to expand. As we have seen, there are many setup decisions that must be made in order to design a sampling algorithm for a specific case or state. It is essential to measure the sensitivity of results to model design, tuning, and data perturbation. Determining conditions for reliable and robust findings is perhaps the most important open question in this space.

5 CONCLUSION: STILL NOT A SOLVED PROBLEM

The overarching goal of MCMC methods for redistricting is to generate ensembles of alternative maps that can put a proposed plan in context of the full universe of possibilities. Setting this up requires hard work to certify that we are drawing samples according to a clear rule for weighting some more highly than others. In order to avoid making a test that only computers can pass, we need to know that our ensembles are sufficiently diverse to account for the many ways that benign but unspoken principles can make people's maps different in subtle ways from computer outputs. Ground truth is hard to come by in redistricting, but we should take it where we can, and use it to calibrate our tests.

Redistricting pushes mathematical knowledge to the research frontier, but the scientific consensus is crystallizing around powerful and efficient methods of analysis. Or in other words...

[17] In many cases, the dual graphs are planar and mostly triangulated (as in Figure 6), but when the units are disconnected (as precincts quite often are) the combinatorics can get substantially worse.

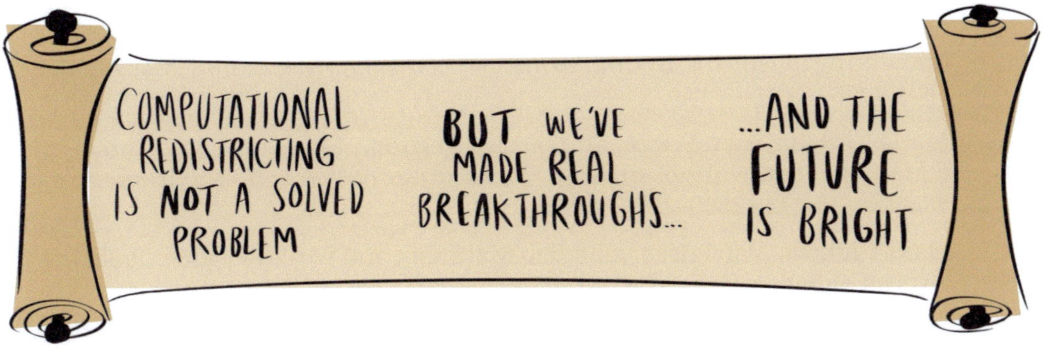

REFERENCES

[1] Robert Eckhardt. Stan Ulam, John Von Neumann, and the Monte Carlo Method. *Los Alamos Science*, pages 131–136, 1987.

[2] David A. Levin, Yuval Peres, and Elizabeth L. Wilmer. *Markov Chains and Mixing Times.* American Mathematical Society, Providence, R.I, December 2008.

[3] David Aldous and James Allen Fill. Reversible Markov chains and random walks on graphs, 2002. Unfinished monograph, recompiled 2014, available at http://www.stat.berkeley.edu/~aldous/RWG/book.html.

[4] Charles J. Geyer. Introduction to Markov chain Monte Carlo. In *Handbook of Markov chain Monte Carlo*, Chapman & Hall/CRC Handb. Mod. Stat. Methods, pages 3–48. CRC Press, Boca Raton, FL, 2011.

[5] N. Metropolis, A.W. Rosenbluth, M.N. Rosenbluth, A.H. Teller, and E. Teller. Equations of state calculations by fast computing machines. *Journal of Chemical Physics*, 21:1087–1092, 1953.

[6] W. K. Hastings. Monte Carlo sampling methods using Markov chains and their applications. *Biometrika*, 57(1):97–109, 1970.

[7] F. Sullivan and I. Beichl. The Metropolis Algorithm. *Computing in Science & Engineering*, 2(1):65–69, 2000.

[8] Daryl DeFord, Moon Duchin, and Justin Solomon. Recombination: A family of Markov chains for redistricting. *arXiv:1911.05725*, 2019.

[9] Elle Najt, Daryl DeFord, and Justin Solomon. Complexity of sampling connected graph partitions. *arXiv:1908.08881*, 2019.

[10] Benjamin Fifield, Michael Higgins, Kosuke Imai, and Alexander Tarr. A New Automated Redistricting Simulator Using Markov Chain Monte Carlo. *Working paper: Princeton University*, page 55.

[11] Gregory Herschlag, Han Sung Kang, Justin Luo, Christy Vaughn Graves, Sachet Bangia, Robert Ravier, and Jonathan C. Mattingly. Quantifying Gerrymandering in North Carolina. *arXiv:1801.03783 [physics, stat]*, 2018. arXiv: 1801.03783.

[12] Maria Chikina, Alan Frieze, and Wesley Pegden. Assessing significance in a Markov chain without mixing. *Proceedings of the National Academy of Sciences*, 114(11):2860–2864, 2017.

[13] Maria Chikina, Alan Frieze, Jonathan Mattingly, and Wesley Pegden. Practical tests for significance in Markov chains. *arXiv:1904.04052*, 2019.

[14] Julian Besag and Peter Clifford. Generalized Monte Carlo significance tests. *Biometrika*, 76(4):633–642, 1989.

[15] Peter J. Cameron. *Combinatorics: Topics, Techniques, Algorithms*. Cambridge University Press, 1994.

[16] Moon Duchin and Bridget Tenner. Discrete geometry for electoral geography. *arXiv:1808.05860*, 2018.

[17] Sarah Cannon, Moon Duchin, Dana Randall, and Parker Rule. A reversible recombination chain for graph partitions. *preprint*, 2020.

[18] Daryl DeFord and Moon Duchin. Chapter 12 repo. *GitHub repository*, 2021.

[19] Max Palmer. Expert Report for *bethune-hill v. virginia state board of elections*, 2017.

[20] Benjamin Fifield, Kosuke Imai, Jun Kawahara, and Christophe Kenny. The essential role of empirical validation in legislative redistricting simulation. *preprint*, 2019.

[21] Daryl DeFord and Moon Duchin. Redistricting reform in Virginia: Districting criteria in context. *Virginia Policy Review*, 12(2):120–146, 2019.

[22] Daryl DeFord, Natasha Dhamankar, Moon Duchin, Varun Gupta, Mackenzie McPike, Gabe Schoenbach, and Ki Wan Sim. Implementing partisan symmetry: Problems and paradoxes. *Political Analysis*, to appear, 2021.

[23] Sophia Caldera, Daryl DeFord, Moon Duchin, Sam Gutekunst, and Cara Nix. Mathematics of nested districts: The case of Alaska. *Preprint: https://mggg.org/uploads/Alaska.pdf*, 2019.

[24] Daryl DeFord, Moon Duchin, and Justin Solomon. A Computational Approach to Measuring Vote Elasticity and Competitiveness. *Preprint*, pages 1–30, 2019.

[25] Wasserman D. Bycoffe A., Koeze E. and Wolfe J. The atlas of redistricting, 2018. `https://projects.fivethirtyeight.com/redistricting-maps/`.

[26] Mira Bernstein, Moon Duchin, Tommy Ratcliff, and Stephanie Somersille. Study of voting systems for Santa Clara, California. *MGGG Technical Report*, pages 1–16, 2018.

[27] Ruth Buck, Moon Duchin, Dara Gold, and JN Matthews. Community-Centered Redistricting in Lowell, Massachusetts. *MGGG Technical Report*, pages 1–8, 2020.

[28] Hakeem Angulu, Ruth Buck, Daryl DeFord, Moon Duchin, Howard Fain, Max Hully, Maira Khan, Zach Schutzman, and Oliver York. Study of Reform Proposals for Chicago City Council. *MGGG Technical Report*, pages 1–31, 2020.

V

On the ground

Chapter 18

Making maps: A practitioner's perspective

MEGAN GALL, KARIN MAC DONALD AND FRED MCBRIDE

CHAPTER SUMMARY

For all the algorithms in the world, it's still people who draw the maps. This chapter features first-hand accounts of the balancing act required when building districting plans in the real world. Three redistricting practitioners tell us how it looks to reconcile complex, sometimes vague, and sometimes conflicting priorities, while keeping communities in view.

1 INTRODUCTION

The actual construction of redistricting plans happens through many different processes. In most cases, the authority to draw state-level plans rests with the legislature. Sometimes state legislators from the dominant party do the map-drawing in-house, and sometimes they contract with consultants and iterate through carefully constructed plans in private meetings. A few states, like California and Arizona, have independent commissions and conduct state-level redistricting with full transparency and public exposure (currently, more states are following suit).

Sometimes courts despair of the politics of the process and appoint outside experts, or "special masters," (officials appointed by a judge) with specific mandates and short turnaround times to produce new maps. Other states, like Florida and Utah, have solicited citizen map submissions, thereby offering at least a gesture toward

M. Duchin, O. Walch (eds.), *Political Geometry*, https://doi.org/10.1007/978-3-319-69161-9_18

greater transparency alongside educating voters on the redistricting process. Advocacy groups and other stakeholders try to lift their voices in the process, often by commissioning their own maps with the assistance of nonprofit civil rights organizations. Litigation shops (groups of lawyers, usually with a particular specialization) themselves are often called upon to propose remedial plans in case of successful court challenges.

Drawing fair districting plans that can withstand public scrutiny and constitutional challenges is an intricate issue. Practitioners must equalize population and follow a range of traditional redistricting principles, while keeping sight of relevant court decisions, state requirements, and interests of residents. This chapter focuses on the challenges for mapmakers in applying these many overlapping factors in redistricting, focusing for the most part on local maps.

WHO WE ARE

Megan Gall
Megan started her career in voting rights working with the Lawyers' Committee for Civil Rights Under Law where she focused on quantitative analyses for Voting Rights Act (VRA) compliance and litigation including racially polarized voting statistics and redistricting. She also worked as the in-house researcher and lead scientist with the NAACP Legal Defense and Education Fund (LDF) and the Leadership Conference Education Fund. She recently launched Blockwell Consulting, LLC. In addition to creating new redistricting plans, she's often called on to evaluate potential VRA litigation by evaluating maps, creating alternative maps, and examining racially polarized voting patterns.

Karin Mac Donald
Karin is the director of California's Statewide Redistricting Database, housed at UC Berkeley. She ran the team that drew the maps for the California Redistricting Commission after the 2010 Census that were ultimately enacted for CA Congressional, Senate, and Assembly districts. She's been a leader on interpreting Communities of Interest districting criteria since 1998.

Fred McBride
Fred used to work for the nonpartisan advocacy group FairVote, and then worked at ACLU for 13 years, where he was often deployed in the field for grassroots work to create maps that reflected people's neighborhoods and communities. Since 2019 he has been serving as a Policy Specialist at the Lawyers' Committee for Civil Rights Under Law. He has worked all over the country, but has particularly deep community mapping experience in the south, along with racially polarized voting analysis.

Megan and Fred hold PhDs in political science and are skilled with GIS. Karin specializes in data access and implementation, and all three of us are professional data wranglers. We have certain values in common, including the importance of collaboration with civil rights groups to ensure access to data and public input. None of us uses algorithmic assistance to draw maps, nor do we consider partisan data when redistricting—except for VRA compliance, which we'll describe below.

2 THE RAW MATERIALS OF A MAP: TOOLS AND DATA

To even get started in map evaluation or creation, there's an entry barrier: assembling the relevant data. Chapter 13 has a more detailed look at the geospatial tools and data; here, we want to give a sense of how we use it.

2.1 CENSUS PRODUCTS

The most basic raw materials for a mapmaker are geographic units matched to official population counts. The principal data sources are various Census products. Most crucial is the decennial release called the Redistricting Data Summary File, or PL 94-171, which is designed to facilitate redistricting. Next, we frequently want to see the district lines from the previous cycle, which are also provided by the Census (at least for state-level redistricting). Some mapmaking makes use of block groups or voting tabulation districts (VTDs; the approximate precincts provided by the Census). We frequently employ a statistical step to take data from larger geographies like these and disaggregate them to the smallest units, *census blocks*, because blocks are the standard pieces from which districts are made.

Like most things in redistricting, there are always exceptions to the rule of census blocks. For example, Louisiana requires that larger units—precincts—be the building blocks of plans, and only allows the precincts to be changed or split under limited circumstances.[a]

[a](LA Rev Stat §18:532.1 2011, §18:532 2013)

The Census gives you data on race down to the block level, and there's a delicate balance in how you use that information. On the one hand, race cannot predominate in the map-drawing, but the VRA means it has to be in the conversation.

How you use race data is complicated—even though districts typically balance total population, VRA case law requires using *voting age population* and *citizen voting age population*, which you'll have to merge into your database.

For citizenship data, we look to another Census product, the American Community Survey, which is based on a far more detailed sample survey and so can provide richer data in its five-year releases. While the ACS data are fundamental, we use it with the understanding that it is less authoritative in legal settings because it is based on a sample rather than on a full enumeration.

Getting all your data matched to the same units is not a trivial task. A redistricting authority would hire a consultant to create a fused database with all of this information prepared. A civil rights organization or community group might have an in-house data wrangler, but the data preparation can be an obstacle to some of their important work. And even though Census data are free and publicly available, you'll need software that lets you interact with the geography and the data attributes to draw your map. Most people use commercial options, like the ones described in Chapter 13, but the costs create an obstacle to transparency. These days, public software options are proliferating.[1]

2.2 STATE AND LOCAL DATA

Census data are important but only part of the picture. To comply with the VRA, you may need to go beyond the Census to get data on the race, ethnicity, age, citizenship, and prior voting patterns of registered voters. For the last of these, it

[1]The Redistricting Data Hub is one option that makes matched data available (redistrictingdatahub.org), and MGGG-States is another (github.com/mggg-states).

is often necessary to turn to individual states to get election results, the election geography, and the voter registration data from prior years.

Voter registration data are mostly useful not necessarily for making maps but as a method to make projections of electoral outcomes. A question often arises as to the likelihood of remedial plans affording minority groups a realistic opportunity to elect candidates of choice.

The voter file can be an important source if you need to use actual turnout figures to make your projections. Some states, like Georgia and North Carolina, have a race field in the voter file (self-identified when residents register to vote), but this is rare. Often, experts will pull registered voters' names from the voter file and conduct a statistical surname analysis to estimate levels of registration or turnout in the Latinx or Asian communities as part of a VRA assessment. Use of these data varies across mapmakers.

Gathering election results themselves can be easy or hard depending on the circumstance. If you're lucky, election results will be maintained at the state level, but often you have to go to individual counties or even municipalities. This is complicated further in that the reporting format is uneven. For example, in 2010, Georgia passed a new law requiring that early and absentee votes be reported by precinct. Prior to that, early and absentee ballot votes could not be included in a racial polarization analysis of the kind discussed in Chapter 7. By contrast, in Alaska, early and absentee votes are still only reported at the Assembly district level and not by precinct.

And then there is geography. If you are *extremely* lucky, the state may have a complete precinct shapefile that is linked to the election results. For instance, the Texas Legislative Council and the Massachusetts Secretary of the Commonwealth offer this publicly.[2] In California, counties are required to report their precinct geography to the state's redistricting database at the time of each election. In Pennsylvania, on the other hand, legislative leaders filed a statement with the court stating that they had no way of discovering the current precinct boundaries in time to comply with a court order.[3]

Even though this is surprising coming from the state redistricting authorities, it's certainly true that learning current precinct geography is a very difficult task. It might require calling individual counties and even digitizing paper maps.

[2] California and North Carolina and now Virginia release election data officially disaggregated down to census blocks. See Statewide Database [1].

[3] See pubintlaw.org/wp-content/uploads/2017/06/2018-01-26-Order-in-Furtherance-of-this-Courts-January-22-2018-Order.pdf, where "a 2010 Census block equivalency and ESRI shape file" is requested in Order 4, and pubintlaw.org/wp-content/uploads/2017/06/2018-01-31-Turzais-No-Answer-Letter-re-ESRI-shape-files.pdf, in which Speaker of the Pennsylvania House of Representatives, Michael C. Turzai, states that he "has no data or documents responsive to [the Order]."

In California, Precincts frequently change with each election. This makes it more important to collect them in a timely fashion for each election to be able to analyze data over time. If they are not collected between elections, they are frequently overwritten by the Election Management Systems as the counties move to a new set of geography. This is a big part of the ongoing work that is done at the Statewide Database.

I was working on an investigation for a Native American community in a rural, cash-strapped county in South Dakota. County election officials were happy to help, but all they could offer by way of precinct geography was a photo of a wall map with precincts shaded with highlighters.

So, electoral geography (precinct shapefiles) and election results are often each hard to come by. Then once you have them, they need to be joined, which is a new puzzle.

Even once you have a precinct shapefile, you still have to hope for a good match in the precinct names, or do some extra work to find the "crosswalk" between the place names in the spreadsheets and the place names in the shapefiles. There are cases where I've spent weeks just trying to do this step in the data wrangling.

This is a small glimpse at some of the difficulty of spatializing vote data. Despite this difficulty, spatializing vote data is crucial: if you don't know where the votes are, it becomes prohibitively difficult to do a VRA analysis.

2.3 COMMUNITY DATA

There are times when you might find yourself scouring websites or interviewing locals to get the information you need. For example, incumbency rules vary greatly across states, but several states have rules that disfavor "double-bunking," or putting two incumbents in the same new district. But if you want to avoid pairing incumbents, you need to know where they live.

I will attempt to get the addresses of incumbents—that's honestly not easy! I might have to dig through candidate bios and campaign material, and I pay it less attention when the information seems less reliable and not useful, like if I can only find office addresses and P.O. Boxes.

Information for drawing a map can also come from a community directly. Communities and advocacy groups can give you a great deal of (valuable) anecdotal information that you can't get from a spreadsheet alone.

Sometimes residents of one area tell me that they don't identify with or want to be grouped with another area nearby, and this may come with a story that speaks to different representational interests. For example, where there's a history of violence, there may be well-known but unwritten lines that delineate "no go" zones, or you may learn about blatant racial appeals in campaign material.

In addition to providing information for a map, community input can also build support for it.

In California's 2011 redistricting, "Unity maps" helped the commission to draw better informed districts. These maps were composed by a collaborative of Voting Rights and Advocacy organizations and ensured that groups' interests were not traded off against each other. The commission was open to accepting input from everyone who wanted to participate and specific knowledge and feedback were appreciated.

18.1 A NEW YORK "UNITY MAP"

The New York Unity Map provides a meaningful example of community input in redistricting. During the 2010 redistricting cycle, mappers, activists, and lawyers—from the Asian American Legal Defense & Education Fund, the Center for Law & Social Justice at Medgar Evers College, LatinoJustice PRLDEF, and the National Institute for Latino Policy—collaborated to present a Unity Map. It was the culmination of months of mapping scenarios, community meetings, legal discussions, and consultation with minority elected officials. The Unity Map was used to demonstrate the combined voting strength of Black, Asian, and Latino populations in three of the city's boroughs.[a]

According to Esmeralda Simmons, Executive Director of the Center for Law and Social Justice at Medgar Evers College, "The creators of the Unity Map understand that, during the redistricting processes, many would like to pit one community of color against another, rather than give all protected groups their fair share of representation. Thus, we have created these redistricting maps to prove by example that, by exercising mutual respect for community ties and population shifts, fair redistricting in compliance with the Voting Rights Act can be achieved."

Figure 1: The New York Unity Map

This combined group presented their unity plan to the New York State task force responsible for redrawing the state's legislative districts.

Unity mapping, in which minority groups (racial, ethnic, language, or otherwise) combine efforts to propose redistricting plans, is not entirely new. However, it is rapidly spreading throughout the country as collectives propose redistricting plans to their respective redistricting commissions and legislative bodies.

[a]See `https://perma.cc/5ARY-7WJ2` and `https://www.aaldef.org/press-release/civil-rights-groups-present-revised-unity-map-redistricting/`, retrieved 18 August, 2019.

3 DRAWING THE MAPS

3.1 LOCAL MAPS MATTER

Most of this book is focused on the largest and highest-profile cases: state congressional and legislative districts, which are done at the state level. But the country is huge and complicated, and we also want to turn some attention to the local texture of redistricting. As of 2017, there were 90,126 federal, state, and local governments in the U.S. [4]. Many kinds of elected bodies are responsible for a piece of the civic puzzle: city councils, county commissions, school boards, railroad commissions, and an array of others. These bodies sometimes elect at-large, but they often use districts or some combination of both systems. Accordingly, there are many thousands of local districts in the nation that are prone to some of the same complications as congressional and legislative districts.

Districts in local jurisdictions often elude public scrutiny, but they have an enormous impact on our daily lives. They can control schools and regulations, local policing issues, local parks and recreation facilities, and utilities like water and trash removal. Even more than for larger jurisdictions, detailed data and information on local redistricting can be difficult to locate in smaller and rural areas. One reason for this is that many budget-strapped and volunteer-staffed local election offices don't have the resources or know-how needed for GIS, or even for good electronic record-keeping. Smaller jurisdictions might also try to fly under the radar when they redistrict; sometimes, they don't even realize they *need* to redistrict regularly (after every decennial Census). However, community demand for transparency and fairness in local redistricting efforts is growing rapidly, and local redistricting is increasingly attracting attention.

Thus, in the following discussion on drawing maps, we want to note that we are drawing on experience pertaining to local districts as well as to state level ones.

3.2 GOALS AND STARTING POINTS

Plans might be made fresh for enactment, or might be part of a remedial court process to replace an invalidated map. Sometimes these are presented in court as a "map that could have been," so that a jurisdiction has more pressure to explain questionable features in its adopted plan.

In 2010, a plan on behalf of the Georgia Black Legislative Caucus and the Southern Coalition for Social Justice was introduced in the House—it was never seriously considered by the majority party and certainly never made it out of subcommittee, but it helped to set the terms for the redistricting conversation that year.

The choices that face you as a mapmaker can feel infinite, so you need to know where to start, and there are several different strategies for this. One possibility (sometimes even preferred officially in the rules) is to try to make the least change from the last map.

> I often focus on making the least change from the previous map. For remedial maps that are designed to replace invalidated plans, they are typically required to narrowly fix the problems, not start from scratch.

Another possibility if you're starting with a blank slate is to split up a redistricting problem into zones or pieces.

> California is huge and complex, so in 2011, we began by splitting it into four areas, each with a different lead mapper to become an 'expert' on.

There are other documented examples of "modularizing" a big problem into smaller ones. For instance, when Bernard Grofman served as the special master in the recent Virginia redistricting, his report for the court began with a description of dividing the state four ways. He then gave options for redistricting each zone, so that they could be assembled to a full plan.[4]

Another well known example comes from North Carolina, where the state constitution has extremely specific guidance about which counties must be clustered in a legislative plan (See Carter et al. [5] for a complete analysis of optimal clusterings).

We're all conscious that the final product may depend heavily on a decision of where to start. But in the end, if a map is made with the right principles, it's going to inspire confidence. The following sections discuss the principles and practices mapmakers consider when drawing districts.

3.3 THE BASICS: POPULATION, CONTIGUITY, COMPACTNESS

Population balance, contiguity, and compactness are among the most common requirements when drawing districts. Population balance is a straightforward constraint for mapmakers, but only if you know which population count you are trying to equalize, and how much deviation is allowed in order to meet your other goals. For now, the total population count in the decennial Census is still the

[4] "The illustrative maps I present to the Court are what I refer to as "modularized" maps. To facilitate Court review, and to provide the Court with options for alternative ways to provide a narrowly tailored constitutional redrawing, I partitioned the unconstitutional districts into four geographic regions..." [3]

standard for which population is to be equalized, but there are clear moves in some parts of the country to change the basis for population count (see Chapter 23).

A population adjustment made in some states concerns the location of incarcerated people. (This issue, known as *prison malapportionment* or *prison gerrymandering,* is discussed in more legal detail in the previous chapter.) In January 2020, New Jersey became one of a growing number of states to pass a law requiring that its population database be made with an attempt to use the last residence for incarcerated people.

In California, for the first time in 2021, we will use administrative records to adjust the census block population by reallocating incarcerated persons back to their last known residential address.

Contiguity is usually straightforward as well, although it can be complicated by water. You can try not to have to make any decisions about water and just to rely on Census geography to guide you about what's "next to" what. Sometimes, this means familiarizing oneself with the local ferry routes that may connect the mainland to an island from a point that is not the closest in proximity. You'll probably still have to make some calls about islands and such, but in our experience this rarely feels like a high-stakes enterprise.

There are recent cases of maps under consideration by courts being challenged for water contiguity, such as in the case that no physical bridge is present. But in reality there's very seldom a clear-cut state rule about what counts as adjacent across water, so mapmakers get reasonable flexibility. Some state guidelines caution against point contiguity (where a district could be disconnected by removing a single point).

I've worked on maps in South Carolina and Florida with lots of water, and I'll ask groups what makes sense when the bridges and ferries don't tell an obvious story of what is "next to" what.

Mathematical-looking compactness scores are often taken most seriously by mapmakers working from afar, with nothing but commercial software as a tool. As you build districts, you can get live-updated scores of your Polsby-Popper, Reock, Schwartzberg scores, and so on. Even if you're trying not to take these scores too seriously, the software can make it easy to use compactness to break a tie. If a city has to be split, say, you may be able to be score-conscious about how you do so. This isn't always the case, though: some jurisdictions specify what compactness means and that definition may not align with the compactness measures that are usually included in redistricting software packages.

The idea of being ruled by these kinds of scores makes those of us who work closely with community groups shudder. There are many more features that matter to mapmaking than compactness. Often, an eyeball test suffices. In public meetings, Karin often poses the question whether it's more important to have a 'good' map or a map that scores highly on an arbitrary compactness standard.

I will work to get a better score if I can, but it's not a deal-breaker because there's never a threshold or limit, or even a definitive choice of measure.

3.4 COMMUNITIES OF INTEREST

This concept is sometimes more loosely defined than other redistricting principles— it tells us to try to keep geographically recognizable communities with "shared interests" together. Shared interests can range widely, including trade, environmental hazards, resource needs, and so on. A community might be made up of, for instance, a significant number of farmers, coal miners, or people living in a historic neighborhood, aggregated in some semblance of shared space.

Courts have made quite clear that race alone cannot justify a community of interest, and race is handled in other ways in redistricting law. To get a flavor of the rules for Communities of Interest (COIs), consider the Vermont state statute. It states that the preservation of COIs is the "recognition and maintenance of patterns of geography, social interaction, trade, political ties, and common interests" (17 V.S.A. §1903. 2012). This is quite broad. The Supreme Court has ruled that race and voting blocs are not COIs in themselves, but other than that there is little guidance in case law.

Informally organized communities, including those that are not officially recognized in government designations (like a city or even a census designated place is recognized), have traditionally not been significant players in redistricting. When communities would find themselves split by a district boundary, they had little to no recourse, because challenging district maps is an expensive endeavor.

Locating communities of interest can therefore be difficult. If the community happens to be a classification defined by the Census, then good data exist. For example, when working in Indian Country, spatial data on the location of reservations and tabular data with race and languages spoken can be particularly valuable. However, when dealing with populations like local farmers or residents organizing around a historic area, quality data from government sources will be sparse if not non-existent. Boundaries will often only exist in qualitative or narrative form, or they might be anchored by landmarks like community centers or RV parks. And that's if you know in advance what kinds of communities you are looking for and which issues people organize around.

These challenges illustrate why it can be essential for mapmakers to have on-the-ground knowledge of the jurisdictions they draw, and why a solid outreach program to solicit input can result in better maps. Local knowledge and context are qualitative data points that can be critical to drawing a plan that all parties find acceptable.

I talked to so many people in Georgia that I got a sense of what constituencies want a voice. After enough conversations, you understand not to mix up the peanut farmers with the downtown community groups!

Let's turn again to California, where districting criteria mandate that: "[t]he geographic integrity of any city, county, city and county, local neighborhood, or local community of interest shall be respected in a manner that minimizes their division to the extent possible without violating the requirements of any of the preceding subdivisions. A community of interest is a contiguous population which shares common social and economic interests that should be included within a single district for purposes of its effective and fair representation" (California Constitution Article XXI Sec 2 (d) (4)). California takes this really seriously and has held dozens of live-streamed public hearings to solicit testimony that names and locates relevant communities.

In recent redistrictings on the state and local levels in California, communities took advantage of their newfound access to make their interests and electoral needs known in public hearings. Those ranged from neighborhood to historic preservation groups, communities that shared watersheds or a propensity for wildfires, environmental concerns, and access to government services.

There was also an increase in advocacy from people who had never been a part of redistricting in previous cycles. These included advocates for linguistically isolated populations, those fighting for better air quality and against offshore drilling, along with those advocating for land rights, restrictions on certain kinds of development, and the interests of institutionalized persons.

But what if you do not have the time and resources to identify communities well through large, state-sponsored sessions? Then you might reach out to local officials and organizers to help set up small or informal meetings and use those to solicit input about what matters in residents' lives that calls for coherent representation. In this cycle, many thousands of community maps will be collected using free online mapping programs that offer people the ability to map and describe their communities, thereby connecting narrative information with geography to make it easier to synthesize COIs into usable data for legislatures and commissions.[5]

I have run easily over a hundred community mapping meetings over the years, with anywhere from five to thirty people, or more. It's a combination of taking in information about their communities and teaching them "nuts and bolts" about the redistricting rules and issues near them. I like to pull up a current plan and ask the community members themselves how they would change it. "No, not that way—include this other neighborhood instead." This leads to really useful conversations about shared interests—it's an iterative process.

3.5 SPLITTING AND NESTING

As we heard above, plans can be made of tiny pieces—census blocks. Some states have requirements or strong preferences that larger pieces be preserved, such as whole precincts in Louisiana and whole counties in Iowa and North Carolina. But most states have language requiring that the preservation of subdivisions should be maintained only "to the extent practicable." This creates wiggle room.

Trying to preserve counties, cities, and other localities is not always clear-cut for a mapmaker. For instance, the mapmaker may be left to decide if it is more expedient to split fewer jurisdictions more times or to split more jurisdictions fewer times.

[5]Districtr, Representable, and the Draw My CA Community COI tool in California are several of the options that will be most widely used.

I was drawing a demonstration plan for Texas and Harris County that far exceeded the ideal district size. So while most of the map grouped districts by entire counties, I had to decide how to handle Harris County, which had two and a half times the ideal population. I could split Harris multiple ways and avoid splitting other counties or I could split Harris fewer times but split additional counties. I drew multiple maps to land on the best option.

If I have to split a county, I'll split it just once if possible, and try to keep city cores intact. Remember that scale matters. You wouldn't want to split Atlanta more than 2–3 ways in a Congressional plan, if possible. But for a state legislative plan, for example, Atlanta has to be split so much anyway due to the smaller district sizes that all bets are off.

If I have to split a jurisdiction then I ask the community for input on where it might best be split. Sometimes the decision is to try and split it in half, other times a different solution makes more sense, considering the areas surrounding the jurisdictions' boundaries.

"Nesting" refers to the process of incorporating smaller districts (like for the state House) seamlessly together to make up a larger district (like for the state Senate). As you can learn in the Introduction, eight states have 2-to-1 nesting rules in their laws, and an additional two states have 3-to-1 nesting.

As with all districting criteria, the rule is motivated by some conceptions of good governance practices. Let's unpack that in the case of nesting. A lot of good theoretical arguments can be made for why nesting is a worthy goal for mapmakers, including:

1. Voters might have an easier time figuring out which districts they reside in if there are fewer boundaries to consider;

2. Nesting could provide extra geographic constraints on gerrymandering (and more redistricting efficiency) by limiting mapmakers' ability to fine-tune districts for candidates or for political parties;
3. Nesting two lower house seats in one upper house seat might increase the collaboration between the representatives;
4. Election administrators might prefer fewer ballot groups, which is a possible effect of nesting.

In practice, though, it is clear that nesting can be a difficult criterion to fulfill. The usual qualifier that accompanies it, "as practicable," is necessary to avoid significant problems with the implementation of other, mostly higher ranked, criteria. The following section turns to the issue of balancing these different criteria.

3.6 TRADEOFFS AND "SHARING THE PAIN"

All of these considerations must be balanced, and the balancing act is different for every jurisdiction because every jurisdiction has a special amalgam of priorities and demands. Splitting counties and political subdivisions is often unavoidable in redistricting. Likewise, tolerating lower compactness measures (for those who use them) and compromising on other geography-based redistricting principles is often deemed necessary. The state of California's 2011 redistricting commission called the process of balancing these tradeoffs "Sharing the Pain."

To see tradeoffs in action, let's go back to nesting. Above, we outlined four reasons why nesting rules might be popular or attractive. At the end of the day, nesting not only constrains the room to gerrymander, but also constrains the ability to fully comply with other redistricting criteria.

Nesting is a great example of a rule that sounds good, but creates constraints that make it difficult to impossible to meet other, higher ranked, criteria. We've studied criteria trade-offs in California. We found that nesting, if prioritized, constrains the ability to draw majority-minority districts and leads to unnecessary city and county splits.[a]

[a] See *The Implications of Nesting in California Redistricting* by Cain et al. [2]

The California Citizen Redistricting Commission (CRC) conducted perhaps the most significant "replication" of the Mac Donald–Cain study[2] on the effect of nesting on other criteria when it conducted its 2011 redistricting. The CRC tried to combine Assembly seats into Senate seats and found that very few districts could be nested without violating higher-ranked criteria.

Moreover, nesting can have consequences outside of the scope of traditional redistricting criteria. It can lead to a lack of collaboration as each representative tries to distinguish themselves and set themselves up for election to the higher seat. It can also have negative consequences on party politics when two members from the same party run for the same seat.

Remember too that tradeoffs may not only be present among the written rules, but also might look like horse-trading between elected officials and other powerful stakeholders. Sitting representatives often have strong ideas about what it will take to keep them happy. In one recent example, a sitting Atlanta City Council member wanted to add a certain neighborhood to her own district, and she initially got a majority of the Council to back her plan. However, the cost was steep—the proposed plan would have resulted in a drop of more than five percentage points in the Black population of another district. Since this was in the preclearance era, worries about VRA compliance ultimately persuaded the Council to drop the "retrogressing" plan.[6]

At the end of the day, you've got to sacrifice some criteria for others. In order to prioritize community, I might have to allow additional county or municipality splits, and I might have to pair incumbents. I have to sometimes tell a group that I don't see a problem with the adopted plan. Sometimes I will argue that there's a pretty good opportunity here, given all the obstacles in our way.

4 TRANSPARENCY AND SECRECY

A final theme to consider in understanding redistricting in practice is the openness of the process. Transparency and public access in redistricting can be viewed on a spectrum. The least transparent processes tend to be overseen by legislative bodies and the courts. The most transparent and accessible processes are often those implemented by Independent Redistricting Commissions (IRCs).

[6](Atlanta City Council Action Minutes, 5 December, 2011 http://citycouncil.atlantaga.gov/legislation/city-council-meeting-minutes retrieved 24 August, 2019)

18.2 CALIFORNIA'S COMPLICATED COMMISSION SELECTION PROCESS

In 2008, Californians voted to turn their legislative, congressional, and Board of Equalization redistricting over to the Independent Citizen Redistricting Commission (CRC). They voted on a process that establishes a qualified, bi-partisan board. The selection process is implemented by the California State Auditor (CSA), an agency independent of the California Legislature. It is part of the executive branch of government but not subject to its oversight. The CSA initiated the application process for the 2021 CRC in July of 2019 by accepting initial applications via a web portal. The initial application is designed to select those that meet the minimum requirements while weeding out those with obvious conflicts of interest.

Applicants:

- must have been continuously registered to vote with the same registration status (i.e., the same political party or no party affiliation) since July 1 of 2015
- must have voted in at least 2 of the last 3 statewide elections
- cannot be and can have no immediate family members that have been appointed, elected, or that have been a candidate for a California legislative or congressional seat
- cannot be a lobbyist, or have served as a paid consultant, employee or officer of a political party, or the campaign committee of any candidate for legislative or congressional office in California.
- complete a supplemental application that assesses analytical skills, impartiality, and an appreciation of California's diverse demographics and geography
- must provide detailed information about themselves and their family members
- must submit three letters of reference
- must authorize the posting of their application publicly

A trio of state auditors, one from each of the two major parties and one who is not a member of those parties, were selected to oversee the application process and evaluate all applicants. The Applicant Review Panel (ARP) selects the 40 most qualified applicants for three subpools (Democrat, Republican, neither), and interviews these 120 Applicants. This process is webcast. From the pool of 120, the ARP identifies 20 from each subpool to advance to the next stage of the selection process.

The applications of the remaining 60 applicants are then sent to the 4 Legislative leaders who can each remove up to 2 applicants from each subpool. This is the only point of legislative involvement in the redistricting process.

Once the legislature has exercised its option to reduce the pool, the remaining names are returned to the CSA for a random drawing of the first 8 commissioners. In 2010, this process involved a bingo draw machine and it was broadcast live over the internet. The first 8 commissioners are then seated and will select an additional 6 to round out the Commission. The first 8 referred to themselves as the 'lucky ones' and the additional 6 were known as the 'chosen ones' in 2010. The final commission consists of 14 members, 5 each registered with one of the two major parties and 4 that are not and could either be nonpartisan or registered with a minor party.

On the state level, legislative redistricting is mostly shielded from public view, not widely promoted, and consequently doesn't attract a lot of participation. These processes are mostly done behind closed doors. In some states, the majority party drawing the map does so without the consultation of the minority party, whereas in some states the minority party is invited to construct a bi-partisan deal. When parties do this work in isolation, the public, good-government groups, and voting rights and advocacy groups are often unheard. States in which legislators are in charge of redistricting have been shown to produce political gerrymanders.

Legislatures generally must hold hearings on redistricting to give the public an opportunity to comment. At these hearings, depending on the stage in the process at which they happen, maps may be presented or public input may be gathered. However, among the public, there are few who know how to participate in a process even when an opportunity arises. Without maps or concrete questions that members of the public may be able to respond to, such as "where is my Community of Interest?" input is often very general. Often, public comment at this stage focuses on the district the commenter is familiar with, and likely resides or works in, and not on the overall plan. In these settings, legislators can mitigate negative comments and highlight positive comments as public support.

Additionally, it often takes legal action to unearth information about how the maps were drawn and by whom. At hearings where maps are presented, discussion focuses on the local jurisdictions contained in the respective districts and where the boundaries are. Thus, public input in legislatively managed processes frequently only provides comments on the plans already developed by the parties. And plans that were rejected by the legislatures do not have the benefit of public evaluation.

Sometimes a court will invalidate a legislative map and appoint special masters to create new ones. When this happens, transparency is not always part of the equation. Courts, used to operating with a maximum of privacy, have largely extended that right to their appointed mapmakers, leaving the public ignorant as to how their final maps were constructed.

In special master proceedings, for various reasons (e.g., time constraints due to impending elections), quite frequently only the legislative record is consulted to construct the new districts. These records often lack the specific public comments necessary to inform mapmakers of not just 'lawful,' but perhaps 'better' districts. This lack of public access can effectively foreclose the possibility of improved representation.

Examples of relatively less transparent, but not completely closed, processes include local-level districts in California as of 2011. At that time, the relatively few jurisdictions (e.g., some special districts, school districts, and county boards of supervisors) that elect by-district and not at-large had made marginal efforts at transparency. (This has begun to change thanks to the California VRA and the FAIR MAPS Act which is modeled on the California Voter's First Act and mandates a certain number of hearings, ranked criteria and transparency in the process). Overall, though, the processes mimic those of partisan legislatures. Litigation and advocacy groups used the Department of Justice list of proposed electoral changes to understand where and how jurisdictions were changing rules. It is important to note that jurisdictions previously required to preclear electoral changes under Section 5 of the VRA would have hearings to demonstrate their proposed plan(s) and allow public comment.

Advocacy organizations have pushed for more transparency. Groups including the League of Women Voters and Common Cause have pushed for redistricting processes to be moved from legislatures to IRCs for many years. The rationale is that not only do IRCs eliminate the conflict of interest that legislators have when drawing their own districts, but they also are a way to insert transparency into how lines are drawn. Since the early 2000s, the push for increased transparency has gone hand-in-hand with options for increased public participation. Both have begun to gain traction throughout the United States.

For instance, Arizona passed Proposition 106, which established a bi-partisan IRC in 2000. This IRC opened the process to the public by holding hearings throughout the state, accepting public input on districts and suggested maps, and deliberating in public.

Subsequently, the State of California moved to an IRC. This took place because of a narrow victory on Proposition 11, the Voters First Act, which amended the State's Constitution in 2008. The initiative's language required transparency and public access in every part of the redistricting process. Later regulations promulgated by the California State Auditor (the implementing agency) further enshrined the intent of the initiative to create a transparent, accessible, and inclusive process. The initiative further confirmed that redistricting data must remain public and that access to tools to use these data must be made available to the public. The first California Commission also accepted a wide range of public comments, including maps. This commission constructed every district in public and streamed every meeting live over the internet. They then posted meeting transcripts, including all public comments, on their website.

Despite 'sunshine' laws that govern governmental processes on all levels, the highest levels of transparency are generally found when redistricting is performed by independent commissions—even in local jurisdictions. The independent redistricting commission for the city of San Diego in 2001 was honored with a Significant Achievement Award from the Public Technology Institute for its successful implementation of a highly transparent and accessible process. The 2011 California Citizen Redistricting Commission received Harvard's Innovation in Government Award for the same reasons.

5 RECOMMENDATIONS

Below, we provide recommendations that commissions, legislatures, and communities can implement. These recommendations will both create more buy-in to the redistricting process and result in a better understanding and appreciation by the public of the difficulty of creating statewide plans.

1. Redistricting Databases should be maintained over time and not constructed at the last minute. Rushing to build redistricting databases with election data

that are collected and merged at the last minute is more error prone and limits opportunities for data quality and robustness checks.

2. Redistricting Databases are paid for with tax dollars and should be public. This would more appropriately focus attention on the lines, not the data. Making data available increases the userbase and enables the public to debug it: it is much better to find a data error earlier in the process than later. It is also a critical step to create buy-in to the redistricting process.
3. Officials can encourage public input and consider it! Quite frequently, members of the public have information about an area that is not known to mapmakers, and ideas about how a district can be constructed that meets everyone's needs. Public input can be a type of crowdsourcing to elicit information that is otherwise difficult to obtain. Officials should accept public input in every way it arrives: via mail, email, fax, phone, and/or testimony at a hearing.
4. Jurisdictions can accept all map proposals as part of public input, irrespective of whether these consist of partial districts or single whole districts. Most lay participants will be experts on their immediate surroundings but know nothing about areas farther away.
5. Map-making tools should be freely available to the public. Even tools with minimal functionality are useful, and they are sometimes better than overly complicated software with too many bells and whistles. Ensure that tools are easy to use, quick to learn, and available not just for people who have their own computers and internet. Libraries are a great place to set up public workstations that can be used by people who do not have access to digital technology.
6. Groups can create easily understandable materials that allow the public to participate in the process. Redistricting is complicated, but it can be simplified to equalize the playing field. As part of creating understanding, groups should translate and disseminate their materials into, minimally, the languages for which their jurisdiction is covered under Section 203 of the Federal Voting Rights Act.
7. Officials can provide translations and interpreters at public hearings.
8. Officials can hold hearings in multiple locations throughout the jurisdiction; ensuring that hearing locations are accessible to differently-abled persons. Hearings should be near public transportation and have parking. Programs like Zoom and other internet viewing options, as well as call-in options, not only aid accessibility, but increase opportunity for public input.
9. Mapmakers should visit the jurisdictions they are planning to redistrict. When travel isn't possible, they should use the internet and other research tools to learn about the area and the people who live there. Talking to communities on the ground will always result in better maps.
10. Community members can utilize the resources available from advocacy groups working on fair districts, and can attend workshops, town halls, trainings, and information sessions on redistricting.

11. The COVID-19 crisis makes it blazingly clear that public participation in the process may be difficult, with unequal barriers to participation. But most district boundaries remain in place for ten years, so it is fundamentally important that we design and advocate for new methods to facilitate public input, while investing in outreach to be sure we reach the communities who have the most to lose when their voices are excluded from the process.

6 CONCLUSION: DEMOCRACY TAKES WORK!

In *Rucho v. Common Cause*, Chief Justice Roberts went out of his way to (approvingly) cite reform measures. Although the verdict of the case opposed intervention by the federal courts in partisan gerrymandering, this comment signified broad support of independent redistricting commissions, additional state redistricting criteria for mappers, and prohibitions against partisan favoritism. Even so, these discussions of redistricting must not miss the point: the question is not simply who draws, but how.

Redistricting is constantly evolving. Quantitative measures, involving assessing a district's performance, are available, along with qualitative measures, such as researching communities. Mapping technology, statistics, and the availability of election data have created enormous opportunities for evaluating plans based on past elections. A number of mapmakers and expert witnesses in voting rights cases are now suggesting how candidates would've performed in previous elections under various "proposed plans." This information and these methods have the potential to produce redistricting plans worthy of robust discussion, input, edits, and reconsideration in the democratic process.

The laws and jurisprudence around redistricting are complex, intertwined, and often have competing goals. Stakeholders, including politicians, citizens, and advisory boards, create a complex tapestry of demands. This forces mapmakers to exercise a remarkable amount of discretion, even amid the federal, state, and local rules that constrain us. This balancing act is one reason why human judgment is still needed for map drawing, even in the age of algorithms. Redistricting requires prioritization and thoughtfulness. Many of the mapping criteria demand that we know the qualitative, contextual picture on the ground. To create effective plans, mapmakers must balance principles and rules alongside the choir of voices.

Democracy takes work! How districts are drawn is ultimately as important as who runs for office in determining the quality of representation. The rules of the game should ensure equal voting power, proper racial and ethnic consideration, and a lot of basic common sense when amassing groups of people into areas to elect people to represent their interests. Electing officials to serve the people is the fundamental mechanism of representative democracy. Drawing the districts they are to serve is a crucial and delicate part of the process.

REFERENCES

[1] Statewide Database: Redistricting. (n.d.). Retrieved 25 May, 2020, from `https://statewidedatabase.org/redistricting.html`

[2] Cain, B. E., Mac Donald, K., Boyle, N., Palazzo, W., Russian, N., & Bonovich, N. (2007). The Implications of Nesting in California Redistricting. Institute of Governmental Studies, University of California at Berkeley.

[3] Grofman, Bernard. Report of the Special Master. Golden Bethune-Hill et al., v Virginia State Board of Elections. 7 December, 2018. `https://perma.cc/Z7LU-RPM3`

[4] US Census Bureau. (25 September, 2019). Census of Governments. Retrieved from `http://www.census.gov/programs-surveys/cog.html`

[5] Daniel Carter, Zach Hunter, Dan Teague, Gregory Herschlag & Jonathan Mattingly (2020) Optimal Legislative County Clustering in North Carolina, Statistics and Public Policy, 7:1, 19–29.

19 *Interview: Drawing for the courts*

MOON DUCHIN AND OLIVIA WALCH

NATHANIEL (NATE) PERSILY
JAMES B. MCCLATCHY PROFESSOR OF LAW
STANFORD LAW SCHOOL

Nathaniel (Nate) Persily has been involved in drawing plans for courts since the 2000 redistricting, sometimes as a "special master," tasked with direct action on behalf of the court. In one form or another, he has assisted courts in crafting Congressional or legislative districting plans in Georgia, Maryland, Connecticut, New York, North Carolina, and Pennsylvania, as well as drawing for numerous local governments.

In 2005, Persily wrote a "primer" on court-drawn plans in a law review article [2], noting:

> "In the many redistricting struggles that now follow each census, plaintiffs routinely turn to the courts not only to strike down plans as illegal, but also to draw remedial plans to take their place. Courts are not mere referees of the redistricting process; they have become active players, often placed in the uncomfortable role of determining winners and losers in redistricting, and therefore, by consequence, elections."

In the time since then, courts have possibly turned to Persily more than any other expert to play this role. We asked him about the process.

Q. What's the backstory—how did you come to be drawing maps for courts?

The first time I was called upon to draw a map for a state was in 2002 for the New York congressional redistricting. The legislature had deadlocked and the federal court had appointed a Special Master, former District Court Judge Fred Lacey, to come up with a redistricting plan. His staff reached out to Heather Gerken, then a Professor at Harvard, now Dean of Yale Law School.

The year before I had done a redistricting simulation for Heather's class, in which I brought in people from Caliper Corporation, who make Maptitude for Redistricting, the software package that is used by most jurisdictions for redistricting. (I had gotten to know the folks from Caliper from previous meetings of the National

M. Duchin, O. Walch (eds.), *Political Geometry*, https://doi.org/10.1007/978-3-319-69161-9_19

Conference of State Legislatures where they did simulations for state legislators.) When Heather got the call, she recommended me and the rest is history.

At the time, I was an Assistant Professor at University of Pennsylvania Law School, on the verge of finishing my political science dissertation at U.C. Berkeley. I had no partisan affiliation or reputation and no track record when it came to redistricting—these turned out to be pluses, not minuses, for me to be appointed as the assistant to the special master, who naturally did not want someone who could be seen as siding with the Democrats or Republicans.

Since this was my first opportunity, though, I wanted help and suggested that Bernie Grofman of U.C. Irvine and Marshall Turner, who had just retired from the Census Bureau, should join me in drawing the map. (I had edited a small volume for the Brennan Center about two years earlier on "The Real Y2K Problem: Census 2000 Data and Redistricting Technology" [3] in which Bernie had a chapter, so I got to know him then.)

We worked with the Special Master and the New York Legislative Task Force on Redistricting (which staffed the legislature) to create a map for New York's congressional districts. Those first maps I drew are now framed on my wall! We had hearings, we drew a plan, but the plan endangered too many incumbents so the legislature rejected it when the Court gave it a second chance to overturn it. But the plan was considered to be politically fair, and that helped build my reputation drawing for courts; I then did legislative plans for Maryland and Georgia that in the end were perceived to be mildly pro-Republican. Sometimes I was working directly with the court, sometimes with a special master. That led me to being one of a small group of people who get appointed by courts.

It helps that I never registered with a party, gave money to candidates, or worked with parties directly. Those three early cases then led me in 2012 to be appointed by the Connecticut Supreme Court as a Special Master for the Connecticut congressional redistricting and then once again to assist a special master for the New York congressional redistricting, and then to serve as Special Master later in the decade in North Carolina to solve a racial gerrymandering problem identified by the federal court there.

Most recently, I was brought in by the state Supreme Court in Pennsylvania in 2018 to assist in their remedying the partisan gerrymander of the congressional districts that they struck down under the state Constitution. The map that emerged was mildly pro-Republican but Democrats still picked up three seats—moving from a 12-6 congressional delegation to 9 to 9, which for a purple state like Pennsylvania seems pretty fair.

Given how politically contentious that map inevitably was, I worry that I might now be out of the business of drawing maps for courts. But I am still doing maps for local jurisdictions, such as a recent one in Lowell, Massachusetts, that arose out of a settlement agreement from a voting rights lawsuit. Doing this kind of work really transforms your view of redistricting. It's a totally different world once you're weighing the fates of voters, candidates, and communities.

Q. What are the ways that a special master is at a disadvantage drawing lines? How might a lack of local knowledge impact the outcome, and how do you try to compensate for that?

In most situations, I get involved because the political process has broken down and a plan needs to be drawn in emergency fashion so that elections can take place. This is never the ideal environment in which to draw a plan. Nevertheless, even in an expedited process, either the Court or I can hold hearings to get participation from the litigants or interested parties.

You learn an enormous amount about community interests and on-the-ground politics in these cases. It brings out the best and worst in our political system. I remember in the first New York congressional redistricting that we moved Jamaica Bay out of Congressman Anthony Weiner's district. He sent the court a letter saying that he wanted his swamp back. No people lived in that census block, but he had ongoing environmental projects that he was supervising there. But we moved that large water block to ensure the adjacent district was more compact and contiguous. So, he never got his swamp.

I use that example when I teach Reynolds v. Sims—especially the line in that opinion that says trees and acres don't vote, people do. Well, apparently swamps do too. Justice Breyer wrote about this example in his Vieth v. Jubelirer dissent to demonstrate that political considerations are not always nefarious.[1]

And then there are always instances where groups of people either want to be in a different district from other groups or joined with their allies. Sometimes these are pretextual arguments for partisan- or incumbency-related considerations. But sometimes they are authentic expressions about how people define the boundaries of their community. For instance, I remember in the Maryland redistricting that obeying the boundary line that separates Baltimore City from Baltimore County ended up splitting the Orthodox Jewish community that straddles that border. But the court case that gave rise to the redistricting centered around protecting political subdivision lines so that community ended up divided.

When I am appointed as an expert or special master, most of what I do is acquaint myself with the community-defining characteristics of the state or locality. Often the parties to the litigation present those arguments. Sometimes the court will hold a hearing that allows nonparties to make submissions. Often I will release a draft map (as I did in North Carolina, for instance) that allows the parties to express their disagreement and suggest revisions based on nonpartisan criteria. However, it is always important to be careful because community of interest arguments are the most ripe for manipulation by partisans and incumbents.

Q. So what does the actual process look like, and how long might it take? Who puts together the raw data materials? Do you seek to uphold principles not

[1] "The use of purely political considerations in drawing district boundaries is not a 'necessary evil' that, for lack of judicially manageable standards, the Constitution inevitably must tolerate. Rather, pure politics often helps to secure constitutionally important democratic objectives. But sometimes it does not" [1].

specifically named in law or by the court that appoints you, or do you view the task very narrowly?

I don't build the data. Often I will have the jurisdiction provide it. Some things sound obvious but are in fact tricky, like identifying the boundaries of municipalities. It can be tempting to use data products from the Census, like their MCDs (minor civil divisions), which mostly nest in counties. I got burned once on this issue, in North Carolina, where it was argued that I had the wrong boundary files for cities. Similarly, you can try to shortcut community identification with CDPs (Census-Designated Places)—but this never matches up to the real conception of community. However it is collected, the parties must agree to the data and it must be transparent and public. Caliper provides the census data and boundary files, however, as used in the Maptitude software program.

The first ten hours (or 20, or 30) is me just staring at maps and trying to understand how the pieces fit together. Most of the time when I'm appointed, the legislature has failed to come up with a plan, so I need to work fast. Two months would be normal turnaround time. For most states, you can rough-draft a congressional plan in a day—but just because you can do it fast doesn't make it good! The number of stakeholders and the number of iterations can increase complexity a great deal. Georgia in some ways was the most intense process, because we drew over 300 districts for both the state Senate and the General Assembly, repeatedly.

My article *When Judges Carve Democracies* [2] discusses the role of courts and special masters in judicially supervised redistricting processes. Precedent exists discussing how aggressive the courts can be in designing a remedy for a given constitutional violation. A remedy ordinarily must be narrow: just solve the problem that gave rise to the court's involvement. If only two districts are malapportioned, for instance, you only adjust those two districts.[2] Most of the time, when I get involved, the Court directs me to remedy a particular problem in a particular set of districts. However, if the plan as a whole is legally infirm (for instance, if the state needs to transition from 30 to 28 districts or the plan as a whole is determined to be an illegal partisan gerrymander), then the remedy will be more comprehensive, perhaps requiring a plan drawn completely from scratch.

Q. Can you talk more about how you end up balancing the traditional districting principles, from compactness to political boundaries to communities of interest?

Let's start with compactness. Judges tend to like compactness measures because they have the feel of objective criteria against which you can evaluate whether one plan is "better" than another. But as you, more than anyone, know, these measures are highly contested and in tension with one another. Generally speaking, judges are also struck by the aesthetics, but they lean on the measures whenever possible. And so, speaking hypothetically here, if you can beat all the submitted plans on all metrics, that would be very persuasive. There is plenty of sleight of hand that

[2] *Perry v. Perez* 2012

lawyers use here. Part of my job is to identify that. It's more challenging for judges to deal with the quantitative data than many political scientists or mathematicians might realize.

Political subdivision splits are another way that courts evaluate redistricting plans, but here too lawyers sometimes try to pull the wool over judges' eyes. For instance, is it better to split one county into five pieces or two counties each into two pieces? There is no neutral way to answer that question.

Similar questions arise with communities of interest. Some communities want to have their influence spread among many districts, others want to be concentrated in one district. I remember in my first redistricting of New York that we drew a nice circular district around Buffalo, which previously had two halves of two districts. But then the headline was—"Buffalo loses congressional district"—when in reality they moved from two halves to one whole.

Compactness often comes with other costs, however. I once drew a gorgeous square district in northern Georgia—it looked like Wyoming! But a mountain range went right down the middle of it, so you had to exit the district if you wanted to travel from one side to the other. Even though it was great on the screen it did not make sense for the population.

In North Carolina, I drew a district around Greensboro that was mocked because of its "buzzsaw" shape. But it had ridges because of town boundaries themselves. That's kind of common: municipal boundaries are often crazily shaped and even noncontiguous due to annexations. People tend not to realize it, so when you draw a district around a city it might be weirdly shaped.

The plan we drew in Georgia paid great attention to political subdivision lines, but not precincts. As a result, hundreds of precincts needed to be redrawn to administer elections under the new districts. I am told the Secretary of State wept. But these are the tradeoffs when you maximize along one dimension without considering others. The impact that redistricting has on election administration is systematically underappreciated.

Q. This is already very complicated. Now how about the Voting Rights Act and racial fairness?

As you know, the law regarding race and redistricting is now, as has been true for thirty years, in flux. The upcoming redistricting is the first since the 1960s that will not be governed by Section 5 of the VRA, given the Court's opinion in Shelby County. In addition, line drawers need to navigate the competing demands of Section 2 of the VRA, which require attention to race, and the Shaw v. Reno line of cases, which caution against using race as the predominant factor in the construction of a district.

It is easier for courts to comply with these conflicting legal pressures than it is for legislators. Because a court need not pay attention to the same concerns about incumbency or partisanship that preoccupy a legislature, it is easier to draw compact districts that ensure minorities have an equal opportunity to elect a candidate of their choice. When legislators do so, they often draw contorted districts that attempt to satisfy, not only the legal requirements, but the many other concerns that lead a district to snake from one area to another.

In my most recent work drawing a plan for Lowell, Massachusetts, I'm appointed pursuant to a settlement agreement between the city and a group of Latino and Asian American plaintiffs who threatened a Section 2 VRA lawsuit. The process has really been eye-opening and inspiring. The litigants and the community submitted maps. I drew a framework map based on what I heard. We then held three public hearings—one in English, another in Spanish and one in Khmer—in which anyone could offer comments. And now the districts are in place for the first time for both City Council and School Board.

It doesn't always need to be contentious! Sometimes the redistricting process makes you have added faith in democracy.

REFERENCES

[1] Stephen Breyer, Dissenting Opinion in *Vieth v. Jubelirer*. Available at law.cornell.edu/supct/pdf/02-1580P.ZD2.

[2] Nathaniel Persily, *When Judges Carve Democracies: A Primer on Court-Drawn Redistricting Plans.* George Washington Law Review 73, 1131–1165, (2005).

[3] Nathaniel Persily, *The Real Y2K Problem: Census 2000 Data and Redistricting Technology* (Brennan Center 2000).

20 *Explainer: Ranked choice voting*

THOMAS WEIGHILL AND MOON DUCHIN

In the subject called *social choice theory*, we regard electoral systems (like plurality voting in districts) as sets of rules that convert ballots from voters into a choice of winners or outcomes. The rules matter—the exact same ballots can be subjected to a different set of rules and produce a different outcome, as we'll see below.

But even before we get to the rules for aggregating the votes, a important element of an electoral system is the form of ballots that voters can cast in the first place. On a standard plurality ballot, you have a number of choices and you select just one candidate. On an approval ballot, you can decide whether you approve or disapprove of each candidate, so you can select more than one. On a limited approval ballot, you can select up to a certain maximum number of candidates. And so on.

Systems matter.

In this explainer, we'll explore *ranked choice voting* (RCV), which is the name for all the electoral systems that use a ranked ballot. In other words, voters see a number of choices and they indicate their first choice, their second choice, and so on—though typically they are allowed to stop early and don't have to include every candidate in the rankings.

Even once we have settled on ballots of this form, there are still (infinitely) many ways to devise rules to combine the ballots of a group of voters into a choice of winners. In the United States today, the two most common forms of RCV under discussion for practical use are *instant runoff voting* (IRV) and *single transferable*

M. Duchin, O. Walch (eds.), *Political Geometry*, https://doi.org/10.1007/978-3-319-69161-9_20

vote (STV). We'll explain those below—they are essentially the same system, but used to elect one person or several.[1]

Supporters of RCV tout it as a way to make democracy work better by giving voters more say in their representation, by allowing ballots that better describe their preferences and policy views. (This is sometimes called "expressiveness.") The tallying algorithm may be complicated, sure, but voters are not required to know anything about how it works to participate fully in the process. Detractors say that more complicated ballots are nevertheless a barrier to participation.

We'll look at some simple examples of the IRV and STV forms of ranked choice voting and then discuss some other properties of the systems.

SINGLE-WINNER (IRV)

Why is it so hard for a third party to gain traction in the United States? There are likely many reasons, but one of the most frequently cited is the problem of third-party "spoilers." Generally, a (weak) spoiler is a candidate who didn't have a chance of winning but nonetheless impacted the outcome among the stronger candidates. This can happen when a newly formed political party ends up hurting the party it most closely aligns with by siphoning enough votes to tilt the election. Let's look at a toy example to see how this works. It's a mayoral election, so there can only be one winner, and there are three candidates: two dogs (the St. Bernard and the Shih Tzu) and one cat. Each animal runs on its own unique platform, but there is a lot of agreement between the dog candidates. Both dogs, for example, are in favor of more fire hydrants for convenient public relief, while the cat would prefer that infrastructure dollars go to scratching posts.

The election takes place, each voter selects just one candidate, and nine ballots are cast (see Figure 1). Clearly the dogs' general platform gained a lot of traction because most voters selected a dog. And yet, it's a cat who won with the most votes overall. The Shih Tzu is branded a "spoiler" because their presence in the race diverted votes away from an ally, the St. Bernard.

Figure 1: Electing a mayor through "first past the post": the animal with the most votes wins, namely the tabby cat. The Shih Tzu was accused of being a spoiler.

Ranked choice voting provides a fix for this predicament. Let's run the election again, this time with rankings. The ballots are shown in Figure 2. We first look at

[1]There are many other forms of combining ranked ballots, like Borda count and Schulze beatpath and on and on, each with passionate aficionados.

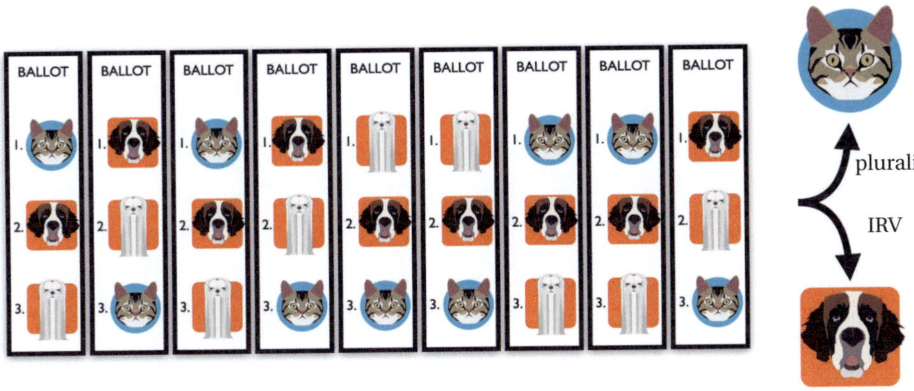

Figure 2: The same plurality votes we saw above might look like this if voters ranked the candidates. If we are electing a mayor through ranked choice voting, we get a different outcome from the initial plurality race: now, once the Shih Tzu is eliminated, their support *transfers* to the voters' next preference and ultimately identifies a majority coalition for the St. Bernard.

only the first place votes: 4 for the tabby cat, 3 for the St. Bernard, and 2 for the Shih Tzu (just like last time). We check to see if any candidate has a majority of these votes (in this case, five or more); if they do, then they win. In our election, no candidate passes that threshold, so we eliminate the candidate with the fewest first place votes, namely, the Shih Tzu. The votes of those who put the Shih Tzu first are not wasted, though, as we now redirect their votes to their second-choice candidate. New totals: 4 for the tabby cat and 5 for the St. Bernard. We check again for a majority and see that the St. Bernard does indeed have a majority of votes. The St. Bernard wins! Note that even though everyone's first-choice votes remained the same, the fact that voters were able to indicate their other preferences secured victory for St. Bernard.

MULTI-WINNER (STV)

In the previous example, we were electing a mayor. Ranked choice voting can easily be extended to elections for a larger number of winners, like for a city council or simply for sending several members to a large governing body using multi-member districts. The number of representatives to be elected from one voting geography is called the *magnitude* of the district. Electing more than one winner opens up a new question: to what extent does the makeup of the representatives reflect voter preferences? Will a city where 30% of voters are deeply concerned with affordable housing policy elect a city council where that reform stance has 30% share? This is a form of *proportionality* that we might ask of our systems of election, and supporters of STV frequently argue that this is the key strength of this system.

To see this play out, we'll go back to the example of a town electing cats and dogs, and we'll suppose a high degree of bloc voting: most voters can be classified as dog-lovers or cat-lovers, and they vote accordingly. Let's say three-eighths (37.5%)

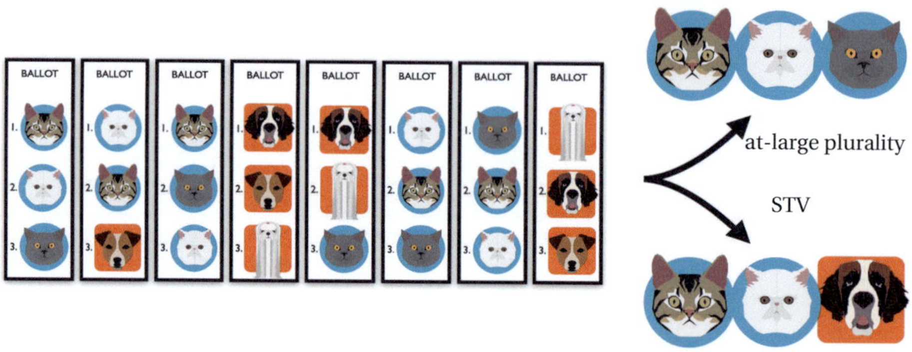

Figure 3: Electing a three-animal council. If we just count who has the most votes on ballots (unranked), we get a cat sweep. But single transferable vote (STV) lets the dog-lovers' votes consolidate on the St. Bernard, and we get a more proportional outcome.

of the town's residents are dog-lovers, and the rest are cat-lovers. The town wants to elect a three-animal town council. In this case, the most proportional outcome would be two cats and one dog. But whether that occurs depends heavily on the system.

Initially, the town used an at-large plurality system, where every voter selects three candidates (to seat the full town council), and the animals with the most overall votes cast for them are elected to the council. Eight ballots were cast, shown in Figure 3, and indeed, cat-lovers showed a lot of cohesion in support of cats. Three cats received five votes, and no dogs did, so the three cats are elected. (Here, we're not paying attention to the rankings, but just to who appears on voter ballots at all.) In other words, bloc-voting by cat-lovers secures a complete sweep of the town council seats.

The dog-lovers might reasonably feel unfairly excluded. What if the exact same voter preferences were taken into account in a ranked way? Let's step through the STV algorithm for tabulating the rankings. We begin with first-place votes and check to see if anyone is over the threshold of election. The threshold is the smallest whole number T of first-place votes that couldn't have too many people clear that level. In this case, four candidates could have two votes each, but you couldn't have four candidates with three first-place votes each, so we set $T = 3$. Initially, no animal hits the threshold, but three animals have two votes each. So we eliminate the little Jack Russell terrier, who has no first-place vote support. Next, the Shih Tzu and the gray cat with spooky eyes are eliminated, because they only have a single first-place vote each. Now the voters that ranked those candidates first see their support transfer to the next eligible candidate on the ballot, which gives an extra vote to the tabby cat and the St. Bernard, putting them at the threshold, so they are elected. The method continues in this way: you check who's over the threshold, and if anyone is over, you mark them as being elected and distribute their excess votes.[2] If nobody is over the threshold, you eliminate those with the least support,

[2]To be precise, there are several different ways to manage the redistribution of excess votes. The Wikipedia page on STV `en.wikipedia.org/wiki/Single_transferable_vote` is a good place to start

and redistribute their votes. Continue until you've elected your full complement of representatives!

PROS AND CONS

There's a lot to like about RCV. In our single-winner example, we saw that RCV can reduce the barriers for additional parties to enter into an election, which many Americans view as favorable, at least at the national level.[3]

In our multi-winner example, we saw that RCV can promote proportional representation in governing bodies. This phenomenon generalizes beyond the very simple example shown here. If you have a sizeable bloc that gives cohesive support to a slate of candidates, they are likely to secure representation roughly proportional to their share of the electorate.[4] Note that this is proportionality promoted by the structural properties of the system, rather than the arranged party proportionality that is popular in other countries through *party list* voting (see Sidebar 2.3 in Chapter 2). This observation is fueling renewed interest by some reform groups in using RCV to promote minority representation across the nation.

It has long been known that at-large plurality voting has a strong tendency to exclude minorities from representation. Localities that decide to move away from at-large plurality (or who are being forced to do so by a Voting Rights Act lawsuit) have traditionally looked to single-member districts as the leading alternative. However, RCV is increasingly being considered as another remedy, such as in Lowell, MA, where the at-large plurality system was challenged in court.[5]

So more proportional representation for minority groups is a huge point in favor of RCV. If that's a pro, what are some cons? The most frequent criticism of ranked choice voting is that it increases the burden on voters by requiring them to make more choices and to absorb more information. To fully complete a City Council ballot in Cambridge, MA, in 2019, a voter had to assess twenty-two candidates (Figure 4), a daunting task for all but the most avid followers of local politics.

A second point against STV is that it can allow for all the representation to come from a small area or neighborhood, so it lacks the geographic distribution that is promoted by the use of a larger number of districts.

There are quite a few other questions around RCV that are hotly debated, like whether the shift to a new form of voting causes changes in voter behavior. See for example the seething polemic—by academic standards—between Donald Horowitz (who argues that RCV reduces racial polarization) and Jon Fraenkel and Bernie Grofman (arguing that this reduction isn't universal). As is the case with a lot of questions about RCV, there is very little data available to address this question,

---

if you want full details.

[3] `news.gallup.com/poll/244094/majority-say-third-party-needed.aspx`

[4] This is a well-known "folklore" fact about STV. It is proved as a lemma in an article that we co-authored [2], where we also demonstrated it in several vote models for approximately realistic voting conditions.

[5] Voters then got a chance to vote on a voting system! City-wide ranked choice was an option, but voters narrowly turned it down; see Chapter 21 for more on that example.

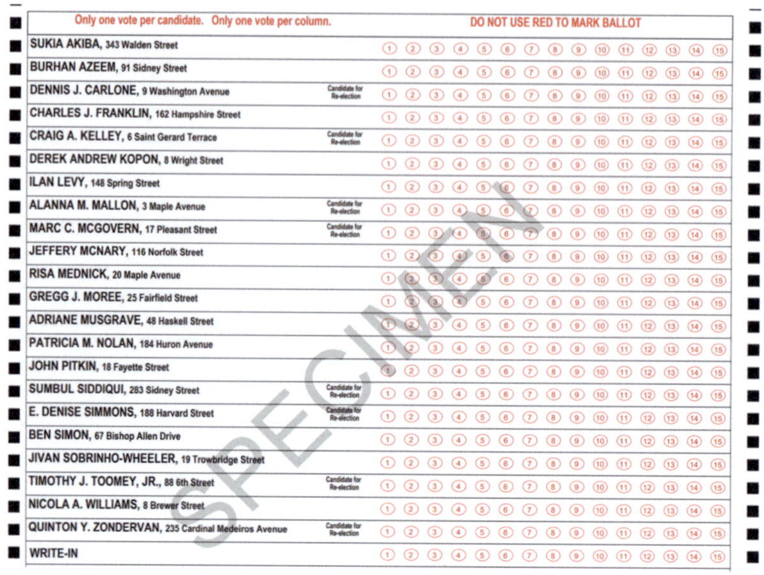
Only one vote per candidate. Only one vote per column.
DO NOT USE RED TO MARK BALLOT

SUKIA AKIBA, 343 Walden Street
BURHAN AZEEM, 91 Sidney Street
DENNIS J. CARLONE, 9 Washington Avenue — Candidate for Re-election
CHARLES J. FRANKLIN, 162 Hampshire Street
CRAIG A. KELLEY, 6 Saint Gerard Terrace — Candidate for Re-election
DEREK ANDREW KOPON, 8 Wright Street
ILAN LEVY, 148 Spring Street
ALANNA M. MALLON, 3 Maple Avenue — Candidate for Re-election
MARC C. MCGOVERN, 17 Pleasant Street — Candidate for Re-election
JEFFERY MCNARY, 116 Norfolk Street
RISA MEDNICK, 20 Maple Avenue
GREGG J. MOREE, 25 Fairfield Street
ADRIANE MUSGRAVE, 48 Haskell Street
PATRICIA M. NOLAN, 184 Huron Avenue
JOHN PITKIN, 18 Fayette Street
SUMBUL SIDDIQUI, 283 Sidney Street — Candidate for Re-election
E. DENISE SIMMONS, 188 Harvard Street — Candidate for Re-election
BEN SIMON, 67 Bishop Allen Drive
JIVAN SOBRINHO-WHEELER, 19 Trowbridge Street
TIMOTHY J. TOOMEY, JR., 88 6th Street — Candidate for Re-election
NICOLA A. WILLIAMS, 8 Brewer Street
QUINTON Y. ZONDERVAN, 235 Cardinal Medeiros Avenue — Candidate for Re-election
WRITE-IN

(ranking bubbles 1–15 for each candidate)

SPECIMEN

Figure 4: Sample ballot from Cambridge City Council election, 2019. Obtained from `cambridgema.gov/Departments/electioncommission`

partly because so few places have adopted RCV. But as more jurisdictions adopt ranked choice, we'll have more and richer data to go on.

The MGGG Redistricting Lab has modeled the effect of RCV on minority representation in localities from large cities like Chicago and Cincinnati to small counties like Yakima County, WA and Terrebone Parish, LA. in each case, we have found positive expected impacts for the representation of people of color.[6]

PAST, PRESENT, AND FUTURE

Outside of the U.S., ranked choice voting is used in national elections in Australia, Fiji, Malta, Papua New Guinea, Northern Ireland, and Republic of Ireland.

Within the United States, ranked choice voting has a very interesting history. Single transferable vote used to be quite popular across the country in the early 20th century, but it got stamped out in mid-century—some have argued that its successes getting Black and Communist candidates elected were the trigger for backlash during the Red Scare [1]. The only holdout has been Cambridge, Massachusetts, a city of about 100,000 that has used ranked choice continuously since initial adoption in 1941.

Today, ranked choice is a reform on the upswing.[7] About twenty U.S. cities adopted some form of ranked choice between 2000 and 2020, most commonly to elect the

[6] You can check out the growing collection of case studies on ranked choice here: `mggg.org/RCV`.

[7] The Ranked Choice Voting Resource Center maintains a list of RCV localities, past and present, here: `rcvresources.org/where-its-used`.

mayor by instant runoff. The most prominent recent addition is New York City, which is now phasing in IRV in primaries and special elections for mayor, city council, and other offices. Needless to say, that will provide a huge and rich source of ranking data.

In 2018, Maine voters adopted ranked choice for all their statewide elections, but only in IRV form. In 2020, Alaska voters narrowly moved to ranked choice: primary elections will now be nonpartisan, and the top four vote-getters will advance to the general election, where IRV will be used to identify the winner.

Multi-winner ranked choice, though it is the form that has proportionality benefits, is still lagging badly behind single-winner ranked choice in reform uptake. Besides Cambridge, only one city council is elected in a multi-winner way: Eastpointe, Michigan, adopted STV in 2019. But there are some new developments. With the increasing hostility in the Supreme Court to the Voting Rights Act, several states have adopted similar new frameworks in state law, which often leave room for ranked-choice remedies (and not just "designer districts"). As the projected twilight of the VRA forces advocates to rethink their approach to electoral opportunity, ranked choice systems may become an indispensable new tool, avoiding both the fence-out effects of plurality systems and the gerrymanderable pathologies of districts.

REFERENCES

[1] Douglas J. Amy, *The forgotten history of the single transferable vote in the United States*, Representation, 34(1): 13–20 (1996).

[2] Gerdus Benade, Ruth Buck, Moon Duchin, Dara Gold, and Thomas Weighill, *Ranked choice voting and minority representation*, available at papers.ssrn.com/sol3/papers.cfm?abstract_id=3778021.

Chapter 21

Reform on the ground in Lowell, MA

IVAN ESPINOZA-MADRIGAL AND OREN SELLSTROM

CHAPTER SUMMARY

Where you vote matters, but the way you vote matters too. This is a case study of litigation using a novel coalition claim to change the system of election in Lowell, MA.

1 INTRODUCTION: THE VRA AS A TOOL

The federal Voting Rights Act of 1965 (VRA) is a powerful tool for advocacy and litigation. It allows communities of color and voters of color to challenge a broad range of systemic and structural practices that deny or abridge their right to vote. Although the Supreme Court has recently limited certain aspects of the VRA, many of its provisions remain as critical to the racial justice struggle as they have been since the VRA's passage in 1965.

Due to its breadth and flexibility, the VRA is an especially potent tool for challenging structural mechanisms that operate to keep an entrenched "old guard" in office, especially in communities that are increasingly diversifying and experiencing demographic transitions. In particular, the VRA has often been used to successfully challenge "at-large" electoral systems that dilute the right to vote.

This chapter examines the continuing relevance of the VRA, in the context of a recent challenge to the at-large electoral system of Lowell, Massachusetts, one of the largest cities in New England.

M. Duchin, O. Walch (eds.), *Political Geometry*, https://doi.org/10.1007/978-3-319-69161-9_21

The VRA was enacted at the height of the civil rights movement in the 1960s. It has been rightly called one of the "crown jewels" of civil rights laws passed during this era. The VRA broadly prohibits any "standard, practice, or procedure" that "results in a denial or abridgement of the right of any citizen of the United States to vote on account of race or color." Two features are evident from the text of the law. First, the VRA is broad in its reach. Overt practices that have historically been used to limit the right to vote of communities of color – such as poll taxes or literacy tests – are clearly prohibited. But the Act reaches further than that: it extends to any practice or procedure that limits the right to vote on account of race, whether that is a discrete voting requirement or a broader structural challenge. Moreover, the obstacle need not impose an outright ban on voting; any "abridgement" of the right to vote is also vulnerable under the VRA.

Second—and perhaps most importantly—the law focuses on the *impact* of such obstacles, not the *intent* behind them. The VRA outlaws practices that "result in" denial or abridgement of the right to vote, whether or not such practices are imposed because of an intent to discriminate against communities of color. This distinction between "intent" and "impact" is critical in civil rights law. Proving intentional discrimination is notoriously difficult – and is made all the more challenging when the defendant is a governmental entity. Since governmental entities are composed of individual actors, who each may have a wide range of motivations, attributing "intent" (much less a nefarious one) can be a daunting task. Moreover, the search for motive often makes little sense from a practical standpoint: whether or not a governmental policy or practice is intentionally discriminatory, the on-the-ground harm to those affected is often the same.

For these and other reasons, civil rights advocates usually prefer to focus not on what a government entity "intends," but rather on the "disparate impact" of governmental action. However, in many areas of law (outside of the VRA), federal courts over the years have cut back on the ability of civil rights plaintiffs to do so. The U.S. Supreme Court, for example, has held that disparate impact is insufficient to prove certain violations of the Equal Protection Clause of the U.S. Constitution. Instead, only intentional discrimination will suffice.

The VRA, however, stands as one of the remaining areas of law that does *not* require plaintiffs to prove intent. As explained further below, analysis under the VRA focuses instead on the impact of the voting practice being challenged. Even following a recent U.S. Supreme Court decision that weakened other provisions in the VRA, this provision remains intact.[1]

[1] In the 2013 *Shelby County* case, the Supreme Court struck down provisions of the VRA that required certain state and local jurisdictions to obtain federal "pre-clearance" before making changes to their voting procedures. Although this ruling essentially gutted the law as it pertains to federal pre-clearance (Section 5 of the VRA), it did not affect the provision of the VRA that allows for affirmative challenges to voting rights practices (Section 2 of the VRA). For a more thorough introduction to the VRA and its case law, see Chapter 6 and Chapter 7

2 AT-LARGE ELECTORAL SYSTEMS: A PRIME TARGET

One of the most common VRA challenges over the decades has been to "at-large" electoral systems. In an at-large system, every voter in the jurisdiction votes for all open seats. This can be contrasted with district or ward elections, in which voters cast their vote only for the candidates running in their particular district or ward.

The problem with at-large electoral systems is that they have the potential to dilute the vote of communities of color, particularly if the majority votes as a cohesive bloc and in opposition to those in the minority. When that happens, 51% of the voters can capture 100% of the seats in 100% of the elections, and effectively leave the minority with no voice at all.

To give an example outside the electoral context, imagine a family with five kids that has "TV night" at home every Friday night. Three of the children always prefer sci-fi movies, while two always prefer cartoons. Most parents who want to keep the peace and come up with a fair system would devise a schedule of alternating sci-fi movies and cartoons from week to week (perhaps with slightly more sci-fi movies to reflect the 3-2 split). But what sensible parents would not do would be to tell the kids: "We're going to vote each week, and we'll watch whatever gets the most votes that week." Because in that case, the family would end up watching sci-fi movies 52 weeks a year. The cartoon fans would rightly feel as if they had no say in the matter. They might check out and stop voting altogether.

For this reason, at-large elections have been banned for federal Congressional seats for years. In 1967, a law was enacted requiring that members of the House of Representatives must be elected by district (in states with more than one Representative). Today, we think of this as natural. A single state, for example, may have 20 Representatives in Congress. But all residents of the state do not vote for all 20 of these Representatives. Rather, the state is split up into 20 districts and residents vote for their own Representative.

But at the local level, at-large elections persist. Because of the potential for inequity, however, federal courts have often found them to violate the VRA. Over the years, guided by the U.S. Supreme Court, federal courts have developed a framework for analyzing such claims. To prevail on a Section 2 claim, Plaintiffs must first satisfy what is known as the "*Gingles* preconditions" (after the name of a U.S. Supreme Court case) by showing that: (1) the minority group is "sufficiently large and geographically compact to constitute a majority in a single-member district;" (2) the minority group is "politically cohesive" in that it tends to vote together in support of particular candidates; and (3) the majority votes "sufficiently as a bloc to enable it . . . usually to defeat the minority's preferred candidate[s]."

Demonstrating the *Gingles* preconditions is largely a map-drawing and mathematical exercise based on publicly available election and U.S. Census Bureau data. The first precondition is proven by demonstrating, typically through expert demographers, that the jurisdiction can be divided into districts where the plaintiff group comprises a majority in at least one district. The second and third preconditions require statistical evidence detailing the voting patterns of communities of color in past elections typically based on publicly available population data and election results.

If these three *Gingles* preconditions are satisfied, plaintiffs must then show that, under the "totality of the circumstances," the minority group has less opportunity than other members of the electorate to elect representatives of its choice. Courts evaluate the "totality of the circumstances" via what is known as the "Senate factors," which include: any history of voting-related discrimination in the political subdivision; the extent of racially polarized voting within the political subdivision; whether voting practices or procedures—such as unusually large election districts—tend to enhance the opportunity for discrimination against the minority group; whether minority group members bear the effects of past discrimination in areas such as education, employment, and health; any lack of electoral success for members of the minority group; and whether elected officials are unresponsive to the particularized needs of the members of the minority group.

3 LOWELL, MASSACHUSETTS: A CASE STUDY

The potency of the VRA to challenge discriminatory at-large electoral systems, and the ways in which such cases can be proven, can best be illustrated by looking at a real-life example: federal litigation against the City of Lowell, Massachusetts. The authors' organization, Lawyers for Civil Rights, represents a coalition of Asian-American and Latinx residents of Lowell who challenged the city's at-large election system under Section 2 of the VRA.

3.1 BACKGROUND

Lowell is a city of approximately 110,000 residents, located roughly 22 miles northwest of Boston. It is a richly diverse city. At the time the lawsuit was filed, in May 2017, the city was approximately 51% White — and on the cusp of becoming a minority-majority community — with more than 49% of Lowell's residents identifying as people of color: 22% Asian-American, 18% Latinx, and 7% Black. A sizeable percentage of the Asian population has Cambodian roots, with many families having moved to Lowell in the 1980s and 1990s as refugees from Cambodia fleeing the Khmer Rouge and its killing fields. This growing and thriving Cambodian community is now the second largest in the United States. Today, Lowell continues to experience demographic transition with vibrant and dynamic Latinx and African populations.

Yet, the city's elected bodies traditionally have not in any way reflected the community's rich diversity. At the time the lawsuit was filed, Lowell's City Council and School Committee were both virtually all-White, and had been for nearly their entire existence. Many qualified candidates of color had run over the years, but very few had been elected. In fact, only *four* candidates of color had *ever* been elected to the City Council. And only *one* person of color had *ever* been elected to the School Committee, even though two-thirds of the children in the Lowell Public Schools are students of color.

The electoral system that had long been in place was a critical reason for this significant power imbalance and racial disparity: all nine City Councilors and all six School Committee members were elected at-large. The top nine and top six vote-getting candidates were elected to the City Council and School Committee respectively. Candidates did not need a majority of all votes cast to win a seat. With an active, unified, and predominantly White majority voting bloc representing approximately 51% of the city's population, all the winners could easily come from — and represent the interests of — one dominant community. This electoral system, with voting patterns that were profoundly racially polarized, suppressed people of color and diverse voices.

For example, in 2013, two Cambodian-American candidates ran for the Lowell City Council. Despite heavy support from Asian-American and Latinx voters alike, neither candidate won a seat on the City Council. Expert analysis of election results indicated that these two candidates were strongly favored by both Asian-American and Latinx voters above all other candidates, ranking as those voters' first- and second-choice candidates. In contrast, they were seventeenth and eighteenth — out of a total of eighteen candidates — among the predominantly White majority voting bloc. Year after year, the White voting bloc consistently elected all nine of its top candidates to the City Council effectively suppressing and diluting the votes of their diverse neighbors.

Interestingly, Lowell did not always use an at-large plurality voting system. From 1943 until 1957, Lowell employed proportional representation voting in municipal elections. In contrast to present-day Lowell, voters in this time period consistently elected a diverse City Council with representatives from the major ethnic groups living in the city at the time, including the Irish, French, Polish, and Greek. Historians credit this system for democratizing the city and increasing the political power of groups that had formerly played a limited role in city politics.

Lowell switched from proportional representation to at-large plurality voting through a city-wide referendum in 1957. Contemporary accounts show that proponents of the shift explicitly stated that moving to an at-large plurality scheme would promote "majority rule" and would limit "minority rule" of the city's various ethnic and national groups. The local paper characterized the "most objectionable feature" of proportional representation as "the opportunity [that] minority groups are given ... for representation" because it purportedly led to "minority representation strictly on a racial or national basis" and motions or decisions based on "racial extraction ...."

The referendum passed in 1957, and Lowell moved to its current at-large system. The effect predicted soon came to be realized as minority-group participation in city politics was significantly curtailed. Over the years, cities across the country and in Massachusetts have moved away from at-large electoral systems — either voluntarily, by court order, or under threat of litigation. Lowell, however, clung to its at-large system. Until the lawsuit, it was the last Massachusetts city with over 100,000 residents to maintain an exclusively at-large plurality electoral system.

3.2 COALITION CLAIMS: THE NEW FRONTIER

The VRA challenge against Lowell was brought by Lawyers for Civil Rights with *pro bono* co-counsel from the law firm of Ropes & Gray in Spring 2017 on behalf of Asian-American and Latinx residents of the city. In so doing, these courageous plaintiffs became part of an emerging trend in voting rights litigation: a coalition lawsuit.

When the VRA was first passed, the paradigmatic voting challenge was a claim brought by African-American voters (often in the South), alleging that a White majority voting bloc was impermissibly suppressing or diluting their vote. As the country has become increasingly diverse, however, and as voting rights litigation outside the South has become increasingly commonplace, this paradigm has shifted. Voting rights cases have been brought on behalf of Latinx residents, Native Americans, and Asian-American communities.

Increasingly, coalition claims are also being brought – that is, claims raised by different communities of color that band together to challenge an electoral system that harms them collectively. The first such cases tended to be African-American and Latinx coalitions. The Lowell case is believed to be the first case brought by a coalition of Asian-American and Latinx voters. Such coalition lawsuits present a valuable opportunity for communities of color to join forces to take on structural systems that harm them in similar ways.

The majority of circuit courts to consider Section 2 coalition claims have recognized them as valid.[2] In 1987, the U.S. Court of Appeals for the Fifth Circuit, which hears cases from Texas, Louisiana, and Mississippi, became the first circuit court to recognize that minority groups could be aggregated for affirmative suits under the VRA.[3] The Fifth Circuit again reached the same result one year later in *Campos v. City of Baytown, Tex.*[4] Analyzing the text of the VRA, the *Campos* court concluded that:

> "There is nothing in the law that prevents the plaintiffs from identifying the protected aggrieved minority to include both Blacks and Hispanics. [Section 2] protects the right to vote of both racial and language minorities. . . . If, together, they are of such numbers residing geographically so as to constitute a majority in a single member district, they cross the *Gingles* threshold as potentially disadvantaged voters."[5]

The court further noted that to prove vote dilution under *Gingles*, a minority coalition must also show that the minority groups "actually vote together and are im-

[2]The U.S. Supreme Court has yet to decide whether Section 2 permits coalition claims, but has assumed without deciding that such claims are cognizable. See *Growe v. Emison*, 507 U.S. 25, 41 (1993) ("Assuming (without deciding) that it was permissible for the District Court to combine distinct ethnic and language minority groups for purposes of assessing compliance with §2, when dilution of the power of such an agglomerated political bloc is the basis for an alleged violation, proof of minority political cohesion is all the more essential.").

[3]See *LULAC v. Midland Indep. Sch. Dist.*, 812 F.2d 1494, 1500 (5th Cir.), *vacated on state law grounds*, 829 F.2d 546 (5th Cir. 1987) (approving of the manner in which African-Americans and Latinx were joined together as a compact minority group "capable of carrying a district")

[4]840 F.2d 1240 (5th Cir. 1988)

[5]*Id.* at 1244

peded in their ability to elect their own candidates by all of the circumstances, including especially the bloc voting of a white majority that usually defeats the candidate of the minority."

The Eleventh Circuit, which hears appeals from Alabama, Florida, and Georgia, adopted the Fifth Circuit's view on coalition districts in *Concerned Citizens of Hardee County v. Hardee County Board of Commissioners.* They found that "[t]wo minority groups (in this case [B]lacks and [H]ispanics) may be a single [VRA] minority if they can establish that they behave in a politically cohesive manner."[6] The Ninth Circuit, which hears appeals from numerous Western states, also recognized coalition districts in *Badillo v. City of Stockton, California.*[7] And in *Bridgeport Coalition for Fair Representation v. City of Bridgeport,* the U.S. Court of Appeals for the Second Circuit (covering Connecticut, New York, and Vermont) also assumed that coalition claims are covered under the VRA. [8] Only the Sixth Circuit, which hears appeals from Michigan, Ohio, Kentucky, and Tennessee, has concluded that coalition claims are not viable.[9]

In the Lowell case, the city attempted to obtain an early ruling from the federal district court hearing the case that coalition claims are improper under the VRA. However, the federal court ruled in favor of the Asian-American and Latinx voters, stating that allowing coalition claims "properly [serves] Section 2's legislative intent of curing past discrimination...." The Lowell ruling makes sense based on the well-established VRA precedent, discussed above, from courts across the country. It also makes sense from a practical perspective as a matter of legal efficiency, particularly to conserve resources and to expedite judicial proceedings. It would be burdensome on the court if voters were to bring separate lawsuits raising identical claims based on the same set of facts against the same responsible parties.

Notably, intersectionality[10] lies at the heart of coalition claims. This is particularly important for voting rights work in cities and towns that have rapidly diversified in the past decade — and that will continue to experience significant demographic shifts and transitions. In the 21st century, natural population growth and migration are changing the faces of communities across the country. As population growth transforms communities beyond the historical Black/White divide, voting rights advocates will confront far more complex social and community dynamics. Even in communities with intensely segregated residential patterns, shared experiences and interests are increasingly common among various racial and ethnic groups vis-à-vis the dominant White voting bloc. Already, it is becoming more challenging to find racial disparity along a single axis of racial or ethnic identity. We predict that this dynamic will continue to consolidate, resulting in intertwined voting experiences across distinct communities of color. This will make coalition claims far more common, relevant, and useful to tackle systemic and structural practices that deny or abridge the right to vote in the next decade.

---

[6] 906 F.2d 524, 526 (11th Cir. 1990)

[7] 956 F.2d 884, 886 (9th Cir. 1992) (assuming that a combined group of African-American and Latinx voters met the first *Gingles* precondition)

[8] *See* 26 F.3d 271, 275–276 (2d Cir. 1994), *vacated on other grounds,* 512 U.S. 1283 (1994)

[9] *See Nixon v. Kent Cty.,* 76 F.3d 1381, 1386–1393 (6th Cir. 1996) (en banc).

[10] See Crenshaw, "Mapping the Margins: Intersectionality, Identity Politics, and Violence against Women of Color" in Stanford Law Review

Coalition claims are, therefore, highly viable under the VRA, allowing communities of color to band together to challenge at-large electoral systems that harm them collectively. As courts approving coalition claims have noted, such coalitions must still satisfy the *Gingles* preconditions (*i.e.*, show that it is possible to draw a majority-minority district and demonstrate racially polarized voting). In the Lowell example, as explained above, such proof is readily apparent.

Coalition claims can be tricky, however. Even when various racial groups share a community and common interests, they may not be used to working together. In Lowell, for example, the Asian and Latinx communities shared many concerns about how their city and school district were run. Both communities had long sought more services for English Language Learners in the schools and both had intense interest in policies concerning how the city cooperated with federal immigration officials. The two communities sometimes intersected around these issues, but not always. Although voting patterns indicated that the two communities were politically cohesive when it came to the ballot box, this had not always translated to on-the-ground coordination in advocacy.

Interestingly, the litigation itself proved to be a galvanizing moment that brought the two communities together. As the plaintiffs and their allies began considering a lawsuit, and then proceeding with it, many evenings were spent in community members' living rooms, over potluck dinners of *tamales* and traditional Cambodian food, learning more about each other's lived experience and rich culture; their concerns about the unequal provision of services in Lowell; and about their shared hopes and dreams for their children and the future of the city.

These conversations were often inter-generational, simultaneously looking backward at often painful and racially charged moments in the city's history and projecting forward to a more equitable and inclusive future. Elders in each community who had lived through decades of being shut out of the political process — and largely relegated to the shadows of civic life more generally — spoke movingly about the weight of that history and how it had created barriers to success. At the other end of the spectrum, young families questioned how the lack of political representation would affect their children in the future. They feared not only that their concerns would be given short shrift by political leaders, but also that their children would grow up not seeing pathways to representation and leadership visibly open to them.

In some cases, the dominant power structure may have been able to remain entrenched, in part, by pitting communities of color against each other—divide and conquer—one of the oldest tricks in the playbook, but unfortunately often successful. In Lowell, for example, members of both the Asian and Latinx communities discussed how over the years the powers-that-be had sometimes held up one community as an example at the expense of the other, whether that was decrying a wave of recent Cambodian refugees to the Latinx community, or subtly signaling to the Asian community that "you're not like them" when referencing the Latinx community.

As these cross-cultural and inter-generational conversations grew deeper and more intense, the commonalities between the Asian and Latinx residents of Lowell became even more clear. So, too did the need to band together to work toward a common solution. They coalesced around meaningful representation in public institutions as a matter of community empowerment and dignity.

3.3 REPRESENTATION AND DIGNITY

Beneath the numbers and maps that are critical to any VRA challenge lies the day-to-day reality of what it means to live in a city or town without enjoying equal representation. The types of proof required for VRA claims — and in particular the "Senate factors" — allow marginalized voters to draw upon their lived experiences in making their case.

The Lowell litigation again provides a compelling example of how this can be done. For the plaintiffs and other residents of color in Lowell, the lack of adequate representation in city government has deep ramifications. Simply put, far too often, the needs of Lowell's communities of color had traditionally been ignored. For example, in the years surrounding the lawsuit, Lowell was considering whether to renovate or move its high school, which was located downtown. This was a critical issue for all residents in Lowell, but particularly for communities of color (as noted above, approximately two-thirds of Lowell's student body are children of color). Many families of color favored renovating the existing infrastructure to make sure their children remained within walking distance of after-school programs and activities. Yet until late in the process, there was little outreach or notice to communities of color about this issue, and little attempt to engage diverse communities in relevant discussions. In this manner, communities of color were largely — and often — excluded from the city's affairs. Meanwhile, students of color faced persistent achievement gaps and disparities in school discipline that had long gone unaddressed.

Similarly, a citizen petition had asked the Lowell City Council to limit local law enforcement cooperation with federal immigration authorities — an issue that had taken on intense importance for many immigrant communities across the country under the Trump Administration. However, the Lowell City Council summarily declined to enact any such policy with little recognition of the fact that this issue presented significant concerns for communities of color, particularly at a time when federal immigration enforcement had been growing increasingly aggressive and intrusive. In a climate of fear and uncertainty, this institutional neglect left many families of color feeling compromised and vulnerable.

Lack of community representation in Lowell had also resulted in unequal distribution of basic city resources, services, and amenities. For example, it took years — and a major push — just to get lights turned on in Lowell's Clemente Park, a park frequented by many children and families of color. Meanwhile, White families in other parks were never left in the dark.

The lack of representation and respect permeated all levels of Lowell's city government. To aggravate matters, there were few, if any, translation services at City Council or School Committee meetings, meaning that non-English speakers were, literally, left out of the loop on major decisions. And Lowell had neglected key entities, including the city's Diversity Council, which was created years ago — only to sit moribund with no appointed members when the lawsuit was filed.

As these examples illustrate, elected officials are simply unaccountable and unresponsive to communities of color under a racially polarized at-large system, with little motivation to be responsive to those communities' needs or concerns. This triggers a cascade of neglect, impoverishing the services that are offered to residents in communities of color. Alarming accounts of services and resources being delayed — or withheld altogether — are not uncommon. This is not just a matter of fundamental fairness; it is also linked to the health and well-being of our democracy. Neglecting constituents based on their race, identity, or zip code erodes political participation and engagement. It generates tension and distrust between communities of color and public officials. It truly is a dangerous and slippery slope from unfilled potholes to victims and witnesses of crime who do not trust the police.

At an even more fundamental level, our plaintiffs and the communities they represented were profoundly aware that Lowell's elected bodies simply did not look like them. They did not see themselves reflected in the halls of power. In a representative democracy, having elected bodies that, in fact, do not represent the rich diversity of the city was an affront to the dignity of communities of color. As our clients emphasized throughout the litigation, they were not suing because they were antagonistic to the city. On the contrary, they sued because they loved their city and were an integral part of it. They just wanted to democratize the electoral system so that it would fairly and equally allow them to be part of all aspects of civic life — including the city's elected bodies.

3.4 REMEDIES

It is not surprising that communities of color must often resort to a lawsuit in order to change entrenched at-large electoral systems. After all, the elected officials — the ones who could change the system voluntarily if they wished — got there through the very system being challenged, and for that reason are often reluctant to undertake reform and nervous to change the status quo. Putting the question of change to the voters as a ballot measure suffers from the same obvious problem: trying to fix a broken system through the system itself.

The Lowell case again provides a stark example of how difficult electoral reform can be:

- For example, in 2009, a proposal to change the city's election system was on the ballot. Asian-American and Latinx residents overwhelmingly voted in favor of the referendum. However, the majority voting bloc voted approximately 2-1 against it, thereby defeating it.
- In 2010 and 2011, the City Council was asked to consider changing the election system, but no reforms occurred.
- In 2016, a City Councilor once again proposed discussing whether the current electoral system should be changed. No other City Councilor seconded that motion, and the matter died without even a discussion of the issue.

This history illustrates why it often takes the outside pressure of a lawsuit, or a court ruling, to effect change in local electoral systems.

And if change is forced through the courts, what are the remedies? Once liability is found, the typical remedy in a voting rights challenge is injunctive relief, meaning an order from the court enjoining — or stopping — the governmental entity from continuing the challenged electoral system. To determine what system gets put in place instead, federal courts often turn first to the defendant, or to the parties jointly, for an alternative system. Courts are typically reluctant to impose a court-ordered remedy, without first asking the parties for proposals. Alternative systems can take many forms. Often the jurisdictions will move to a district-based system, which by its nature does not suffer from the same dilutive flaws as an at-large system. Frequently, the result will be a hybrid system: where certain seats are still elected at-large, but others are elected by district or ward. Proportional voting may also be a viable remedy. Depending on the specific needs of a given community, each of these alternative systems can have its pros and cons.

4 CONCLUSION: A SETTLEMENT IN LOWELL, AND NEXT STEPS

The Lowell lawsuit successfully settled in 2019, with the City agreeing to abandon its at-large method of electing City Council and School Committee members. At the national level, the settlement was groundbreaking: rather than imposing a particular alternative electoral system, the settlement intentionally and deliberately placed the issue back on the community to decide what alternative system to adopt.

The settlement set forth four different types of alternatives that the city could choose from – all of which were deemed to be acceptable alternatives under the VRA that would fully remedy the vote dilution problem of the old at-large system. This "menu of options" included:

1. An all-district system;
2. Several "hybrid" systems that combined at-large and district seats;
3. An at-large but ranked choice voting model; and
4. A three-district ranked choice voting model.

Throughout the summer of 2019, the community debated which of the models would work best for their City. Just as the litigation had brought together communities of color in discussions of the problem of political exclusion, now those same communities came together in a conversation about solutions to the problem. This time, though, they knew that they had the force of a federal court Consent Decree behind them: that the question was no longer *whether* the City would change but *how* it would do so. And they knew that it had been their communities, working together through the lawsuit and beyond, that had created the opening for change to happen. The sense that the City was entering a new chapter in its history was palpable.

Pursuant to the Consent Decree, in September 2019, the City Council narrowed the options to two: a) the at-large ranked choice option; and b) the 8-3 hybrid option (eight district seats and three at-large seats). The options were placed on the ballot in November 2019, as a non-binding initiative, to let voters express their preference. Under the Consent Decree, this was accompanied by a comprehensive trilingual public education campaign in English, Spanish, and Khmer. Ultimately, voters expressed their preference for the 8-3 hybrid system, which the city then officially adopted in December 2019.

Jurisdictions that have moved away from at-large electoral systems to one of these fairer systems have typically enjoyed a number of benefits. First, residents become more engaged when they know that their vote really counts. This increases integrity in the system and reduces alienation and voter apathy. Second, representatives elected through fairer electoral systems become more responsive to the needs of communities of color. When elected officials know that they need your vote to win, they have more incentive to be responsive to your needs once in office. Third, it is more affordable for candidates to run in a discrete district or ward than to run city-wide. Reducing the costs and barriers to entry often results in more diverse and nontraditional candidates who simply would not have the resources to run city-wide.

Finally, when elections move to fairer systems, elected bodies often become more diverse. And diversity on elected bodies carries with it a host of benefits: from increasing the comfort level of constituents of color in approaching their elected leaders to allowing children of color growing up in a community to see that pathways to leadership are open to them as well. Everyone in diverse communities such as Lowell should see themselves reflected in the halls of power. Notably, local elected positions are also often stepping-stones to higher office, thus creating a diverse pipeline for positions of even greater prominence.

At-large election systems carry with them the inherent potential for inequity. A cohesive majority is not only able to rule, but able to sweep the table every time in every election. Particularly when the vote of communities of color is diluted in this way, the VRA can be a potent tool for forcing change. The law remains broad and flexible, allowing increasingly diversifying communities to take on established power structures—and win.

ACKNOWLEDGMENTS

The authors extend their deep gratitude and appreciation to the plaintiffs in this lawsuit, who continually inspired us with their courage and commitment (Chanmony Huot, Vladimir Saldana, Lianna Kushi, Thoeun Kong, Denisse Collazo, Sue J. Kim, Soady Ouch, Tooch Van, Carmen Bermudez, Kei Kawashima-Ginsberg, and Fahmina Zaman) and to the pro bono attorneys at Ropes & Gray LLP who worked tirelessly on this case (including Robert Jones, Scott Taylor, Daniel Fine, Matthew Mazzotta, Martin Njoroge, Jonathan Magaziner, Lilly McCullough, Ioana Moldovan, and Amin Danai). The impressive results achieved by this diverse team working together will reverberate in Lowell and beyond for generations to come.

22 Explainer: Race vs. party

ARUSHA GORDON

The Voting Rights Act (VRA) is a tool that is discussed throughout this book, and it safeguards the ability for minority groups to elect candidates of their choice. This would not be necessary if there were no systematic differences in preference between the minority and the wider society.[1] On the national scale, recent presidential elections provide a way to examine racially polarized voting in our country. The sidebar below explores this polarization in exit polls from the two most recent Presidential elections.

22.1 VOTING POLARIZATION TODAY

These sex-by-race figures on presidential support come from CNN exit polls. They show interesting patterns around the country. (In blank cells, the number of people polled from that group was judged to be too small to produce a reliable estimate.)

National	White women	White men	Black women	Black men	Latina women	Latino men	All other
Clinton '16	43	31	94	82	69	63	61
Trump '16	52	62	4	13	25	32	31
Biden '20	44	38	90	79	69	59	58
Trump '20	55	61	9	19	30	36	38

This shows that Trump improved his relative standing in nearly every group from 2016 to 2020, while losing the popular vote by a larger margin. This is possible because White voters were estimated at 67% of the 2020 electorate, down from 71% in 2016.

AL	White women	White men	Black women	Black men	Latina women	Latino men	All other
Biden '20	19	23	93	82	-	-	-
Trump '20	80	74	7	18	-	-	-

CA	White women	White men	Black women	Black men	Latina women	Latino men	All other
Biden '20	51	51	-	75	77	73	68
Trump '20	47	47	-	21	22	24	28

MI	White women	White men	Black women	Black men	Latina women	Latino men	All other
Biden '20	49	39	95	88	-	-	66
Trump '20	51	60	5	11	-	-	30

[1] To bring a VRA case, plaintiffs must show that voting patterns are racially polarized, with the minority cohesively supporting one set of candidates while the majority has a different, and prevailing, preference—these are the 2nd and 3rd Gingles criteria discussed elsewhere in the book.

M. Duchin, O. Walch (eds.), *Political Geometry*, https://doi.org/10.1007/978-3-319-69161-9_22

What's going on here? Party polarization itself is very high, among both voters and legislators.[2] The preferences of people of color around the country are, in the main, very Democratic. Some authors have used the term "conjoinment" or "conjoined polarization" to refer to the tight correlation of race with party preference. With the conversion of the "Solid South" from Democratic to Republican now complete, the degree of race/party conjoinment may well be at a 50-year high. As political scientists Bruce Cain and Emily Zhang put it: "Since the migration of Southern White conservatives to the Republican Party, party identification has become more consolidated and consistent. As the parties have become more distinct from each other, they have also become more internally ideologically consistent. This assortative political sorting has been accompanied by the strengthening of racial partisan identification, leading to a conjoined polarization of party, ideology, and race. Conjoined polarization complicates and undermines the efforts of an earlier time to protect minority voting rights, most notably through the passage of the Voting Rights Act [3]."[3]

Let's look at how that has actually played out in some recent court cases.

In his dissenting opinion in a 2017 North Carolina racial gerrymandering case, Justice Samuel Alito noted that "partisan and racial gerrymandering can be made to look much the same on a map."[4] Since racial gerrymandering is unconstitutional but partisan gerrymandering is not, this might allow defendants to disguise impermissible predominance of race over other principles, or can at least muddy the waters and make it hard to discern intent. However, as Justice Kagan wrote in this same North Carolina decision, "[t]he sorting of voters on the grounds of their race remains suspect even if race is meant to function as a proxy for other (including political) characteristics." In other words, no matter the focus in a redistricting case, a decision is suspect if the map-drawers considered race in determining how constituencies would vote.

In practice, the challenge of distinguishing between race-based versus party-based voting is a vivid one for advocates working on the ground to advance equal voting

[2] Many authors have tried to assess the impact of simple partisan polarization on the work of legislative bodies (for just two examples, see Andris et al. and Dimock et al. [1, 4]).

[3] Trends in racially polarized voting have historically been particularly strong in jurisdictions previously covered by the Voting Rights Act. As Ansolabehere et al. explain in *Race, Region, and Vote Choice in the 2008 Election* [2], White voters in previously covered jurisdictions voted distinctly more Republican that year than those in the noncovered jurisdictions. Only 28% of White respondents in jurisdictions previously covered by the Voting Rights Act said they voted for the Democratic nominee—fourteen percentage points lower than their counterparts in the noncovered jurisdictions, where 42% of Whites on average reported voting for Democratic nominees. This is thirty-three percentage points lower than Democratic nominees' average vote share among Latinos (61%) and fifty-six percentage points lower than the average among African Americans (84%) in the covered jurisdictions. Regardless of whether they live in covered or noncovered jurisdictions, racial minorities, in contrast, were not found to differ substantially in the share that reported voting for Democratic nominees.

[4] Alito continued, "This phenomenon makes it difficult to distinguish between political and race-based decision-making. If around 90% of African American voters cast their ballots for the Democratic candidate, as they have in recent elections, a plan that packs Democratic voters will look very much like a plan that packs African American voters. '[A] legislature may, by placing reliable Democratic precincts within a district without regard to race, end up with a district containing more heavily African American precincts, but the reasons would be political rather than racial.'" *Cooper v. Harris*, 137 S.Ct. 1455, 1488 (U.S.N.C., 2017) (citing *Easley v. Cromartie*, 121 S.Ct. 1452, 1455, 532 U.S. 234, 235 (U.S.N.C., 2001)). The tension between partisan and racial claims is discussed further by Charles and Spencer in Chapter 9.

opportunities.[5] After plaintiffs make an argument that the Gingles preconditions are satisfied, including a showing of racially polarized voting, defendant jurisdictions often respond by arguing that voting trends are based on party allegiance, rather than race. For instance, San Juan County, Utah was sued multiple times because of discriminatory voting practices making it harder for its Navajo residents to vote, particularly by cutting down in-person voting to a single (poorly located) polling place and providing inadequate language support for voting materials. This minority group had its electoral preferences blocked by polarized voting. But defendants argued in a brief that voting trends were explained best by the alleged fact that "Navajo [residents] vote along party lines." The county asserted that "political party affiliation among Navajo voters in San Juan County is so strong that they will vote for a non-American-Indian Democratic candidate rather than a Navajo Republican candidate" and that "non-Navajo Democratic candidates prevailed over Navajo Republican candidates."[6] This case was ultimately settled with an agreement to maintain at least three polling places close to Navajo Nation and to provide increased translation and interpretation support for voters.

Similarly, in a Lawyers' Committee case challenging Alabama's method of electing judges to a number of the state's courts, the Middle District of Alabama found that, while "there is a significant correlation between race and voting behavior in Alabama," the real question was why that was the case. The court queried, "[i]s it on account of race, as condemned by § 2 of the VRA, or on account of some other cause or causes, such as partisan politics?"[7] In answering this question, and ultimately ruling against plaintiffs, the court pointed to a number of factors—other than a race-based unwillingness to vote for people of color—contributing to White bloc voting. For instance, the court noted that the relative weakness of the Alabama Democratic Party "makes it [] harder for any Democratic candidate — white or black — to get elected." The court also noted the fact that "appellate judges must run under a party banner" and the prevalence of straight ticket voting (voting for a single party up and down the ballot) as additional evidence that "judicial election results are driven [ ] by the party of the candidate, not the race of the candidate." Again, race/party conjoinment was used to undermine a VRA case.

The court's decision in the Alabama case reflected the finding in *LULAC v. Clements*, a case challenging a single-district system of electing state trial judges in Texas. In considering the Gingles preconditions, the Fifth Circuit found that:

> "The race of the candidate did not affect the pattern. White voters' support for black Republican candidates was equal to or greater than their support for white Republicans. Likewise, black and white Democratic candidates received equal percentages of the white vote. Given these

[5] Compare *Easley v. Cromartie*, 532 U.S. 234, 239 (2001) with *Hunt v. Cromartie*, 526 U.S. 541, 550 (1999) (struggling to determine whether North Carolina District 12 was a racial gerrymander due to "a strong correlation between racial composition and party preference"); *Bush v. Vera*, 517 U.S. 952, 968 (1941) (O'Connor, J., principal opinion) ("If district lines merely correlate with race because they are drawn on the basis of political affiliation, which correlates with race, there is no racial classification to justify"); Richard L. Hasen, Race or Party? [5].

[6] *Navajo Nation Hum. Rts. Commn. et al v. San Juan County et al*, 2:16-cv-00154-JNP D. Utah, Def. Opp. to Pls' Mot. for Prelim. Inj.

[7] *Alabama State Conference of National Association for Advancement of Colored People v. Alabama*, 2020 WL 583803 (M.D.Ala., 2020).

> facts, we cannot see how minority-preferred judicial candidates were defeated 'on account of race or color.' Rather, the minority-preferred candidates were consistently defeated because they ran as members of the weaker of two partisan organizations. We are not persuaded that this is racial bloc voting as required by *Gingles*."[8]

One way out of this bind is to view race-party conjoinment as an *expression* of racially polarized voting, not a confounding factor.[9] That is, in a setting where the parties themselves are associated with racialized messages, the preference of people of color for Democratic candidates should still be understood as bloc voting that is salient to shared interests as a minority group.

REFERENCES

[1] Clio Andris et al., The Rise of Partisanship and Super-Cooperators in the U.S. House of Representatives, 10(4) PLoS One (2015) (available at https://journals.plos.org/plosone/article/file?id=10.1371/journal.pone.0123507&type=printable)

[2] Stephen Ansolabehere, Nathaniel Persily, and Charles Stewart III, Race, Region, and Vote Choice in the 2008 Election: Implications for the Future of the Voting Rights Act, 123 Harv. Law Rev. 1385, 1403 (2010)

[3] Bruce E. Cain and Emily R. Zhang, Blurred Lines: Conjoined Polarization and Voting Rights, Ohio State Law J., https://kb.osu.edu/bitstream/handle/1811/79603/1/OSLJ_V77N4_0867.pdf

[4] Michael Dimock et. al., Political Polarization in the American Public: How Increasing Ideological Uniformity and Partisan Antipathy Affect Politics, Compromise and Everyday Life, Pew Research Service (12 June, 2014) (available at https://www.people-press.org/2014/06/12/political-polarization-in-the-american-public/)

[5] Richard L. Hasen, Race or Party? How Courts Should Think About Republican Efforts to Make it Harder to Vote in North Carolina and Elsewhere, 127 Harv. L. Rev. F. 58 (2014)

[8] *League of United Latin American Citizens, Council No. 4434 v. Clements*, 999 F.2d 831, 879 (C.A.5 (Tex.),1993).

[9] Another obvious way out of this bind is to look for polarization patterns in Democratic primary elections, so that party preference is held constant. But data from contested primary elections is not always available.

Chapter 23

The state of play in voting rights

KRISTEN CLARKE AND ARUSHA GORDON

CHAPTER SUMMARY

So where are we now, and where are we going? Civil rights attorneys Clarke and Gordon recount key history, situate the current litigation landscape, and look to the future in a timely overview of redistricting and voting rights for the nation.

1 HOW WE GOT HERE

Many view the U.S. presidential election as a central determinant of American policy, both at home and abroad. Although the redistricting process will never drum up the same kind of headlines or excitement as a presidential election, it arguably has as significant an impact on policy decisions. Who gets counted in the Census and how district maps are drawn have important implications far beyond the elections that are conducted in those districts.[1] These decisions determine not just who is able to get elected, but can also impact how limited resources such as water and electricity are distributed, which roads get repaired, what is taught in schools, and, in the case of judicial districts with jurisdiction over capital cases, even who gets put to death. Yet, rather than ensuring that these critical decisions are made in a dispassionate fashion, the United States arguably stands alone among democratic nations in allowing self-interested legislators to draw their own districts [19].

[1] In this volume, Buck and Hachadoorian talk more about Census practices, and Gall, Mac Donald and our LCCR colleague Fred McBride give some nitty gritty views on mapmaking.

M. Duchin, O. Walch (eds.), *Political Geometry*, https://doi.org/10.1007/978-3-319-69161-9_23

Because of the Supreme Court's devastating 2013 *Shelby* decision, discussed more below, the redistricting cycle following the 2020 Census marks the first time since the civil rights movement of the 1960s that redistricting will occur without the full protections of the Voting Rights Act (VRA). The *Shelby* ruling has sparked a years-long effort to push Congress to respond to the Court's ruling and restore the full vitality of the Act. As such, it is a particularly appropriate time to examine current issues in redistricting.

In our work for the Lawyers' Committee for Civil Rights Under Law ("Lawyers' Committee"), we bring lawsuits that protect the rights of Black people and others from historically marginalized backgrounds to have an equal opportunity to participate in all stages of the electoral process. Since its founding in 1963 at the request of President John F. Kennedy, the Lawyers' Committee has been at the forefront of the fight for voting rights and has brought many of the most significant cases impacting voting rights in our country. Today, our docket of voting rights lawsuits remains incredibly comprehensive and far-reaching.[2]

The vast majority of lawsuits concerning redistricting include claims under the VRA, a landmark piece of federal legislation from 1965 that has been discussed throughout this book.[3] We will offer a brief recap here, because current voting rights contestation is best understood with a long view of American voting rights history.

1.1 HISTORICAL SIGNIFICANCE

In 1857, the Supreme Court's infamous *Dred Scott* decision held that African Americans could not be U.S. citizens, whether enslaved or free. Black people were constitutionally recognized as full citizens only after the Civil War, via the 14th and 15th Amendments (ratified in 1868 and 1870 respectively). Despite formal citizenship, they faced considerable challenges in running for office or even registering to vote across the Southern U.S. throughout the Reconstruction Era. More systematic repression took hold in 1877, when a deal brokered in Washington removed federal troops from the South and left the new civil rights laws unenforced.[4] The Jim Crow Era—the long period of official anti-Black laws and practices that followed—is often given 1877 as its start date and 1965, the passage of the VRA, as its end.

The VRA came about because of the demonstrations and protests that were carried out by people like the great civil rights leader John Lewis. There was one march in particular during the 1960s—a march from Selma, Alabama to Montgomery, Alabama in March 1965—where peaceful demonstrators were preparing to cross the Edmund Pettus Bridge when they were attacked by police officers armed with billy clubs and dogs.[5] John Lewis was struck across the head and bore scars from

[2] You can find an overview of some of this work here: `https://lawyerscommittee.org/project/voting-rights-project/`

[3] Chapter 6 of this volume gives a quick overview of the VRA's origins and key provisions, and Chapter 7 includes a detailed discussion of its most important legal challenges to date.

[4] For an unparalleled history of the Reconstruction Era, see Eric Foner's books Reconstruction: America's Unfinished Revolution, 1863–1877 (2014) and How the Civil War and Reconstruction Remade the Constitution (2019) [7, 8].

[5] The bridge was built in 1940 and named for an Alabama senator and Klansman.

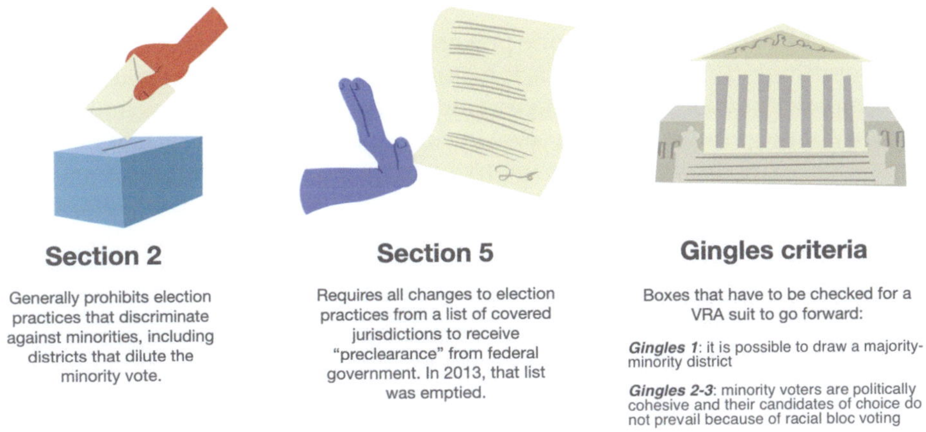

Figure 1: A brief recap of Sections 2 and 5 of the VRA and the Gingles criteria

this incident for the rest of his life. But the painful marches and protests from the Civil Rights Era are what gave rise to the Voting Rights Act. Images of the march across the Edmund Pettus Bridge and other civil rights demonstrations were televised across the globe and became an impetus for President Lyndon B. Johnson to act. The law was passed by the Senate on May 26 of that year and Johnson signed the bill into law on August 6, with Martin Luther King and other civil rights leaders present for the signing ceremony.

The Voting Rights Act banned outright literacy tests, grandfather clauses, and other Jim Crow tools that had been used to disenfranchise minority voters. But the Voting Rights Act contains other strong provisions as well. The two sections you hear about most are Section 2 and Section 5 (see Figure 1 for a brief overview of both, as well as the Gingles criteria). Section 2 applies nationally, prohibiting jurisdictions from states to small localities from putting in place laws that may dilute minority voting strength or deny minority voters access to the polls. Litigators often work with statisticians to use Section 2 as a tool to challenge redistricting plans that fail to provide minority voters with an equal opportunity to elect candidates of choice.

There's another provision of the Voting Rights Act that has sadly been the subject of a lot of controversy in the courts: the Section 5 "preclearance" provision. At the time that this law was put into place, there were some parts of the country where voting discrimination seemed intractable and truly presented a problem that required strong medicine to heal. Alabama, site of the Pettus Bridge attack, was one of those places; Mississippi, Louisiana, South Carolina, parts of North Carolina, Texas, Arizona, parts of California, Florida, parts of New York: all of these states,16 in total, were subject to the enormously important provision of the Voting Rights Act that required federal review before any change could be made to any voting law or procedure. It was intended to make sure that jurisdictions didn't turn the clock back and worsen the position of minority voters. Preclearance helped to block hundreds of discriminatory voting changes, including discriminatory redistricting plans, over the course of the past few decades.

Kilmichael, Mississippi provides one powerful example of how Section 5 operated long after initial VRA passage. This is a small community off the beaten path in Mississippi where 2000 census data revealed that African Americans had become a majority of the population. It's a town governed by a five-member Council and a mayor—all White throughout the town's history up to then—but because of the demographic shift, a number of African Americans decided to run for seats on the council and even the mayoral seat. So the council decided to change the rules of the game: they voted to simply cancel the 2001 election. The DOJ stepped in, the election went forward, and the town elected three of five Black councillors and a Black mayor. This demonstrates the importance of the Voting Rights Act—it is a law that's helped open up access to democracy across our country, from members of Congress to the mayor of Kilmichael.

It is worth highlighting just how involved the U.S. Congress has been over the long life of the VRA. The law was resoundingly passed in 1965, but its "coverage formula" (the list of places that were subject to preclearance) was only supposed to last five years. In 1970, 1975, 1982, and again in 2006, Congress went back to examine whether the VRA and preclearance in particular had served its purpose, and each time they opted for renewal or even extension.

One moment from the 2005-2006 House debate over reauthorizing Section 5 stands out as a vivid visual: Republican Congressman Jim Sensenbrenner was discussing the recent history of voting changes blocked by preclearance (Figure 2). He began to pile the books and files onto a table, showing the volume of evidence amassed by his staff, to the point that it tipped over and books started to fall onto the floor. It was a very powerful illustration of this Congress doing its job, and doing its homework, to really study carefully the need for an important law like this. At the end of the debate, the law was reauthorized 98-0 in the Senate and 390-33 in the House. By an overwhelming bipartisan margin, Congress agreed that Section 5 of the Voting Rights Act was still playing a vital role in our democracy.[6]

1.2 LIFE AFTER PRECLEARANCE

In 2005, just as Congress began to debate the latest VRA extension, the Roberts Court was born. Here is Justice John Roberts in a 2009 case, presenting a rosy view of the world:

> "The historic accomplishments of the Voting Rights Act are undeniable. When the act was first passed, unconstitutional discrimination was rampant, and the racial gap in voter registration and turnout... was great. Today that gap has been dramatically diminished, and most of the barriers to equal voting rights have long been abolished."[7]

Section 5 survived that earlier constitutional challenge mounted by a Texas municipal utility district, but it was an Alabama case, *Shelby County v. Holder*, where it

[6] It is fascinating to watch the CSPAN coverage of the House debate: `https://www.c-span.org/video/?193337-1/house-session`

[7] *Northwest Austin Municipal Utility District No. 1 v. Holder* (No. 08-322) 573 F. Supp. 2d 221.

Figure 2: In the 2006 hearings, Sensenbrenner cataloged DOJ activity under preclearance from 1982 to 2006. Georgia: 91 objections; Texas: 105 objections; Mississippi: 112 objections; Louisiana: 96 objections; South Carolina: 73 objections; North Carolina: 45 objections; Alabama: 46 objections; Arizona: 17 objections. He detailed dozens of voting rule changes that were withdrawn by those states under DOJ pressure and hundreds of federal observers assigned to monitor elections in just the four years prior to this debate. He concluded: "We have put in the work on this. We've done the hearings. The record is replete... let's go down in history as the house that did the right thing."

finally gave way in 2013. The Supreme Court's *Shelby* decision didn't strike down the preclearance provision, but instead nullified the coverage formula which set forth the states and localities that were covered, effectively ending preclearance. *Shelby* has fundamentally changed the field and landscape for voting rights attorneys and advocates. Before the *Shelby* decision, advocates were alerted to changes in the works when a covered jurisdiction sought preclearance from the Department of Justice or the D.C. District Court; this allowed advocates and voting rights attorneys to preemptively work to stop changes that would hurt minority communities. In the aftermath of *Shelby*, changes large and small can be implemented without stakeholders receiving any notice. As a result, the work of voting rights attorneys and advocates has shifted from preventing problematic rule changes to a game of "whack-a-mole," where lawsuits and other advocacy efforts are of a more reactive nature. In practice, a discriminatory change to an electoral process must often be implemented and disenfranchise voters before that harm can be the basis of a court challenge.

This more reactive process is particularly troubling because state governments have moved boldly in the post-*Shelby* world. In the days after the Supreme Court handed down its decision, several states that were previously covered moved swiftly to enact conspicuous changes. Within months, restrictive voter ID requirements were introduced in four states (Alabama, Mississippi, North Carolina, and Texas) and a number of states (including Florida, Georgia and Virginia) carried out mass purges of their voter rolls [3].[8]

Mid-decade redistricting is a sure tell that some states were ready to take advantage of the withdrawal of oversight. Georgia was a particularly bad actor, and in fact their re-redistricting was so egregious that advocates filed suit. Consider District 105 in their 180-member state House. As constituted after the 2010 Census, this district had a White population of 48.6% and a combined Black and Latino population of 51.6%. Its 2012 election was extremely competitive, with a challenger largely backed by minority voters coming within 554 votes of an incumbent backed by White voters, and nearly as close again in 2014. The state went in and carved up the district in 2015, shifting the population to make the district Whiter by about 4%. The 2016 outcome was the closest yet, a margin of just 222 votes for the incumbent, leaving it pretty clear that the incumbent was saved by those race-conscious adjustments.

1.3 RESIDENTIAL SHIFTS

Decades after the initial passage of the Voting Rights Act, the need to keep governments in check has not dissipated, although some conditions on the ground have certainly shifted. To set the stage for today's developments, it's worth looking at changes in human geography. The country is more racially and ethnically diverse today than ever before, and the trend is not slowing. Census statistics tell us that in 1965, just 5% of the U.S. population was born abroad; today, that number has more than doubled to 14%. The Hispanic and Latino population is expected to grow from 18.73% in 2020 to 27% by 2060.[9] In the same timeframe, the Asian population will grow from just over 6% to 9%. Because of this growth, the Pew Research Center estimates that by 2055 no racial or ethnic group will be a majority group in the United States [13].

Where you live is bound up with where you can work, where you attend school, how you are policed, where you can vote, and who's on your ballot. Housing policy, school policy, policing, race, and voting have always been intertwined. VRA practice reminds us of this fundamental role of geography by requiring that plaintiffs show that the minority group is sufficiently concentrated to constitute the majority in a district (Gingles 1).

But the flip side of concentration is segregation. Segregation can make it easy for a group's voting strength to be diluted through a practice known as packing (see Chapter 0, Chapter 2). And even when districts are favorable at one level for communities of color, it may be difficult for minority candidates from a tightly clustered community to be elected to higher office, such as an at-large county commission seat or a larger congressional district.[10]

[8] None too subtle, Texas announced its intended voter ID changes on the very afternoon of the *Shelby* decision.

[9] Projected Race and Hispanic Origin: Main Projections Series for the United States, 2017–2060. U.S. Census Bureau, Population Division: Washington, DC (released Sept. 2018).

[10] We can turn to the major VRA historical survey by Katz et al. to see that courts have noted both effects at work: "[T]he district court in the Charleston County litigation noted severe societal and housing segregation and found that this ongoing racial separation 'makes it especially difficult for

Cities and counties themselves have had their borders constantly made and remade along race and class lines.[11] As human geography is transformed through processes of immigration, gentrification, and resettlement, changes are sometimes accompanied by contortions in electoral and school districts to maintain a racial status quo. As social scientist Meredith Richards explains in her geospatial study of school redistricting: "[l]ike congressional districts, school zones are highly gerrymandered; the gerrymandering of school zones serves to worsen the already severe racial segregation of public schools" [15, 16]. Because public schools are largely funded by property taxes levied by local governments, an intense feedback loop of housing, schooling, and voting can severely exacerbate divisions. We should be vigilant when districting magnifies inequality.

In addition to the political, housing, and educational implications of changing demographics, an explosion in mass incarceration that disproportionately targets certain demographic groups has amounted to a transfer of residential population whose consequences for redistricting we will explore further below.

While the vast majority of Section 2 cases have historically been brought on behalf of African American communities, immigration and demographic growth will likely mean that Latino and Asian plaintiffs become more common in the future.[12] And these groups have different population patterns, molded in part by decades of policy that has circumscribed where people of color are able to live.

Increasingly, counties, cities, school districts, and other jurisdictions may have Black and brown communities making up a majority of the population—meaning that successful Section 2 vote dilution cases may be more likely—but only when one considers these groups collectively (e.g., when one combines Black and Asian populations, or Latino and Native American communities).

2 WHERE WE'RE GOING

The shifting landscape has brought major setbacks but has also opened up promising new frontiers. We'll look at the local level, discuss prison gerrymandering, overview the state of coalition claims, and touch on state-level VRAs.

African American candidates seeking county-wide office to reach out to and communicate with the predominately White electorate from whom they must obtain substantial support to win an at-large elections [sic].' The district court in the Neal litigation likewise concluded that similar segregation meant 'that whites in the County have historically had little personal knowledge of or social contact with blacks....Quite simply, whites do not know blacks and are, as a result, highly unlikely to vote for black candidates'" [12].

[11] Municipal annexation and de-annexation often follow conspicuous racial patterns.

[12] In 2005, just seven Section 2 cases were brought with an Asian American plaintiff, compared to 268 with an African American plaintiff [12]. See also *Diaz v. Silver*, 978 F. Supp. 96, 129 (E.D.N.Y. 1996), aff'd, 522 U.S 801 (1997) (successful § 2 claim by Asian Americans in Chinatowns of Manhattan and Brooklyn); *Chen v. City of Houston*, 206 F.3d 502 (5th Cir. 2000); Chen and Lee, Reimagining Democratic Inclusion [4] (discussing the lack of success of Asian Americans in § 2 claims and proposing reforms); Ingram, The Color of Change (discussing changing demographics and VRA claims) [11].

2.1 LOCAL CHALLENGES

While redistricting challenges have been historically directed toward U.S. congressional maps, state legislatures, county commissions, and local school boards, advocates are increasingly applying these same principles to challenge vote dilution in other electoral bodies. Judicial districts are one new frontier. In 2016, for instance, the Lawyers' Committee filed a lawsuit aimed at ending the discriminatory practices by which judges in Texas are elected. The suit alleged that the state's practice of electing judges statewide to the Texas Supreme Court and the Texas Court of Criminal Appeals (the two highest courts in the state) violated the Voting Rights Act. Latinos comprised 26% of the voting age population of Texas in the 2010 Census, while White residents made up 56.4%. Because voting in Texas is heavily polarized, Latino-preferred judicial candidates have had difficulty getting elected to these two courts. In fact, in the past seven decades, just two of the 48 judges serving on the Court of Appeals have been Latino. Similarly, just five of the 77 judges serving on the Texas Supreme Court have been Latino.[13] Clearly, these numbers are not representative of Texas demographics; more importantly, there is reason to believe that they will not lead to equal justice for Texans. Unfortunately, in 2018, the court found against plaintiffs noting that plaintiffs "failed to satisfy their burden of demonstrating that the lack of electoral success by Hispanic-preferred candidates for high judicial office is on account of race rather than other factors, including partisanship."

Although Section 2 challenges against K-12 school districts are not uncommon, claims challenging districting decisions of bodies governing higher education are a newer development. In 2013, the Lawyers' Committee filed suit in Arizona Superior Court challenging the method used for electing the Governing Board of the Maricopa County Community College District.[14] The lawsuit was initiated after the Arizona Legislature enacted H.B. 2261 in 2010, requiring that two at-large seats be added to the Governing Board, increasing the size of the Board from five to seven, amounting to a new system of election by creating two new seats that would be very difficult for minority-preferred candidates to secure. The lawsuit alleged that H.B. 2261 violated the Arizona State Constitution because it only applied to counties with at least three million residents, effectively singling out Maricopa County because no other county had even one million residents. The suit alleged that H.B. 2261 violated the state Constitution's prohibition against local or special laws and the Constitution's privileges and immunities clause. This lawsuit went from Arizona Superior Court to the state Court of Appeals and finally the state Supreme Court, ultimately ending unfavorably for the plaintiffs.

In addition to extending redistricting claims to judicial bodies and community college districts, voting rights attorneys have also challenged redistricting decisions concerning so-called special districts such as utility districts. In 2000, for instance, the United States Department of Justice filed a lawsuit against the Upper San Gabriel Valley Municipal Water District in Ventura County, California—water

[13] See plaintiffs' brief in *Lopez et al. v. Abbott,* available at `https://lawyerscommittee.org/wp-content/uploads/2016/07/Texas-Courts-Complaint_07-20-16_FINAL.pdf`

[14] *Gallardo et al. v. Arizona.* See `https://www.lawyerscommittee.org/wp-content/uploads/2015/06/0444.pdf`

districts are of crucial policy importance in the drought-ridden Southwest.[15] Although the district was approximately 46% Hispanic at the time the lawsuit was filed, and although nine Hispanic candidates had run for a board position, no Hispanic resident had ever been elected.[16] The United States argued that the water district improperly split the Hispanic population across the five divisions making up the district, "with the result that Hispanics d[id] not constitute a citizen voting-age majority in any of the five Divisions."[17] After the complaint was filed, the District adopted new division borders that no longer diluted Hispanic voting strength, and so the court dismissed the suit as moot.[18]

While the work rooting out discrimination at the federal level and in state legislatures and in county councils is not done, challenging voter suppression as it occurs in electoral bodies that have not traditionally been the focus of vote dilution challenges is equally important. Advocates must continue to think creatively to target discrimination in judicial election processes, community college districts, utility districts, and elsewhere. This won't just be through litigation, but will just as importantly involve candidate recruitment, community organizing, and voter education.

2.2 PRISON MALAPPORTIONMENT

Incarceration rates in the U.S. have grown dramatically in recent decades, from about 150 people per 100,000 in the mid 1970s to 707 people per 100,000 in 2012. Today, approximately 2.2 million people are incarcerated in the United States, up from just 300,000 in 1970 [14, 20].

These staggering incarceration rates have had a disproportionate impact on Black and brown communities: 60% of incarcerated individuals are people of color, even though they account for just 30% of the general U.S. population. While just 1 in every 106 White men are incarcerated, the rates for African American and Hispanic men are drastically higher, with 1 in 15 African Americans and 1 in 36 Hispanic men incarcerated [14, 17].

---

[15] *United States v. Upper San Gabriel Valley Mun. Water District.*, 2000 WL 33254228 (C.D. Cal, 8 September, 2000).

[16] Complaint, *Upper States v. Upper San Gabriel Valley Mun. Water District*, No. CV 00-07903 (C.D Cal, 21 July, 2000), at 3–4 (available at `https://www.justice.gov/crt/case-document/file/1175831/download`). For case overview, see Katz et al. [12].

[17] Complaint, *Upper San Gabriel Valley*, at 5.

[18] *U.S. v. Upper San Gabriel, et al.* 2:00CV07903, (C.D. Cal.), Stipulation and Order by Judge A. H. Matz entered 16 June, 2003 (docket entry 52).

23.1 FIXING PRISON MALAPPORTIONMENT

LITIGATION

Advocates working to address prison gerrymandering have employed varied litigation strategies, bringing lawsuits in numerous states. In Florida, the ACLU sued Jefferson County. According to the 2010 Census, the county had a total population of 14,761, of which 1,157 were incarcerated at Jefferson Correctional Institution (JCI), a state prison. Only nine of those inmates were convicted in the county. Districts for county commission and school board had roughly 2,900 residents each, so JCI made up almost half of a district. The court found that the massive up-weighting of voting strength for the non-incarcerated population of District 3 was "clearly an equal protection violation," ordering defendants to submit a new districting plan.

In a similar case brought in Rhode Island, the First Circuit declined to follow the Florida example, instead finding that "the Constitution does not require [a jurisdiction] to exclude... inmates from its apportionment process" and "gives the federal courts no power to interfere" with a jurisdiction's decision. Given the relatively scarce and substantively scattered case law, advocates must tread carefully when considering litigation on the issue.

LEGISLATION

The most comprehensive approach to fixing prison gerrymandering would require the Census to change how and where it counts prisoners. However, given that such an approach has yet to be implemented federally, various state and local actors have taken steps to address the issue. Since 2010, at least 20 states and more than 200 counties and municipalities have introduced legislation to address prison gerrymandering. As of this writing, more than half a dozen states have passed legislation addressing the issue, including California, Delaware, Maryland, New York, Washington, New Jersey, and Nevada, with Maryland and New York taking steps to address how prisoners were counted prior to the 2020 redistricting cycle.

Maryland's legal fix, broadly similar to many of these states, applies to districts at every level from congressional and state legislative to counties and municipalities. Mapmakers must allocate incarcerated individuals "at their last known residence before incarceration if the individuals were residents of the state." It also requires federal and state correctional facilities to be excluded from population counts.

New York's prison gerrymandering law is somewhat narrower, as it does not include congressional districts and does not require federal prisoners to be reallocated to a previous address for counting purposes. The law requires the New York State Legislative Task Force on Demographic Research and Reapportionment (LATFOR) to "reallocate people in correctional facilities back to their home communities for purposes of drawing state and local districts."[a] In order to make this possible, the State Department of Corrections is required to send information regarding the residential address of offenders prior to their incarceration. The task force then must match the previous residential addresses of incarcerated individuals with the appropriate census block and maintain a database to track this information for use in drawing state legislative districts.

[a] Part XX of Chapter 57 of the Laws of 2010

But what do these high rates of incarceration and the disproportionate imprisonment in Black and brown communities have to do with redistricting? The answer relates back to the fact that the Census counts inmates as residents of the jurisdiction in which they are incarcerated,[19] and states and other jurisdictions then rely on that Census data in redrawing their electoral districts. During the explosion of incarceration rates in the 1980s and 1990s, many new prisons were built in largely rural areas. Prison construction and maintenance followed, creating economic opportunities and jobs in these rural areas.[20] As a result, "fewer than half of all prisons were located in non-metropolitan areas in the 1960s and 1970s," while "rural communities developed hundreds of new prisons during the 1980s and 1990s" with almost two-thirds of new prison development occurring in rural areas by the mid-1990s.[21]

This trend meant that, while urban centers where Black and brown communities are concentrated are disproportionately targeted for arrests, convicted offenders are often relocated to prisons in rural areas with majority-White populations to serve their time. Because incarcerated individuals lose their right to vote in nearly every state,[22] the vast majority of incarcerated people are unable to vote in the jurisdiction in which the Census counts them as a resident.

2.3 COALITION CLAIMS

The shifting demographics we outlined above have created new opportunities to bring coalition claims, in which more than one minority group comes together to plead a VRA violation. Given the changing landscape, lawyers and other advocates fighting for equal voting rights will need to consider the dynamics between different ethnic and racial groups when bringing claims under Section 2. What do these coalition claims look like and what do they mean for future redistricting decisions? We will focus on articulating the broader trends and the best legal approaches being taken in approaching them.[23]

Unlike many other voting rights issues, the legal framework for coalition claims under Section 2 is still being defined, with courts in some circuits more friendly to these claims (e.g., the Fifth Circuit), than others (e.g., the Sixth Circuit).[24] Yet,

---

[19] *Calvin v. Jefferson County Board of Commissioners* 172 F.Supp.3d 1292, 1297 (N.D.Fla., 2016).

[20] Pfaff writes "In Pennsylvania, the state laid off only three guards when it closed two entire prisons in 2013 . . . [M]any legislators and citizens believe that prisons provide vital economic support, even beyond guard salaries, to the disproportionately rural communities in which so many are located" [14].

[21] Michael Skocpol says that "Areas classified as rural are home to 20% of the overall U.S. population but 40% of all prisoners" [18].

[22] Currently, only two states (Maine and Vermont), as well as the District of Columbia, allow incarcerated individuals to vote while serving time.

[23] Hopkins gives an excellent survey of the state of aggregate minority claims under Section 2 as of 2012 [10]. Elmendorf and Spencer find high cohesion in Asian American and Latino communities and argue that this "implies that Asians and Latinos ought to have considerable success bringing 'coalitional' vote dilution claims under Section 2" [6]. The previous chapter features Espinoza-Madrigal and Sellstrom taking an in-depth look at one recent Asian/Latino coalition case.

[24] Friendly Fifth Circuit examples include *LULAC v. Clements*, 999 F.2d 831, 864 (5th Cir. 1993)("if blacks and Hispanics vote cohesively, they are legally a single minority group, and elections with a candidate from this single minority group are elections with a viable minority candidate."); *Campos v. City of Baytown*, 840 F.2d 1240, 1241 (5th Cir. 1988) (emphasizing that voting patterns among the

recognizing changing demographics, advocates have brought a number of lawsuits using coalition claims. For instance, consider *Arbor Hill v. Albany*, a case in which plaintiffs argued that Black and Hispanic voting strength was being diluted by the districting plan. In the decision, the Northern District of New York articulated some conditions for the success of a coalition claim by specifying that "Black and Hispanic groups are politically cohesive when most members of the two groups vote for the same candidates in most elections" and that, in determining whether groups are cohesive, courts should also consider "whether black groups and Hispanic groups have worked together to form political coalitions and promote the same candidates."[25] The court went on to find that plaintiffs in Arbor Hill successfully showed cohesion between Black and Hispanic groups and pointed to evidence including, among other things, the fact that leaders in the Black and Hispanic communities "attest[ed] without contradiction" that the groups "joined together to further each other's political and social interests" by supporting "various events and projects of interest" to the groups, such as sporting events and festivals. In addition, the court noted that the groups "jointly publish a bilingual community newspaper," and that there was anecdotal evidence that "blacks and Hispanics joined to support candidates preferred by one group or the other."

More recently, voting rights advocates have broken new ground and brought coalition claims joining more than two racial minority groups. In *Georgia Conference of the NAACP v. Gwinnett County*, attorneys with the Lawyers' Committee argued that the district maps for the county board of commissions and school board of Gwinnett County, Georgia, violated Section 2 by diluting the voting strength of African Americans, Latinos, and Asian Americans. Together, African American, Latino and Asian American voters comprise approximately 43% of the voting age population of Gwinnett County. However, at the time the suit was filed in 2016, no minority candidate had ever won election to the County Board of Commissioners or Board of Education.

Gwinnett County's maps pack approximately 74.4% of the African American, Latino, and Asian American voters into one of the County's five districts, while splitting the balance of the minority population across the other four districts such that African Americans, Latinos, and Asian Americans do not constitute a majority in any of those districts. The complaint alleged that the districts should be re-drawn to include a second majority-minority district for both the school board and the board of commissioners so that minority voters have a fair opportunity to elect candidates of their choice to those bodies. It also argued that one majority-minority coalition district could be drawn among the four single member County Board of Commission districts (excluding the chair).

The case in Gwinnett was voluntarily dismissed in 2019 after non-White candidates were elected to the Board of County Commissioners and School Board for the first

minority groups are particularly important to a showing of political cohesiveness). A less friendly Sixth Circuit example is *Nixon v. Kent Cnty.*, 76 F.3d 1381, 1393 (6th Cir. 1996) (rejecting the notion that coalitions of more than one racial or ethnic minority can bring a Section 2 claim).

[25] *Arbor Hill Concerned Citizens Neighborhood Ass'n. v. County of Albany*, 2003 WL 21524820, at *8 (N.D.N.Y., 2003) (citing *League of United Latin Am. Citizen Council v. Clements*, 999 F.2d 831, 864 (5th Cir. 1993) (focusing on elections with minority candidates) and *Concerned Citizens of Hardee County v. Hardee County Bd. of Comm'rs*, 906 F.2d 524, 526 (11th Cir. 1990)).

time in the county's history, though similar cases are likely in the future. It is clear that, in order to successfully assert a coalition claim under Section 2, plaintiffs must be sure to include substantial evidence showing cohesion between the various racial minority groups. Multiple courts have rejected coalition claims when evidence of minority group cohesion is slim. For instance, in *Johnson v. Hamrick*, the Eleventh Circuit rejected a claim that the Black and Hispanic communities of Gainesville, Georgia were politically cohesive. Plaintiffs' evidence of cohesion in this case failed to include any "statistical evidence that blacks and Hispanics voted together in any election," and instead relied solely on anecdotal evidence of individuals in the community. In rejecting plaintiffs' claim the court explained that it would "not indulge the presumption that blacks and Hispanics vote together merely because a few have worked together on various, non-electoral, community issues."[26]

Despite the challenges of bringing coalition claims, changing demographics demand that voting rights advocates take this avenue seriously and develop tools for demonstrating cohesion.

2.4 STATE VOTING RIGHTS ACTS

A final development in recent years worth noting is that several states have introduced their own state-level voting rights acts, which sometimes echo the federal VRA (so that they would serve to keep its protections in place even if it is struck down) and sometimes differ in interesting ways.

First on the scene was the California Voting Rights Act (CVRA)[27], passed in 2001. Its key difference from the federal VRA is that plaintiffs must only show racial polarization (Gingles 2–3) and do not need to demonstrate the existence of potential majority-minority districts (Gingles 1) to press a case. The CVRA was designed to dismantle at-large elections for localities around the state, and its impact has been enormous, as cities and counties have scrambled to redesign their elections. In 2016, the California legislature put a "safe harbor" provision in place for 45 days, allowing all localities that moved to create districts in that period of time to be shielded from litigation.[28] A white paper by civil rights organizations cites the research of political scientist Morgan Kousser in enumerating at least 335 localities (school and community college boards, city councils, utilities districts, and so on) that shifted their system of election under the CVRA as of 2018. Kousser's work found major impacts: for instance, affected school districts had a 60% increase in Latino representation in a ten-year span. Interestingly, most of this happened without litigation. Of the cases enumerated in Kousser's study, 12% had a lawsuit as the precipitating event, 25% were triggered by a demand letter (which attorneys use to put localities on notice of a potential lawsuit), and the remaining 63% were preemptive switches.

The rest of the West Coast followed suit, with a Washington VRA and an Oregon

[26] *Johnson v. Hamrick*, 155 F. Supp. 2d 1355, 1368 (N.D. Ga. 2001) aff'd, 296 F.3d 1065 (11th Cir. 2002).
[27] California Voting Rights Act of 2001, Cal. Elec. Code § 14025 (West 2017).
[28] Cal. Elec. Code § 10010.

VRA now on the books as of 2018 and 2019 respectively.[29] The Oregon VRA applies specifically to school districts; the Washington VRA is broader and expressly calls for the consideration of alternative remedies, so that ranked choice options can be considered in addition to districts. Quite a few other states have legislation in various stages of preparation for their own state-level VRAs, including New York.

3 CONCLUSION: WHY IT MATTERS

So why does all of this matter? We believe that fair redistricting has a direct correlation with the quality of people's lives in our country. If you care about issues like the school-to-prison pipeline, then having a school board that fairly reflects the diversity of the community served by that school board is key. If you care about issues like unjustified police shootings of unarmed individuals, particularly of African Americans, then the makeup of your city council is key—city councils sometimes have a say in police chiefs and whether or not those police chiefs are held accountable for how their police departments are run. If we collectively believe in an inclusive democracy, then we want a democracy in which our local governments, our state governments, and our federal government reflects the diversity of the communities they serve. Diverse governing bodies help to increase public confidence that elections reflect the will of the people, and ultimately boost confidence in the work of government.

ACKNOWLEDGMENTS

We thank Lawyers' Committee for Civil Rights Under Law attorneys Ezra Rosenberg and Jon Greenbaum, as well as the Lawyers' Committee's courageous clients.

REFERENCES

[1] ACLU California, Asian Americans Advancing Justice, and MALDEF, The California Voting Rights Act in 2018: The Impact of Model Civil Rights Legislation on Discriminatory Voting Practices. White paper. Available at `https://www.advancingjustice-alc.org/wp-content/uploads/2012/11/CVRA-Fact-Sheet-12-03-2018.pdf`

[2] Clio Andris et al., The Rise of Partisanship and Super-Cooperators in the U.S. House of Representatives, 10(4) PLoS One (2015) (available at `https://journals.plos.org/plosone/article/file?id=10.1371/journal.pone.0123507&type=printable`)

[3] Desmond Ang, Do 40-Year-Old Facts Still Matter? Long Run Effects of Federal Oversight under the Voting Rights Act, HKS Faculty Research Working Paper Series (2018)

[29] Oregon HB 3310 A (2019); Washington SB 6002 (2018).

[4] Ming Hsu Chen and Tauku Lee, Reimagining Democratic Inclusion: Asian Americans and the Voting Rights Act, 3 UC Irvine L. Rev 359, 361 (2013)

[5] Michael Dimock et. al., Political Polarization in the American Public: How Increasing Ideological Uniformity and Partisan Antipathy Affect Politics, Compromise and Everyday Life, Pew Research Service (12 June, 2014) (available at https://www.people-press.org/2014/06/12/political-polarization-in-the-american-public/)

[6] Christopher S. Elmendorf & Douglas M. Spencer, Administering Section 2 of the Voting Rights Act after Shelby County, 115 Columbia L. Rev. 2143 (2015)

[7] Eric Foner, Reconstruction: America's Unfinished Revolution, 1863–1877 (2014)

[8] Eric Foner, How the Civil War and Reconstruction Remade the Constitution (2019)

[9] Richard L. Hasen, Race or Party? How Courts Should Think About Republican Efforts to Make it Harder to Vote in North Carolina and Elsewhere, 127 Harv. L. Rev. F. 58 (2014)

[10] Chelsea J. Hopkins, The Minority Coalition's Burden of Proof under Section 2 of the Voting Rights Act, 52 Santa Clara L. Rev. 623 (2012)

[11] Joanna E. Cuevas Ingram, The Color of Change: Voting Rights in the 21st Century and the California Voting Rights Act, 15 Harv. Latino L. Rev. 183 (2012)

[12] Ellen Katz et al, Documenting Discrimination in Voting: Judicial Findings Under Section 2 of the Voting Rights Act Since 1982, 39 U. Mich. J.L. Reform 643, 706 (2005)

[13] Jeffrey S. Passel and D'Vera Cohn, U.S. Population Projections: 2005–2050. Pew Research Center Report. https://www.pewresearch.org/wp-content/uploads/sites/3/2010/10/85.pdf

[14] Jon Pfaff, Locked In: The True Causes of Mass Incarceration—and How to Achieve Real Reform 2–3 (2017)

[15] Meredith P. Richards, The Gerrymandering of School Attendance Zones and the Segregation of Public Schools: A Geospatial Analysis, American Educational Research Journal 51, 1119–1157 (2014)

[16] Meredith P. Richards. Gerrymandering educational opportunity. Phi Delta Kappan. 99, 65-70. doi:10.1177/0031721717739597. (2017)

[17] Leah Sakala, Breaking Down Mass Incarceration in the 2010 Census: State-By-State Incarceration Rates by Race/Ethnicity, Prison Pol'y Initiative (28 May, 2014), https://www.prisonpolicy.org/reports/rates.html.

[18] Michael Skocpol, The Emerging Constitutional Law of Prison Gerrymandering, 69 Stan. L. Rev. 1473, 1487–1488 (2017)

[19] Nicholas O. Stephanopoulos, Our Electoral Exceptionalism, 80 U. Chicago L. Rev. 769, 772 (2013).

[20] Peter Wagner & Bernadette Rabuy, Mass Incarceration: The Whole Pie 2017, Prison Pol'y Initiative (14 March, 2017), https://www.prisonpolicy.org/reports/pie2017.html

Chapter 24

Epilogue: The view from 2022

MOON DUCHIN

This book was begun in 2018 and has come together slowly over the few years since then. Meanwhile, once the Decennial data was released in August 2021, states started redistricting immediately (and some didn't even wait for the new data!). Now the book is on the verge of publication in February 2022, and most states have new plans in place and ready for the midterm elections in November.

So, what have we seen this time?[1]

New commissions. There were new redistricting commissions of some description in Michigan, Utah, New Mexico, Missouri, Ohio, Colorado, Virginia, and more—running the gamut from being fully independent to playing a merely advisory role to being dominated by partisan elected officials and behaving just as self-interestedly as the legislature. So this year has provided a crash course in all the promise and limitations of alternative configurations of decisionmakers. It was not without drama. There were tears in Michigan; commissioners threatened to quit in Virginia, and one actually quit and walked out in Utah. Some commissions gridlocked and refused to draw, or re-draw, maps, throwing the process to the

[1] For full disclosure, work of the MGGG Redistricting Lab from 2020 through Feb 2022 includes the following: Collecting public input in the form of districting plans and/or community mapping for the Michigan Department of State, the Wisconsin Department of Administration, the New Mexico Citizens Redistricting Committee, the Pennsylvania Governor's Office, and the Alaska Redistricting Board. Grassroots and/or local community map collection in Ohio, Missouri, Florida, Minneapolis, MN, Dallas, TX, San Diego, CA, and so on—approximately 100 localities altogether. Support for mapping competitions in Missouri and Indiana. Two presentations for the Arizona Independent Redistricting Commission. Support for analytical work of line-drawing bodies (e.g., partisan fairness or racial polarization analysis) for the Wisconsin People's Maps Commission, the Maryland Citizens Redistricting Commission, the Utah Independent Redistricting Commission, and the Massachusetts State Senate. And I have done expert work in litigation in state court in Wisconsin, North Carolina, Pennsylvania, and South Carolina, and in federal court in Alabama.

M. Duchin, O. Walch (eds.), *Political Geometry*, https://doi.org/10.1007/978-3-319-69161-9_24

courts. Some commissions drew maps and were fully ignored by the legislatures that were supposed to act on them. It will take some time for the dust to settle and to sort out all the lessons learned!

Lots more attention to Communities of Interest. Many line-drawing bodies launched major campaigns to collect community of interest testimony in the form of digital mapping data. Some of the campaigns were extraordinarily successful, with COI submissions in the hundreds or even thousands! But... then what? Next up, we need ideas for persuasive accounting of how all that data was taken into account.

All the litigation. Every ten years comes the Census, the maps, and the lawsuits! This cycle has been no exception. There have already been initial decisions in a bunch of states, but in most cases things are still ping-ponging between various courts and panels. It's too early for a postmortem, but one thing we can say is that state courts look highly willing to step in.

Leaching out competition. One hot-take narrative that is emerging about new maps in several states is that they are strikingly uncompetitive. In some cases this seems to be done expressly, in classic incumbent-favoring gerrymander fashion. But in other cases, it is possibly a byproduct of trying to create a map that has good features: good efficiency gaps, good proportionality, and so on. After all, if you're a state like Minnesota that has 8 seats and a typical vote that's in spitting distance of 50-50 between the major parties, then one way to get a map that looks shiny to lots of the metrics is to lock 4 districts down for each party. If you make a map with more swing and responsiveness, then it could give unexpected results in an individual election and bring unwanted attention. In a moment with so much scrutiny, predictable is safe.

Dark clouds for the VRA. In early January of 2022, a district court held a week-long hearing about whether Alabama has a VRA obligation to create a second effective district for Black voters, out of seven Congressional districts. They found for the plaintiffs, requiring the state to re-draw the districts; when the state refused, a special master was set to take up the pen in early February. But at the eleventh hour, the U.S. Supreme Court stepped in with a "stay," putting the brakes on the re-draw. The Supreme Court majority indicated that the outcome is not sure when they finally hear the case.... next year. So the VRA is in a serious kind of limbo for something like 15 months, and it's cued up for reshaping by a hostile court. Especially alluring to this court might be the long shot implication by Alabama that maybe key provisions of the VRA should be read race-blind.

So this book, which aimed to cover all of the tools for redistricting in its messy interdisciplinary richness, might now especially be read to help you understand four things:

- What does it look like to take citizen input seriously?

- What is this VRA thing I keep hearing about, and what are the stakes if it goes away?
- How do algorithms come in to the picture, and what do they tell us about how districts work when they're run "blind"?
- Yikes, if districts can't do the work we want them to do, then **what else can we do**?

It's a dispatch from within a turbulent moment... but it has all the tools for the hard democracy work that comes next.

Correction to: Political Geometry – Rethinking Redistricting in the US with Math, Law, and Everything In Between

Moon Duchin and Olivia Walch

Correction to:
M. Duchin, O. Walch (eds.), *Political Geometry*,
https://doi.org/10.1007/978-3-319-69161-9

The original HTML version of the book was inadvertently published with the incorrect author names attributed to the chapters. This has now been amended in each chapter.

The updated version of the book can be found at
https://doi.org/10.1007/978-3-319-69161-9

M. Duchin, O. Walch (eds.), *Political Geometry*, https://doi.org/10.1007/978-3-319-69161-9_25

Author Biographies

Amariah Becker is a data scientist who worked at the MGGG Redistricting Lab in 2019-2021. Her interests include algorithm design, combinatorial optimization, and computing for social change. Amariah's background is in graph algorithms.

Mira Bernstein is a math and statistics educator and consultant, with a PhD in algebraic geometry. She was one of the co-creators of MGGG in 2017. She also chairs the Board of Directors for Canada/USA Mathcamp, a unique summer program for high school students.

Ruth Buck is a graduate student in Geography at Penn State University studying population, scale, and redistricting. She previously worked as a geographer at the MGGG Redistricting Lab.

Guy-Uriel Charles has recently moved to Harvard Law School as the inaugural Charles J. Ogletree, Jr. Professor of Law and the faculty director of the Charles Hamilton Houston Institute for Race and Justice. He is an expert in constitutional law, election law, campaign finance, redistricting, race, and politics.

Kristen Clarke is a civil rights attorney with extensive experience in government and the nonprofit sector. She is the former president & executive director of the National Lawyers' Committee for Civil Rights Under Law and has also worked at NAACP-LDF, in the Civil Rights Bureau for the New York State Attorney General's office, and in the Civil Rights Division of the U.S. Department of Justice.

Daryl DeFord is an assistant professor of Data Analytics at Washington State University. His research applies algebraic and combinatorial methods to problems in data analysis, with a special focus on redistricting. As a postdoc with Justin Solomon's group at MIT, he was affiliated with the MGGG Redistricting Lab and was one of the key developers of GerryChain.

Moon Duchin is a mathematician at Tufts University, where she founded the interdisciplinary Science, Technology, and Society program and runs the MGGG Redistricting Lab at the Tisch College of Civic Life. Her background is in geometry, topology, and dynamics, and her current research focus is data science for civil rights.

Iván Espinoza-Madrigal is the Executive Director of Lawyers for Civil Rights, whose focus is on extending legal protections for people of color and immigrants. Previous positions include legal director for the Center for HIV Law and Policy and immigrant rights attorney at MALDEF (the Mexican American Legal Defense and Education Fund).

Chris Fowler is a geographer at Penn State, where he serves as Director of The Peter R. Gould Center for Geography Education and Outreach. He is currently building methods for representing neighborhood change in complex, multiscalar contexts

M. Duchin, O. Walch (eds.), *Political Geometry*, https://doi.org/10.1007/978-3-319-69161-9

and developing a line of research that explores the increasing neighborhood-scale diversity in U.S. cities.

Keith Gåddie is the President's Associates Presidential Professor of Architecture and Executive Faculty Fellow of the University of Oklahoma. His specialties are redistricting, voting rights reform, and the built environment of democratic systems.

Megan Gall's career as a civil rights practitioner focuses on quantitative analyses for Voting Rights Act compliance and litigation. Her specialties include racially polarized voting statistics and redistricting. She's a certified GIS Professional and holds a PhD in Political Science.

Arusha Gordon is an attorney at the Lawyers' Committee for Civil Rights Under Law, where she serves as Associate Director of the James Byrd Jr. Center to Stop Hate. She was formerly counsel for the Lawyers' Committee's Voting Rights Project.

Larry Guth is the Claude Shannon professor of mathematics at MIT, with specialties in metric geometry, harmonic analysis, and extremal combinatorics, as well as a longstanding commitment to pedagogy and mentorship.

Lee Hachadoorian is an assistant professor of instruction in Temple University's Department of Geography & Urban Studies. His research is broadly concerned with applying geospatial technologies to urban and demographic analysis; his teaching and research heavily emphasize the use of free and open-source software.

Ellen Katz, the Ralph W. Aigler Professor of Law at the University of Michigan, writes and teaches about election law, civil rights and remedies, and equal protection. Her scholarship addresses questions of minority representation, political equality, and the role of institutions in anti-discrimination law.

Karin Mac Donald directs the Election Administration Research Center at Berkeley Law, as well as the Statewide Database, the redistricting database for the State of California. Her work and research center on access to elections focusing on voting rights and representation via election administration and nonpartisan, transparent districting processes.

Fred McBride is a Redistricting and Voting Rights Policy Specialist at the Lawyers' Committee for Civil Rights Under Law, where he engages in quantitative and qualitative research in redistricting and voting rights law. He previously held positions at the ACLU and FairVote, and holds a PhD in Political Science from Clark Atlanta University.

James M. Murphy is an assistant professor of mathematics at Tufts University, with interests in applied harmonic analysis, mathematics of data science, and applications to image, signal, and network processing.

Garrett Dash Nelson is the Curator of Maps and Director of Geographic Scholarship at the Leventhal Map & Education Center at the Boston Public Library. His research investigates how geographic units are historically formed and contested. His methods combine qualitative research on the politics, culture and landscape of cities and regions with spatial analysis and critical cartography.

Ari Nieh is a mathematician, professional singer, and game designer. She was a founding member of MGGG, specializing in math pedagogy and communication. She co-led a cycle of four Educator Training workshops with a focus on gerrymandering and geometry.

Jonathan Rodden is a professor in the political science department at Stanford who works on the comparative political economy of institutions. His specialties include federalism, political geography, and the historical origins of political institutions.

Heather Rosenfeld is an environmental geographer currently teaching at Smith College, formerly staff geographer at the MGGG Redistricting Lab and instructor in the Science, Technology, and Society Program at Tufts University. Heather's research examines the co-construction of citizen/activist science and political-economic structures, using feminist and critical cartographic methods.

Oren Sellstrom is the litigation director of Lawyers for Civil Rights in Boston, MA. He oversees the organization's litigation and advocacy work in all areas including education, economic justice, employment, police accountability, immigrants' rights, and voting rights.

Justin Solomon is an associate professor at MIT, where he heads the Geometric Data Processing Group in CSAIL (the Computer Science and Artificial Intelligence Laboratory). His specialties include discrete differential geometry and numerical analysis, with applications to computer graphics and machine learning. He co-founded MGGG's Voting Rights Data Institute summer program.

Douglas M. Spencer is an associate professor of law at the University of Colorado. He is an election law scholar whose research addresses the role of prejudice and racial attitudes in voting rights litigation, as well as campaign finance and election dynamics.

Alma Steingart is an assistant professor of history at Columbia University, researching the interplay between politics and mathematical rationalities. Her current book project examines how mathematical thought and computing technologies have impacted electoral politics in the United States in the twentieth century.

Olivia Walch is a mathematician, app developer, and cartoonist. Apps developed by her have been downloaded more than half a million times, and her comics have appeared in *The Washington Post* and *The Nib*. She is currently CEO of Arcascope, a company that makes software for tracking sleep and circadian rhythms.

Thomas Weighill is an assistant professor of mathematics at the University of North Carolina at Greensboro, following a postdoc at MGGG. His research is in topology and geometry and their applications to data science, focusing on ways in which Census and election data can be harnessed for social good.

MIX
Papier aus verantwortungsvollen Quellen
Paper from responsible sources
FSC® C105338

If you have any concerns about our products,
you can contact us on
ProductSafety@springernature.com

In case Publisher is established outside the EU,
the EU authorized representative is:
Springer Nature Customer Service Center GmbH
Europaplatz 3, 69115 Heidelberg, Germany

Printed by Libri Plureos GmbH
in Hamburg, Germany